THE PHYSICISTS

THE PHYSICISTS

The History of a Scientific Community in Modern America

Daniel J. Kevles

With a New Preface by the Author

HARVARD UNIVERSITY PRESS
CAMBRIDGE, MASSACHUSETTS, AND LONDON, ENGLAND

Published by arrangement with Alfred A. Knopf, Inc.

Portions of this book appeared in different form in the journals *Isis*, *Minerva*, *The Physics Teacher*, and *Technology & Culture*, and *Nineteenth Century American Science*, edited by George H. Daniels, The Northwestern University Press, 1972.

Grateful acknowledgment is made to the following for permission to reprint previously published materials, as indicated:

M.I.T. PRESS: Excerpts from an essay by Daniel J. Kevles, in Helen Wright, Joan N. Warnow, and Charles Weiner, eds., *The Legacy of George Ellery Hale*, M.I.T. Press, Cambridge, Mass., 1972.

MILITARY AFFAIRS: An article by Daniel J. Kevles in *Military Affairs*, December 1969, pp. 374–84.

Library of Congress Cataloging-in-Publication Data

Kevles, Daniel J.
    The physicists / the history of a scientific community in modern
America / Daniel J. Kevles.
        p.   cm.
    Originally published: New York : Knopf, 1977. With new pref.
    Includes bibliographical references and index.
    ISBN 0-674-66656-9
    1. Physics—United States—History.   2. Physicists—United States.
3.   Science—United States—History.   I. Title.
QC9.U5K48   1995
530′.0973—dc20   94-23708
                                                                              CIP

For Bettyann, Beth, and Jonathan,
who have little known this
dancer from this dance

Contents

Preface, 1995

The Death of the Superconducting Super Collider
in the Life of American Physics

This book accounts for the generation of American physicists who changed the world by forging atomic weapons in the laboratories of World War II, notably at the famed installation on the mesa at Los Alamos, New Mexico. It explores the roots of their revolutionary achievements in the efforts of their predecessors toward building American physics from the Civil War onward, addressing how they overcame the obstacles to the practice of pure science in the American democratic culture to win world standing in their discipline and recognition as assets to American society. It is also occupied with how, after the atomic bombings of Hiroshima and Nagasaki a half century ago, physicists of the Los Alamos generation became a kind of secular establishment—with the power to influence policy and obtain state resources largely on faith and with an enviable degree of freedom from political control.

What brought them to power is, to a considerable degree, what kept them there for most of the last half century—the identification of physics with national security. During World War II, a physicists' war, physicists not only devised the atomic bomb but also crafted numerous other technical miracles, including radar, rockets, and proximity fuses. Throughout the Cold War, they were crucial figures in maintaining American superiority in arms, advising on defense policy in relationship to technical possibilities, training students who joined the weapons laboratories, and carrying out basic research under military contracts. A number of them also fought to slow or halt the arms race, contributing importantly to the movements that led to the Nuclear Test Ban Treaty and the Strategic Arms Limitation Treaties, as well as energizing opposition to President Ronald Reagan's Strategic Defense Initiative. Whichever side they took on issues of arms control and defense, physicists remained honored and empowered because they remained essential in determining the shape and capabilities of American national security.

They were also valued for their role in the development of the high-

For help in preparing this account of the SSC, I am grateful to the Andrew W. Mellon Foundation for research support; to Kathy J. Cooke, Janet Jenks, and Ingeborg E. Sepp for indispensable assistance; to Linda R. Cohen and Gretchen A. Kalsow for providing me with an early draft of their analysis of the congressional votes on the SSC; to Steven E. Koonin, Diana Barkan, Thomas E. Everhart, Peter Galison, and Lillian Hoddeson et al. for critical readings; and to David Salzman for conversations.

technology postwar economy. Their contributions—made both indirectly, through military spinoffs, and directly, through academic and industrial research—have been essential in myriad fields, including transistors, computers, lasers, and fiber optics, areas that in recent years about one-third of physics Ph.D.s entered a short time after receiving their degrees. State officials, their eyes on the regional economy, established academically connected centers of technological innovation to exploit pertinent areas of physics, among other fields. Politicos at every level extolled research and training in the sciences as requirements of competitiveness in the international marketplace.

Under the circumstances, pure physics prospered handsomely in the United States, receiving abundant support for the pursuit of studies in esoteric areas of knowledge that were mainly conducted either in academia or in federally supported installations such as the Brookhaven National Laboratory, on Long Island. The federal budget for basic physics rose steadily through the late 1960s, turned down, then started rising again in the 1980s. A similar pattern characterized the American production of physics Ph.D.s, although senior members of the field found it disturbing that a steadily increasing fraction of the new recruits—42 percent in 1985–86—came from foreign countries. A good deal of basic research and training was conducted in small groups and concerned the physics of condensed matter, a branch of physics that is related to such practical arenas as semiconductors and superconductivity but that has its own basic conundrums to be explained.

A fresh demonstration of the value of such research was provided when, in 1986, several scientists at the IBM research laboratory in Zurich, Switzerland, reported a dramatic development in superconductivity. Discovered in 1911, superconductivity is the ability of certain materials to conduct electrical current with no resistance when they are cooled to within a few degrees of absolute zero, which is almost 460 degrees below zero on the Fahrenheit scale. The development in the 1960s of new alloys such as niobium-titanium, which permitted the maintenance of large superconducting currents at temperatures as high as 10 degrees above absolute zero, opened the door to the practical exploitation of superconductivity, primarily in the development of superconducting magnets. Such devices could achieve very powerful magnetic fields at high currents with no loss of energy. The scope of such exploitation was limited by several factors, however, including the cost of cooling the alloys to the extremely low temperatures at which they become superconducting. The IBM scientists devised a new compound that achieved superconductivity at a much higher temperature—30 degrees above absolute zero—and in 1987 physicists at several universities in the United States created still other compounds that would superconduct at 90 degrees above absolute zero. The results were scientifically exciting—at the American Physical Society meetings in March 1987, more than a thousand physicists came to hear talks on what was rapidly called high-temperature superconductivity—and the economic implications of the results were declared to be breathtaking by President Reagan himself.

The type case of Big Science was elementary-particle physics, a field in which some 10 percent of American physicists (about 4,000 practitioners) absorbed themselves in the 1980s and whose essential experimental tool was the particle accelerator. The first accelerators were devised in the early 1930s to explore the atomic nucleus; they operated at energies in the range of tens of millions of electron volts, which is characteristic of nuclear reactions. (An electron volt is the energy that an electron gains by crossing a difference in electric potential of one volt.) In the postwar decades, particle accelerators left nuclear physics behind, moving into the high-energy region necessary to probe the elemental structure of matter and forces. The accelerators successively designed for the task were increasingly sizable machines costing hundreds of millions of federal dollars. They ran at billions of electron volts and were exploited by large groups of researchers. High-energy physicists came to represent an influential subfield composed of overlapping groups: physicists who designed and built the accelerators; physicists who did experiments with them; and physicists who theorized about the meaning of the data they produced.

High-energy physicists were among the most prominent members of their profession—key figures in the nation's strategic defense and science policymaking councils and winners of many of the Nobel prizes awarded in physics to Americans. When they spoke, the American government tended to listen, at least about policy for basic physics. One of the leading rationales for the policy that gave abundant funds to particle physics was a reading of history: seemingly impractical research in nuclear physics had led to the decidedly tangible result of the atomic bomb; thus, research in particle physics had to be pursued because it might produce a similarly practical surprise. In the context of the Cold War, particle physics provided an insurance policy that if something important to national security emerged unexpectedly, the United States would have the knowledge ahead of the Soviet Union.

In the mid-1960s, high-energy physicists won authorization to build a still more powerful accelerator, to be located at the new Fermi National Accelerator Laboratory in Batavia, Illinois—despite widespread objection to proceeding with "the expensive irrelevance of a 200 billion electron volt accelerator to any real present national problem," as the *New York Times* editorialized, noting the troubles besetting the country as a result of the Vietnam War and the social tensions of the cities.[1] The Batavia accelerator was completed on time, within budget, and with a top energy of 500 instead of just 200 billion electron volts, making it the most powerful accelerator on earth. In the early 1980s, high-energy physicists urged the construction of a new, gargantuan accelerator—the Superconducting Super Collider, commonly called the SSC. It would be far more energetic than the original machine at Fermilab (as the Batavia installation was known) and would encircle an area 160 times as great. Nothing better symbolized the continuing power and influence of high-energy physicists in American society than the

---

[1] *New York Times*, July 16, 1967, p. 12.

serious consideration that Congress began, in 1985, to give the SSC project, which was then estimated to cost some $4 billion to build and several hundred million dollars a year to operate.

Yet now, just nine years later, the SSC is dead, having been killed in the House of Representatives in October 1993, partly in response to angry opposition from physicists themselves. A high official of the American Physical Society called the Super Collider project "perhaps the most divisive issue ever to confront the physics community."[2] The turn of events sent the nation's high-energy physicists reeling, but bad times have suddenly hit virtually every area of American physics. The sharp change in fortunes no doubt derived in part from the recent recession and the ongoing sluggishness of the economy. But far more important was the singular event of recent years—the end of the Cold War. In the post–Cold War environment, the death of the SSC expressed more than a setback for high-energy physics. It symbolized the end of an era for physics in the United States, especially its high-energy branch, and its relationship to the federal government.

Readers of this book will learn that the fate of physics in recent years was roughly adumbrated a century ago, when hard times overcame the earth sciences in the United States. During the years following the Civil War, federal support of research in the earth sciences had expanded enormously, supplying unprecedented patronage to disciplines relevant to one of the major national missions of the era: the exploration, settlement, and economic development of the Far West. Yet the degree of expansion in federal science generated suspicion among fiscal conservatives that the government was spending too much money for seemingly impractical work and among populist-oriented congressmen who did not see why funds should be spent for research on the slimy things of the earth when human beings were earning too little to keep their farms. During the depression of the 1890s, the conservatives and reformers formed a coalition that sharply reduced the government's support of impractical science and forced the federal scientific agencies onto bare-bones budgets. The depression was the occasion for the cutbacks, but the geographical frontier had closed, the country was emphasizing the agenda of its urban industrial order, and the earth-sciences agencies were no longer at the top of it.

A similar coalition formed in the 1960s, holding in one or another of its quarters that physics was too great an absorber of tax dollars, too little attentive to social issues, and too much a creature of the military and the war in Vietnam. It was this coalition that forced the leveling in the growth of federal funds for physics. The turn provoked much more far-reaching effects than the cutbacks of the 1890s, when federal patronage of science had been largely confined to support of work carried out directly by federal agencies. By the mid-1970s, in constant dollars, the federal budget for research and

---

[2] Steven Weinberg, *Dreams of a Final Theory: The Search for the Fundamental Laws of Nature* (New York, 1992), pp. 54–55.

development was 20 percent lower than it had been in 1967, but the number of physicists was higher. Since the federal government was the primary supporter of basic physics research everywhere it was practiced, the contraction adversely affected virtually the entire enterprise of the physical sciences in the United States, making jobs in academic physics, the center of basic research in many areas of the subject, particularly hard to find. High-energy accelerators were being shut down, research programs terminated.

The trend was well advanced by the time I finished writing this book, in the late 1970s, and it prompted me to conclude that American physicists had undergone a degree of disestablishment. Yet shortly thereafter, the disestablishment appeared to ease. The country was said to have been made militarily vulnerable by the reductions in spending for defense research and development (R&D) and by the weakening of the academic base for technical preparedness. It was declared to be economically vulnerable to vigorous foreign competition, especially from Japan, not only in the world's but even in the nation's own technological markets. Such concerns prompted a boost in federal research expenditures under President Jimmy Carter that continued under President Ronald Reagan, despite the budget slashing that marked the early Reagan years. By 1983, in constant dollars, federal R&D expenditures had reached the level of 1967. The largest share of the increase went to defense, many of whose research programs tended to be directed at a variety of physics-related subjects, including semiconductors, optics, lasers, integrated circuits—subjects that can yield results both of robust economic and of sensitive military utility. A then-recent Ph.D. in quantum electrodynamics, surprised to find herself engaged in defense-connected work at the Texas Research Institute, remarked that "all roads seem to lead to the Pentagon."[3] Support for high-energy physics followed the budgetary rise in the physical sciences, providing the high-energy community with means enough to initiate, in 1977, construction of a powerful new accelerator, called Isabelle, at Brookhaven that would use superconducting magnets to keep the beam on course. Funds also became available to upgrade the main existing machines, including the Stanford Linear Accelerator and the one at Fermilab whose energy would be doubled—to one trillion electron volts (TeV), making it a Tevatron—by similar use of superconducting magnets.

Still, enthusiasts of high-energy physics worried that resources remained inadequate to maintain American leadership in the field. They rightly argued that Europe, which supported the grand multinational accelerator installation CERN (for Conseil Européen de Recherche Nucléaire), on the French-Swiss border, was spending twice as much on high-energy research relative to GNP as was the United States. (Indeed, American investment in research in all the physical sciences was, in proportion to its population and wealth, similarly low.) In 1982, Fermilab had money to operate at only about a quarter of its capacity, while CERN was running at almost a three-quarters level. More-

---

[3] Bruce Schechter, "Beyond the Ivory Tower," *Physics Today*, 39 (June 1986), 36.

over, European accelerators were beginning to outclass their American counterparts in the significance of the experimental evidence they were producing and the energies they were seeking to reach. Sidney Drell, the deputy director of the Stanford Linear Accelerator, plaintively remarked: "The quality of a society is indicated by the questions it asks. One of these questions is, What is man made of? The answer is matter, and it is the nature of matter that is the domain of high-energy physics. The society that doesn't ask this question is a suffering society."[4]

The drive for more powerful accelerators was symbiotically tied to the development and testing of elementary-particle theory, which in the 1970s had achieved a formal, overarching structure called the Standard Model. The model seeks to account for three of the four known forces in nature: the electromagnetic force, which acts on ordinary charged particles such as electrons and protons; the weak force, which is involved in radioactive decay; and the strong force, which holds together the particles in the atomic nucleus. (The fourth force, gravity, has so far remained beyond the reach of any accepted theoretical model.) The Standard Model holds that all matter is formed of particles called quarks and leptons, that the existence and behavior of these particles is governed by different types of force fields, and that the interactions of these fields are mediated by the exchange of elementary particles. Some of these exchange particles tend to be very massive. Since mass is the equivalent of energy, they can represent the compaction of an enormous quantity of energy, an amount rarely found concentrated in single reactions in the contemporary universe. However, they can be—and many had been—produced in the high-energy reactions that occur in accelerators, adding weight to the evidentiary foundation of the Standard Model, which by 1980 included the detection of all the leptons and quarks (except of the "top" quark) whose existence it predicted. In one of its major triumphs, the Standard Model also unifies the electromagnetic and the weak forces, convincingly holding that at high energies a deep symmetry characterizes both of them and they operate as a single "electroweak" force. And it has accomplished plausible though not entirely satisfying unions of the strong force with the electroweak one in a so-called Grand Unified Theory.

A number of particle theorists exploited the Standard Model to understand the behavior of the universe close to the time of its origin in a Big Bang, when enormous energies were concentrated in a very small volume. The Big Bang hypothesis had been bolstered by several arresting classes of observational evidence, particularly detection of a low-energy microwave background radiation that theory predicted should be present throughout the contemporary universe as a residue of its colossal birthing explosion. Using the Standard Model, physicists speculated about cosmological processes back to the first few minutes of the universe, even to its first tiny fractions of a micro-second. High-energy accelerators go some distance toward reproduc-

[4] Bruce Schechter and Gary Taubes, "Battle of the Big Machines," *Discover* (April 1982), p. 68.

ing the energies that were present at those early moments; thus, together with the Standard Model, they provide a window directly onto some of the phenomena of the early universe—strongly suggesting, for example, that as the universe cooled, the deep symmetry of the electroweak force was broken in a way that generated the electromagnetic and weak forces. Grand Unified Theory, by conceptually analyzing phenomena at still higher energies, reaches back theoretically to the behavior of the universe at still earlier moments. It permits physicists to ask not only what the properties of the universe are but why it possesses them. At the beginning of the 1980s, several groups of physicists showed that the Grand Unified Theory could plausibly account for some of those properties. Exploiting the theory, they generated a line of analysis that many physicists found compelling, partly because it solves several conundrums about how the universe came to be the way it is, but also because it provides an entry to the early universe that allows many—if not all—of its features to be calculated rather than posited as arbitrary initial conditions.

The accelerators of the 1970s were inadequate by any measure to test all the facets and assumptions of the Standard Model (and no earthly machine could conceivably reach the enormous energies—a trillion times the designed energy of the Tevatron—necessary to test the Grand Unified Theory). At the opening of the 1980s, electroweak unification theory had been experimentally confirmed indirectly but awaited direct confirmation of one of its essential points—that the electromagnetic force and the weak force are mediated by the photon plus three massive particles from a class called bosons, specifically, the Z-zero, the W-plus, and the W-minus. (The designation "W," a long-standing commonplace in theoretical speculations, stood for weak, whereas the name "Z" had been coined by Steven Weinberg, then at Harvard University, who independently co-devised electroweak theory in 1967, and who would share the Nobel Prize in physics in 1979 for his contributions to it. Weinberg, who in 1983 moved to the University of Texas at Austin, says that he picked "Z" as the name for the W's new sibling partly "because Z is the last letter of the alphabet, and I hoped that this would be the last member of the family."[5]) It was an ambition of high-energy physicists in the United States to beat the Europeans to the punch in observing the particles, using one of the accelerators whose upgrading was then under way. Problems with the development of the necessary superconducting magnets had slowed the enhancement of the accelerator at Fermilab, however, and had put the Isabelle project completely on hold. And then, in January 1983, a team at CERN announced that they had detected the two W particles, and in June, that they had found the Z. In an editorial, the *New York Times* twitted the country's high-energy community: "Europe 3, U.S. Not Even Z-Zero," adding, "The 3–0 loss in the boson race cries out for earnest revenge."[6] What American high-energy physicists were resolved upon was not revenge but a restoration of preeminence—via the Superconducting Super Collider.

[5] Weinberg, *Dreams*, pp. 119–120.
[6] *New York Times*, June 6, 1983, p. 16.

Their eagerness for the SSC was prompted in significant part by the intellectual exigencies of elementary particle physics. A particle accelerator operating beyond a trillion volts would reveal phenomena that must have occurred in the early moments of the universe, when electroweak unity came to be broken; and in any case certain essential theoretical problems connected with the Standard Model could be illuminated only at accelerator energies of at least five to ten trillion electron volts. High-energy physicists were particularly interested in probing for evidence of what they call the Higgs force field—named after Peter Higgs, of Edinburgh University, who had most clearly postulated it in 1964—which was believed to play a role in the shattering of electroweak unification and was considered necessary to explain why the particles in electromagnetic and weak interactions possess the masses they do; indeed, why they have any mass at all. On theoretical grounds, it was expected that the Super Collider would reveal the presence of a new particle called the Higgs boson. Leon Lederman, the director of Fermilab, attempted to explain the Higgs boson's relationship to the behavior of the particles that come out of electroweak unification by telling a Senate hearing to think of a group of extraterrestrials watching a soccer game who are somehow incapable of seeing the ball: "They see a lot of people running around seemingly at random in a chaotic disorganized activity, but if someone postulates the existence of a soccer ball, then the whole thing becomes clear and simple and elegant."[7] Theory predicted that the Higgs soccer ball is a particle with a mass equivalent to an energy of up to a trillion electron volts, which is about a thousand times the mass of the proton.

Lederman was one of the principal spokesmen for the SSC—in congressional hearings an unbridled advocate of its merits, which he advanced with colloquial and often comic directness. An accomplished high-energy experimentalist, he had made Nobel Prize–winning contributions to the development of the Standard Model during the 1960s, although the prize itself did not come until 1988. For some time, along with other physicists, he had been dreaming of building a huge, multinationally sponsored accelerator powerful enough to reveal the Higgs particle, but at a meeting of high-energy physicists in Snowmass, Colorado, in mid–1982, he had advanced the idea of the United States's recapturing leadership in the field by building a super, predominantly American, machine.[8] In July 1983, the High Energy Physics Advisory Panel to the Department of Energy, the agency that funds almost all high-energy machines in the United States, issued a formal recommendation for the SSC, stressing its essential importance to further progress in elementary-particle physics.

The proposed machine, a circular accelerator, would operate at perhaps

[7] U.S. Congress, Senate, *Joint Hearing before the Committee on Energy and Natural Resources and the Subcommittee on Energy and Water Development, Importance and Status of the Superconducting Super Collider,* 102nd Cong., 2nd Sess., June 30, 1992, p. 25.

[8] Adrienne Kolb and Lillian Hoddeson, "The Mirage of the 'World Accelerator for World Peace' and the Origins of the SSC, 1953–1983," *HSPS: Historical Studies in the Physical and Biological Sciences,* 24 (Part 1; 1993), 117–120.

ten trillion electron volts, an energy a million times greater than that of the accelerators in the 1930s and high enough to reveal phenomena in the Higgs region. It seemed technically feasible: the superconducting magnets for Isabelle and for the Fermilab enhancement had by now been successfully developed; they could be scaled up and would serve to keep the SSC particle beam on its curving track. The machine was estimated to be costly, but the project seemed so important to the high-energy community that many—though not all—of its policymaking members were willing to scrap Isabelle to get it. (Even though its magnet problems had been solved, Isabelle would not be powerful enough to explore the energy region where Higgs phenomena would manifest themselves. Nick Samios, the director of Brookhaven, nevertheless called the scrapping of Isabelle "one of the dumbest decisions ever made in high energy physics."[9]) In November 1983, the Department of Energy halted work on Isabelle and obtained authority to redirect its funds to research on the SSC.

Extensive technical studies of the proposed machine followed, and by 1986 the SSC had taken detailed conceptual shape. It would accelerate two beams of protons, each in the opposite direction from the other, through a circular tunnel some fifty-two miles in circumference to an energy of twenty trillion electron volts. Because they would be rotating contrary to each other, the two proton beams could be made to collide with an energy of forty trillion electron volts. (Such an energy was needed to explore phenomena in the Higgs range—that is, of several trillion electron volts—because a great deal of the acceleration energy is shared among the constituents of the proton, which do not participate in the interactions of interest, leaving only a fraction of that energy available for the particles that do.) The SSC's acceleration energy would be sixty times greater than the CERN collider's and twenty times greater than that of Fermilab's upgraded machine. Allowing for inflation, it would cost roughly $6 billion to construct over ten years. It would be by far the most powerful proton accelerator in the world, could be ready by the 1990s, and would restore the United States's preeminence in high-energy physics.

The price might have been high, but to the devotees of particle physics it unquestionably merited payment. To many of them, particle physics was a transcendent pursuit made holy by its quest for a theory of physical nature at its deepest level and for how that theory might illuminate the origins and development of the universe. Some of them likened the great particle accelerators to modern-day cathedrals. Indeed, their devotion to the newly fashioned mixture of particle physics and cosmology resonated with popular educated culture of the period, in which particle-physics theories of the early universe were prompting an avalanche of quasi-religious treatises. (In *Roger's Version*, the novelist John Updike expressed the gist of the outpourings in the

---

[9] The report is reprinted in U.S. Congress, House, *Hearings before the Committee on Science, Space, and Technology: Superconducting Super Collider,* 100th Cong., 1st Sess., April 7, 8, 9, 1987, pp. 59–132 (hereafter, House, *SST Hearings,* 1987); Weinberg, *Dreams,* p. 265.

remarks of an upstart computer hacker to the divinity teacher Roger Lambert: "Dr. Lambert, aren't you excited by what I've been trying to describe? God is *breaking through*. They've been scraping away at physical reality all these centuries, and now the layer of the little left we don't understand is so fine God's face is staring right out at us."[10]) Lederman would dip into the cultural trend by referring to the Higgs boson as "the God Particle" in a book that he published in 1993 under that title. The book, which illuminatingly recounted the history of atomic and particle physics, especially the experimental side of the high-energy epoch, amounted to a historical brief for the SSC, including a rendering of what Lederman semimockingly called "The Very New Testament": "And the Lord came down to see the accelerator, which the children of men builded. And the Lord said, Behold the people are unconfounding my confounding. And the Lord sighed and said, Go to, let us go down, and there give them the God Particle so that they may see how beautiful is the universe I have made."[11]

Steven Weinberg declined to indulge in such notions. In a book on the very early universe, he had written that "the more the universe seems comprehensible, the more it also seems pointless"—by which he meant in part that the more the fundamental principles of the universe were revealed, the less they seemed to have to do with us. When given an opportunity in a congressional hearing to comment on whether the SSC might reveal the face of God, he maintained a prudent silence.[12] What Weinberg preferred to emphasize was that physicists were "desperate" for the SSC because they were "stuck" as physicists in their progress toward what he called "a final theory" of nature—a complete, comprehensive, and consistent theory that accounted for all the known forces, fields, and particles in the universe.[13] In eloquent testimony before Congress and elegant prose for the public—in a book called *Dreams of a Final Theory*, published in 1992—he explained the intellectual content of the Standard Model, including the questions concerning it that needed to be explored at the energy level of the Higgs field and that the SSC would address. The SSC was a sure bet, Weinberg stressed, not because it would reveal the deity or enhance American prestige, but because even if it did not find the Higgs boson it would expose the existence of new forces and phenomena that would bring the achievement of a final theory closer.

For the most part, conventional religious implications had no bearing on the particle-physics community's eagerness for further knowledge. They located themselves in the traditional drive to understand nature that had originated with the ancients and that—in the view of both Lederman and Weinberg—the United States might break faith with only at its peril. In 1985, in an article on the SSC, Lederman and Sheldon Glashow, a co-winner of

---

[10] John Updike, *Roger's Version* (New York, 1986), p. 20.

[11] Leon Lederman, with Dick Teresi, *The God Particle: If the Universe Is the Answer, What Is the Question?* (New York, 1993), p. 24.

[12] Steven Weinberg, *The First Three Minutes* (updated edition; New York, 1988), p. 154; Weinberg, *Dreams*, pp. 253–254, 243–244.

[13] Weinberg, testimony, House, *SST Hearings*, 1987, pp. 243–244.

the Nobel Prize with Weinberg for his role in the development of elec-troweak theory, averred that "high-energy physics must go in this direction or terminate the 3000-year-old quest for a comprehension of the architecture of the subnuclear world," adding, "If we forgo the opportunity that [the] SSC offers for the 1990s, the loss will not only be to our science but also to the broader issue of national pride and technological self-confidence. When we were children, America did most things best. So it should again."[14]

Such arguments received a friendly reception in congressional hearings on the progress of the SSC planning program, where the question of whether God was to be found in the particles cropped up only occasionally, but where more than one congressman reminded scientific witnesses that the SSC might be an unaffordable luxury. At a House hearing in 1985, Congressman Joe L. Barton of Texas asked the physicist Alvin W. Trivelpiece whether, high-en-ergy physics being an international enterprise, the United States should build the SSC by itself. Trivelpiece, the director of the Office of Energy Research in the Department of Energy and an enthusiast of the SSC, was working hard on its behalf. He had to say, nevertheless, that "a project of this sort is almost certainly going to be an international activity one way or another," continu-ing, "The idea or the luxury that this would be done entirely within the United States exclusively by U.S. scientists with exclusive U.S. support is unrealistic."[15] To the congressmen, the costs of the project had to be closely counted, as always, but especially now that the passage of the Gramm-Rud-man-Hollings Act had committed both Capitol Hill and the White House to deficit reduction.

President Reagan's science adviser, the physicist George Keyworth, was on record that if the SSC were built elsewhere it would be "a serious blow to U.S. scientific leadership."[16] Trivelpiece persuaded Secretary of Energy John S. Herrington, a California attorney who had come to his post in January 1985 freely admitting that he knew nothing about energy issues, nuclear or otherwise, to support the SSC. Herrington, who was close to Reagan, lobbied hard for the project, but he faced opposition from hard-nosed officials who saw no need for it and worried about its impact on the budget. In a showdown at the White House, President Reagan, having heard the arguments on both sides, issued his decision in the form of an anecdote about the Oakland Raiders star quarterback Kenny Stabler. Taking a card from his pocket, Reagan read a poem by Jack London that began: "I would rather be ashes than dust / I would rather that my spark / Should burn out in a brilliant blaze / Than it should be stifled in dry rot"—and ended: "I shall

[14] Sheldon L. Glashow and Leon M. Lederman, "The SSC: A Machine for the Nine-ties," *Physics Today*, 38 (March 1985), 37, 34.

[15] U.S. Congress, House, *Hearing before the Subcommittee on Energy Development and Ap-plications of the Committee on Science and Technology: Status and Plans of the United States and CERN High Energy Physics Programs and the Superconducting Super Collider [SSC]*, 99th Cong., 1st Sess., Oct. 29, 1985, p. 20.

[16] Gary Taubes, "The Atom," Collision over the Super Collider," *Discover*, July 1985, p. 62.

use my time." According to Reagan, Stabler, when once asked about the poem, said that it meant "Throw deep," which Herrington took to mean that he should go for the SSC.[17]

The arguments being made for the SSC in and out of the White House, including in newspapers, magazines, and congressional hearings, indicated that there was more than one intended receiver—not only the intellectual adventurers of high-energy physics but also their prospective allies in the American political economy. Enthusiasts of the SSC held that it would pay considerable practical dividends. The outcomes of cutting-edge scientific endeavors being largely unpredictable, they could not be very specific about the future; they thus enlisted the historical record of particle physics, pointing to its past spinoffs and extrapolating from them to sketch the SSC's practical promises. Once accelerators had moved beyond the relatively low energies of nuclear interactions to the higher energies of elementary-particle research, the knowledge of nature that they revealed was, in and of itself, no longer practically relevant. Elementary-particle research had produced many highly trained physicists, however, a number of whom migrated from the field and successfully deployed their skills in other branches of science and technology. And since the first inventions of particle accelerators, a series of useful dividends had come from the development and operation of the machines themselves.

For example, accelerators running in the range of tens of millions of electron volts supply radiations used in the processing of foods and materials and in the treatment of cancer. (At a House hearing in April 1987, Lederman declared that "one person in eight in this room will at one point in their life be treated in a hospital by an accelerator, generally in a beneficial manner."[18]) Accelerators at the level of hundreds of millions to several billion electron volts provide sources of powerful light beams that can etch integrated circuits onto semiconductor chips at much greater densities than could otherwise be achieved. And most contemporary high-energy accelerators rely on computerized methods and sophisticated technologies to screen and analyze the superabundance of data they generate that have been exploited in many other fields. The drive to develop machines operating at or near a trillion volts—the push for Isabelle and then the Doubler at Fermilab—had produced significant advances in the technologies of superconducting magnets. In 1991, Lederman testified to the House Budget Committee that these advances had "enabled" the deployment of the "powerful medical diagnostic tool called magnetic resonance imaging," continuing, "Some 25 companies are making these things in a new industry that is approaching $1 billion a year."[19]

Advocates of the SSC declared that it, too, would assist in the battle

---

[17] Ibid., p. 62; Irwin Goodwin, "Reagan Endorses the SSC, a Colossus among Colliders," *Physics Today*, 40 (March 1987), 48.

[18] House, *SST Hearings*, 1987, p. 263.

[19] U.S. Congress, House, *Hearing before the Task Force on Defense, Foreign Policy and Space, Committee on the Budget, Establishing Priorities in Science Funding*, 102nd Cong., 1st Sess., July 11 and 18, 1991, p. 78 (hereafter, House, *Hearing, Task Force*, 1991.)

against cancer: a low-energy accelerator used to inject particles into the super machine would produce excess protons; they would be diverted to cancer treatment in a facility that the University of Texas Southwestern Medical Center, based in Dallas, would establish on the site. Yet nothing in the expected spinoffs of the SSC received more attention than the stimulation it was—or was said to be—providing to advances in superconducting technologies. In the future, spinoffs from the project's development of superconducting magnets would contribute to innovations in power generation and transportation in the form, for example, of magnetically levitated trains. In the here and now, the project's demand for large superconducting magnets was said to be yielding important improvements and price reductions in the niobium-titanium wires needed to carry the enormous superconducting currents. Deputy Secretary of Energy W. Henson Moore, III, a lawyer and former congressman from Louisiana, permitted enthusiasm to take him much further than Lederman when he indicated to a congressional committee that magnetic resonance imaging had been made possible by the work on superconducting magnets for the SSC itself.[20]

Lederman and Glashow pointed out that the construction of the collider would no doubt result in the improvement of tunneling technologies, but it did not take a physicist to recognize the direct and proximate dividends of the SSC's $6 billion price tag. The project involved boring a 52-mile circular tunnel 14 feet in diameter and rifling it with steel pipes and more than 8000 superconducting magnets that would be energized by 625,000 miles of superconducting cable. It required a dozen refrigeration plants supplied with 300 Goodyear blimps worth of helium to cool the magnets and two subterranean experimental halls comparable in size to a football stadium that would each house a particle detector weighing more than a battleship. The SSC would produce an abundance of industrial contracts and, as one congressman put it, "an awful lot of jobs." The circumference of the collider circle approximated that of the Washington Beltway. In the circles of the concrete industry, the SSC came to be known as "the big pour."[21]

Some five to eight thousand jobs were estimated to come to the locality where the SSC would be built. Trivelpiece remarked that there were "lots of physicists, politicians and businessmen out there lusting after it."[22] More than half the states in the Union took steps to enter the site-selection competition, which began on April 1, 1987. That month, at the first congressional hearing on the project after the Reagan administration had endorsed it, dozens of statements were entered into the record, including declarations of support

[20] U.S. Congress, Senate, *Hearing before the Subcommittee on Energy Research and Development of the Committee on Natural Resources, on the Department of Energy's Superconducting Super Collider Project*, 102nd Cong., 1st Sess., April 16, 1991 (Washington, D.C., 1991), p. 12 (hereafter, Senate, *Hearings, Energy, SSC*, 1991.)

[21] Taubes, "The Atom," pp. 62–63; Irwin Goodwin, "Amazing Race: The SSC Contest Generates Disorder and Discord," *Physics Today*, 41 (May 1988), 69.

[22] Irwin Goodwin, "Race for the Ring: DOE Reacts to Congress's Anxieties on SSC," *Physics Today*, 40 (Aug. 1987), 47.

for the SSC and the qualifications of their states from some ten governors, lieutenant-governors or their representatives, and twenty-five congressmen. Would-be competitors had to satisfy several technical criteria for the site, including geological soundness and access to sufficient electrical power, and several budgetary ones, too, including a willingness to provide the federal government with some 16,000 acres of land gratis for the SSC's tunnel and laboratory installations. The *New Republic* called the site-selection process an invitation to "quark barrel politics."[23]

On November 10, 1988, the day after George Bush was elected to the presidency, Secretary Herrington announced at a press conference that the winner was Waxahachie, Texas. ("You wax a car; you live in [Walks-a-hatchy]," Congressman Joe Barton, who represented the area, said, distinguishing the pronunciation of the name.[24]) A town of 18,000 people about 25 miles southwest of Dallas, Waxahachie was best known, to the extent it was known at all, for the turn-of-the-century Victorian ambience and pink granite courthouse that provided the setting for some two dozen films, including *Bonnie and Clyde* and *The Last Picture Show*. Now it would achieve a different kind of attention as the home of the facility that Herrington declared would be named the "Ronald Reagan Center for High Energy Physics." The mayor of Waxahachie revealed his constituents' overall enthusiasm for the project, declaring at the end of a windmill speech, "And this nation must be the first to find the Higgs scalar boson!"[25]

Herrington insisted that "there are no politics in this" and that neither Reagan nor president-elect Bush had been involved in the decision. Waxahachie was, in fact, a compelling site independent of politics. It had been ranked outstanding on every major criterion by a selection committee of the National Academy of Sciences, which had submitted a short list of eight qualifying sites in as many states to the Department of Energy in January 1988. For example, the site straddled a large geological formation called Austin Chalk, which is soft enough to permit rapid tunneling but strong enough on its own to hold shape after having been hollowed out. And Texas, which would provide 200 square miles of land for the facility, had also promised the project $1 billion, a sweetener offered by no other state. Congressman Barton declared at the press conference, "We're celebrating in Texas this afternoon because we are the best."[26]

Observers could not help noticing, however, that the president-elect called Texas home and that the Texas congressional delegation was a powerhouse that included Speaker of the House Jim Wright and Senator Lloyd

[23] Ibid., p. 48; Robert Bazell, "Quark Barrel Politics," *New Republic*, 196 (June 22, 1987), pp. 9–10.

[24] U.S. Congress, House, *Hearing before the Committee on Science, Space, and Technology: The Superconducting Super Collider Project*, 103rd Cong., 1st Sess., May 26, 1993, p. 108 (hereafter, House, *SST Hearings*, 1993).

[25] Lederman, with Teresi, *The God Particle*, p. 380.

[26] J. Michael Kennedy, "For Waxahachie, It Was a 'Super' Day"; Robert Gillette, "Texas Selected . . . ," *Los Angeles Times*, Nov. 11, 1988, p. 19.

Bentsen, who had just run for the vice-presidency on the Democratic ticket, and Phil Gramm, a co-author of the deficit-reduction act and Waxahachie's congressional representative before his election to the Senate. (Gramm, ecstatic like his fellow Texans at the choice of their state, dubbed the SSC "another Spindletop oil gusher," the harbinger of a new high-technology economic revolution called "High Tex."[27]) It was not unreasonable to think that the collider might need the help of such allies. In 1987 Congress, wondering where funds for the SSC would come from, had flatly refused to appropriate any money to initiate its construction. In the spring of 1988, it had slashed the administration's budgetary request for the project by 75 percent, allowing only $100 million to keep research on the collider moving ahead. By then, all but eight states had been eliminated from the site competition, which presumably reduced the other forty-two's interest in the project, and Department of Energy officials had to admit that they had no firm commitments of any significant foreign assistance.

In 1989 Congressman Tom Bevill of Alabama, the chairman of the House Appropriations Committee, threatened to hold up construction funds again without evidence of substantial foreign contributions. According to press reports, however, a White House meeting with Bush and a letter from his secretary of energy, James Watkins, assured Bevill that Japan would soon contribute between half a billion and a billion dollars in cash or in kind to the SSC.[28] In any case, that year first the House and then the Senate, where Gramm and Bentsen had lined up sixty colleagues, voted decisively to fund the construction of the SSC, agreeing in the end on a total appropriation to the project of $225 million for 1990, only $25 million less than the Bush administration had requested, and accepting a total cost for its construction of $5.9 billion. Two years later, in the summer of 1991, Congress appropriated more than $450 million for the collider, and a move in the Senate to kill the project failed by a margin of almost 2 to 1.

The SSC appropriations were to be an add-on to the existing high-energy budget, which amounted annually to more than half a billion dollars and had remained steady in constant dollars. By 1991, the add-on had resulted in the award of more than $100 million in grants and contracts for SSC research to scientists and engineers at 90 universities and institutes in roughly 30 states. The flood of money exemplified an upward trend in appropriations for basic physics—and, indeed, for all basic academic research, which had continued to rise, reaching some 20 percent more in constant dollars than the pre-downturn peak, in 1967. Jobs for physicists were plentiful. Some 30,000 physics Ph.D.s were working in the United States, and a shortage was predicted for the 1990s, with accelerator physicists becoming particularly scarce. By 1990 some 150 of them were hard at work in Waxahachie, part of

[27] Irwin Goodwin, "DOE Picks Texas for 'Gippertron' amid Political and Managerial Collisions," *Physics Today*, 42 (Feb. 1989), p. 96.

[28] David Lindley, "SSC Is Brought Another Step Nearer Reality," *Nature*, 340 (July 6, 1989), 4; Mark Crawford, "House Approves SSC Construction," *Science*, 245 (July 7, 1989), 25.

an overall SSC workforce that was moving past 1,000 people under the direction of Roy Schwitters, who was appointed head of the SSC in January 1989. Schwitters, in his mid-forties, was a distinguished high-energy experimentalist, a former member of the Harvard faculty, and a project leader at Fermilab. He was urbane, articulate, and appeared to be politically savvy as well as technically competent. He knew his business and was a very good bet to turn the gargantuan project into the reality of an operating accelerator.

Although physicists, like other Americans, have embraced political engagement in arenas of technological policy such as arms control, they have tended to resist it on behalf of their science, fearing that it would undercut their social authority, not to mention their self-image, if they behaved like just another interest group in American society. However, the SSC inflamed resentments among physicists who did not work in elementary particles that had long been simmering against the power, authority, and budgetary leverage of those who did. Once the collider became a serious public-policy initiative, opposition to it from within the physics community was openly expressed in a variety of forums, especially hearings before the House Committee on Space, Science, and Technology. One drumbeat of criticism came from Rustum Roy, a distinguished materials scientist at Pennsylvania State University who in 1985 had co-authored a book—*Lost at the Frontier*—critical of American science policy since World War II for giving too much support to esoteric fields of research. Roy considered high-energy physicists "spoiled brats" for wanting a multibillion-dollar accelerator when the country was running up $200 billion annual deficits.[29] Yet Roy was regarded as something of a noisy eccentric in American science policymaking. More telling dissent came from within the scientific establishment—from distinguished scientists who worked in fields such as condensed matter and superconductivity. They included former presidents of the American Physics Society and Nobel laureates. Most respected and admired particle physics, but like the laureate J. Robert Schrieffer, who called himself a "loyal opponent" of the initiative to build the machine, none thought it a justifiable use of public resources at its multibillion-dollar price tag.[30]

The opposition fire intensified after the passage, in 1990, of the Omnibus Budget Reconciliation Act, which imposed caps on defense and discretionary domestic spending, including research and development. The caps made R&D funding into a zero-sum game and sent a frisson of apprehension through the American physics community. Even the powerful high-energy sector was told in 1991 that the Department of Energy's particle physics programs might well have to be cut by 10 percent. In January of that year, the council of the American Physical Society resolved that, though "the SSC should be built in a timely fashion," the necessary funds "must not be

[29] Therese Lloyd, "SSC Faces Uncertain Future," *The Scientist*, Feb. 23, 1987, p. 8; Rustum Roy, letters, *Physics Today*, 38 (Sept. 1985), 9–10; 39 (April 1986), 11; 40 (Feb. 1987), 13.
[30] House, *SST Hearings*, 1993, p. 89.

[obtained] at the expense of the broadly based scientific research program of the U.S." In April, the Nobel laureate Nicolaas Bloembergen, then president of the society, conveyed the resolution to a Senate hearing, pointedly noting the vulnerability of science to the budgetary caps and warning that, under the current fiscal circumstances, "major new initiatives, whose annual costs are projected to escalate for several years, threaten the already precarious house of Government-funded research."[31]

The budgetary claims of the SSC particularly exercised condensed-matter physicists, who had long been given reason to think that at least some particle theorists considered their work intellectually second class. While the high-energy community proclaimed that it quested ever more deeply after the fundamental laws and constituent particles of the universe, condensed-matter physicists dealt with laws that govern matter as it exists in the messy aggregate—for example, matter in the solid state, such as semiconductors—that human beings devise or encounter. It rankled in solid-state quarters that Murray Gell-Mann, the brilliant particle theorist and Nobel laureate, had mocked the field as "squalid-state" physics.[32] Perhaps the leading antagonist of the particle theorists' position was Philip Anderson, a Nobel Prize winner for his work in condensed-matter physics, who had been challenging the merits of their claim to deep fundamentality at least since the early 1970s and who repeated his arguments in several hearings on the SSC. Big and complex aggregates of matter "often behave in new ways and according to new laws," he typically explained. "These new laws don't contradict the laws the elementary-particle people discover; they are simply independent of them, and I would argue they are in no way any less . . . fundamental."[33]

Robert Schrieffer stressed that elementary-particle physics cast no light on the behavior of ordinary matter even in its disaggregated forms, which meant that it was irrelevant to most of atomic and molecular physics and chemistry and as such to any science with utilitarian potential. Anderson told Congress that fields such as condensed matter served society at far lower cost and with far greater payoffs than did elementary-particle research. It cost one-third to one-sixth as much to train a graduate physicist in condensed matter as in high energy. "Dollar for dollar," Anderson testified, articulating the conviction of many of his colleagues, "we in condensed-matter physics have spun off a lot more billions than the particle physicists . . . and we can honestly promise to continue to do so."[34]

A number of physicists were angered by the spinoff benefits that had been claimed for high-energy accelerators, especially the decisive contribu-

[31] Senate, *Hearing, Energy, SSC,* 1991, p. 43; *Physics Today,* 44 (March 1991), 79.

[32] David Berreby, "The Man Who Knows Everything: Murray Gell-Mann," *New York Times Magazine,* May 8, 1994, p. 27.

[33] House, *Hearing, Task Force,* 1991, p. 64; Silvan S. Schweber, "Physics, Community, and the Crisis in Physical Theory," *Physics Today,* 46 (Nov. 1993), 36.

[34] House, *SST Hearing,* 1993, pp. 89–90; U.S. Congress, Senate, *Hearing before the Subcommittee on Energy Research and Development of the Committee on Energy and Natural Resources, Department of Energy's Fiscal Year 1990 Budget Request for the Office of Energy Research,* 101st Cong., 1st Sess., Feb. 24, 1989, p. 135.

tions to the development of magnetic resonance imaging (or MRI) that had been implied by Lederman and explicitly declared by Deputy Secretary Moore. If the truth be told, Theodore Geballe, a condensed-matter physicist at Stanford who had spent part of his career at the Bell Telephone Research Laboratories, informed a congressional committee, the SSC's demand for superconducting wire had "at most, caused a blip on the market price."[35] Nicolaas Bloembergen testified in 1991 that not superconducting magnets, the superconducting magnet industry, nor magnetic resonance imaging had come primarily from the development of accelerators: "As one of the pioneers in the field of magnetic resonance, I can assure you that these are spinoffs of small-scale science."[36] In a follow-up letter to an official at Fermilab that was entered into evidence in a congressional hearing in the spring of 1992, Bloembergen allowed that superconducting wire technology had, in fact, greatly benefited from the work at Fermilab and for the SSC and that this benefit had improved the equipment in magnetic resonance imaging. But he called Moore's testimony that such imaging was therefore a spinoff of Fermilab and the SSC "unwarranted, and . . . ill-advised," averring that "MRI would be alive and well today even if Fermilab had never existed."[37]

To opponents of the SSC, its advocates appeared to be promulgating gross misrepresentations—not least in hard-sell symposia sponsored by the Texas National Research Laboratory Commission, which was devoted to promoting the collider—of what the machine would do for the world. Philip Anderson reckoned in one hearing that the dilution of the message about the SSC "in the Department of Energy and in the political rhetoric" produced "claims that particle physics did everything from MRI and the computer revolution to the television screen and sliced bread." To Anderson, "The saddest sight of all is to see officials of the Department responsible for our energy supply deliberately misleading Congress and the public with these false claims, and to see my particle-physics colleagues, many of whom I admire and respect, sitting by and acquiescing in such claims."[38]

Particularly galling to condensed-matter physicists, high-energy research was taking a disproportionately large share of the federal basic physics research budget. By Anderson's calculation, particle physics received some ten times more money per capita than did other fields. James Krumhansl, a condensed-matter physicist and a former president of the American Physical Society, estimated that perhaps one thousand physicists might be served directly by the SSC—which meant that the machine's cost of $5 billion would amount to a ten-year grant of $5 million to each particle physicist, and this

[35] House, *SST Hearing,* 1993, p. 65.

[36] Senate, *Hearing, Energy, SSC,* 1991, p. 43.

[37] Bloembergen to Richard A. Carrigan, Jr., May 21, 1992, in U.S. Congress, Senate, *Joint Hearing before the Committee on Energy and Natural Resources and the Subcommittee on Energy and Water Development, Committee on Appropriations, Importance and Status of the Superconducting Super Collider,* 102nd Cong., 2nd Sess., June 30, 1992, p. 12. On later reflection, Lederman said that Fermilab could take "a modicum of credit" for the MRI industry. Lederman, with Teresi, *The God Particle,* pp. 233–234.

[38] House, *Hearing, Task Force,* 1991, p. 65.

at a time when the average grantee in other fields was struggling to obtain $40,000 for one year.[39] Eric Bloch, an electrical engineer and a former vice president at IBM who had recently stepped down as head of the National Science Foundation, told a House hearing that, although he considered the SSC a valuable scientific instrument, he ranked it well below other projects. "Let it wait," he added. "We have more important things to fund within the R&D budget—people, instrumentation, facilities, economic competitiveness . . . A rich country can afford [the SSC]. We're not as rich anymore as we were at one time."[40]

Not as rich, and yet being asked to pay more. In 1988, a report from the Congressional Budget Office noted that the costs of three of the four accelerator projects of the 1980s had increased significantly and concluded that therefore the price of the SSC might also escalate. In fact, by 1989 it was becoming evident that the machine's superconducting magnets might be inadequate to their task, requiring modifications in the design of the accelerator that could increase costs by as much as 30 percent. Deputy Secretary of Energy W. Henson Moore reportedly told the project's program managers that "either we're going to build it for the figure of 5.9 [billion dollars] or we're not going to build it at all."[41] Nevertheless, a redesign of the magnets, an increase in the size of the accelerator circle to fifty-four miles, and several other enhancements to the research facilities were proposed by the physicists and accepted by the Bush administration. The changes would raise the quality and reliability of the machine. They would also increase its total cost—to 8.249 billion 1990 dollars, according to the official estimate of the Department of Energy, although another estimate that it commissioned put the figure at $11.8 billion.

The cost escalation provoked congressional critics of the project. Some began to raise questions about the managerial competence of the SSC project, especially since its director, Roy Schwitters, had declared reassuringly that the original magnet design had needed only fine tuning. The managerial issue was exacerbated when, in the spring of 1992, congressional investigators reported that more than $20,000 had been spent the previous year on office plants and plant care and that additional thousands had been spent on entertainment, including liquor.[42] Many more congressmen turned a searchlight on the promises of foreign assistance to the SSC. The Reagan and Bush administrations had assured Congress that fully one-third of the total construction costs would come from nonfederal sources—which now meant, at the elevated price of the machine, $2.7 billion. A billion dollars would come from Texas, leaving $1.7 billion to be provided by foreign countries. Yet by

[39] Senate, *Hearing, Energy, SSC*, 1991, p. 35.

[40] Daniel S. Greenberg, "Former NSF Head Says SSC Should Be Delayed," *Science and Government Report*, July 1, 1991, p. 3.

[41] R. Jeffrey Smith, "Problems May Lead to Cuts in Collider Project," *Los Angeles Times*, Nov. 19, 1989, p. 27.

[42] Kim A. McDonald, "University Consortium Charged with Mismanaging Work on Supercollider," *The Chronicle of Higher Education*, May 6, 1992, p. A30.

1990 nothing had been pledged from abroad except $50 million of in-kind contributions by India, which Lederman said he had personally obtained through the science adviser to Prime Minister Rajiv Gandhi. The Department of Energy kept reassuring Congress that a hefty subvention would be forthcoming from Japan. "We are not asking for money, but a partnership," Deputy Secretary Moore told a reporter, noting that Japanese scientists and students would be welcome at the SSC.[43] But Japanese physicists, who had their own funding problems, opposed helping out the collider, and the Japanese government apparently did not mind abiding by the opinion of its scientists. Several American delegations to Japan were unable to extract anything for the SSC. President Bush did not even raise the subject when he visited Prime Minister Kiichi Miyazawa in January 1992, nor did President Bill Clinton when he met with him in Washington in April 1993 and then in Tokyo in July of the same year.

By the spring of 1992, amid the deepening economic recession, the attacks against the SSC—especially its budgetary expansion and the absence of foreign pledges to pay for it—were drawing blood in the House of Representatives, where on June 17 a coalition of the project's active critics introduced an amendment to an appropriations bill to terminate it. The chief sponsors were the Democrats Howard Wolpe, Dennis Eckart, and Jim Slattery, and the Republican Sherwood Boehlert. Wolpe was an unabashed liberal from Michigan who made a point of siphoning appropriations from the Sun Belt into his own district, including its heavily black west side. He was sure that a dollar spent on the SSC would be a dollar unavailable for education or the environment, among other social welfare programs. Eckart, from a northern Ohio region of declining smokestack industries and ethnic minorities, was a more moderate liberal, and Slattery, from Topeka, Kansas, hewed to the middle of the road. Both Eckart and Slattery were members of the Energy and Commerce Committee, whose chairman, John D. Dingell, was Eckart's good friend. As chair of his own Subcommittee on Oversight and Investigations, Dingell was much practiced in, and much feared for, his scourging of defense contractors, university cost-accounting practices, and scientific fraud. Eckart, pointing to the lack of foreign contributions and the project's escalating budget, remarked on the House floor, "The SSC is committing suicide. It commits suicide by failing to be honest with us and with the taxpayers as to how much it will cost."[44]

Sherwood Boehlert, from the Oneida district in upstate New York, was a moderate Republican of independent mind and sharp tongue. In the mid-1980s he had urged that land on the New York–Canadian border would be an excellent location for the SSC, and some said that his opposition to the

[43] David E. Sanger, "U.S. Asks Japan to Accept Role in Supercollider," *New York Times*, June 1, 1990, pp. 1, C9.

[44] U.S. Congress, *Congressional Record*, 102nd Cong., 2nd Sess., House, June 17, 1992, p. H4811 (hereafter, *CR*, 102nd, House).

project derived from a sour-grapes resentment that the Department of Energy had ruled out any transborder site. Boehlert said that, on the contrary, his opposition had originated in an encounter at Cornell with the physicist James Krumhansl, who detailed the reasons against the SSC. Whatever the case, Boehlert derided the SSC as a medley of endlessly increasing costs, threats to other sciences, and unwarranted predictions of spinoffs for competitiveness. "Contrary to all the hype, the SSC will not cure cancer, will not provide a solution to the problem of male-pattern baldness, and will not guarantee a World Series victory for the Chicago Cubs," Boehlert declaimed.[45] It was also a creature of shell-game misrepresentations by the Department of Energy, not least because the department was negotiating sole-source contracts for SSC components from countries such as South Korea, where wages were low, and intending to count as foreign contributions to the project the difference between buying the components in Korea and purchasing them in the United States. Above all, Boehlert doubted, as he told the House, "that the most pressing issues facing the Nation include an insufficient understanding of the origins of the universe, a deteriorating standard of living for high-energy physicists, or declining American competitiveness in the race to find elusive subatomic particles."[46]

The defense of the SSC was floor-managed by Waxahachie's congressman, Joe Barton. A smart arch-conservative Republican—just a week earlier he had helped lead the unsuccessful fight for a balanced-budget amendment to the Constitution—Barton ably contended that the project was on time, within its $8.249 billion budget, and well worth every cent. (Congressman Lawrence J. Smith, an outspoken liberal Democrat from Florida and an enemy of the SSC, could not help gibing that Barton, the budget balancer, was "obviously a contortionist, being on two opposite sides of fiscal policy at the same time.") Barton was supported in the debate by several fellow Texans who extravagantly extolled the spinoffs to come from the SSC, as did Congressman Robert L. Livingston, a Republican from suburban New Orleans, who praised high-energy research for producing, among other things, "virus remedies, maybe even for AIDS."[47] Yet Barton's most significant ally was Congressman George E. Brown, a Democrat from southern California and the chairman of the Committee on Science, Space, and Technology; he was a good friend of science in general, an enthusiast of the SSC in particular, counting its intellectual aims meritorious and its development an advantageous way of keeping scientists and engineers employed as the country converted from its Cold War footing. One of the most respected members of Congress on R&D policy, Brown was also politically shrewd and well aware that the lack of foreign budgetary participation in the project rankled in the House. He co-introduced an amendment, which was quickly accepted, that would deny money to the SSC beginning a year hence unless the president

[45] Ibid., p. H4820.
[46] Ibid., p. H4823.
[47] Ibid., pp. H4814, H4820.

certified that the project would obtain at least $650 million from international sources.

Nevertheless, on the night of June 17, 1992, the House voted to terminate the SSC by the hefty margin of 232 to 181, stunning its advocates and sending them into a frantic effort to reverse the decision in the Senate.[48] On June 24, the Executive Board of the American Physical Society issued a statement deploring the cancellation of the project. The next day 40 physicists, including 21 Nobel laureates, sent a letter expressing shock and dismay at the House's action and insisting on the importance of the SSC for American scientific prowess to President Bush, to House members who had switched from "for" to "against" between 1991 and 1992, and to every member of the Senate. Within three weeks their letter was endorsed by more than 1,700 other American scientists plus some 300 from foreign countries. The National Association for the Superconducting Super Collider—a lobby that included high-technology and construction companies—arranged for scores of physicists to buttonhole senators and their staffs.

Undaunted, Dale Bumpers of Arkansas, the SSC's chief nemesis in the Senate, moved to end the project. A moderate to liberal Democrat with a gift for the down-home phrase, Bumpers was unfriendly toward a number of federal big-technology enterprises, including the Strategic Defense Initiative, the Space Station, and revamped Trident missiles. Although he knew nothing about particle physics, he did know that inadequate funds were available for Head Start as well as prenatal and neonatal care; that thirty-five million people lacked health insurance and that ten million people were unemployed in the United States; and that the National Institutes of Health did not have enough money to fund more than a quarter of the meritorious grant applications received in the biomedical sciences. He also knew that the SSC would not add to American competitiveness, except perhaps to make the United States the world's leading supplier of super colliders—because Philip Anderson, Nicolaas Bloembergen, James Krumhansl, and Rustum Roy, among others, had said so, in testimony and documents that he presented to the Senate, remarking, "These are not people who just fell off the turnip truck."[49]

However, Bumpers faced the powerful pro-SSC phalanx of Senators Gramm and Bentsen of Texas, who mounted an educational campaign for their colleagues in which they were joined by J. Bennett Johnston of Louisiana. Johnston chaired the Energy and Natural Resources Committee as well as the Energy and Water Development Appropriations Subcommittee, both of which had jurisdiction over the collider project. Originally an opponent of the SSC, Johnston had turned into its most formidable friend in the Senate after General Dynamics committed itself to producing superconducting mag-

[48] Irwin Goodwin, "2100 Physicists Use a Democratic Process for the SSC," *Physics Today*, 45 (Aug. 1992), 59.

[49] U.S. Congress, *Congressional Record*, 102nd Cong., 2nd Sess., Senate, Aug. 3, 1992, pp. S11145–S11146 (hereafter, *CR*, 102nd, Senate).

nets for the accelerator at a large factory in Hammond, Louisiana. He was a Senate insider, a charming, knowledgeable, and effective legislative dealer, an able defender of his state's oil and gas industry who did not mind telling people that Louisiana received more than one billion dollars a year in defense contracts. He was also a man of independent and energetic intellect, an outspoken opponent of the Strategic Defense Initiative. He counted the collider as important to the post–Cold War high-technology economy, and he had also developed a genuine intellectual enthusiasm for the quest after the Higgs boson. In June he presided over a hearing on the SSC during which a stellar lineup of scientific witnesses rang the changes on its virtues. (Bumpers later noted that Johnston almost never invited critics of the project to testify.) Leon Lederman unashamedly bore witness that it would be "crazy" to build the SSC for its spinoffs, insisting, "We build it because we are . . . insatiably curious, and have an unquenchable determination to know."[50]

Johnston and his allies prevailed, producing a Senate vote on August 3, 1992, of almost 2 to 1 in favor of continuing the project, although with an appropriation $100 million less than the Bush administration had requested. The SSC was not "just a toy for the amusement of American physicists," Senator Bentsen had insisted, but an instrument of American technical leadership in the post–Cold War world.[51] On September 15, a Senate-House conference agreed to the essentials of the Senate vote. The House promptly upheld the conference report, no doubt because it was an election year and Johnston had made the SSC part of a measure appropriating $22 billion for energy and water projects all over the country.

Early in 1993, Washington insiders were saying that, with a new Congress and a new administration in office, the prospects of the SSC's surviving another year were problematic. Voters had sent 113 new members to the House, refreshing more than a quarter of that body with the message to cut spending. Although President Clinton had endorsed the SSC during the campaign, his vice president, his secretary of agriculture, and, most important, his budget director, Leon Panetta, had all resisted the SSC during their days in Congress. In early February, the new secretary of energy, Hazel O'Leary, remarked in her first meeting with reporters that she did "not feel passionately" about the project, adding that the question was whether the country could afford it.[52] In January, Governor Ann Richards of Texas had reportedly persuaded Clinton to keep the SSC alive despite Panetta's opposition, but Clinton's budget called for stretching out the project by an additional three years. The maneuver would reduce its annual cost, but it would raise the total cost to almost $11 billion, according to a report from

[50] U.S. Congress, Senate, *Joint Hearing before the Committee on Energy and Natural Resources and the Subcommittee on Energy and Water Development, . . . Importance and Status of the Superconducting Super Collider,* 102nd Cong., 2nd Sess., June 30, 1992, pp. 25–26.

[51] CR, 102nd, Senate, Aug. 3, 1992, p. S11156.

[52] Thomas L. Friedman, "Collider and Space Station Saved from Budget Ax," *New York Times,* Feb. 6, 1993, p. 7.

the General Accounting Office, in May 1993, which declared the SSC behind schedule and already over budget.[53]

Secretary O'Leary, by now a declared enthusiast of the accelerator, pronounced the report mistaken, and in mid-June the president, in a letter to the chairman of the House Appropriations Committee, urged that it be continued, because abandoning it "would signal that the United States is compromising its position of leadership in basic science." But in late June, shortly before the House was to vote on the collider, its critics leaked a report from the Energy Department's own inspector general sharply critical of the project's management: of the half billion dollars spent by SSC contractors, 40 percent was found to be "unnecessary, excessive," or the result of "uncontrolled growth."[54]

In the House, Eckart and Wolpe had both retired, but Slattery remained and so did Boehlert, who that spring remonstrated in a hearing on the accelerator that the session deserved the title "The Night of the Living Dead," explaining that, even though the House had killed the SSC, "it keeps rising from the dead to suck up budget dollars."[55] On June 24, Boehlert and Slattery introduced an appropriations amendment to slay it once again, adding the adverse findings of the General Accounting Office to their prior indictments of the project. Boehlert summarized the case against it: "In short, the costs are immediate, real, uncontrolled, and escalating; the benefits are distant, theoretical, and limited. You don't have to be an atomic scientist to figure how that calculation works out. We can't afford the SSC right now."[56]

Congressman Barton again avidly defended the project, insisting that money taken from the SSC would go not to other sciences but into more water projects; reporting that former President Bush, in a letter to him, encouraged all Republicans to vote for the SSC; and contending that the accelerator was on time, within budget, and well managed. (Just for good measure, Barton introduced an amendment, promptly accepted, stipulating that no funds provided in the appropriations bill for Department of Energy facilities could be used for "food, beverages, receptions, parties, country club fees, plants or flowers.") Barton's case was strengthened by allies from California, hard hit by defense cutbacks, and nearby districts in Texas who pointed out that the SSC had already provided hundreds of millions of dollars for defense conversion, creating thousands of jobs and awarding some 20,000 contracts to businesses in most states of the Union, more than 10 percent of them to firms owned by women or members of minority groups. Carrie P. Meek from Miami, Florida, and Eddie Bernice Johnson from the Dallas

[53] Victor S. Rezendes, Director, Energy and Science Issues, General Accounting Office, House, *SST Hearing*, 1993, pp. 177–178, 181–182, 189–190, 195.

[54] Clinton to William Natcher, June 16, 1993, reprinted in U.S. Congress, *Congressional Record*, 103rd Cong., 1st Sess., Senate, Sept. 29, 1993, p. S12692 (hereafter, *CR*, 103rd, Senate); Kim A. McDonald, "House Kills Superconducting Super Collider by 130 Votes; Space Station Scrapes through by a Margin of 1," *The Chronicle of Higher Education*, June 30, 1993, p. A25.

[55] House, *SST Hearings*, 1993, p. 2.

[56] U.S. Congress, *Congressional Record*, 103rd Cong., 1st Sess., House, June 24, 1993, p. H4057 (hereafter, *CR*, 103rd, House).

area—both black and both newly elected to the House—praised the SSC, with Meek declaring, "It gives us a chance, the minorities in this country . . . to get into jobs that are developed by technology and science."[57]

The House nevertheless voted once again, on June 24, to end the SSC, by a strongly bipartisan and decidedly lopsided vote of 280 to 150. On June 30, Congressman John Dingell went after the SSC in a hearing before his subcommittee, lambasting Universities Research Association, the project's overall manager, for some of the worst contract "mismanagement" his committee had seen, for a remarkably "high level of arrogance and . . . intolerance towards government oversight," and for "lavish spending of taxpayers' money on luxuries and entertainment." Secretary O'Leary chimed in that the project was characterized by "lax management" and attitudes of high "self-importance," and that she intended to correct both deficiencies.[58]

Anticipating trouble, advocates of the SSC had been lobbying in defense of it since the spring, the Texas National Research Laboratory Commission having quietly paid for public-interest forums on its behalf in several parts of the country, and the National Association for the Superconducting Super Collider having footed the bill for a public-relations offensive that marshalled a press conference with seven Nobel physicists, who then met with Vice President Gore and congressional staff. After the House vote, SSC enthusiasts came out swinging. In Senate hearings in August, Schwitters insisted that the project was on time and even under budget, declaring, while displaying pictures of the Waxahachie site, that superconducting technology was now installed on "a piece of land that 3 years ago was described as being so poor that you could not even raise armadillos on it." Secretary O'Leary backed Schwitters's assessment but also announced a major managerial shake-up: authority over the construction of the SSC would be removed from Universities Research Association and turned over to an industrial contractor. Steven Weinberg questioned the close oversight of the project—eight audits had been under way in the spring—noting that previous accelerator projects had overrun their estimates by only a few percentage points. He suggested, "It seems to me that there is an effort being made to replace the system that gave us Fermilab with the system that gave us the B-1 Bomber." In a House hearing in May, Weinberg had testified that if the SSC was killed, "you may as well say good-bye to any responsible program of high-energy physics, and with it . . . any hope in this country in our time of discovering a final theory of nature." Now he warned that killing the SSC could also begin "the killing of support for basic science in this country."[59]

Still, a pre–Labor Day poll by the *Dallas Morning News* found only eleven

[57] Ibid., pp. H4052, H4063–H4064.
[58] U.S. Congress, House, *Hearing before the Subcommittee on Oversight and Investigations of the Committee on Energy and Commerce, Mismanagement of DOE's Super Collider,* 103rd Cong., 1st Sess., June 30, 1993, pp. 1–2, 43.
[59] U.S. Congress, Senate, *Joint Hearing before the Committee on Energy and Natural Resources and the Subcommittee on Energy and Water Development, Superconducting Super Collider,* 103rd Cong., 1st Sess., Aug. 4, 1993, pp. 102, 78, 54; House, *SST Hearings,* 1993, p. 59.

senators saying they would vote for the project, while six of those who had supported it in 1992 said they would vote against it. On September 29, Bumpers weighed in with a renewed attempt to kill the SSC by terminating its appropriation. During the summer hearings he had declared, noting that Johnston accused him of being "president of the Flat Earth Society," that he preferred membership there than in "the Flat Broke Society"; now he expatiated on why the SSC was "an outrageous waste of money" that could be used for better purposes.[60] Johnston not only rang all the changes on the SSC's merits but provided the Senate with a rare moment of attempted instruction in theoretical physics, outdoing himself with a celebration of the high importance of the elementary-particle research that the collider would make possible. He explained that knowledge to come from the SSC would bear upon whether the universe would continue to expand or "turn around and begin to collapse upon itself," adding the reassurance, "Indeed, beyond the lifetime of any of us here. Not to worry." As he understood it, the questions that the SSC would address "lie astride the common boundary between theology and science." Suffice it to say, Johnston reported, "that many scientists see in the patterns, the complexities, the symmetries, and yet the simplicity of matter and quarks and leptons and the way they are put together, the hand of God. And to the extent that I am given a peek at what they have to say, I agree with that."[61]

When Bumpers's amendment to kill the SSC came up, on the morning of September 30, it was rejected by a bipartisan majority of 57 to 42 (partly because, in a shrewd maneuver, the management issue had been defused by a prior amendment stipulating that all spending on the project was to be halted after ninety days unless its managerial deficiencies had been resolved). The SSC thus cleared the Senate with its full appropriation, imbedded again in a multibillion-dollar energy and water appropriations bill. It was agreed to on October 15 in a House-Senate conference—to which no representative who had voted against it in June had been appointed—although the conferees, Congressman and conferee Tom Bevill later told the House, "expressed their intention" to keep the cost of the project under $11 billion.[62]

Boehlert and Slattery were, to put it mildly, irritated by the slanted composition of the conferees from their side of Capitol Hill and by the conference's seemingly high-handed indifference to the House's overwhelming rejection of the SSC. When the conference report came to the House for action, on October 19, Bevill said that the conferees had spent more time on the SSC than on any other issue, but that the delegation from the Senate had adamantly refused to recede. Boehlert insisted, "If we accept this action from the conferees then the House amounts to nothing more than a very expensive version of Boys' State—just going through the motions of govern-

[60] Senate, *Joint Hearing . . . Energy and Natural Resources*, Aug. 4, 1993, p. 5; *CR*, 103rd, Senate, Sept. 29, 1993, p. S12725.

[61] *CR*, 103rd, Senate, Sept. 29, 1993, p. S12705.

[62] *CR*, 103rd, House, Oct. 19, 1993, p. H8109.

ing." The House was not about to accept, its prior objections to the SSC having been strengthened by what John Dingell's hearing had appeared to reveal—"huge cost overruns, and fraud, and abuse, and waste in this, the largest pork project in this Government," according to the declamation of Sherrod Brown, a freshman Democrat from Ohio and a member of Dingell's subcommittee. The House rejected the conference report by the overwhelming vote of 282 to 143, insisting that a reconvened conference cut the SSC, which it did two days later, agreeing to use the funds allocated to the project for its orderly termination. "The SSC has been lynched, and we have to bury the body," Johnston snapped.[63]

Johnston, like a number of analysts, blamed the execution on the House freshmen, typically describing them as "the product of an angry electorate that wants to cut projects and cut perks."[64] True enough, the 113 House freshmen voted against the collider almost 3 to 1. And true enough, the House was in a budget-cutting mood, slashing out money for several smaller physics projects. Civilian research and development, which accounted for 11 percent of the domestic discretionary budget, was a natural fiscal target under the circumstances of the budget-cap rules. In the charged economizing atmosphere, it did not help matters that, despite repeated promises from both the Reagan and the Bush administrations, not a single yen had been pledged for the SSC. Yet the incumbent House voted against the collider by a margin of 200 to 111, almost 2 to 1. And from the beginning the House, as well as the Senate, had been of divided mind on the issue of foreign cost sharing—on the one hand wanting the money but, on the other, not wanting to relinquish any of the project's jobs or control of its technological spinoffs to the nation's economic competitors. Besides, as Johnston pointed out several times, killing the SSC would not make much of a dent in the deficit, since the year's appropriation for it amounted to only a tiny fraction of the overall federal budget. The Congress of the United States is selective in its economizing, tending to be tolerant of expenditures for high national purposes, especially if they are reinforced by important local political and economic interests. Far more important than the freshman effect or the foreign deficiency in shaping the fate of the SSC was the fact that the SSC failed to qualify on the national or local ground.

Missing at the national level was what had made physics, including its high-energy branch, so important since World War II—real or imagined service to national security. Both the House and the Senate debates several times made the point advanced at summary length in a report that the General Accounting Office prepared for Senator John Warner, a conservative Democrat from Virginia, who presented it to Congress on May 18, 1993: the SSC had no direct bearing on national security, though its indirect benefits,

[63] Ibid., pp. H8101, H8109, H8117; "Congress Pulls the Plug on the Super Collider," *Los Angeles Times*, Oct. 22, 1993, p. 20.

[64] Clifford Krauss, "Knocked Out by the Freshmen," *New York Times*, Oct. 26, 1993, p. B10.

such as more powerful superconducting magnets and conversion awards to defense contractors, could assist the military indirectly. The promise of indirect benefits, including the desire to maintain a world-leading elementary-particle community, had assisted in the creation of Fermilab, whose builder, the physicist Robert Wilson, frankly assured an inquiring senator in 1969 that the new accelerator would do nothing for national defense except to help make the United States worth defending. In 1993, Congressman Ralph M. Hall of Texas wondered wistfully whether the SSC might not "bring us back one more time to the financial position that we had in the early 1950s and the geopolitical strength that we had." The SSC tended to receive support from the minority of House members who, following a more specific but similarly wishful preference, voted for the Strategic Defense Initiative. The SSC was disadvantaged, however, by the general outlook that went almost without saying but that was made explicit by Senator Dave Durenberger, a Minnesota Republican: "If we were engaged in a scientific competition with a global superpower like the former Soviet Union, and if this project would lead to an enhancement of our national security, then I would be willing to continue funding the project. But . . . we face no such threat."[65]

Dissociated from national security, the SSC was subject to the play of domestic politics, presidential as well as congressional. In July 1992, after the SSC's first setback in the House, the Bush administration lobbied hard for the project; President Bush himself welcomed a delegation of pro-SSC physicists to the White House, invited undecided senators there for a pep talk, and visited Waxahachie, where he praised the collider as "a big part of our investment in America's future," adding that for basic research "this is the Louvre, the Pyramids, Niagara Falls all rolled into one."[66] In 1993 the influence of the Texas congressional delegation had been weakened by the loss of House Speaker Jim Wright and Lloyd Bentsen's elevation to the cabinet, while Texas's adopted son in the White House had been replaced by Bill Clinton. Clinton added a final-hours phone call to Tom Bevill on behalf of the SSC to his letters endorsing the project, but his support struck insiders, including his Arkansas ally Dale Bumpers, as tepid, and it did not begin to match the all-out jawboning effort he made on behalf of the space station, which the House approved by a margin of only one vote in June 1993, the day before it rejected the SSC.

The SSC's political-economic muscle did not compare with the space station's, which, with a price tag more than twice that of the collider, had commitments of some $8 billion in foreign financing, the heavyweight support of the aerospace industry, and a reported 75,000 jobs created to its credit. Although SSC expenditures reached into almost every state, Slattery pointed

[65] House, *SST Hearing*, 1993, p. 120; *CR*, 103rd, Senate, May 18, 1993, pp. S5996–S5997, S6000–S6003; *CR*, 102nd, Senate, Aug. 3, 1992, p. S11165; Linda R. Cohen and Gretchen A. Kalsow, "Who Killed the Superconducting Supercollider?" unpublished ms., 1994, p. 22.

[66] Irwin Goodwin, "Good News for the SSC as Senate Approves Funds and Magnets Work," *Physics Today*, 45 (Sept. 1992), 55.

out to the House that most states would pay far more for the project than they would receive from it. Indeed, from October 1989, when construction began, through April 1992 the vast majority of procurement contracts had gone to only five states—Massachusetts, New York, Illinois, California, and, of course, Texas, where some four times as much procurement money was spent as in California, which ranked second, or Illinois, which ranked third.[67] Boehlert summarized, with only slight exaggeration, the political dynamic of the SSC: "My colleagues will notice that the proponents of the SSC are from Texas, Texas, Texas, Texas, and Louisiana, and maybe someone from California. But my colleagues will also notice that the opponents are . . . from all across the country."[68]

Apprehension that the SSC would reduce the funds available for all types of science figured significantly in the congressional debates and likely colored the House vote, as did worries that it might jeopardize even high-energy physics outside Texas. Besides, SSC grants and contracts were not distributed evenly throughout the congressional districts in any state, including the four outside Texas that led in the procurement sweepstakes. In 1993, although the Texas congressional delegation voted for the SSC 26 to 1, the delegations in those four states voted overwhelmingly to kill it. The Illinois delegation cast 18 of its 20 votes against the collider, the majority including the representatives from the districts of Fermilab and the Argonne National Laboratory, near Chicago, who perhaps feared for the future of those installations. The California delegation split 34 to 18 against the SSC, with 16 of its 18 freshmen—including the congresswoman from the district that included the Stanford Linear Accelerator—voting with the majority.

Above all, many congressmen declared the SSC not worth its cost when measured against the country's other needs. The spinoff arguments were uncompelling to congressmen such as Peter Hoagland, a Democrat from Omaha, Nebraska, who instructed the House, "If we need a cancer therapy machine, let us invest in a cancer therapy machine, or a super computer, or a magnetic resonance imaging machine. Let us invest directly in those machines." A number praised the SSC's intellectual purposes, some in the vein of Senator Herb Kohl, a Wisconsin Democrat who thought it would be fine to learn what happened "moments after—and perhaps even before—the Big Bang."[69] Others declared that it did not matter whether the origins of matter were discovered now or in the distant future; it would not change how people lived. Respectful of the science or not, the opponents of the SSC considered the project simply too expensive.

Yet the opponents were not all simply economizers as such. The congressional debates revealed that, although many wanted to kill the collider solely for the sake of cutting the budget, Dale Bumpers was not alone among

[67] Holly Idelson, "House Denies Atom Smasher Its 1993 Expense Account," *Congressional Quarterly*, June 20, 1992, p. 1783.
[68] CR, 102nd, House, June 17, 1992, p. H4820.
[69] CR, 103rd, House, June 24, 1993, p. H4062; 103rd, Senate, Aug. 3, 1992, p. S11167.

the SSC's enemies in insisting that expenditures for it were unwarranted when appropriations for social programs such as medicare, nutrition, vaccination, education, and inner-city redevelopment were being cut. Analysis of the 1993 House SSC vote in light of the voting record of incumbents on other issues shows that its opponents comprised a coalition of conservatives and, in greater proportion, liberals. Its defenders included a higher proportion of conservatives, a tendency echoed by the vote in the Senate that year, where the collider won only a bare majority of Democrats but prevailed among Republicans by more than 2 to 1. In the end, the collider resolved into a creature of Cold War conservatism when the majority of Congress—both liberals and conservatives—was undergoing a transformation to a post–Cold War political order.

As many pro-SSC physicists saw it, the collider's fate defined one of the new order's chief convictions—which was, as Schwitters said, that "curiosity-driven science is somehow frivolous, and a luxury we can no longer afford." Leaders of American physics variously declared the collider's death to mean that high-energy physics had no future in the United States, that the country was relinquishing its role as a scientific leader, and that the half-century-old partnership between science and the federal government was ending. At the level of grand interpretation, Murray Gell-Mann called the cancellation "a conspicuous setback for human civilization." At the level where scientists worried about jobs and opportunities, the killing of the collider was proclaimed in a letter to *Physics Today* to have sent a clear message: "Physics and physicists are not valued in this country! Enter this profession at your peril!"[70]

They had certainly been devalued in Waxahachie, Texas, where a local newspaper, published in nearby DeSoto, called the collider's opponents "lower than pond scum" and compared the action of Congress to "the technological equivalent of Pearl Harbor."[71] The SSC staff, now totaling two thousand people—including two hundred physicists who had relocated from Europe as well as the United States—had almost gotten used to the disruptive uncertainties of the mid-year appropriations season, though many had hedged their bets by renting rather than buying homes. Now they were revising their résumés while Secretary O'Leary set up severance procedures and the government began negotiations with the state of Texas, which estimated that it had invested more than $400 million in the project, over the disposition of the site and its facilities. "What we need now are some lawyers, not physicists," the head of the Texas National Research Laboratory Commission noted.[72]

[70] Malcolm W. Browne, "Pondering Supercollider's Demise," *New York Times*, Aug. 9, 1994, p. 15; Colin Macilwain, "SSC Decision Ends Post-War Era of Science-Government Partnership," *Nature*, 365 (Oct. 28, 1993), 773; Robert J. Reiland, letter, *Physics Today*, 47 (March 1994), 91.

[71] "Supercollider Outlasted a Texas Town's Illusions," *New York Times*, Oct. 28, 1993, p. 8.

[72] Robert L. Park, *What's New (in My Opinion)*, Newsletter, The American Physics Society, Nov. 5, 1993. In July 1994, the Department of Energy agreed to pay Texas $145 million in cash and contribute $65 million toward the conversion of the proton injector accelerator into a cancer treatment facility. Christopher Anderson, "DOE and Texas Settle SSC Claims," *Science*, 265 (July 29, 1994), 600.

The death of the SSC exacerbated a broad contraction of opportunities in physics that had begun with the defense cutbacks and economic downturn around 1990. The predictions that a shortage of physicists would develop during the 1990s now appeared a cruel joke. By every measure, the supply of physicists exceeded demand in most fields and in every sector—government, industry, and academia. The employment outlook was uneven across the profession, with high-energy physicists finding themselves worse off than those in industrially related fields such as condensed matter, lasers, and optics. Still, the predictions now were that prospects would worsen across the spectrum of fields, as new physics Ph.D.s continued to pour out of the graduate schools and émigré Russian physicists sought work in the United States. Young physicists went into holding patterns as postdoctoral fellows and applied by the hundreds for single faculty positions, even at liberal arts colleges with limited research programs. Those who did land jobs reported that competition for funds was so intense that they spent more time trying to raise money, often without success, than doing research. Some physicists contemplated career shifts—for example, into designing video games; others left physics to deploy their analytical skills on Wall Street. Asked about the job market in 1994, one young physicist called it about average: "worse than last year, but better than next year."[73]

The turn of events provoked bitter resentment. Some physicists accused the country of bad faith, of callously disregarding the investment in science it had encouraged them to make. Others complained that the autonomy of science was being increasingly undermined by bureaucratic intrusion, that all the national laboratories had to cope with hundreds of prescriptive directives, regulations, and audits. In a postmortem on the SSC, Wolfgang K. H. Panofsky, the former director of the Stanford Linear Accelerator, who considered the collider's death a "senseless killing," attacked the sheer volume of oversight to which it had been subjected and which had cost senior laboratory personnel enormous amounts of time. "We should be devoting ourselves to completing this machine as rapidly and cheaply as possible, and getting on with the real science," Schwitters had told a reporter early in 1993, adding impoliticly, "Instead, our time and energy are being sapped by bureaucrats and politicians. The SSC is becoming a victim of the revenge of the C students."[74]

The complaints of the physicists were commonplace throughout the American scientific community, where chasing after dollars had become the order of the day. Although federal funds for basic research had increased by 20 percent over the levels of the late 1960s, the number of research scientists in the United States had doubled. Many scientists spoke wistfully of the autonomy and the opulence that had characterized science in the United States in the quarter century after World War II. Lederman called explicitly

[73] Gary Taubes, "Young Physicists Hear Wall Street Calling," *Science*, 264 (April 1, 1994), 22.
[74] Irwin Goodwin, "After Agonizing Death in the Family, Particle Physics Faces Grim Future," *Physics Today*, 47 (Feb. 1994), 87; Malcolm W. Browne, "Building a Behemoth against Great Odds," *New York Times*, March 23, 1993, p. B9.

for restoration of that golden age, urging in 1991 a doubling of the funding for all of academic science, which meant enlarging its annual budget by ten billion federal dollars.[75]

The physicist Walter E. Massey, director of the National Science Foundation in 1991, observed a "growing perception that the research community considers itself exempt from the pressures of competition and accountability and 'entitled' to public funding." The impression of entitlement left by high-energy physicists—their tendency to measure the quality of society by how generously it supported their enterprise—irritated a number of people while it infuriated some, including Rustum Roy, who was gleeful at the death of the SSC and told a *New York Times* reporter that "this comeuppance for high-energy physics was long overdue." During the 1970s, observers had warned that exponential growth in physics, measured by Ph.D. production or any other indicator, could not continue indefinitely; the warnings had been forgotten amid the defense-driven resumption of expansion in the 1980s. Now Frank Press, who had been President Carter's Science Adviser and was president of the National Academy of Sciences, reminded Lederman and his allies that "no nation can write a blank check for science" and that, if the number of scientists had doubled in twenty years, there was no reason why taxpayers should come to the rescue or why science should take precedence over other meritorious demands on the federal treasury.[76]

A major trouble, Massey observed in 1993 at a roundtable on the outlook for physics, was that people drew selectively on the past of the science to predict its future, that they started "with World War II as if there was no science in the world or in America" before then, believing that "the only standards" available for "quality of life" are those that prevailed during "the last forty years."[77] In fact, as this book shows, in the half century before 1940, physics in the United States had thrived in dependence on money provided for basic research by state legislatures, industrial corporations, individual donors, and philanthropic foundations, especially the Rockefeller Foundation, which operated as a national research agency during the interwar period.

In a sense, the reduction in per-capita federal research dollars was forcing physics now, along with most other sciences, to cope with pre-1940-like circumstances, at least to a limited degree; in recent years a similar mixture of nonfederal public and private patrons has been providing an increasing fraction of support for science in universities. But the federal government remains the most generous single patron of American physics—indeed the dominant one for all of science in the United States, providing in 1993 roughly 60 percent of all monies spent on academic research. As Senator Bill Bradley of New Jersey explained his vote to kill the SSC, in 1990 the federal

[75] Leon M. Lederman, *Science: The End of the Frontier?* (Washington, D.C., 1991), p. 18.

[76] Malcolm W. Browne, "Supercollider Demise Disrupts Lives and Rattles a Profession," *New York Times*, Nov. 14, 1993, p. 23; Irwin Goodwin, "Funding Gloom: Mood of Foreboding Pervades Forum at Science Academy"; "Distress Call from Three Physicists: Is the Image of Science out of Sync?" *Physics Today*, 44 (Feb. 1991), 76; 44 (June 1991), 94.

[77] "Roundtable: Physics in Transition," *Physics Today*, 46 (Feb. 1993), 39.

government had spent roughly $60 billion on R&D, while industry had spent nearly $78 billion on it—which in Bradley's view added up to quite an investment, thank you, in technological progress for the twenty-first century.[78]

The vote against the SSC was thus not a vote against science or for an end to the longstanding partnership of science and government; rather, it signified a redirection of the partnership's aims in line with the felt needs of post–Cold War circumstances. Emphasis would go to what policymakers were calling "strategic" or "targeted" areas of research—fields likely to produce results for practical purposes such as strengthening the nation's economic competitiveness or its ability to deal with global environmental change. As Congressman George Brown explained the shift, the country had to forgo "the myriad serendipitous" scientific paths it was capable of following in favor of "the strategic paths where we must go if the planet and its increasing population are to survive."[79] Emphasis would also be given to science education and to efforts to diversify the social composition of the scientific professions so that they would better mirror the increasingly multicultural makeup of American society. (American physics remained predominantly white and male, with women accounting for only 10 percent of its yearly crop of doctorates, and blacks and Hispanics less than 2 percent of them.)

Yet virtually no significant policymaker at either end of Pennsylvania Avenue urged that all undirected, untargeted basic research be denied federal largesse. The Clinton administration declared that the advancement of fundamental research was one of the chief goals of its policy for science and technology. Congress maintained appropriations for many areas of basic physics at a substantial level, awarding even high-energy research dispensation for several new initiatives in the same year it killed the SSC. Physics continues to be recognized as a mighty source of innovation and, as such, essential to sustaining a high-technology society.

But not at any price. In a joint newspaper column published shortly after the demise of the SSC, Brown and O'Leary endorsed the importance of attending to the scientific questions that the collider had been designed to address, but added that in the new era the big-science effort required to pursue them had to be genuinely international. During the hearings on the collider, the further internationalization of high-energy physics had been called for by critics such as Anderson and Schrieffer, who remarked, "Not to build the SSC is conceivable. Not to pursue particle physics is totally unacceptable to those who are concerned with and depend upon the health of science."[80]

International collaboration underlay the intellectual triumph announced at Fermilab on April 26, 1994—that 440 scientists, representing the United States, Japan, Canada, Taiwan, and Italy, had accumulated tentative but strong

[78] CR, 102nd, Senate, Sept. 30, 1992, p. S12765.
[79] Roland W. Schmitt, "Public Support of Science: Searching for Harmony," *Physics Today*, 47 (Jan. 1994), 29.
[80] Hazel R. O'Leary and George E. Brown, Jr., "Resuming the Pursuit of Knowledge," *Los Angeles Times*, Nov. 21, 1993, p. M5; House, *SST Hearings*, 1993, p. 91.

experimental evidence for the existence of the top quark, the last particle needed to confirm the validity of the Standard Model. In 1994, high-energy policymakers were giving serious consideration to the United States's joining CERN, if CERN would accept a formal American contingent, and to participating in the development of a new accelerator likely to be built there called the Large Hadron Collider; the machine would smash protons and antiprotons together at only 40 percent of the SSC's energy, but it was thought to have a chance, albeit a small one, of finding the Higgs boson. When Sherwood Boehlert was told about the prospect at a congressional hearing, he responded favorably, calling the idea "a thoughtful, specific blueprint for how to pursue this most basic of basic sciences."[81]

Whether the federal government would commit substantial funds to CERN would be a matter for political decision—political in the best sense that politics is the means by which the state resolves conflicting claims for the allocation of resources. So, similarly, would politics determine the country's mix of investment in targeted and untargeted research. The scarcity of resources for research provoked competing interests in physics to resort to the political process in the SSC controversy, and it will likely prompt them to make a habit of the practice. With the end of the Cold War, the disestablishment of American physics that began in the 1970s has been further extended; its claims to a share of the public purse are no longer taken largely on faith or dispensed with little obligation to accountability. Physics in the United States has been irreversibly incorporated into the conventional political process, making it a creature of political democracy, its fortunes, like those of other interest groups, contingent on the outcome of the fray.

---

[81] Walter Sullivan, "Panel Urges U.S. to Join European Collider Project," *New York Times*, May 31, 1994, p. B8.

Preface to the First Edition

A number of years ago I met the American Nobel laureate physicist I. I. Rabi. Veterans of Los Alamos during World War II remember how Rabi would get off the train in the hot Albuquerque sun wearing rubbers and carrying an umbrella; Columbia University students recall how he would bustle about the laboratory humming, sometimes Mozart. Now Rabi remarked to me in his freewheeling fashion that there were few histories of modern science or biographies of its leading figures. Why doesn't someone write about my generation of physicists? he asked in his twinkling manner: "After all, we changed the world."

In an important way, Rabi's question stimulated the writing of this book. There was—and has continued to be—no history of his generation of America physicists, and I set out to study the scientists who came to professional maturity after World War I, mastered the atom, then built the bomb and rushed the world for better or worse into a fundamentally new era. But the more I proceeded with my work, the more I realized that to understand Rabi's generation I had to go back to the post–Civil War period. And the more I studied the Rabis and their predecessors, the more I learned that not only they changed the world, but the world had changed them—their opportunities and institutions, responsibilities and attitudes, their power, status, and expectations.

To understand this process of mutual change, it seemed essential to examine the American physicists' larger community: the individuals and institutions—academic, cultural, industrial, and governmental—that linked them to their society and helped make possible the pursuit of their discipline. I thus decided to deal in this book with two broad topical strands. The first is American physicists in their own right; the second is the general world of the sciences, especially the physical sciences, in the United States. Neither strand has ever been wholly independent of the other. And both have been profoundly affected by the social, cultural, and political developments around them. Consequently, from the post–Civil War years to the present, this book is an intertwined history of American physicists and of their contextual community, set within the turbulent evolution of the nation at large.

Above all, this is a book about scientists as human beings. A few scientists—James D. Watson and Stanislaw Ulam come readily to mind—have revealed the human dimension of the scientific life. Yet books like theirs are the exception rather than the rule. Having come to know many scientists well in the course of my research and my teaching at the California Institute of Technology, I can suggest one reason for this literary reserve: despite commanding remarkable, often awesome, powers of mathematical exposition, many scientists consider themselves incompetent to write about anything except the content of science itself. But a number of them can write about people and nonscientific subjects with admirable ease and grace. The most important reason why so few scientists give us books like Watson's *Double Helix*, I think, is the scientific community's convention of impersonality. The convention is commonly manifest in the requirement that scientific papers must be written in the passive rather than the active voice. The convention essentially declares: The actors in research are much less important than their processes and results. Only the truth about nature is worth having. Knowledge of the human beings who seek it is relatively unimportant.

Perhaps some scientists adhere to the convention the more firmly because of the same psychological disposition which in the first place turned them away from the contemplation of self to the study of things. Many abide by it, I suspect, because, by requiring them to submerge their individuality in an indubitably glorious collective quest, the convention adds a further touch of nobility to their pursuits. And their persistent loyalty to it even beyond the conduct of science proper quite likely arises from a fear that to reveal the human reality of their activity will somehow render it vulnerable before a suspicious public.

Yet the convention of impersonality is merely that—a convention. Scientists know how human an enterprise science is, if only from the strong interest they feel in the award of credit for priority of discovery or from their community's practice of naming theories, effects, and experiments after particular people. Science is shaped in the endlessly unpredictable way that social and family environment work upon individual temperament and imagination and the way that professional circumstances meld with the ordinary drives for money, status, and power. For this reason, I have in this book treated the human story as such of the people who have practiced physics, interpreted it to the larger society, politicked for it, built its institutions, and responded to its revolutionary impact upon the world.

I must finally stress that this book is not in any way an exhortation about what physicists should do to improve their lot in America, or what Americans must do to cope better with the torrentially onrushing advance of science. It remains above all a history, guided by the historian's personal quest for detachment and balance. Before transferring to full-time historical studies, I spent much of my period of higher education in preparation for a scientific career. While neither pro- nor anti-physicists in any sense, I have approached their history with considerable admiration for their achievements and a recognition that, like all other human beings, they can provoke dismay. This is an unabashedly interpretive history, especially with respect to certain key issues raised by the pursuit of science in the American democracy. These issues may at first appear simple to resolve if only they are approached with intelligence and generosity. Yet I came to understand in writing this book, as I hope you will in reading it, that the more they are pondered, the more tortuous they become.

AUTHOR's NOTE

Manuscript and records collections are cited in the footnotes in code-letter form—for example, FDR for the Franklin D. Roosevelt Papers. An explanatory list of these coded abbreviations is to be found in the glossary of manuscript citations which follows the text. The complete body of records and papers on which this study is based is discussed in the essay on sources.

THE PHYSICISTS

I

The Many Wants of Science

S hortly before the nation's centennial, Joseph Henry, the physicist and secretary of the Smithsonian Institution, white-haired and wise, was stirred to reflect: The surprise is not "that science has made comparatively *little* advance among us, but that . . . it should have made *so much*."[1]

In the post-Civil War decade the circles of American science included gentlemanly friends of natural philosophy along with amateur students of nature, and totaled perhaps two thousand people. About five hundred of them were serious researchers, centered in New England and the Middle Atlantic states, and mainly employed in colleges or governmental agencies.[2] At the top of American science was a small, internationally respected group that included Asa Gray, the masterful analyst of botanic specimens; the Swiss import Louis Agassiz, a brilliant student of rocks and fossils; James Dwight Dana, an authority on volcanic cones and coral structures; the mathematician Benjamin Peirce, his mind at once playful and implacable, his work ranging from linear algebra to planetary per-

---

[1] Henry to the Committee of Arrangements . . . , Feb. 3, 1873, in *Proceedings of the Farewell Banquet to Professor Tyndall . . . Feb. 4, 1873* (New York, 1873), p. 19.

[2] Nathan Reingold, "Definitions and Speculations: The Professionalization of Science in America in the Nineteenth Century," in Alexandra Oleson and Sanborn C. Brown, eds., *The Pursuit of Knowledge in the Early American Republic: American Learned and Scientific Societies from Colonial Times to the Civil War* (Baltimore, 1976), pp. 38–39, 52, 57; Daniel J. Kevles, "The Study of Physics in America, 1865–1916" (unpublished Ph.D. dissertation, History, Princeton University, 1964), Appendix III.

turbations; the patrician chemist Wolcott Gibbs, who untangled the structures of complex acids; and the geophysicist Alexander Dallas Bache, Benjamin Franklin's great-grandson, an authority on terrestrial magnetism. Physicist Henry, the orphaned son of a cartman in Albany, New York, had pursued electrical experiments while teaching at the Albany Academy and discovered electromagnetic induction independently of the great British scientist Michael Faraday.

In concert and separately, these men had shaped the course of American science since the 1830s, when Henry had returned from a kudos-filled tour of Europe avowing to Bache: "the real working men . . . of science in this country should make common cause . . . to raise their own scientific character."[3] Soon Agassiz arrived and built the Museum of Comparative Zoology at Harvard. At Yale, Dana edited the nation's principal organ for research, the half-century-old *American Journal of Science and Arts*. Bache directed the U.S. Coast Survey, expanding it into the largest employer of physical scientists in the nation. Henry, gracing the Princeton faculty before assuming the secretaryship of the Smithsonian, operated as a key link between the nation's academic and federal scientists. From these strategic positions Henry and his allies politicked, maneuvered, cajoled, and exhorted to strengthen science in America. Their chief goals: to exclude amateurs from the control of science, or, in the terms of a later generation, to professionalize research; to improve the condition of science in institutions of higher learning; and to enlarge its role in the federal government.

In the 1840s Henry had complained, "Our newspapers are filled with puffs of quackery and every man who can . . . exhibit a few experiments to a class of young ladies is called a man of science." Any dabbler who collected natural specimens or fiddled with an electrical battery might expect to deliver his results before a scientific meeting or publish them in one of the scores of local scientific journals. Even Dana's *American Journal* depended upon subscriptions from numerous amateurs, "men of general rather than scientific intelligence," Dana aptly observed. Yet, as Henry once remarked, scientific opinions had to be "weighed not counted," and neither meetings nor publications could serve the special interests of the scientific few if they had to remain agreeable to the unscientific many.[4]

By the postwar years, the general, and gentlemanly, phrase "natural philosopher" was giving way to the vocational word "scientist"—it had been coined by the Englishman William Whewell in 1840—and pro-

[3] Henry to Bache, Aug. 9, 1838, quoted in Nathan Reingold, ed., *Science in Nineteenth-Century America: A Documentary History* (New York, 1964), p. 85.
[4] Henry to ———?, Feb. 27, 1846, quoted in Howard S. Miller, *Dollars for Research: Science and Its Patrons in Nineteenth-Century America* (Seattle, 1970), p. 7; Dana to A. Gould, Feb. 16, 1848, quoted in Robert V. Bruce, "Science in Civil War America," unpublished manuscript, Chap. IV, p. 9; Henry to C. Dewey, Nov. 7, 1859, quoted in Robert V. Bruce, "Democracy and American Scientific Organizations in the Mid-Nineteenth Century," unpublished manuscript, pp. 13–14.

fessionals were playing a more prominent, and dominant, role among the traditional mixture of dabblers and researchers. Evidently bowing to complaints from Henry and others, Dana had made the *American Journal of Science and Arts* into a professional organ (which may have been why it was losing money and had no more subscribers than when it was founded in 1818). And Henry and his colleagues had created a national professional organization, the American Association for the Advancement of Science (AAAS). Though membership was open to any interested layman willing to pay the dues, the constitution virtually guaranteed the election of professional scientists to the governing offices, including a standing committee whose approval was required for the presentation of a paper at the meetings or its publication under AAAS auspices.

In higher education, science had come a long way since the early nineteenth century, when theology, the classics, and moral philosophy dominated the curriculum, ministers were one of the main college products, and many scientists got their training by taking medical degrees. In 1828 the Yale faculty sensibly concluded, in a report that set the coming standard for higher educational policy: "As knowledge varies, education should vary with it." More important, the Yale professors emphasized that one requirement of liberal education was to encourage in students "a proper *balance* of character," a balance that could be achieved only by exposing them to the different branches of science as well as literature.[5] By the 1860s at the leading private colleges the prescribed curriculum introduced students to botany, chemistry, astronomy, geology, and physics. The required physics courses now covered light, heat, electricity, and magnetism besides the traditional Newtonian mechanics, and they incorporated a degree of mathematical exposition and analysis.[6]

By this time, Washington had become a major scientific center. The red gothic towers of the Smithsonian Institution jutted skyward less than a mile from the White House. Henry administered the Smithsonian with the primary object "of stimulating the talent of our country to original research—in which it has been lamentably deficient—to pour fresh material on the apex of the pyramid of science, and thus to enlarge its base."[7] There was also the National Academy of Sciences, a private organization with a federal charter, created in the middle of the Civil War to provide expert advice to the government. And there were the regular scientific agencies, created over the years within the executive departments

[5] "Original Papers in Relation to a Course of Liberal Education," *American Journal of Science*, XV (1829), 325, 301.

[6] Stanley M. Guralnick, "Science and the American College, 1828–1860" (unpublished Ph.D. dissertation, Education, History, University of Pennsylvania, 1969), pp. 128–29, 193, 195–96.

[7] Henry to J. B. Varnum, June 22, 1847, draft, quoted in Wilcomb E. Washburn, "Joseph Henry's Conception of the Purpose of the Smithsonian Institution," in Whitfield J. Bell, ed., *A Cabinet of Curiosities* (Charlottesville, Virginia, 1967), p. 111.

to aid commerce, defense, and agriculture. The Coast Survey mapped the shoreline. The Naval Observatory gathered astronomical data for navigation. The new Department of Agriculture, headed by Commissioner Isaac Newton, appropriately hired a few scientists to study plants and insects.

In the post-Civil War years, with settlement thrusting toward the edge of the plains, the army sponsored various surveys of the geographic and economic possibilities of the far western states and territories. Among the most significant was the Geological and Geographical Exploration of the Fortieth Parallel, which followed the general route of the transcontinental railroad and was headed by Clarence King. Handsome, well connected, and in Henry Adams' judgment "the best and brightest man of his generation," King was only twenty-five when in 1867 the secretary of war appointed him to lead the Fortieth Parallel survey, a post which more than one major general coveted.[8] King, who had studied at Yale under some of the country's best geologists, staffed the exploration with first-rate scientists: topographers, geologists, botanists, and even an ornithologist. His men collected rock specimens and fossils, set up barometric stations, studied the structure of the great formations. The work was fruitful scientifically. Much of it was also economically pertinent, especially the geological investigation of the region in the neighborhood of the rich Comstock silver lode. In 1869 King returned to Washington, where he won a special appropriation to continue his survey—and, in effect, to cement the marriage of science with western exploration.

A few years later, Henry understandably enthused that a "great change" had occurred over the previous quarter century in the public's "appreciation of abstract science as an element in the advance of modern civilization."[9] Yet many professional practitioners of science were discontented.

The professionals distinguished between "abstract" and "practical" science, just as a later generation would distinguish between "basic" and "applied" research, or between "science" and "technology." To Henry's generation, "abstract" science meant the study of nature for the sake of understanding its substance, its working, its laws. "Practical" science meant, generally, the exploitation of nature and nature's laws for the sake of material development. The object of abstract science was knowledge; of practical science, processes, metals, crops, machines, and gadgets. Henry's group took no less pride than did other Americans in the nation's practical scientific achievements, but in the abstract realm they felt themselves without much reason to be proud.

It was largely in the earth and biological sciences—geology, topography, paleontology, botany, and zoology—that Americans had earned

---

[8] Quoted in William H. Goetzmann, *Exploration and Empire: The Explorer and the Scientist in the Winning of the American West* (New York, 1966), p. 431.

[9] *Annual Report of the Board of Regents, Smithsonian Institution, 1874,* p. 8.

the respect of Europeans. In physics, chemistry, and astronomy they had published only one third as much work since the revolution as the French Academy of Sciences and the Royal Society of London. In physics the bulk of American research had dealt with such geophysical subjects as meteorology or terrestrial magnetism. The main advances in heat, light, electricity, and magnetism had been the work of Europeans. Apart from Benjamin Franklin and Henry himself, few American physicists had ever won foreign acclaim. In the 1870s no more than seventy-five Americans called themselves physicists. The entire group published research at the decidedly low average rate of about one article per physicist every three years, and very few members of the group, Henry lamented, were doing significant work.[1]

Almost all American physics was experimental rather than theoretical. The colleges typically taught little more than the rudiments of the essential tool of theory, mathematics; students often did not reach calculus until the senior year, if ever. More important, the nation's scientific tradition was the offspring of Francis Bacon. Bacon had of course assaulted mere scholastic hypothesizing and urged instead "keeping the eye steadily fixed upon the facts of nature." He had also asserted that empirical knowledge would provide a "vision of the footsteps of the Creator" and inevitably lead to "power."[2] Perhaps few American scientists actually read Bacon, but the circumstances of their culture and environment—the strength of Protestant religion, the accessibility of nature, the challenge of mastering it—all encouraged them to be Baconian. Naturalists who laid flora and fauna away in specimen cabinets and astronomers who recorded the positions of the stars and planets were accumulating a factual variorum of the Creator's imprint on the universe. Benjamin Franklin, the American prototype in so many ways, was the prototypical American physicist, with his deism, his attachment to useful knowledge, and his penchant for experimentation.

Over the years American scientists grew much less concerned with reading God's imprint in nature. But the propensity for stressing empirical knowledge, handed down as the legacy of Bacon from teacher to student, remained vigorous. Physicists of Henry's generation also remained Baconian friends of practical science. Some even spent part of their time dabbling in invention, or advising technological entrepreneurs, or pursuing practical problems in the federal research establishment. Henry and his colleagues did insist that the progress of practical science depended on the advance of abstract science. Without the creation of new knowledge, professional

---

[1] Simon Newcomb, "Exact Science in America," *North American Review,* 119 (1874), 290; Samuel E. Cassino, ed., *The Naturalists Directory* (Boston and Salem; S. E. Cassino, 1878), listed seventy-one physicists in the United States; Henry to Alfred M. Mayer, May 27, 1870, HM. In the four years 1870–73 Americans published 110 physics articles in the *American Journal of Science.*

[2] Edwin A. Burtt, *The English Philosophers from Bacon to Mill* (New York, 1939), p. 23.

scientists argued, there could be no new knowledge to exploit; if only for that reason, abstract science deserved a good deal more prestige and support in the United States.

But these were the boom years of the post-Civil War period, when the golden spike was driven linking East and West in the first transcontinental railroad; when the advance of technology was combining with widespread economic opportunity to spur the economy ahead at a dizzying pace; when thousands were moving westward, drawn alike by cheap land and the increasing miles of railroad track; when the factory system was pouring out a constantly swelling volume of goods; when people seemed unprecedentedly well off. Few Americans distinguished between abstract and practical science—people often called the inventor a scientist—and few understood Henry's claim of the dependence of technology upon scientific progress.

To a considerable extent the industrial expansion was energized by machine-based technology. Products like the sewing machine or barbed wire, or production devices like the steam-powered machine in the brewery that almost seemed capable of drinking the beer—these innovations required mainly mechanical skill and ingenuity, not scientific knowledge and training. The managers of industries more closely related to science—mining, rubber, petroleum, electrical power—sometimes called upon the advice of academic geologists, chemists, and physicists. But on the whole, the industrial machine throbbed ahead without scientists and research laboratories, without even many college-trained engineers. The advance of technology relied on the cut-and-try methods of ingenious tinkerers, unschooled save possibly for courses at mechanics' institutes.

Thomas Alva Edison, the symbolic proprietor of the burgeoning electrical industry, stressed a preference for plain figuring over scientific formulas. "Oh these mathematicians make me tired!" he once gibed. "When you ask them to work out a sum they take a piece of paper, cover it with rows of A's, B's and X's, Y's . . . scatter a mess of flyspecks over them, and then give you an answer that's all wrong." Nonetheless, while Edison's approach to invention was often cut-and-try, it was highly systematic. His laboratory at Menlo Park, New Jersey, was equipped with a rich variety of scientific instruments, and its library shelves included the latest scientific books as well as periodicals. Edison also employed some scientists, including the mathematical physicist Francis R. Upton. But Americans of the day, with no small encouragement from the inventor himself, typically thought of Edison as the practical, unschooled inventor who needed no science. And it was true that neither mathematical nor scientific training necessarily made ordinary mortals a match for Edison's kind of genius. Eminent British physicists, sure that the carbon-filament lamp would draw too much current, scoffed at Edison's plans for low-cost electric lighting. Upton recalled the attack on the electric light: "I cannot imagine why I did not see the elementary facts more clearly. . . . I came to Mr. Edison a trained man, with . . . a working knowledge of calculus and a mathematical turn of mind. Yet my eyes were blind . . .; and I want to say that I had *company!*"[3]

In 1870 a few meteorological physicists managed to get a United States Weather Service organized under the Army Signal Corps. Rural Americans, always threatened by the capriciousness of the climate, frequently attacked the Service because its predictions were not uniformly accurate.[4] Farmers preferred patent remedies for insects, almanac forecasts for the weather. And it was not a scientist but a prosperous Illinois farmer, Joseph F. Glidden, who at a county fair in 1873 noticed an exhibit of the newly invented barbed-wire fencing, thought of a way to improve it considerably, and soon produced the innovation that did so much to open up the fertile grasslands of the Great Plains to agricultural settlement.

In the public colleges, where engineering and agricultural studies were centered, practical considerations dominated the teaching of science. In the older private colleges, where vocational training was generally inadmissible, science was subordinate to the goal of liberal education. Subjects were taught for the sake of "the *discipline* and *furniture* of the mind," in the persistent doctrine of the Yale report. Students were to learn physics in order to become "familiar with facts, with the process of induction, and the varieties of probable evidence," not to become physicists or engineers.[5] The older private colleges preferred to relegate mere vocational training to the country's few technical schools, like Rensselaer Polytechnic Institute, or the Massachusetts Institute of Technology. Harvard kept technical studies segregated in its Lawrence School of Science, Yale in its Sheffield School. For years technical students at Harvard and Yale lived in different buildings from the rest of the undergraduates, went to different lectures, sat apart in the college chapel, and earned degrees that Harvard and Yale proper held in suspicion, if not disdain.

Practical technical studies found more of a welcome in the public colleges, especially those that the Morrill Act had authorized to foster learning in agricultural and mechanical arts. But in this period a good deal of even the most sophisticated engineering knowledge was more a melange of codified experience, empirical relationships, and proved techniques than an analytical science based on the fundamentals of physics and chemistry. More important, the technical curriculum was tailored to

[3] Quoted in Matthew Josephson, *Edison* (New York, 1959), pp. 283, 194; Thomas Hughes, "Edison's Method," unpublished paper.

[4] Donald R. Whitnah, *A History of the United States Weather Bureau* (Urbana, 1965), pp. 43–44.

[5] "Original Papers," *American Journal of Science*, XV (1829), 301.

immediate vocational needs. Rather than studying the thermodynamics of steam engines, engineering students learned how to build the contraptions, and build them swiftly. Agricultural students preferred working the furrows of the college farm to studying the obscurities of soil chemistry. The many Americans who wanted their children back plowing or out earning a living were inclined to agree with the practical journalist who argued that agricultural students ought to be introduced not to a "laboratory and . . . philosophical apparatus" but "to a pair of heavy neat's boots and corduroy pants" and should be taught "how to load manure and drive oxen."[6]

In the federal research establishment, abstract investigations might develop under an unusual director like Clarence King. They could often spring up under an administrator lax enough to let his scientists go their own way (the respected astronomical work of the Naval Observatory had risen willy-nilly behind the cloak of its official purpose, the collection of charts and instruments; the superintendent, usually a naval officer on shore duty, had simply left the management of the work of the Observatory to the scientific staff). But only the Smithsonian Institution made the advancement of abstract knowledge a matter of policy. And Henry, who had expected to have to make only "a few oblations to Buncombe," was forced to spend a sizable fraction of James Smithson's bequest upon a turreted pile of red bricks and the care of a museum; he had precious little money annually, a scant $10,000 in 1865, to dispense for abstract research.[7]

Practical or abstract, governmental research was, compared to railroads, a pauper of national investment, and scientists were scarcely considered indispensable to the conduct of federal business. When Simon Newcomb, a Harvard graduate and one of the navy's leading astronomers, was offered an academic post, the secretary of the navy frankly advised him: "By all means accept the place; don't remain in government service a day longer than you have to. A scientific man here has no future before him, and the quicker he can get away the better."[8] The government rarely asked for the expert advice of the National Academy of Sciences. Federal officials might welcome the Academy's counsel when it reported upon a narrow technical subject, like a method of conveniently measuring the proof of distilled spirits. But Washington politicos reacted quite differently when the Academy meddled in such major matters of public policy as the settlement and use of the remaining public land in the West.

[6] Quoted in Earle D. Ross, *Democracy's College: The Land Grant Movement in the Formative Stage* (Ames, Iowa: State College Press, 1942), pp. 90–91.

[7] Washburn, "Joseph Henry's Conception of the Smithsonian Institution," pp. 111–12, 119, 141–42; *Annual Report of the Board of Regents, Smithsonian Institution, 1865*, p. 107.

[8] Quoted in A. Hunter Dupree, *Science in the Federal Government: A History of Policies and Activities to 1940* (Cambridge, Mass., 1957), p. 185.

Responding to a congressional request in 1878, the Academy recommended the consolidation of the various western surveys into a United States Geological Survey; it also urged reform of the public land system in the arid regions beyond the one-hundredth meridian, where the annual rainfall was insufficient to sustain the traditional homestead. Calling for a breakaway from the 160-acre homestead tract, the report urged the scientific classification and distribution of the land in parcels whose size varied with their suitability for mining, grazing, farming, or irrigation. The report was made the basis of a congressional bill which Representative Abram S. Hewitt, the respected ironmaster and friend of science, commended to the House. The plan came from "the highest scientific authority of the land," Hewitt declaimed, but that argument failed to diminish the vigorous opposition to the measure from representatives of the western states and territories.[9]

Did scientists say that the land was too arid for small farms? Theorists had pronounced the homestead system dead before, cried Montana Territory's congressional delegate, Martin Maginnis; settlers had gone west and, "practical men" all, had "seen the capabilities of this land which had escaped the notice of our scientists and statesmen." Did the proposal enjoy the imprimatur of the highest scientific authority in the land? The National Academy, argued Congressman Thomas MacDonald Patterson from Denver, Colorado, "never published but one work, and that was a very thin volume of memoirs of its departed members. And if they are to continue to engage in practical legislation," he said to the laughter of the House, "it would have been very well for the country if that volume had been much thicker."[1]

In the end, the Congress created the U.S. Geological Survey but left the public land system intact. Congressman Dudley C. Haskell, a Yale graduate from Lawrence, Kansas, did not mind giving science its "little appropriation." "Now, if you want a geographical survey, if you want a lot of astronomical figures, if you want a lot of scientific material, then organize your geographical surveys and authorize them to get out there and dig and hunt bugs and investigate fossils and discover the rotundity of the earth and take astronomical observations. But if you please, while you are there acting in the interest of science and in the interest of professional bug-hunting, leave the settlers upon our frontier alone. . . ."[2]

In the post-Civil War United States, the only reliable home for abstract science was the leading private colleges. But since the private college curriculum was designed to produce the intellectually well-balanced,

[9] U.S. Congress, House, *Congressional Record*, 45th Cong., 3d Sess., Feb. 11, 1879, p. 1207.
[1] *Ibid.*, Feb. 11, 1879, p. 1202; Feb. 18, 1879, p. 1564.
[2] *Ibid.*, Feb. 11, 1879, p. 1211.

liberally educated man, students were allowed too little curricular opportunity—at Yale, typically, less than 20 percent of total course time—to probe deeply into physics and chemistry. Most scientific courses rarely went beyond the elementary level. Since the curriculum included science primarily as a tool for training the mind, professors of science were not expected to specialize in any one subject. Quite the contrary: One man usually taught two scientific disciplines; sometimes he taught them all, including astronomy, physics, chemistry, botany, zoology, mineralogy, geology, and physiology. Many of the professor's students, the able chemist Frank W. Clarke complained, came to class "miserably prepared, caring little for what he considers important, and regarding his instruction as so much an impediment between them and their degrees."[3] With some exceptions, science was taught by the methods of lecture and recitation. Students rarely entered a laboratory to question nature but learned in a book what nature had to say.

What with the limited pedagogical need for adequate laboratory facilities, professors in public and private colleges were often like the physicist at Brown who, the president of the college recalled, occupied the "damp dark basement rooms of Rhode Island Hall . . . only at the risk of his health." Public colleges and technical schools generally ignored the challenge of discovering knowledge. Private college administrators may have genuflected before the ideal of scholarship, but most agreed with the assertion of the Yale report that elementary principles were not taught best by those "whose researches have carried them so far beyond these simpler truths, that they come back to them with reluctance and distaste."[4]

Whatever the avowed intellectual aim of the private colleges, they displayed little zest for the advancement of knowledge and a good deal of concern for the development of moral, which tended to mean religious, character. Most private college presidents were clerics operating under predominantly clerical boards of trustees. Reminders of religious piety usually insinuated themselves into the curriculum, and science was generally presented not as an ongoing intellectual enterprise but as though it were a completed body of knowledge exemplifying God's mind and benevolence. Private college administrators expected their faculty to meet a test of religious soundness, sometimes in a sectarian sense, usually by showing agreement with the fundamentals of Christian doctrine. Moral character counted for more in the rating of college faculty than scholarly accomplishments. "While we seek the highest scholarship," the presi-

[3] Frank W. Clarke, "American Colleges vs. American Science," *Popular Science Monthly*, 9 (Aug. 1876), 470. The Yale curriculum devoted 0.71 academic years out of 4 to science. *Catalogue of Yale College, 1866–67.*

[4] The Brown president is quoted in Walter C. Bronson, *The History of Brown University, Providence, 1764–1914* (Providence, 1914), p. 389; "Original Papers," *American Journal of Science*, XV (1829), 306.

dent of Amherst put it, "we also seek to make Christian faith its top stone and cornerstone."[5]

Graduate training was limited and desultory. Yale, the first American institution to offer the Ph.D., awarded just two doctorates in the 1860s. (In 1871 graduate enrollment in all subjects in the United States totaled 198 students.) A Harvard official recognized the implications: Asa Gray, Benjamin Peirce, and Louis Agassiz were "all going off the stage and their places cannot be filled with Harvard men, or any other American that I am acquainted with. This generation cannot match them. These men have failed to train their successors."[6] Speaking in 1869 as the retiring president of the American Association for the Advancement of Science, the Harvard astronomer Benjamin A. Gould summarized the well-known deficiencies of abstract science in America, especially "the want of access to such implements of research as are beyond the reach of most private men; the want of time and energy to spare from the 'grand struggle for existence;' and above all, the want of competent scientific counselers and guides for the beginners in scientific research." But Gould, one of the younger and more impatient members of Joseph Henry's circle, added emphatically: Whatever the difficulties past and present, the duty of his generation was clear—to hasten the time when the United States should awe all other nations by "leading the science of the world."[7]

[5] Quoted in Thomas Le Duc, *Piety and Intellect at Amherst College, 1865–1912* (New York, 1946), p. 45; Richard Hofstadter and Walter P. Metzger, *The Development of Academic Freedom in the United States* (New York, 1955), pp. 283–84. A standard college text was William Paley, *Natural Theology: or Evidence of the Existence and Attributes of the Deity, Collected from the Appearances of Nature.* See, for example, *Princeton University Catalogue, 1864–65*, p. 18; *Catalogue of Yale College, 1866–67*, pp. 36–37.

[6] The Harvard official was Charles William Eliot, quoted in Henry James, *Charles W. Eliot: President of Harvard University, 1869–1909* (2 vols.; London, 1930), II, 12–13.

[7] *Proceedings of the American Association for the Advancement of Science, 1869*, pp. 2, 37.

II

Ennobling and Purifying the Mind

During the winter of 1872, in many major cities of the eastern seaboard, crowds jammed lecture halls to hear the British physicist John Tyndall, a lean, bony-faced evangel of science. The New York *Tribune*'s special edition of the talks sold more than fifty thousand copies. As his tour progressed, Tyndall met a crescendo of acclaim that reached its peak in a glittering farewell banquet at the New York restaurant Delmonico's.[1]

Despite the general disregard for science in the post-Civil War United States, it clearly enjoyed considerable favor with one group. Its devotees were the prominent citizens of their communities, some of them landed gentry, others patrician aristocrats, and many more professionals or businessmen of the well-established mercantile type. The social cream of every city—the most "cultivated and intelligent people," in the phrase used to describe Tyndall's Manhattan audience—flocked to the talks.[2] Americans of cultivation had long provided an audience for scientific lecturers. Now, college graduates in an era when most Americans scarcely thought of college, they were paying unprecedentedly keen attention to what they considered science, its methods as well as theories. Charles Francis Adams recalled his response to an essay of John Stuart Mill's:

---

[1] A. S. Eve and C. H. Creasy, *Life and Work of John Tyndall* (London, 1945), p. 172; *Popular Science Monthly*, 2 (Feb., March 1873), 499, 625–26; New York *Tribune*, Feb. 5, 1873; Frank Luther Mott, *History of American Magazines* (4 vols.; Cambridge, Mass.: 1939–57), III, 107.

[2] *Popular Science Monthly*, 2 (Feb. 1873), 499.

"That essay . . . revolutionized in a single morning my whole mental attitude. I emerged from the theological stage in which I had been nurtured and passed into the scientific."[3] *Harper's*, *Scribner's*, and the *Atlantic Monthly*, widely read in the best circles, devoted numerous pages to natural science; ministers sermonized upon its lessons; many lecturers in addition to Tyndall explained it to the uninitiated. The most influential journal of the new diffusion was the *Popular Science Monthly*, edited by the self-taught chemist Edward Livingston Youmans.

"Deeper than all the questions of Reconstruction, Suffrage and Finance," Youmans insisted, "is the question, *What kind of culture shall the growing mind of the nation have?*" Youmans established the *Monthly* in 1872 to give the unequivocal answer: a scientific culture. Knowledgeably written, purposefully edited, its circulation soon reached ten thousand, as high as the readership of the influential *Nation*. The elder Oliver Wendell Holmes told Youmans that *Popular Science Monthly* "comes to me like the air they send down to the people in a diving bell. I seem to get a fresh breath with every new number."[4]

What stimulated Americans like Holmes was not the current work of professional scientists but the popularization of the main scientific triumphs of the mid-nineteenth century, notably the mechanical theory of heat, the conservation of energy, and the theory of evolution. In the 1840s the British physicist James Prescott Joule had decisively overturned the prevailing view that heat consisted of an imponderable fluid called "caloric." In a brilliant series of experiments, Joule showed that the motion of a paddle wheel in water raised the temperature of the liquid in a quantitatively predictable fashion. The implication: Heat was not a fluid but the sensible effect of mechanical motion.

What scientists would later call the theory of the conservation of energy had been most notably enunciated by the great German scientist Hermann von Helmholtz of Heidelberg University, who came to physics by way of physiology. In the 1840s, while serving as a doctor in the Prussian army, he did a good deal of research into the physical and chemical nature of life processes, particularly into the relations of metabolism to muscular activity. No animate body, he became sure, could do work without some metabolic expenditure. The assertion of an analogous physical assumption was the starting point of his classic 1847 paper, *On the Conservation of Force*: "it is impossible to create a lasting motive force out of nothing. . . ."[5] In a closed system, Helmholtz developed the argument, energy that disappeared in one form had to reappear in another;

[3] Charles Francis Adams, *Charles Francis Adams, 1835–1915: An Autobiography* . . . (Boston, 1916), p. 179.

[4] Edward L. Youmans, ed., *The Culture Demanded by Modern Life* (New York, 1867), p. v; Holmes is quoted in John Fiske, *Edward Livingston Youmans: Interpreter of Science for the People* (New York, 1894), p. 315.

[5] Quoted in Charles C. Gillispie, *The Edge of Objectivity: An Essay in the History of Scientific Ideas* (Princeton, 1960), p. 387.

it could neither be created nor destroyed. The principle of conservation had run through the work of many scientists in a vague, unspoken way; Helmholtz made it elegantly explicit. Physicists quickly realized that the theory of the conservation of energy was a fundamental postulate of limitless fruitfulness.

Youmans called the conservation of energy "the most important discovery of the present century" and gave all the advances of physics abundant space in the *Popular Science Monthly*.[6] Physics was interpreted —and celebrated—to magazine and lecture audiences in the accessible language of simple mechanisms, including Tyndall's description of heat as so many "modes of motion," a phrase echoed on both sides of the Atlantic. The conservation of energy, the Harvard scientist John Trowbridge was ready to announce, represented a "great guiding principle, which . . . will conduct us as Moses and the Israelites were once conducted to an eminence from which we can survey the promised scientific future."[7]

Charles Darwin, with his explosive doctrine of evolution, added to the intellectual appeal of science the sharp fillip of theological heresy. *The Origin of Species*, published in 1859, had skirted direct inquiry into the beginnings of humanity. *The Descent of Man*, which appeared in 1871, treated it outright. "Man," Darwin declared in sledgehammer words, "is descended from a hairy, tailed quadruped, probably arboreal in its habits, and an inhabitant of the Old World."[8]

Through the 1860s and 1870s, most Protestant ministers and theologians vented their wrath on the Darwinian heresies. Youmans, lecture-touring in upstate New York, came to Freeport, where "there was a protracted meeting in full blast in every church in town except the Episcopal, and a general feeling of pious rage at my appearance. . . . The next night was scarcely better." In tract after tract, Darwin's sentence about the origin of man was punned upon, ridiculed, and damned. The assault only amplified the attention Darwinism received. "I have never known anything like it," Youmans exulted. "Ten thousand *Descent of Man* have been printed and I guess they are all gone. . . . The progress of liberal thought is remarkable."[9] The prestigious *North American Review*, cautious about Darwinism in the late 1860s, was soon giving it enthusiastic coverage. In 1876 Thomas Henry Huxley came from England to lecture on evolution, and his reception matched Tyndall's. The evolu-

---

[6] Edward L. Youmans, ed., *The Conservation and Correlation of Forces* (New York, 1869), p. v.

[7] "Recent Progress of Natural Science," *Popular Science Monthly*, 2 (March 1873), 600; C. W. Siemens, "Science in Relation to the Arts," *ibid.*, 22 (Nov. 1882), 55; Trowbridge, "What Is Education," *ibid.*, 26 (Nov. 1884), 77.

[8] Charles Darwin, *The Origin of Species . . . and the Descent of Man . . .*, (New York, Modern Library edition), p. 911.

[9] Quoted in Fiske, *Youmans*, p. 244, and in Richard Hofstadter, *Social Darwinism in American Thought* (Boston, 1955), p. 27.

tionary analogies of Herbert Spencer made his writings a must for any American who considered himself *au courant*. At Dartmouth in 1873, Whitelaw Reid, the editor of the New York *Tribune*, observed: "Ten or fifteen years ago, the staple subject here for reading and talk . . . was English poetry and fiction. Now it is English science. Herbert Spencer, John Stuart Mill, Huxley, Darwin, Tyndall have usurped the places of Tennyson and Browning, Matthew Arnold and Dickens."[1]

To applaud science was to set oneself apart socially in a country so exuberant over mere gadgets and machinery. To discuss it was to mark oneself as a cultivated man. Patricians and landed gentry, professionals and mercantile businessmen—most of the patrons of science liked to be called cultivated; a description of pleasingly lofty connotations, it signified their sense of distance from a country they found downright philistine. Raised in comfortable homes, they disdained post-Civil War America's encomiums to prosperity. The more the country measured progress in tons of steel and miles of track, the more a Henry Adams deplored its failure to contribute to the higher thought of the world. The more the nation celebrated the advance of technology, the more its cultivated citizens insisted that its inventors were, after all, merely drawing on the treasury of science. "I took to pieces the claims of their practical men," Tyndall remarked about one of his American lectures. "I was as plain as I could be." His audience repeatedly interrupted with "loud approval."[2]

If cultivated Americans found the country's preference for the machine over the intellect annoying, they considered its taste in leadership repugnant. They were accustomed to doing the leading. Now many men in public office were like George Washington Plunkitt of Tammany Hall, who announced: Colleges "don't count in politics."[3] Citizens of cultivation were generally white, Anglo-Saxon, and Protestant; they resented the ascendant political power of Irish Catholics in the North and the enfranchisement of Negroes in the South. They assumed that the country should defer to their judgment. Yet it was applauding a different breed, the new industrial leaders, the rough politicos, a disturbing number of whom seemed to have little sense of what the salons called "public responsibility." In the era of Ulysses S. Grant, officials on every rung of the political and business ladders seemed to be indulging in bribes and boodling.

The trouble seemed to come down to a question of standards. Americans of cultivation believed that in every activity self-interest ought to give

[1] Quoted in Mott, *History of American Magazines*, III, 105.

[2] Henry Adams to Daniel Coit Gilman, Nov. 17, 1875, DCG, Adams file; Tyndall is quoted in Eve and Creasy, *Tyndall*, p. 171.

[3] Quoted in Richard Hofstadter, *Anti-intellectualism in American Life* (New York, 1963), pp. 187–88.

way to public responsibility. Sure that they lived by this standard, they were alarmed that the Grants and Vanderbilts did not. The expedient ethics of the new leadership seemed to be poisoning standards everywhere. Even higher education, a special sanctum of cultivated concern, appeared infected with Grantism. In some states the sales of college land grants were questionable enough to generate talk of "an agricultural college 'ring.' "[4]

Cultivated Americans worried still more about the intellectual thrust of the entire college system. They considered higher education crucially important to the nurture of public responsibility and now feared for the values and professional integrity of college graduates. A good deal of their anxiety stemmed from the attitude of the colleges toward that signet of cultivation, science. Here were the private colleges, offering only elementary science, and the public colleges, smothering science under a blanket of practical studies. Men of only practical training, like the railroad builders, seemed satisfied with shabby technical standards. In the haste to finish the Central Pacific, miles of track were laid on the frozen winter ground of the Sierras. When the earth thawed, the tracks sank in an ooze of mud.[5] Professional responsibility had apparently been forsaken in the rush for profits. Cultivated Americans, fascinated by the intellectual content of science, thought that it should have a more substantial place in the curriculum. They were also sure that teaching more science to engineers would improve their professional knowledge and standards.

The cultivated public accused the religion-dominated colleges of dogmatism. Against dogmatism as such, they particularly opposed the kind of dogmatism evident when prominent clergymen adapted theology to the defense of wealth against poverty. A characteristic example was the famous exhortation of the Reverend Russell Conwell in his *Acres of Diamonds*: "I say that you ought to get rich, and it is your duty to get rich. The richest people are generally those of the best character. It is wrong to be poor."[6] As enthusiasts of science, the cultivated public felt sure that science was the antithesis of dogmatism. Yet even the prominent private colleges sometimes practiced outright intellectual suppression, including the dismissal of a professor of biology at Amherst who had sought to inculcate knowledge of the clam through a course in dissection and whose successor was told to teach the subject "as an absolutely de-

[4] Charles W. Eliot, in *Proceedings of the National Education Association, 1873*, p. 111.

[5] Oscar Lewis, *The Big Four: The Story of Huntington, Stanford, Hopkins, and Crocker, and of the Building of the Central Pacific* (New York, 1945), pp. 83, 345. See also William Z. Ripley, *Railroads, Finance and Organization* (New York, 1920), pp. 45–46; Andrew D. White, "Scientific and Industrial Education," *Popular Science Monthly*, 5 (June 1874), 182.

[6] Quoted in Sidney Fine, *Laissez-Faire and the General Welfare State* (Ann Arbor, 1956), p. 119.

pendent product of an absolutely independent and spiritual creator."[7] More important, the splenetic clerical opposition to Darwinism called into obvious question the commitment of the private colleges to scientific subjects and to the scientifically open mind.

At the Tyndall banquet, one speaker had in mind the posture toward science of both the public and private colleges when he declared: Something is wrong with education that breeds "devotion to the mere husks and rinds of good . . ."; something is wrong with training that encourages "struggle for mere place and pelf . . . faith in mere material comfort and wealth."[8]

Charles William Eliot emphatically agreed. A product of Boston's Beacon Street aristocracy, Unitarian enthusiast of Darwinism, the ultimate cultivated American, Eliot assumed the presidency of Harvard in October 1869. At the inaugural in the bare, wooden meetinghouse across from the college Yard, he received the ancient seal, keys, and charter from the governor of Massachusetts, listened politely to some mutterings about the dangers of certain kinds of science, and in a gentle but determined way let the gathering know that he intended to set things right in Cambridge.

Eliot had a patrician's belief in public responsibility and a patrician's resentment against the crudely self-made man. He thought the "vulgar conceit that a Yankee can turn his hand to anything . . . preposterous and criminal."[9] Bookish from childhood, the son of a Harvard treasurer, educated at Harvard himself, he was convinced of the tempering influence of higher education. But he had small respect for the typical American college. If a merely "polytechnic"—meaning vocational—spirit dominated a college, it fell below its "true ideal." If narrowly cultural training and ministers dominated it—"a really learned minister is almost as rare as a logical sermon"—it could scarcely pretend to educational competence.[1] In this industrial age Harvard's new president, whose father had helped create the Lawrence School of Science, believed that the college ought to encourage professional studies, and in a way that was not merely vocational. In an age when new conditions were challenging the old orthodoxies, he thought that the college ought also to nurture an antidogmatic cast of mind. In Eliot's program, "The university will hold

---

[7] Thomas Le Duc, *Piety and Intellect at Amherst College, 1865–1912* (New York, 1946), pp. 83–86. See also Frederick Rudolph, *The American College and University: A History* (New York, 1962), p. 347; George P. Schmidt, *The Liberal Arts College: A Chapter in American Cultural History* (New Brunswick, N.J., 1957), p. 67.

[8] Andrew D. White, "Science and Public Service," *Popular Science Monthly*, 2 (April 1873), 738–39.

[9] Charles William Eliot, *Educational Reform* (New York, 1898), p. 11.

[1] Eliot, "The New Education," *Atlantic Monthly*, 23 (1869), 215.

high the standards of public spirit and will enlarge that cultivated class which is distinguished, not by wealth, merely, but by refinement and spirituality."[2]

Eliot, who had taught chemistry at Harvard and MIT, considered science the most powerful weapon in his educational arsenal. Engineering studies based on science would make for more competent engineers. More important, the replacement of the "dogmatic and overbearing" with the "receptive" mind would be "a most beneficent result of the study of natural history and physics." The study of nature, Eliot assumed, encouraged habits of "candid, fearless truth-seeking. . . ."[3] Young men, tempered by science to a high-minded devotion to truth, would carry high-mindedness beyond the daily practices of their lives to their outlook upon the world. Give science more weight in the curriculum, in short, and Harvard's graduates were bound to become virtuous citizens.

But the science that Eliot espoused was not the old science, learned by rote, committed to memory as a completed body of fact—that science led straight to dogma. Eliot had pioneered laboratory work in his days as a junior chemistry professor; he was sensitive not only to the study of nature but to the nature of study. The student must discover for himself, his judgment governed by a scrupulous regard for the facts. What better way to encourage that habit than to have him study science in a laboratory? There he would learn the scientific method, the nineteenth century's "great addition . . . to the idea of culture." In the laboratory, as Eliot assessed the process, the student "scrutinizes, touches, weighs, measures, analyzes, dissects, and watches things. By these exercises his powers of observation and judgment are trained."[4]

Along with the emphasis on science, Eliot intended to provide a wider variety of courses in the arts—economics and politics, modern languages and modern history—and to experiment with a limited introduction of the elective system, so that students could pursue advanced work. He would of course maintain the traditional studies, and religion need not fear for a place in the college. "A worthy fruit of academic culture is an open mind . . . penetrated with humility," Eliot told the inaugural gathering. "It is thus that the university in our day serves Christ and the church."[5]

At Princeton University, President James McCosh was advancing the cause in a somewhat different way. A Scotsman educated at Glasgow and Edinburgh universities, a highly respected author of philosophical treatises before his arrival at Princeton in 1868, he was a minister whom Charles William Eliot could not fault for want of erudition. McCosh joined in the

[2] Eliot, *Educational Reform*, p. 44.

[3] Eliot, *American Contributions to Civilization* (New York, 1897), p. 238; Eliot, *Educational Reform*, p. 227.

[4] William A. Neilson, ed., *Charles W. Eliot: The Man and His Beliefs* (2 vols.; New York, 1926), I, 191; quoted in Henry James, *Charles W. Eliot: President of Harvard University, 1869–1909* (2 vols.; London, 1930), I, 358.

[5] Eliot, *Educational Reform*, p. 8.

Darwinian wars, arguing that it was "useless" to denounce evolution.[6] One of the first Protestant theologians in America to attempt a reconciliation of Darwinism and religion, McCosh never went so far as to accept *The Descent of Man.* But his reluctance added to the authority of his basic admonition: Religion need not quaver before the facts of nature. Instead, strengthen religion by emphasizing the religious lessons of science.

In McCosh's view, the colleges had to prevent the scientific approach to knowledge from overwhelming all others. An intuitionist as a philosopher, an opponent of John Stuart Mill's thoroughgoing empiricism, McCosh held that "fundamental truth," as he called it, could be perceived without experiments (he once argued that he had been prepared to "receive" the theory of the Conservation of Energy because it followed "directly from a doctrine laid down by me . . . in my work on 'The Method of the Divine Government'"). In academic terms, McCosh counseled scholars to "fight determinedly" to hold the curricular fort of Greek, Logic, and Ethics. In all, Princeton would preserve the well-educated man. As McCosh expressed his dissent from too broad an application of the elective system: "The objection is, that it would nurture specialists without a general or comprehensive culture."[7]

But as an enlightened conservative, McCosh believed firmly in the colleges as a source of public leadership, equally firmly that the "radiating power [of science] is especially needed in our day . . . to counteract the coarseness, the earthliness, the clayiness" spreading across the land. Of course, with regard to "religious truth . . . no uncertain sound [would be] uttered within these walls."[8] McCosh issued no paeans to laboratory work, and he was no great enthusiast of practical scientific training. Whatever the caveats, he did favor a limited elective system and a school of science, which would include engineering. Of course, scientific students at Princeton would have to study Latin. In McCosh's defense of the requirement, "We seek to make [them] . . . educated gentlemen, and not mere scientists."[9]

Different in their emphases but alike in their overall public purpose, McCosh and Eliot together gave the thrust of institutional leadership to cultivated America's enthusiasm for the study of the cosmos.

"We are going to have new times here at Harvard," a young devotee of science cheered Eliot's inaugural. "No more old fogyism, I hope."[1]

[6] Quoted in *Popular Science Monthly,* 16 (Feb. 1880), 556.
[7] James McCosh, *Christianity and Positivism* (New York: R. Carter, 1871), p. 14; McCosh, *What an American University Should Be* (New York: J. K. Lees, 1885), p. 8; McCosh, *Course of Study in Princeton College, Report to the Board of Trustees, Nov. 8, 1877,* pp. 6–7.
[8] *Inauguration of Reverend James McCosh . . .* (Princeton, 1868), pp. 19–20; quoted in Laurence R. Veysey, "The Emergence of the American University: A Study in the Relations of Ideals and Institutions" (unpublished Ph.D. dissertation, History, University of California, Berkeley, 1961), p. 192.
[9] McCosh, *Twenty Years of Princeton College* (New York, 1888), pp. 26–27.
[1] John Fiske, quoted in James, *Eliot,* I, 228.

An era of educational reform was under way. It was as much a part of the attack against corruption as were the calls for civil service, tariff, and currency reform that coalesced in the liberal political movement of the early 1870s. Edwin L. Godkin's liberal-minded *Nation*—a large fraction of its readers had an interest in the colleges, Godkin noted—carried a good deal of educational material as well as denunciations of political practices. Writers in *Popular Science Monthly* matched their admiration for the triumphs of science with indignation at high tariffs and paper money. Reformers could warmly agree with the enthusiast who proclaimed: "the chief influence of modern science upon . . . education will be its ethical influence."[2]

A fresh generation of college presidents laid down bold policies for newly founded Cornell and the University of California, attacked the customs of established Minnesota and Michigan, assailed the certitudes of venerable Yale and Columbia. By birth, background, and conviction, all were cultivated Americans. Like the antebellum college presidents, all believed higher education to be a wellspring of public spirit and were devoted to making the college curriculum suitable to its public purpose.[3] Like McCosh and Eliot, they disagreed among themselves about the proper place in the curriculum of the classics, modern languages, and social studies and debated the extent to which the elective system should be introduced. But for all the differences, they were united—and set apart from their presidential predecessors—by the degree to which their educational attitudes were shaped by the challenge of science.

There was Noah Porter at Yale, McCosh's chief ally, who welcomed the expression of "Christian truth" in the forms of science; James B. Angell at Michigan, who argued for science as a way of developing "the faculties of observation, imagination, and reasoning"; William Watts Folwell at Minnesota, who proposed the "extension of the scientific method, till it shall become the guide of conduct in the every day life"; Frederick A. P. Barnard at Columbia, who called for reading the scrip-

---

[2] E. L. Godkin to Daniel Coit Gilman, April 19, 1866, DCG, Godkin file; William P. Atkinson, "Liberal Education in the Nineteenth Century," *Popular Science Monthly*, 4 (Nov. 1873), 23.

[3] Veysey, "American University," p. 375; Andrew D. White, "Inaugural," in *Inauguration at Cornell University* (Ithaca, 1921), pp. 13–15, 28–29; White, "Scientific and Industrial Education," *Popular Science Monthly*, 5 (June 1874), 186–87; Carl Becker, *Cornell University: Founders and the Founding* (Ithaca, 1943), p. 194; John S. Brubacher and Willis Rudy, *Higher Education in Transition: An American History, 1636–1956* (New York, 1958), pp. 159–60; Fabian Franklin, *The Life of Daniel Coit Gilman* (New York, 1910), pp. 2–4, 14, 28–29; William Warren Ferrier, *Origin and Development of the University of California* (Berkeley: Sather Gate Book Shop, 1930), pp. 339–40; William Watts Folwell, *University Addresses* (Minneapolis; H. W. Wilson, 1909), pp. 16–21, 27–28, 39, 50, 52; Folwell, *William Watts Folwell: The Autobiography and Letters of a Pioneer of Culture*, ed. Solon J. Buck (Minneapolis, 1933), pp. 4–8, 256–57; Shirley W. Smith, *James Burrill Angell: An American Influence* (Ann Arbor, 1954), pp. 6, 14; *Exercises at the Inaugural of President Angell . . .* (Ann Arbor, 1871), pp. 18, 22, 24; *Proceedings at the Inauguration of*

tures by the "torch" of science.[4] All of these new educational leaders intended to promote science from its subordinate role in the traditional college and to endow it with equal curricular opportunity. Some even intended to make science central to the intellectual spirit and outlook of the whole academic enterprise.

The new scientific persuasion gradually spread to the less prestigious institutions. Reformers in even the most pietistic sectarian schools could argue for the ideas of McCosh without fear of offending orthodox authorities. Reformers in the secularly minded colleges were strongly influenced by Eliot. Public college administrators, bound to justify apparently impractical science to their legislatures, listened in particular to Andrew D. White at Cornell. As much the aristocrat as Eliot, White was also the head of a mixed private and land-grant institution. He defended federal aid to education, which Eliot opposed, and displayed less interest than did his Harvard colleague in using science to develop antidogmatic habits of mind. White once charged that "millions on millions of dollars have been lost by the employment of half-educated engineers"; he emphasized rather the value of science to technical training. The tone given science at Cornell softened the hauteur of Eliot's arguments and made them more appropriate to the public colleges.[5]

In the University of California at Berkeley, Daniel Coit Gilman was astir. A friend of White's, scion of a prominent eastern family, the organizer of the Sheffield School at Yale, Gilman was a type-case advocate of science. He no sooner introduced a cultivated curriculum at the University of California than the State Grange and the Mechanics' Deliberative Assembly issued demands for more bread-and-butter courses. The administration, the social reformer Henry George charged in 1873, had "perverted" Berkeley from a college of industry into "a college of the classics and polite learning." Gilman, "infinitely disgusted," was eager to leave, and two years later he accepted the presidency of the newly founded Johns Hopkins University in Baltimore.[6]

At Gilman's inaugural in 1876, Thomas Henry Huxley delivered the oration and the ceremonies closed without prayer. Keep young students

President Frederick A. P. Barnard . . . (New York: Hurd and Houghton, 1865), pp. 10, 47, 68, 69; Frederick A. P. Barnard, *Memoirs of Frederick A. P. Barnard*, ed. John Fulton (New York, 1896), pp. 1, 39, 83, 142, 355; Barnard to Eliot, May 27, 1871, CWE, Box 67; Noah Porter, *The American Colleges and the American Public* (New York, 1878), pp. 71, 94, 179, 224; *Addresses at the Inauguration of Professor Noah Porter . . .* (New York, 1871), pp. 28, 45–46, 48, 52–53; Ralph Henry Gabriel, *Religion and Learning at Yale: The Church of Christ in the College and University, 1757–1957* (New Haven, 1958), p. 183; Porter to James McCosh, November 3, 1868, JM.

[4] *Inauguration of . . . Porter*, p. 48; *Inaugural of President Angell*, p. 18; *Inauguration of . . . Barnard*, p. 51; Folwell is quoted in *Science*, II (Aug. 24, 1883), 228.

[5] White, "Scientific and Industrial Education," *Popular Science Monthly*, 5 (June 1874), 182; Rudolph, *American College and University*, p. 267.

[6] George and Gilman are quoted in Franklin, *Gilman*, pp. 146–47, 155.

away from Hopkins, a Presbyterian weekly warned: "Honor here was refused to the Almighty."[7] No honor went to general education either. Gilman, who had coped enough with undergraduate matters in Berkeley and who had once savored the atmosphere of the German universities, proposed a different way of promoting science for cultivated purposes.

The goal of Hopkins would be "the encouragement of research"—not for practical ends but for its own sake. Gilman had to satisfy the more undergraduate-minded trustees and could not manage a complete breakaway from the traditional college. The best teachers, he deftly argued, were those "free, competent, and willing to make original researches."[8] But his program directly countered what the cultivated public was deploring: the fact that, for all America's material wealth, her contributions to higher learning had been minuscule. By the time Hopkins opened, various other colleges had inaugurated nominal graduate work, but none had given the sheer advancement of knowledge so prominent and serious a place in their programs. A bright beacon for cultivated Americans everywhere, the Hopkins idea was the capstone of the movement for educational reform.

As the pressure for change worked its way through the college system, science, in the words of President Ezekiel G. Robinson of Brown University, "got a foothold in the curriculum which it is never likely to lose."[9] Liberal arts colleges created degrees in science and engineering. Established engineering schools everywhere made fundamental scientific studies curricular requirements; almost every one of the fifty-five new technical institutions opened in the two years after 1870 gave physics and chemistry special emphasis.[1] Increasing numbers of schools, technical and nontechnical, taught science in the laboratory. Professors of science acquired enhanced status and the academic leverage that went with it.

By the mid-1870s, cultivated America's patronage of science was giving the subject an enormous academic boost. But among the better-educated and higher-status public of the day, the expected result of the academic shift overshadowed the benefits to the scientific disciplines themselves. For beneath the pleasure in the transformation was the dogged assumption of an Eliot: Science "ennobles and purifies the mind."[2]

[7] Quoted in Hugh Hawkins, *Pioneer: A History of the Johns Hopkins University, 1874–1889* (Ithaca, 1960), p. 71.
[8] Quoted in Richard Hofstadter and Walter P. Metzger, *The Development of Academic Freedom in the United States* (New York, 1955), p. 377, and in Abraham Flexner, *Daniel Coit Gilman* (New York, 1946), p. 58.
[9] Quoted in Walter C. Bronson, *The History of Brown University, Providence, 1764–1914* (Providence, 1914), p. 388.
[1] Charles R. Mann, *A Study of Engineering Education* (New York, Carnegie Foundation for the Advancement of Teaching, Bulletin #11, 1918), p. 6; *Report of the Investigation of Engineering Education, 1892–93* (2 vols.; Pittsburgh: Society for the Promotion of Engineering Education, 1930–34), I, 818–19.
[2] Quoted in James, *Eliot*, I, 64.

III

The Flaws of
American Physics

The founding of Johns Hopkins University was an event to excite the young physicist Henry A. Rowland. Descended from a line of Yale-trained ministers, Rowland was sent to Andover Academy, where he found the science inadequate and complained of the "horrid" boys who swore. In college he announced that he had no ambitions for a mere industrial career; success in business was no more important than business itself. Rowland decided to devote himself to science, to the kind of research that brought "not . . . filthy lucre but good substantial reputation."[1] He closeted himself in his boarding-house room, hunched over a book—it was often one of Faraday's works—from his personal scientific library, and scrawled notes of research problems in contemporary physics. His experiments in electromagnetism identified him as a highly promising scientist. Shortly before Hopkins opened, Daniel Coit Gilman, who liked to bet on young men, appointed Rowland, then twenty-seven, the first professor of physics at the new university.

In the post-Civil War decades the revolution in the colleges created new opportunities in science for young men with the upper-status background and attitudes of Rowland, but there was no general democratization of access to the physics profession. In the America of 1880 only eleven secondary schools offered a physics course with laboratory work, only four a course lasting a full academic year. In higher education the

[1] Rowland to Anna Rowland, May 29, 1865; June 9, 1873, HR.

number of scholarships and fellowships available was minute; most young people could not afford an undergraduate education, let alone graduate school or study in Europe.[2] Then, too, as a Cornell scientist aptly expressed a widespread belief: "In this country, men devoted to science purely for the sake of science are and must be few in number. Few *can* devote their lives to work that promises no return except the satisfaction of adding to the sum of human knowledge. Very few have both the means and inclination to do this." American physicists of the day tended to come from the narrow fragment of society that called itself cultivated; most were the sons—or married the daughters—of well-to-do merchants, gentry, lawyers, ministers, or teachers; almost all were white Anglo-Saxon Protestants.[3]

By the early 1890s only some 200 Americans were practicing the discipline of physics. The large majority were employed in the academic world. About one-fifth of the profession published research regularly. The work of these productive men, who were more likely than members of the profession at large to have taken a Ph.D. and to have studied abroad, was distributed among all the principal fields of contemporary interest on both sides of the Atlantic, including heat, light, electricity, and magnetism.[4] Some of the new American physicists earned the respect of their European colleagues. A few, including Rowland and Albert A. Michelson, won considerable acclaim.

In 1876 Rowland went abroad for a year to study and to purchase apparatus for the Johns Hopkins physics laboratory. While working in Helmholtz's laboratory in Berlin, he succeeded, where others had failed, in demonstrating experimentally that an electrically charged rotating disc would produce a magnetic field. Back at Hopkins, he meticulously obtained an authoritative measure of the ohm and an improved value for the mechanical equivalent of heat. Then, in 1882, Rowland revolutionized the study of light spectra, one of the new and increasingly important fields of late-nineteenth-century physics and astrophysics.

Every chemical element was known to emit light in a unique spectrum of colors, or combination of wavelengths. Analyze the spectrum of star-

---

[2] Albert E. Moyer, "The Emergence of the Laboratory Approach in the Teaching of Secondary-School Physics in Late Nineteenth-Century America" (unpublished Master's thesis, History of Science, University of Wisconsin, 1974), pp. 3–4; Geraldine Joncich, "Scientists and the Schools of the Nineteenth Century: The Case of American Physicists," *American Quarterly*, XVIII (1966), 677–85.

[3] William A. Anthony, in *Proceedings of the AAAS, 1887*, p. 70; Daniel J. Kevles, "The Study of Physics in America, 1865–1916" (unpublished Ph.D. dissertation, History, Princeton University, 1964), Appendix V.

[4] Daniel J. Kevles and Carolyn Harding, "The Physics, Mathematics, and Chemical Communities in the United States, 1870 to 1915: A Statistical Survey," California Institute of Technology, Social Science Working Paper No. 136, March 1977, Tables 1, 3. In 1888, the number of fellows in Section B, the physics section, of the AAAS was 151. *Proceedings of the American Association for the Advancement of Science, 1888*, pp. lxvi–lxxxvi.

light and you could determine both the chemical composition of the star and whether it was in a solid, liquid, or gaseous state. To disperse the different colors mixed in a beam of light, physicists used a flat grating—a reflective plate closely ruled with thousands of fine lines per inch. Both the purity and accuracy of the resulting spectrum depended on the regularity of the interlinear spacings, which depended in turn on the perfection of the screw in the ruling engine. Rowland contrived a way to machine a screw of almost flawless pitch. No less important, he used the screw to make a new type of grating, by ruling the lines not on a flat but on a spherically curved surface.

Rowland's curved grating advantageously focused the entire spectrum on a circle. That property of the grating eliminated the need for glass lenses, which absorbed part of the spectrum. It also vastly reduced the time necessary to take an extended series of spectral measurements. When Rowland demonstrated his apparatus in London in 1883, he was received, a fellow American physicist marveled, "like the Yosemite, Niagara, Pullman parlor car; far ahead of anything in England." In France a student of spectra was heard to mutter, "We must begin all over again."[5]

Rowland also remapped the spectrum of the sun. His published tables of solar wavelengths, which were ten times as accurate as their best predecessors, became standards; his gratings, which he supplied at cost to other physicists, became stock equipment in the major laboratories of Europe and America. Rowland provided his own testament to the importance of his gratings by having his ashes interred in the wall of the Hopkins laboratory where he had ruled the revolutionary instruments.

When Rowland invented his grating, Albert A. Michelson was already on his way to the first Nobel Prize in physics awarded an American. Born in a small town on the Prussian-Polish border, Michelson grew up in remote Murphy's Camp, a mining town in Calaveras County, California. His father, like many Jewish immigrants in the West of that day, ran a drygoods store. His mother, the daughter of a Jewish businessman, dreamed of better things for her talented son. In rough Murphy's Camp, Michelson spent a good deal of his time reading books and playing the violin. When the family moved to booming Virginia City, he was sent off to a high school in San Francisco. The principal, with whom he boarded, encouraged the young man at the violin and in his studies—especially science—taught him to box, and urged him to complete his education at the Naval Academy.

Michelson, who failed to get the Annapolis appointment allotted to his Nevada congressman, went to Washington to try for one of the ten at-large places awarded by the President. The congressman helpfully

[5] John Trowbridge to Daniel Coit Gilman, Nov. 30, 1882, Nov. 12, 1882, HAR, Box T.

wrote to Grant: The boy's father was highly influential, and the boy himself something of a "pet" among the Jews of Virginia City; the Jewish community was "a powerful element in our politics"; an appointment by the President would do a great deal "to fasten these people to the Republican cause."[6] Grant, who had already filled the at-large places, stretched the rules and made Michelson the eleventh appointee.

At the Naval Academy Michelson excelled at boxing—he was the school champion—and at physics. At commencement, Superintendent John L. Worden, the commander of the *Monitor* in its famed engagement with the *Merrimac*, told Midshipman Michelson, "If you'd give less attention to scientific things and more to your naval gunnery, there might come a time when you would know enough to be of some use to your country." After Michelson completed his two-year postgraduate cruise, the navy found a use for him as an instructor of physics and chemistry at Annapolis. Michelson, who considered the teaching easy—"all I had to do was to keep a few textbook pages ahead of the cadets"—had ample time for the Academy social life and for research.[7] In time he met Margaret Heminway, a niece of his superior officer and the daughter of a wealthy Wall Street broker; the couple was soon married in the family's Episcopal church. Soon, too, with the aid of $2,000 from his father-in-law, Michelson performed his first significant experiment, a measurement of the speed of light to unprecedented accuracy. A few years later, in France, the twenty-nine-year-old physicist walked into a laboratory and was asked, Are you the son of the famous Michelson?[8]

Michelson was study-touring in Europe that academic year, 1880–81, and in Berlin, while working in Helmholtz's laboratory, he began the most significant investigation of his career—the attempt to measure the relative motion of the earth and the "luminiferous ether." It was a fundamental hypothesis of nineteenth-century physics that the ether was the medium which transmitted the undulatory motions of light waves. Most physicists assumed that the ether pervaded all space; some, that it remained entirely stationary despite the passage of matter through it. In a letter to an officer at the *Nautical Almanac*, the great British physicist James Clerk Maxwell had called attention to the importance of empirically demonstrating the presence of the ether. Michelson, who read Maxwell's letter, reasoned that if the earth in its orbital motion was plowing through this undulatory medium, an ether wind would sweep the planet. Light, like a boat in a river, would travel more slowly against the wind than

---

[6] Quoted in Dorothy Michelson Livingston, *The Master of Light: A Biography of Albert A. Michelson* (New York, 1973), p. 25.

[7] The superintendent is quoted in *The New York Times*, May 10, 1931, p. 3; Michelson, in Bernard Jaffe, *Michelson and the Speed of Light* (Garden City, N.Y., 1960), p. 46.

[8] Livingston, *The Master of Light*, p. 87.

across it. With funds provided by Alexander Graham Bell, whom Simon Newcomb had probably told about the proposed experiment, Michelson constructed an ingenious and extraordinarily sensitive optical instrument to measure the expected difference in velocity. But he could not detect any such difference. The hypothesis of a stationary ether, meaning one not dragged along by the earth moving through it, he boldly concluded, was "erroneous."[9]

All the same, the optical instrument that Michelson had developed for the experiment was adaptable to numerous other investigations with light. Eventually known as the Michelson interferometer, it became a standard item, like the Rowland grating, on the apparatus shelves of the world's physics laboratories.

Michelson's failure to detect the ether decidedly troubled theorists of physics. An ether seemed essential to understand how light waves traveled from place to place. After Maxwell, who in his grand synthesis of electricity and magnetism had proposed that light was an electromagnetic oscillation, the same ether also seemed to an increasing number of physicists, including most in the United States, the seat of all electrical and magnetic forces. Michelson typically declared that by taking on "various modes of motion," the ether might even account for "all the phenomena of the physical universe," including heat, light, electricity, and magnetism.[1]

Like many of their brethren in Europe, American physicists tended to believe that all physical phenomena could be reduced to mechanical models; a few even held that these mechanisms were not merely models of reality but reality itself. In the view of the mechanists, the ether was some kind of elastic solid that obeyed the laws of dynamics. (In the interpretation of a small group of antimechanist physicists on the Continent, it was fundamentally electromagnetic.) Typically for an American physicist, Rowland dismissed the idea that magnetism could act across "perfectly vacant space" as an "absurdity." He did an elaborate series of experiments to show the usefulness of "Faraday's method of lines of magnetic force." Intellectually, American physicists had always been closest to the British school. Now they assumed with Maxwell that the light spectrum given off by a gas must be produced by the vibrations of its molecules, which somehow set up vibrations in the luminiferous ether. They paid respectful attention to the atomic theory of the commanding British physicist William Thomson, who once declared that he could not understand anything unless he could make a mechanical model of it. Atoms, Thomson proposed, were a kind of vortex motion in the ether,

---

[9] Michelson, "The Relative Motion of the Earth and the Luminiferous Ether," *American Journal of Science*, XXII (1881), 128.
[1] Michelson, *Light Waves and Their Uses* (Chicago, 1903), p. 162.

like smoke rings. Colliding vortices would rebound; vibrating vortices, produce spectra.[2]

But such mechanical reductionism made John B. Stallo restless. The product of a German gymnasium and a Jesuit college, once briefly a scientist and now a prominent Cincinnati lawyer, Stallo was not only cultivated but erudite; he had more than an amateur's continuing interest in physics. In 1882, in his *Concepts and Theories of Modern Physics*, Stallo attacked the prevalent notion that the "mechanical explanation" of nature was "not only unquestionable, but absolute, exclusive, and final." "The truth is," he asserted, "that neither mass nor motion is substantially real"; both, like every other physical concept, including space, were merely tools of understanding. In fact, the path to knowledge was paved with a series of "logical fictions." While these fictions were both legitimate and indispensable to thought, they were no more than "symbolical representations" of the real world.[3]

Somewhat later, Albert Einstein, with his special theory of relativity, would eliminate the ether, mechanical or otherwise, from the house of physics. Much later, views like Stallo's would find a receptive audience among the post-quantum-mechanics generation of American physicists. But in the 1880s the respected and influential astronomer Simon Newcomb pronounced Stallo's book unworthy of critical notice; even philosophical reviewers generally dismissed the work as an ignorant presumption.[4] In the 1880s, too, British physicists eagerly endorsed further efforts by Michelson, now a professor at the Case Institute of Technology in Cleveland, to detect the ether experimentally. Delayed by marital troubles and a nervous breakdown, Michelson finally repeated the experiment in 1887 with a colleague at Case, Edward W. Morley. Done with painstaking exactitude, the Michelson-Morley experiment revealed no evidence of an ether wind. Michelson's conclusion: The earth, as it swept around the sun, must drag an envelope of ether with it.

In the late nineteenth century, American physicists, along with many of their European colleagues, held tenaciously to the hypothesis of a mechanical ether, insisted that all of physics could be reduced to matter

[2] *The Physical Papers of Henry Augustus Rowland* (Baltimore, 1902), pp. 56, 127; A. F. Chalmers, "The Limitations of Maxwell's Electromagnetic Theory," *Isis*, 64 (Dec. 1973), 473. Robert H. Silliman, "William Thomson: Smoke Rings and Nineteenth Century Atomism," *Isis*, 54 (Dec. 1963), 461, 465, 469–71; Edmund Whittaker, *A History of the Theories of Aether and Electricity* (rev. ed.; 2 vols.; London, 1951), I, 294.

[3] J. B. Stallo, *The Concepts and Theories of Modern Physics*, ed. Percy W. Bridgman (Cambridge, Mass., 1960), pp. 56, 170, 302.

[4] *Ibid.*, pp. vii–lx; Simon Newcomb, "Speculative Science," *International Review*, XII (April 1882), 336. Stallo's book was in fact an extravagant attack which often caricatured the ideas actually prevailing among Anglo-American physicists and sometimes rested on a misunderstanding of the physics. See John L. Heilbron, "A History of the Problem of Atomic Structure from the Discovery of the Electron to the Beginning of Quantum Mechanics" (unpublished Ph.D. dissertation, History, University of California, Berkeley, 1964), pp. 39–40.

in motion, and doggedly declared with Joseph Lovering of Harvard: "The great problem of the day is, how to subject all physical phenomena to dynamical laws. With all the experimental devices, and all the mathematical appliances of this generation, the human mind has been baffled in its attempt to construct a universal science of physics. But nothing will discourage it."[5]

If the reduction of phenomena to dynamical laws was a task for theorists, in the United States physical research remained predominantly experimental. The new stress on laboratory work and the introduction of graduate training had given the experimental tradition considerable institutional buttressing. Yet the Yale faculty, bristling against too much science, as Eliot once mused, like "a porcupine on the defensive," included Josiah Willard Gibbs.[6] Gibbs was a Yale professor's son, Yale A.B., Yale Ph.D.—and also one of the major theoretical physicists of the nineteenth century.

Professor Hubert A. Newton encouraged young Gibbs's uncommon talent for geometry and mathematics, but since the Yale physics curriculum was so elementary in the early 1860s, Gibbs took his doctorate in engineering. He got his real professional training during three years in Europe, studying mathematics and physics at the universities of Paris, Berlin, and Heidelberg. In 1871 Yale offered him the new professorship of mathematical physics, which Newton had probably persuaded the administration to create. Yale, in its grudging way, provided no salary for the post. Gibbs, who had independent means and an umbilical attachment to New Haven, accepted. Still something of an engineer, Gibbs briefly indulged in the development of a governor for steam engines, then began a long-term pondering of the principles of thermodynamics.

Scientists might persistently stress the debt of technology to science, but thermodynamics owed much to the invention of the steam engine. The marriage of steam with mechanical power had fathered a good deal of study into the relations among heat, temperature, volume, pressure, and work. By the 1860s thermodynamics was solidly founded upon two laws. The first, whose meaning had been illuminated by Joule's demonstration of the mechanical equivalent of heat, expressed the interchangeability of heat and work. The second law declared that the transformation of heat and work was subject to limitations, because heat does not flow naturally from cold to hot bodies. To express this basic characteristic of thermodynamic change, the German physicist Rudolph Clausius, who

[5] *Proceedings of the American Association for the Advancement of Science, 1875*, pp. 34–35.
[6] Charles W. Eliot to Daniel Coit Gilman, March 9, 1880, quoted in Laurence R. Veysey, "The Emergence of the American University: A Study in the Relations of Ideals and Institutions," (unpublished Ph.D. dissertation, History, University of California, Berkeley, 1961), p. 252.

formulated the second law, introduced the concept of "entropy," after the Greek for "transformation." Mathematically defined as the ratio of heat exchanged to the temperature at which the process occurs, entropy expressed in a sense the degree to which the energy in a thermodynamic system was unavailable for use. It was ultimately recognized to be as fundamental a property of thermodynamic systems as energy itself.

Gibbs recognized the full theoretical significance of Clausius' concept. In 1873 he used it to treat the way the physical states of a chemically homogeneous substance—its solid, liquid, and gaseous forms—varied with thermodynamic conditions. Gibbs represented the thermodynamic properties of a substance by a surface in a space whose three dimensions were volume, energy, and entropy, and, in a tour de force of geometrical reasoning, he demonstrated that the relationships among the physical states of the substance could be deduced from the nature of the surface. Gibbs noted that this approach would work not only for chemically homogeneous but also for chemically mixed materials. That insight was the seed of two extensive papers published in 1876 and 1878, which, together, constituted his masterpiece, "On the Equilibrium of Heterogeneous Substances."

Gibbs set down a remarkable basic equation for the states of the system, asserted that it was to be solved in accord with a postulate about energy and entropy, then proceeded to a deductive discourse of some three hundred pages. His logic was tight and his argument mounted in a crescendo of ramifications. Gibbs's "phase rule" was to prove the most celebrated. A simple mathematical formula, it predicted the number of different physical combinations in which the components of the mixture could coexist. Yet equally important was the analysis of the properties of fluids and solids, chemical solutions and catalysts, capillarity, diffusion, and electrochemical phenomena. In Maxwell's opinion, Gibbs had laid "a massive foundation for a splendid structure."[7]

In the America of his day Gibbs was *sui generis*; he was not only a theorist but a particular kind of theorist. In constructing his theories he preferred not to rely on particular mechanical models of natural phenomena. The fundamental equation for equilibrium, he remarked, applied "independently of any theory in regard to the internal constitution of the mass." In the 1880s Gibbs ventured a few papers on the theory of light; characteristically, he argued that if one simply accepted light as an electrical flux, tortuous assumptions of stress and strain in a mechanical ether were unnecessary to predict the results of experiment.[8] No less a mechanist than his contemporaries in the Anglo-American school, Gibbs did believe that reduction to mechanics was the goal of physics. But he

[7] Maxwell to Simon Newcomb, May 31, 1879, SN, Box 31.

[8] *The Collected Works of J. Willard Gibbs*, ed. Henry A. Bumstead and Ralph G. Van Name (2 vols.; New Haven, 1948), I, 63; II, 210, 246.

insisted upon avoiding the construction of unnecessary hypotheses, mechanical or otherwise. As a theorist he possessed a ruthless sense of economy. It led him to treat physical theory more as a tool of analysis than as a portrait of nature, an approach that set him apart both from the Anglo-American and to a degree from the dominant Continental school.

Gibbs's economical approach was nowhere more evident than in his climactic treatise of 1902, the *Elementary Principles of Statistical Mechanics*. He was at work on it in the 1880s, and it stemmed from the then recent efforts of physicists, notably James Clerk Maxwell and the Austrian Ludwig Boltzmann, to apply probability theory to the behavior of systems of molecules in motion. In the programmatic preface to his book, Gibbs declared that, since framing "hypotheses concerning the constitution of material bodies" led to the "gravest difficulties," he would not attempt to "explain the mysteries of nature." Instead, he would be content with "the more modest aim of deducing some of the more obvious propositions relating to the statistical branch of mechanics [itself]."[9] Following out his program with razorlike logic, Gibbs arrived at equations that were analogues of the prime equations of thermodynamics and that threw the light of statistical interpretation upon such prime thermodynamic terms as entropy. Historically, the *Statistical Mechanics* paralleled the *Analytical Mechanics* of Joseph Lagrange, who at the end of the eighteenth century had appropriated an elegant algebra to rationalize Newtonian physics. A powerful general treatment, Gibbs's book achieved a similar end for the thermal and statistical physics of Clausius, Maxwell, and Boltzmann, and its methods ultimately dominated the field.

Rowland, all too aware that "mathematical physics is so little cultivated in this country," was proud of Gibbs, but for years few American scientists knew about him.[1] His papers on equilibrium appeared in the *Proceedings* of the Connecticut Academy of Arts and Sciences, hardly a show window of research. He published almost nothing on statistical mechanics before the full book. Gibbs began to win recognition in the United States only after Maxwell called attention to his work. Even then, few American physicists grasped his importance, if only because most were unprepared to cope with his mathematics. Besides, Gibbs, who could write with punch and wit when he wished, frequently tortured his scientific papers with sentences of Germanic length.

He accepted his obscurity without complaint. The man was shy, retiring, "one who seldom looks," in the doggerel of his schoolmates, "with playful countenance from off his books."[2] He rarely attended

[9] *Ibid.*, II, ix–x.

[1] Quoted in Lynde P. Wheeler, *Josiah Willard Gibbs: The History of a Great Mind* (New Haven, 1951), p. 87.

[2] Quoted *ibid.*, p. 20.

scientific conferences outside of New Haven, and spent the bulk of his life within the few blocks bounded by his house, his office, and his ultimate burial place.

Had Willard Gibbs been more gregarious, mathematical physics might have taken root in late-nineteenth-century America. But although he was eager enough for reputation to distribute reprints of his papers throughout the scientific world, Gibbs had no ambition to generate a school at home. He had few graduate students. He did not invite his small band of acolytes to cooperate in the development of his work; they learned the finished theory, not the process by which he forged it. In any case, Gibbs would have had a hard time promoting theoretical physics at Yale. The college complimented him with a salary of $2,000 only after a decade of service—and an offer of $3,000 from Johns Hopkins.

William James once caught the irony in the career of Yale's withdrawn genius. "Wait till we're dead twenty years. Look at the way they're now treating poor Willard Gibbs, who during his lifetime can hardly have been considered any great shakes at New Haven."[3]

Rowland, Michelson, Gibbs—yet, on the whole, American physicists hardly led the world in research. It was true that, however small the number of physicists, the profession in America compared favorably in size to its counterparts in the major nations of Europe. And if American physicists were strapped for funds and laboratory facilities, so were Europeans. Yet European physicists produced proportionately more research than their American brethren, who were publishing about the same number of articles in the early 1890s as in the early 1870s.[4] In the United States, academic physicists, along with professors in every discipline, faced an administrative obstacle to the advancement of knowledge. "The prime business of American professors in this generation," Eliot had said in his inaugural, "must be regular and assiduous class teaching."[5] Eliot and his fellow college presidents had reformed higher education principally for cultural purposes. They respected scholarship more than their old-time predecessors; they believed, with the president of MIT, that "Our aim should be: *the mind of the student*, not scientific discovery, not professional accomplishment."[6] The main object of the university was

[3] Quoted in Muriel Rukeyser, *Willard Gibbs* (Garden City, N.Y., 1942), p. vii.

[4] Paul Forman, John Heilbron, and Spencer Weart, *Physics Circa 1900: Personnel, Funding and Productivity of the Academic Establishments* (*Historical Studies in the Physical Sciences*, vol. V; Princeton, N.J., 1975), pp. 6, 116, 119, 121. The publication data is based on a statistical analysis of the physics articles in the *American Journal of Science* (hereafter, "AJS Statistical Analysis"). Part of the remainder of this chapter previously appeared in George H. Daniels, ed., *Nineteenth-century Science: A Reappraisal* (Evanston, 1972) and is reprinted here in different form with the permission of Northwestern University Press.

[5] Quoted in Henry James, *Charles W. Eliot: President of Harvard University, 1869–1909* (2 vols.; London, 1930), II, 18.

[6] Francis A. Walker to Alpheus Hyatt, Aug. 29, 1889, HM, Walker file.

to develop character by diffusing science and its way of thinking, not by stressing its advancement.

Presidential policy increasingly conflicted with the rising research aspirations of American scientists. Like Rowland, some scientists disliked having to teach "the ABC's of physics" to intellectual striplings.[7] In any case science was growing more complex, and in response scientists were becoming more specialized. In discipline after discipline they were forming their own professional societies, starting to publish their own journals. The directors of the magazine *Science* had hoped to win a professional and lay readership by publishing the results of the latest research in a form intelligible to any educated man. The number of scientific readers kept declining; the journal was losing money. "We have found it impossible to make *Science* the ideal paper we desire," one of its publishers lamented, "because all scientists are specialists. . . ."[8]

Specialists could in principle still diffuse science, but just as teaching failed to pay professional dividends, so did popularization. The pre-Civil War generation of scientists had happily traveled the lyceum lecture circuit; the "later generation," Andrew D. White complained, tended "to underrate everything except minute experiments." Enough scientists were now contemptuous of popularization to make Youmans' *Monthly* bristle: "In order that the relations between science and the age may be what they ought to be, the world at large must be made to feel that science is, in the fullest sense, a ministry of good to all, not the private possession and luxury of a few, that it is the best expression of human intelligence and not the abracadabra of a school, that it is a guiding light and not a dazzling fog."[9]

In the universities, Eliot and his presidential peers had all the power they needed to keep science a guiding light. The American university president, Lord Bryce noticed, held an "almost monarchical position." He might confer with his faculty; he exercised virtually autocratic power over salaries, appointments, promotions, and policy.[1] With the exception of Gilman at Hopkins, the major presidents used their power to stress teaching over research in the use of both funds and appointments. Rowland's laboratory was magnificently equipped for research; physics laboratories elsewhere were better suited to pedagogical purposes. On the whole, academic physicists had to wangle money for equipment where

[7] Rowland to Daniel Coit Gilman, April 20, 1876, HAR.

[8] Gardiner Hubbard to the Directors of the Science Company, Feb. 13, 1885 [1886] SN, Box 38.

[9] White to D. C. Gilman, July 24, 1878, quoted in Fabian Franklin, *The Life of Daniel Coit Gilman* (New York, 1910), p. 344; *Popular Science Monthly*, 38 (Nov. 1890), 121.

[1] Bryce is quoted in George W. Pierson, *Yale: College and University, 1871–1937* (2 vols.; New Haven, 1952), I, 129; Joseph Ben-David, "The Universities and the Growth of Science in Germany and the United States," *Minerva*, 7 (Autumn 1968), 19–20; Hugh Hawkins, *Between Harvard and America: The Educational Leadership of Charles William Eliot* (New York, 1972), p. 62.

they could, often paying for apparatus from their own pockets. "Scientific men," Alexander Agassiz once wryly noted, "rarely spend their money as their wives wish."[2] However they spent their money, they could not usually persuade the university president to ease the pedagogical demands on their time. The president ran the university so that the advancement of knowledge paid no dividend in advancement through the college hierarchy. In fact, those who did no research often seemed to get ahead more swiftly. Eliot exaggeratedly acknowledged that research in physics or any other discipline required "fanatical zeal."[3]

A number of physicists in the United States did research anyway, of course, but their work earned them few garlands from Europeans. While, much as in literature or art, it is difficult to specify what constitutes quality in science, one may call attention to certain prominent features of merit. In theoretical physics, quality has frequently consisted of formulations like Gibbs's on heterogeneous substances that throw explanatory light on a general category of phenomena. On the experimental side, it has often been marked by the prosecution of such key tests of theory as Michelson's ether drift investigation or by the devising of such apparatus as Rowland's grating which permit the accurate measurement of fundamental physical characteristics. In contrast to the work of Rowland, Michelson, and Gibbs, much of American physical research amounted to narrow, pedestrian experiments; one out of five American publications in the discipline reported upon mere observations of natural phenomena or objects, including the characteristics of meteoric remains.[4] The trouble with American physics was not only its failure to produce more but, with its undue attention to the gathering of insignificant facts, the frequently inferior quality of what was produced.

An explanation for this qualitative deficiency was proposed by Joseph Lovering of Harvard: "Unless our physicists are contented to lag behind and gather up the crumbs which fall from the rich laboratories and studies of Europe, they must unite to delicate [experimental] manipulation the power of mathematics."[5] True, mathematics often won victories where experiment was beaten. Theoretical analysis could also suggest significant tests in the laboratory, help focus attention on important results which might otherwise be ignored. But Lovering was asking America's deficiency in theorists to carry too heavy an explanatory load. Faraday had not relied on abstruse mathematics, and neither did Michelson. Joseph Henry devised significant experiments in accord with his own

---

[2] Alexander Agassiz to Miss Hyatt, August 3, 1900, HM.

[3] Quoted in Samuel Eliot Morison, *Three Centuries of Harvard, 1636–1936* (Cambridge, Mass., 1936), p. 378.

[4] "AJS Statistical Analysis."

[5] Lovering, "Mathematical Investigations in Physics," *Popular Science Monthly*, 6 (January 1875), 321.

understanding of hypotheses, and Rowland had an admirable knack for translating theoretical issues into concrete experiments. None of these physicists needed special help from mathematically sophisticated theorists, and none ever used more than elementary calculus.

The Baconian tradition no doubt encouraged American physicists to pursue an arid form of empiricism; the importance of facts to science was all too easily transmuted into an emphasis on mere fact-gathering. But the British were no less heirs to Bacon; they managed to do good physics. More important, the Baconian tradition was common to all the sciences in the United States. Yet the Geological Survey won the prestigious Cuvier Medal of the French Academy of Sciences for its collective work, and American geology in general enjoyed the respect of Europeans.

Geology done well in the same country where physics was done poorly—a comparison of the two disciplines reveals what was the matter with Rowland's science. There were more geologists than physicists in the United States, but greater numbers do not guarantee greater professional quality.[6] There were also more theorists in geology; geological theorizing did not require the analytic training and mathematical mastery necessary in physics. Still, at the time the principal task in both disciplines was not the construction of new but the elucidation of existing theories— in physics, electromagnetism, light, and thermodynamics; in geology, evolution—through observation or experiment. The significant difference between the two disciplines was institutional. While both were set in the university, physics had nothing like the United States Geological Survey under John Wesley Powell, the famed geologist and explorer who, despite the loss of an arm at Shiloh, had scaled Long's Peak and conquered the canyons of the Colorado River.

His exploring days over, his body paunching but his beard still a bristling red, Powell, the director of the Survey since 1881, fired the entire operation with enthusiasm. He employed topographers, geologists, and paleontologists; he farmed out work to university consultants. Lester Frank Ward wrote part of his seminal book *Dynamic Sociology* on Survey time. Othniel C. Marsh, the paleontologist, president of the National Academy of Sciences, and director of the Peabody Museum of Natural History at Yale, remained a member of the Yale faculty after he joined the Survey. Charles Darwin considered Marsh's work on ancient horses and on birds with teeth the best support of the theory of evolution to appear since the publication of *The Origin of Species*.[7] Marsh, who was independently wealthy and not on the Yale payroll, spent his handsome

[6] As compared with approximately 200 physicists, in 1889 the new Geological Society of America had 187 members; in 1893, 224. See the *Bulletin of the Geological Society*, I (1890), 586, and IV (1893), 450.

[7] Charles Schuchert, "Othniel C. Marsh," *Biographical Memoirs of the National Academy of Sciences*, XX (1939), 22.

Survey salary for research on fossils gathered by its expeditions. The arrangement was good for Marsh, for Yale, for the Survey, for science.

Through its consultantships and field expeditions, the Survey provided important research opportunities for many American geologists. Its access to the Government Printing Office endowed the geology profession with the equivalent of its own major research journal. But more important than the existence of these opportunities for the doing and publishing of research was John Wesley Powell's control of them. By determining the tasks of Survey expeditions and by establishing the editorial criteria of its publications, Powell and his colleagues were able to set standards of what constituted significant work.

Put mediocre scientists in control of a standards-setting institution, and the result will be mediocre standards. But Powell was a good geologist and his associates were even better ones; the standards of the Geological Survey were high. Whoever was paid or published by the Survey was evidently discouraged from the gathering of inconsequential facts. Equally important, with its consultants spread through nineteen states and territories, the Survey made its influence felt in numerous academic localities. Its plentiful bulletins—more than a hundred had appeared by the mid-1890s—and its staff's consistent contributions to the *American Journal of Science* provided models of meritorious work for every American geologist. Under Powell, who found American geologists "always anxious" to fit their "several labors" into "the grand system . . . for the development of knowledge," the U.S. Geological Survey shaped the intellectual thrust of the entire geological profession.[8]

In the physics profession, geniuses like Willard Gibbs needed no intellectual guidance; neither did Rowland or Michelson. But all physicists, like all facts, were not equally important. More than four fifths of the American research published in the discipline between 1870 and 1893 was produced by only one fifth of the practitioners.[9] The rest of the profession, and even many in the productive group, stood to benefit from a system that established the questions to ask of nature, the techniques to use in answering them, the kinds of facts worth gathering.

To set standards in physics did not necessarily require a Geological Survey for the discipline. Standards could be exemplified in school, where young scientists, working as apprentices to the masters in their fields, might acquire the taste to choose the significant instead of the trivial experiment, or more generally to understand what constituted quality in

[8] U.S. Congress, Joint Commission to Consider the Present Organization of the Signal Service, Geological Survey, Coast and Geodetic Survey, and the Hydrographic Office of the Navy Department, with a View to Secure Greater Efficiency and Economy of Administration of the Public Service . . ., *Testimony*, March 16, 1886, 49th Cong., 1 Sess., Sen. Misc. Doc. 82, Series 2345, p. 178.

[9] Kevles and Harding, "The Physics, Mathematics, and Chemical Communities," p. 10.

scientific approach and accomplishment. Standards could be encouraged through local institutions, academic or otherwise, that provided opportunities for the critical discussion of research problems with able colleagues. And standards could be set nationally through private, not only public, organizations. With only Rowland, Michelson, and Gibbs, the American physics profession did not have a large group of meritorious leaders to set standards of excellence in research. Adding to that deficiency, it also lacked an effective institutional framework for the purpose, either academic, local, or national.

The graduate schools did little to set standards, in no small part because relatively few practicing physicists of the day—about one quarter—took Ph.D.'s. Then, too, only six colleges offered adequate preparation for graduate work in physics, and fewer still provided first-rate doctoral training. One Ph.D. candidate went to Princeton and for a year, he recalled, "browsed in the library, played in the laboratory, and deteriorated intellectually." Some American physicists earned doctorates abroad; many more considered a stint of study in Europe essential.[1] Between 1873 and 1890 only twenty-two American physicists took Ph.D.'s at home, mostly at Hopkins, where Rowland claimed to neglect graduate students, or at Yale, where few ever saw Willard Gibbs. Graduate training aside, institutionally, except for some concentration at Harvard, MIT, and Yale, the more competent productive physicists were widely dispersed; they tended to work alone, isolated from their comparably productive and presumably stimulatingly critical peers. Of course, they might find professional companionship at the local scientific society. There were many such societies in the United States, Rowland tartly observed, most of them "dignified by high-sounding names," each with "its local celebrity, to whom the privilege of describing some crab with an extra claw . . . is inestimable."[2]

Nationally, academic physicists had neither a special journal nor a professional society of their own. Their access to foreign journals was somewhat limited by the high tariff—it was 25 percent ad valorem—on the importation by individuals of all recent foreign books and periodicals. "These vexatious duties," a scientific journalist fumed, "protect no American interest. What they do protect, if anything, is the European in-

[1] *Ibid.*; George H. Daniels, "The Pure-Science Ideal and Democratic Culture," *Science*, 156 (June 30, 1967), 1702. The Ph.D. candidate was Henry Crew, quoted in William F. Meggers, "Henry Crew," *Biographical Memoirs of the National Academy of Sciences*, XXXVII (1964), 36.

[2] The Ph.D. statistic is based on M. Lois Marckworth, comp., *Dissertations in Physics: An Indexed Bibliography of All Doctoral Theses Accepted by American Universities, 1861–1959* (Stanford, 1961); employment statistics are in Kevles and Harding, "The Physics, Mathematics, and Chemical Communities," Table 11. Rowland, "A Plea for Pure Science," in *The Physical Papers of Henry Augustus Rowland*, p. 610.

vestigator from equal competition on the part of American workers."[3] American physicists tended to be much more current with research published in the *American Journal of Science* than with what appeared in the leading British, French, or German periodicals.[4] But in the 1870s and 1880s, neither the *American Journal* nor the American Association for the Advancement of Science, which was the only active national scientific society, could effectively set standards in physics.

To maintain the standards of the *American Journal,* James Dwight Dana might refer potential articles to an expert in the field for evaluation. When in 1873 the young and virtually unpublished Rowland submitted a bold experimental study of magnetism, Dana turned to physicists at Yale for an assessment of its merits. Following their advice, Dana rejected the paper, explaining to Rowland that he "needed more study." Rowland sent the paper to James Clerk Maxwell, who was so impressed that he had it published forthwith in the British *Philosophical Magazine*. The physicists whom Dana could ask for judgments were often mediocre or out of touch; after repeated rejections of Rowland's papers, they admitted that they just could not understand his mathematics.[5] Dana evidently tended to treat generously articles submitted by physicists better established and less mathematical than the young Rowland. Compared with the number of papers in geology, few papers in physics came along. Sure that a "purely geol[ogical] j[ournal]" will not pay here," Dana had little choice but to publish mediocre studies. When Rowland proposed that Johns Hopkins sponsor an American journal of physics, Dana, eager to block the venture, assured Daniel Coit Gilman of the "early publication" in the *American Journal of Science* of "any physical paper that may be received."[6]

[3] *Science*, I (June 29, 1883), 589. Books and periodicals imported by such institutions as scientific societies and universities were admitted free. In 1882, Edward D. Cope, testifying for the National Academy of Sciences, protested vigorously the duty on foreign books and periodicals before the U.S. Tariff Commission. So did W. H. Doyle, a wholesale merchant from Minnesota, who, stressing the importance of advancing and diffusing knowledge, urged that all books should be admitted free of duty. "Including the works of Huxley, Darwin, and Herbert Spencer?" asked Commissioner John W. H. Underwood, a former congressman from Rome, Georgia. "Do you think that that kind of literature ought to be encouraged in the United States by putting it on the free list?" Doyle did, but Commissioner Underwood obviously did not. U.S. Congress, House, *Report of the Tariff Commission . . .,* 47th Cong., 2d Sess., 1882, Misc. Doc. 6, part 1, pp. 1074, 2454–56.

[4] Based on a comparison of the frequency and age of the references in physics articles published in the *American Journal of Science* to other articles in the *Journal* and in *Annalen der Physik, Comptes Rendus,* and the *Philosophical Magazine*.

[5] See Rowland's exultant note penned on Dana's letter of rejection, Dana to Rowland, July 2, 1873, HAR. John D. Miller, "Rowland's Physics," *Physics Today,* XXIX (July 1976), 41.

[6] Dana to D. C. Gilman, June 20, 1884, DF. Statistically, Dana showed no special favoritism to Yale physics faculty or Ph.D.'s. Kevles and Harding, "The Physics, Mathematics, and Chemical Communities," Table 13.

The American Association for the Advancement of Science had little ability to set standards for any discipline. By the 1880s, the AAAS had given physicists their own Section B, and at the annual meetings they exchanged ideas and read papers. But Section B was no more influential between meetings than the AAAS itself. The Association published only the proceedings of its annual get-togethers; its officers did little more year-round than plan for the next convention. In addition, it had only a minuscule research endowment, and physics had to compete with every other discipline for the funds.

Whatever its fiscal difficulties, some scientists considered the AAAS inherently unqualified to set standards of excellence. To strengthen control of the Association by professional scientists, the constitution was revamped in 1874 to restrict officeholders to a new, special class of members, or "Fellows," who were accomplished in research. But even if now the AAAS was more of a scientocracy, the critics considered it too democratic. Some ridiculed the Association for its lay membership. Others were dissatisfied for a more fundamental reason: While the AAAS operated in a way that distinguished between professionals and laymen, it made no distinction between good and mediocre professionals. For the critics, an association that allowed any scientist to join and hold office was not in any position to encourage the best research.[7] The dissidents understood that in every field, as in physics, the vast majority of the best science was produced by a small minority of the practicing scientists. The critics preferred an organization founded on the key assumption: Standards of excellence could be set only by an elite of the best scientists.

The advocates of best-science elitism typically found their model scientific tribunals in the exclusive academies of Europe.[8] Many European scientists actually considered the Royal Society of London or the French Academy of Sciences to be useless organizations of superannuaries. Nevertheless, admission was possible only through election by their distinguished memberships, and annual additions to the rolls were limited. Both academies published transactions and memoirs and held frequent meetings. In America, advocates of a best-science tribunal looked to the government's official scientific advisory agency, the National Academy of Sciences.

The leaders of the Academy always hoped that the government

---

[7] Benjamin Gould implied that some scientists cast "ridicule" upon the AAAS because it allowed "all lovers of science" to join. *Proceedings of the American Association for the Advancement of Science, 1869,* p. 36. See also Wolcott Gibbs to Othniel C. Marsh, May 15, 1882, OCM; Simon Newcomb, "Exact Sciences in America," *North American Review,* CXIX (October 1874), 300; Sally Gregory Kohlstedt, *The Formation of the American Scientific Community: The American Association for the Advancement of Science, 1848–1860* (Urbana, 1976), pp. 168–9.

[8] Alexander Agassiz to Simon Newcomb, May 30, 1906; Benjamin Gould to Newcomb, March 13, 1867, SN.

would make good use of what one of its presidents called "the ability, the patriotism, and the disinterestedness of its members."[9] These leaders also evidently expected that the Academy's official advisory connection with the government would in turn help endow it with the prestige necessary to set standards in research. Some of the founders even hoped for a federal subvention. In any case, the charter of incorporation limited the membership to fifty. Louis Agassiz, one of the Academy's originators, rejoiced: "We have a standard for scientific excellence."[1]

Joseph Henry had opposed the creation of the Academy. He suspected that it might be considered "at variance with our democratic institutions" and might become "perverted . . . to the support of partisan politics."[2] When Henry became president of the Academy in 1867, he did not press for federal funds and kept the Academy scrupulously out of politics. The membership ceiling was removed so that five scientists could be added to the rolls each year. Patronizing research by rewarding success with titles and pensions was the European way and unacceptable in America, Henry noted. Still, an "intelligent democracy" could properly bestow honors for achievement and the creation of the Academy had opened in America another "avenue for the aspirations of a laudable ambition."[3]

Despite Henry's modifications the Academy was largely ineffective in the cause of scientific excellence. American scientists were dispersed over the eastern and midwestern United States and, like Professor John Le Conte at Berkeley, Californians felt hopelessly isolated. The National Academy met twice yearly at the most; its sessions were poorly attended. Federal scientists tended to dominate its affairs, particularly elections to the ranks. Scientists in Cambridge and New Haven, wary of the centralization of science, worried that it was developing into a Washington clique.[4] More important, Henry had made the Academy into an honor society, but could a primarily honorific body expect to influence the course of research in the United States?

Not in the opinion of Leo Lesquereux, a paleobotanist imported to America from Europe: "What will you do with a body of scientific men who, in *America!* are expected to work for nothing and what influence

[9] Wolcott Gibbs to Hoke Smith, March 2, 1896, U.S. Congress, Senate, *Report of the National Academy of Sciences, 1896,* 54th Cong., 2d Sess., S. Misc. Doc. 50, Series 3469, p. 15.

[1] Agassiz to Alexander Dallas Bache, May 23, 1863, in Nathan Reingold, ed., *Science in Nineteenth-Century America: A Documentary History* (New York, 1964), p. 210.

[2] Henry to Louis Agassiz, Aug. 13, 1864, *ibid.,* pp. 213–14.

[3] U.S. Congress, Senate, *Report of the National Academy of Sciences,* 1867, 40th Cong., 2d Sess., S. Misc. Doc. 106, Series 1319, p. 3.

[4] Le Conte to A. M. Mayer, Jan. 17, 1873; James D. Dana to Mayer, March 14, 1874, HM; Alexander Agassiz to O. C. Marsh, May 4, 1884, OCM. See also Wolcott Gibbs to Marsh, May 15, 1884, OCM; A. Agassiz to Simon Newcomb, Dec. 20, 1885, and May 30, 1906, SN.

can . . . the so-called highest scientific society of the Land [exert], which cannot even publish a scientific memoir. In Europe, honor conferred is worth more than money but in America the same honor is worth nothing by itself." Since Henry's shunning of the public purse became a hallowed tradition, in the late nineteenth century the Academy had only the income from a few private trust funds to dole out for the support of research.[5] It could afford to publish little save, as the congressman said, the obituaries of its deceased members.

The official status of the Academy was no help. As the government acquired more of its own scientific bureaus, the advisory Academy became increasingly superfluous. By the late 1880s federal requests were so infrequent that some members were urging the leadership to volunteer the Academy's services. President Othniel C. Marsh refused. In his judgment, the Academy would "lose both influence and dignity by offering its advice unasked." As its advisory duties declined, the Academy lost prestige. Few scientists considered membership in the NAS important, Simon Newcomb noted, since it rarely brought "increased consideration in any quarters outside of the Academy itself."[6]

In 1883, in an address as vice-president for physics of the American Association for the Advancement of Science, Henry Rowland, a member of the Academy and an exponent of best-science elitism, considered "what must be done to create a science of physics in this country, rather than to call telegraphs, electric lights, and such conveniences by the name of science. . . . The cook who invents a new and palatable dish for the table benefits the world to a certain degree; yet we do not dignify him by the name of chemist." But the flaws of American physics, Rowland argued, ran deeper than the public disregard of abstract science. The difficulties lay in the American scientific system, which, to Rowland's elitist mind, reflected the "pride" taken in "a democratic country . . . in reducing everything to a level."[7]

Hardly any of the some four hundred institutions of higher learning—Rowland dubbed them "a cloud of mosquitoes, instead of eagles"—deserved to be called genuine universities, where great men, amply funded, could inspire and enjoy the helping hands of many assistant professors and still more students. Only seventeen institutions of higher education had more than twenty faculty; only eight, endowments of more than

[5] Lesquereux to J. P. Lesley, March 21, 1866, in Reingold, ed., *Science in Nineteenth-Century America*, p. 222. By 1895 the Academy had but $94,000 capital in trust funds for the support of research. Howard S. Miller, "A Bounty for Research: The Philanthropic Support of Scientific Investigation in America" (unpublished Ph.D. dissertation, History, University of Wisconsin, 1964), p. 217.

[6] Quoted in Schuchert, "Marsh," *Biographical Memoirs of the National Academy of Sciences*, XX (1939), 30; Newcomb to T. C. Mendenhall, Oct. 22, 1892, TCM, Box 6.

[7] Rowland, "A Plea for Pure Science," *The Physical Papers of Henry Augustus Rowland*, pp. 594, 601.

$1,000,000. And in Europe, Rowland added, the "great academies" impelled every physicist to his "highest effort" and provided him with "models of all that is considered excellent." In the United States, neither the National Academy nor any other scientific society did the same.[8]

Rowland's address was more than a critique; it also outlined a program for American physics. The United States might of course hope to import meritorious practitioners of the discipline from abroad. But whatever the accession of foreigners, Rowland implied in no uncertain terms, the creation of a sustained and outstanding indigenous physics would require the concentration of educational wealth in a few first-class universities. It would require the joining together of the best scientists in a national organizational network committed to the setting and enforcement of standards of excellence. It would require, in short, the construction of an institutional pyramid, oriented toward the advancement of abstract physics, open to the talents at the bottom, and commanded at the heights by a best-science elite.

[8] *Ibid.*, pp. 600–601, 610. Simon Newcomb analyzed the weakness of pure science in the U.S. in much the same vein as Rowland in "Abstract Science in America, 1776–1876," *North American Review*, 121 (1876), 104, 108, 111.

IV

Pure Science and Practical Politics

Rowland called his address of 1883 "A Plea for Pure Science." The word "pure" did more than differentiate science from invention; it added to the traditional descriptive phrase "abstract research" the connotation of high virtue. Most Americans, in the opinion of Youmans' *Monthly*, preferred to "degrade" science to a "low, money-making level." Pure research was uninfected by the motive of profit or ultimate practical use; it was done wholly for the love of truth. The friends of pure science liked to tell the story of the scientist and the businessman. Urged to apply his talents outside research, the scientist retorted: "My dear sir, I have no time to waste in making money."[1]

In Rowland's day the phrase "pure science" was replacing the term "abstract research," especially among academics. Many university scientists spent their postgraduate years in Germany, where patent-minded professors were often criticized by their colleagues and where doctoral candidates foreswore the advancement of science "for gain's sake."[2] Whatever the general belief that the pure scientist was doomed to poverty, in

[1] *Popular Science Monthly*, 22 (April 1883), 847–48; the scientist's retort is repeated in Andrew D. White, "Science and Public Affairs," *Popular Science Monthly*, 2 (April 1873), 736. Frank Jewett, the long-time director of the Bell Telephone Laboratories, succinctly caught the connotation of the phrase "pure science" while explaining his dislike for it: "The word 'pure' implies that all other kinds of research are 'impure.'" Jewett to Vannevar Bush, June 5, 1945, OSR, Entry 2.

[2] See, for example, Max Von Laue, "My Development as a Physicist," in Paul P. Ewald, ed., *Fifty Years of X-Ray Diffraction* (Utrecht: International Union of Crystallography, 1962), p. 288; Carl Barus, "Autobiography," CB, p. 147.

the United States academic scientists, if not rewarded richly, were comfortably paid. On the average, they earned $1,400 a year: 40 percent more than clerical workers, 75 percent more than ministers, almost 300 percent more than common laborers; at such places as Harvard, Yale, and Johns Hopkins, some professors were known to make $4,000 annually. Whatever the salary, professors, as Lord Bryce noted, moved among the "social aristocracy" of the college town.[3] If they often complained about salary, it was because they had to conform to the expensive pace of cultivated society.

Like their cultivated patrons, academic scientists had their quarrels with America, not least because in the 1870s and 1880s people found economic issues, including plummeting farm prices, low wages, and persistent unemployment, more compelling than the abstractions of science. While locusts and grasshoppers spread disaster across the prairies, college agricultural courses languished for lack of students, and when Congress provided funds for entomological studies in 1876, the appropriation amounted to a piddling $18,000. Most businessmen went on agreeing with Andrew Carnegie that higher learning of any sort was useless preparation for the practical demands of industry. While engineering enrollments rose somewhat, every year less than one tenth of 1 percent of the population took degrees in any subject. *Electrical World* summarized the ongoing attitude of the electrical industry. "Edison's mathematics would hardly qualify him for admission to a single college or university . . ., but we would rather have his opinion on electrical questions than [that] of most physicists."[4]

Most physicists, along with other academic scientists, were certain that their work added to the nation's cultural stature, especially since so many Americans paid no attention to it. Physicists advanced knowledge and remained obscure; the Edisons appropriated their findings and became rich and famous. Some academics occasionally consulted for industry; a few ventured into business on the side. Most were remote from crude matters of commercial practicality. Out of curiosity, the National Academy invited Edison to demonstrate his phonograph, but if some scientists held Edison in awe, most scorned him. Academics, devoted—if not limited—to research for its own sake, naturally insisted with Rowland upon the higher dignity of chemists over cooks.

For professionals like Rowland, the phrase "pure science" did more

[3] Walter Metzger, "College Professors and Big Business Men: A Study of American Ideologies, 1880–1915" (unpublished Ph.D. dissertation, History, State University of Iowa, 1950), pp. 27–28; James Bryce, *The American Commonwealth* (2 vols.; London, 1889), II, 533.

[4] Leland O. Howard, *A History of Applied Entomology* (Smithsonian Miscellaneous Collections, vol. 84; Washington, D. C.: Smithsonian Institution, 1930), p. 80; Irving G. Wyllie, *The Self-Made Man in America* (New Brunswick, N.J., 1954), p. 101; Daniel J. Kevles, "The Study of Physics in America, 1865–1916" (unpublished Ph.D. dissertation, History, Princeton University, 1964), Appendix II; *Electrical World*, IV (Sept. 20, 1884), 96.

than identify their work. In the age of Edison it established their worth—and their emerging insistence that science ought to be valued for its own sake, not for whatever practical benefits it might ultimately produce.

Yet Rowland's claim that science was somehow superior to invention annoyed Alexander Graham Bell. The son and grandson of British scientists, a one-time student at the University of London, a former professor of speech mechanics and vocal physiology, Bell had hit upon his way to transmit speech electrically while working with the deaf at Boston University. Granted a basic patent in 1876, Bell and his father-in-law, Gardiner G. Hubbard—Hubbard's deaf daughter had been a student of Bell's—proceeded to commercialize the telephone. Bell, whose patent battles brought him to Washington, D.C. in 1878, kept a foot in the world of science. He became a part-time member of the Johns Hopkins staff, hosted a Wednesday evening salon for scientists in the capital, and was elected to the National Academy. The part-time professor respected science for its own sake; the inventor of the telephone resented the castigations of science for profit.

Bell helped propagate a view different from Rowland's when, in 1883, he and Hubbard philanthropically relaunched the young but failing magazine *Science*. The first editorial declared: "Research is none the less genuine, investigation none the less worthy, because the truth it discovers is utilizable for the benefit of mankind." Had not the inventors of the electric light and the telephone added to the stock of physics? The man who simply appropriated science for invention might stand on a "lower plane" than the discoverer; the inventor who advanced knowledge while on the way to the patent office was superior to them both. Scientists, Bell's magazine admonished, "should cast aside all prejudice against the man of patents and practical devices, and should stand ready to welcome the investigator in whatever garb he appears."[5]

Scientists in the federal service were inclined to agree. Working for the government, they could hardly disdain utility and keep their self-respect. They had to worry continually about Congress. "We are not fomenting science," the director of the Coast and Geodetic Survey (the "Geodetic" was added to the name in 1878) carefully assured a congressional committee. "We are doing practical work for practical purposes."[6] And federal scientists knew from specific instances that work for practical ends often added to the stock of the world's knowledge.

[5] *Science*, I (Feb. 9, 1883), 1–3. For Bell's relationship to the magazine, see Robert V. Bruce, *Bell: Alexander Graham Bell and the Conquest of Solitude* (Boston, 1973), pp. 276–78.

[6] Julius E. Hilgard, in U.S. Congress, Joint Commission to Consider the Present Organization of the Signal Service, Geological Survey, Coast and Geodetic Survey, and the Hydrographic Office of the Navy Department, with a View to Secure Greater Efficiency and Economy of Administration of the Public Service . . ., *Testimony*, March 16, 1886, 49th Cong., 1 Sess., Sen. Misc. Doc. 82, Series 2345 (hereafter, Allison Commission, *Testimony*), p. 54.

Under the mathematician Benjamin Peirce, the Coast Survey had begun a geodetic connection across the continent, which included measuring the curvature of the earth along an arc equal to about a full eighth of the circumference of the globe. A key contributor to this effort was Charles Sanders Peirce, Benjamin's son and a graduate of the Lawrence School of Science, the vain, eccentric, prickly founder of pragmatism whom no university would hire (some said because he had committed the Victorian sin of divorce). During his thirty years with the Survey, Charles Peirce produced a stream of brilliant papers on formal logic and philosophy—he wrote his famous "How to Make Our Ideas Clear" while on the way to an international geodetic conference—and he earned an international reputation for his scientific work, especially the development of highly accurate methods for measuring gravity with a pendulum. At the Nautical Almanac, Simon Newcomb was cooperating with the Naval Observatory to determine the precise position of the stars. No one could fault the scientific activities of the Weather Service and the Geological Survey. Both were contributing to the economic life of the country *and* to the advancement of knowledge.

Military and civilian observers around the country would wire reports of local conditions to the Washington offices of the Weather Service, housed in the quarters of the Army Signal Corps on G Street. There, the next day's predictions were drawn up, then telegraphed across the United States. The reports were published in newspapers and tacked up at local post offices. Many farmers learned of cold waves, sunshine, or storms from signal flags hoisted on a passing train or a public building (the flags would remain a commonplace of the American landscape until the days of radio). By the mid-1880s, the Service enjoyed widespread national support and its annual appropriation had reached almost $1,000,000.[7]

In 1880 Cleveland Abbe, a graduate of City College in New York, an astronomer turned geophysicist, an early advocate of the Weather Service and now head of its small civilian staff, had persuaded the new Chief Signal Officer General William B. Hazen to spend money for the study of meteorology. Hazen was a veteran of duty in the Dakotas, the author of a book attacking railroad promoters for exaggerating the amount of rainfall on the plains, and an enthusiast of weather prediction. At Abbe's urging, Hazen also encouraged the enlistment of college graduates in the Weather Service by exempting them from regular military duties; he added civilian scientists to the staff, who were given the titles of professor and assistant professor. In the musty building on G Street, not far from the White House, Abbe created a Study Room for his professors— there were seven or eight of them, plus ten assistants—and opened a physical laboratory in the basement. Under Hazen's beneficent command,

[7] Donald R. Whitnah, *A History of the United States Weather Bureau* (Urbana, 1965), p. 21.

Weather Service scientists explored the nature of storms and tornadoes, investigated the mysteries of the atmosphere, and, in all, nudged ahead the science of meteorology.[8]

Abbe had Hazen, whom he liked well enough to name his son for him; the Geological Survey had John Wesley Powell. The son of an abolitionist, Powell was a reformer who was aware that most of the good public land, the kind "a poor man could turn into a farm," was already sold. He had shaped the attempt by the National Academy of Sciences to reform the public land system.[9] He remained eager to use geological research as a tool to recast the West into a region of economic opportunity, and he quickly transformed the Survey, which had previously emphasized mining studies, into a wide-ranging inquiry into the physical evolution and economic possibilities of the land.

No disdainer of Washington politics, Powell could usually find a position for the relative of a well-placed congressman, and he distributed the Survey's attractively, sometimes lavishly illustrated publications around the capital. He was also legislatively skilled. The Survey had been created to explore the territories; states'-righters in the Congress opposed the director's bid to extend it over the entire country. In 1882, by deftly slipping into the Sundry Civil Bill the words "and to continue the preparation of a geological map of the United States," Powell won nationwide jurisdiction. Between 1881 and 1884 the Survey's annual budget jumped to almost $500,000, a fivefold increase.[1]

Thus, the federal government, with its Weather Service, Naval Observatory, Coast and Geological surveys, and various other offices and departments, became the home of much of American science, including one out of every six jobs in physics. The Government Printing Office, which since 1870 had issued hundreds of memoirs for the scientific bureaus, was the nation's principal publisher of research. Relative to population, more scientists were working in the capital than in any other city, including Cambridge, Massachusetts. Like academics in Cambridge, New Haven, or New York, Washington scientists had their own congenial community. They gathered at meetings of the various scientific societies

[8] *Ibid.*, pp. 38–39; William J. Humphreys, "Cleveland Abbe," *Biographical Memoirs of the National Academy of Sciences*, VIII (1919), pp. 480–81; Thomas Corwin Mendenhall, "Autobiographical Notes," TCM, vol. V, pp. 149–50; vol. VI, pp. 2–3, 8.

[9] Powell, New York *Tribune*, April 28, 1877, quoted in William C. Darrah, *Powell of the Colorado* (Princeton, 1951), pp. 226–27. See John Wesley Powell, *Report on the Lands of the Arid Region* (2d ed.; Washington, D.C., Government Printing Office, 1879), *passim*; William H. Goetzmann, *Exploration and Empire: The Explorer and the Scientist in the Winning of the American West* (New York, 1966), p 585.

[1] U.S. Congress, House, *Congressional Record*, 47th Cong., 2d Sess., Feb. 22, 1883, XIV, pt. 4, 3127; A. Hunter Dupree, *Science in the Federal Government: A History of Policies and Activities to 1940* (Cambridge, Mass., 1957), p. 212; Darrah, *Powell*, p. 285.

then proliferating in the capital.[2] They conjoined at the Cosmos Club, which Powell, Henry Adams, and friends had founded in 1878 for people devoted to science, art, and literature. Alexander Graham Bell's Wednesday evenings brought them together for good talk and fine cigars.

Washington scientists, constantly faced with the task of winning appropriations from a utilitarian-minded government, had a common bond of experience which few academics shared. Most could agree with the dissent from Rowland's attitude of Thomas Corwin Mendenhall, a physicist working in the dark basement laboratory of the Weather Service, who excoriated the "unfortunate, and perhaps growing tendency among scientific men to despise the useful and practical in science." "The arrogance of genius," he reminded the academics, "is no less disagreeable than that of riches."[3]

Yet whatever the disdain for utility of the Rowlands or the respect for it of the Powells, both groups, like dissident brothers, were still loyal members of the same family. A number of academics had at one time or another worked in the government service; many federal scientists had taught in the universities. They knew each other and appreciated each other's work. The academics might deplore the practice of science for profit; Washington scientists were hardly gathering personal fortunes. Indeed, if they lacked Rowland's hauteur, they still believed strongly in the advancement of knowledge for its own sake.

Both groups proudly assumed that by advancing science they were helping to make possible the material growth of the country. Culturally, both could regard themselves, in the description of the senior Holmes, as "the oracles, the prophets of today," who were "teaching [the laity] to look for truth at first hand, and not to be afraid of it." Intellectually, they exemplified the disinterested scientific mind. Above all, as Rowland once said, they were "a small and unique" community, "an aristocracy, not of wealth, not of pedigree, but of intellect and ideals."[4]

"An aristocracy . . . of intellect and of ideals"—in a practical nation which had a way of scoffing at aristocracies, academic and federal scien-

[2] Daniel J. Kevles and Carolyn Harding, "The Physics, Mathematics, and Chemical Communities in the United States, 1870 to 1915: A Statistical Survey," California Institute of Technology, Social Science Working Paper No. 136, March 1977, Table 11; *Science*, V (April 17, 1885), 328–29; G. Brown Goode, "The Beginnings of American Science, The Third Century," *Report of the United States National Museum*, in *Annual Report of the Board of Regents of the Smithsonian Institution, 1897* (Washington, D.C.: Government Printing Office, 1901), 462–63; J. Kirkpatrick Flack, *Desideratum in Washington: The Intellectual Community in the Capital City, 1870–1900* (Cambridge, Mass., 1975), pp. 60, 62.

[3] *Proceedings of the American Association for the Advancement of Science, 1890*, pp. 11, 14.

[4] Holmes to A. M. Mayer, Jan. 28, 1873 (a letter read at the Tyndall banquet), HM; Rowland, "The Highest Aim of the Physicist," in *The Physical Papers of Henry Augustus Rowland* (Baltimore, 1902), p. 668.

tists did not dwell on their differences, if only because they had too much in common to defend.

In the mid-1880s the Coast Survey, the Geological Survey, and the Weather Service, all of them showpieces of the burgeoning federal research establishment, ran into heavy political fire on Capitol Hill. Some congressmen urged the removal of the Weather Service from the army, charging General Hazen, in the language of Senator Preston B. Plumb of Kansas, with providing too many "easy berths" for army officers and "eminent scientific civilians."[5] States'-rights legislators attacked the extension of the Geological Survey to the states. Secretary of the Navy William E. Chandler, undoubtedly alive to new possibilities of patronage, kept trying to get the work of the Coast Survey shifted to the Navy Hydrographic Office. In July 1884, in response to the turmoil, Congress authorized an investigation of these scientific agencies by a Joint Commission of the House and Senate. The six-man commission was bipartisan and chaired by the GOP warhorse Senator William B. Allison of Iowa, a powerful friend of federal science, who delayed the opening of hearings until after the 1884 presidential election. But in the months following the inauguration of Grover Cleveland, stories circulated in the newspapers that the major scientific bureaus were scandalously corrupt, and in July 1885 the Cleveland administration suspended Julius E. Hilgard from the directorship of the Coast Survey.[6]

Hilgard was widely believed to be a drunkard; his physician later said that he suffered from Bright's disease.[7] Whatever the case, there was "no ground for doubt," in the report of a *New York Times* correspondent, that Hilgard's office operated in a state of "demoralization" and was "to a serious extent, inefficient, unjust, and to some degree disreputable." One salient item of evidence: For years Charles Sanders Peirce had been experimenting with pendulums wherever and whenever he pleased.[8] Cleveland offered Hilgard's post to Alexander Agassiz, who declined, then gave it to a functionary in the Treasury Department. Despite protests from scientists, the President not only stuck by the appointment but also recommended the transfer of coastal surveying to the control of the navy. Simon Newcomb, who acknowledged Hilgard's incapacity, complained: "When a naval officer gets drunk and runs a ship aground, no one ever thinks of concluding that the [whole] navy . . . is

[5] U.S. Congress, House, *Congressional Record*, 47th Cong., 2d Sess., Feb. 21, 1883, XIV, pt. 4, 2040.

[6] Thomas G. Manning, *Government in Science: The U.S. Geological Survey, 1867–1894* (Lexington, Ky., 1967), pp. 127–28; *New York Times*, July 26, 1885, p. 8.

[7] Thomas Corwin Mendenhall, "Autobiographical Notes," TCM, vol. VI, pp. 29–31; E. W. Hilgard, "Julius E. Hilgard," *Biographical Memoirs of the National Academy of Sciences*, III (1895), 337.

[8] *New York Times*, Aug. 7, 1885, p. 3.

demoralized. But hundreds are now ready to conclude that the case against any scientific man being fit to take charge of scientific work is conclusively proved."[9]

Now, too, First Auditor J. Q. Chenowith of the Treasury Department was reportedly investigating charges that even John Wesley Powell had been doing useless research, overpaying favorites, publishing lavish books on irrelevant subjects, buying his way into the National Academy with patronage, and giving away federally owned fossils to Othniel Marsh at Yale.[1] At General Hazen's school for weather observers across the Potomac at Fort Myer, college recruits complained that the food was scrofulous, the methods of martial discipline harshly inappropriate to men of good breeding ("Carry your gun properly . . .," one sergeant said; "you drill like a wooden militiaman"). When a group of enlisted men formally complained to Hazen that their lieutenant repeatedly cursed them, the general ordered the offended soldiers subjected to a barracks court-martial.[2]

Many—perhaps most—scientists had voted for Cleveland and sympathized with the administration's concern for reform. Many wanted the Weather Service removed from the army and placed under civilian control.[3] But the scientific community had long resisted the subjection of the Coast Survey to navy control, and now it collectively fumed with Alexander Agassiz against "ignorant interference with scientific affairs." Were the eminent chiefs of the research agencies "to submit the scientific expenses of these bureaus to the judgment of a clerk in the Auditor's Department?" The AAAS expressed the highest approval of Peirce's measuring gravity with his pendulum and went to the core of the issue in a solemn resolution: The work of the scientific bureaus was best judged by scientific men.[4]

Endorsing that cardinal point, a predominantly academic committee of the National Academy of Sciences proposed to the Allison Commission a radical reorganization of the entire federal scientific effort, preferably the clustering of the government's research bureaus in a department of

[9] *New York Times*, Sept. 27, 1885, p. 8; Oct. 2, 1885, p. 3; *Science*, VI (Dec. 11, 1885), 507; Newcomb to David Gill, Dec. 28, 1885, SN, Box 6.

[1] *New York Times*, Sept. 16, 1885, pp. 1, 4; Nov. 9, 1885, p. 4.

[2] Allison Commission, *Testimony*, p. 501; *New York Times*, Oct. 29, 1885, p. 2; Oct. 31, 1885, p. 2.

[3] See the resolution of the AAAS, Sept. 1, 1880, attached to George J. Brush to Simon Newcomb, Nov. 1, 1880, SN, Box 18. Thomas Corwin Mendenhall typically found it difficult to carry on scientific work "under military discipline and by military methods." Mendenhall, "Autobiographical Notes," TCM, vol. VI, pp. 1, 6.

[4] Agassiz, "The Coast Survey and 'Political Scientists,'" *The Nation*, Sept. 17, 1885, p. 236; *New York Times*, Aug. 29, 1885, p. 2. It is interesting to note that when a few years later Thomas Corwin Mendenhall, then director of the Coast Survey, reviewed the report of the investigation of the agency, he concluded that under Hilgard it had indeed fallen into considerable managerial disarray and some dishonest practices. Mendenhall, "Autobiographical Notes," TCM, vol. VI, pp. 29–31.

science.[5] The capriciousness of Washington politics constantly threatened the life of federal research. Moreover, bureau directors were always uneasy at having to wangle approval for their activities from scientifically illiterate cabinet members. Unlike labor or agriculture, which had millions of votes, or business, which had millions of votes and dollars, science did not enjoy a natural base of political power. Lacking that, federal scientists decidedly wanted—and thought they needed—what the reorganization proposed by the Academy's committee would provide: a considerable measure of self-determination and insulation from philistines in and out of the government.

Powell, who preferred to centralize federal science under the nominally apolitical Smithsonian Institution, told the Allison Commission: Scientists "spurn authority." They are, "as a class, the most radical democrats in society—patient, enthusiastic, and laborious when engaged in [absorbing] work . . . but restive and rebellious when their judgments are coerced by superior authority." The control of science had to be kept out of the hands of "officers or functionaires" whose principal interest was "official position or dignity." The great body of scientific men would not tolerate the promotion of research under a predominantly "political institution."[6]

But by attacking the subjection of science to "political institutions," Powell and his colleagues were calling for more than protection from the improprieties of power-hungry officeholders. To them, politics gave laymen the power to establish the programmatic scope of federal research, to assess the merit of its execution, even to control the dissemination of its results. Since the line between informed lay management and know-nothing restriction was hardly sharp, Powell and his colleagues preferred direct governmental support with only nominal governmental control. Behind their preference lay the assumption that they were a select group who deserved to exercise power with only limited accountability, an assumption which in its approach to government can be identified as political elitism.

Distinct from the best-science elitism of Rowland, this political elitism was widespread among American scientists. It was not in any respect ideological. It bespoke neither a cavalier disregard for the processes of democratic government nor a bald-faced bid for power. Rather, it revealed the scientific community's self-conception as an aristocracy of intellect and ideals. It was reinforced by the general assumption of the cultivated that scientific habits of thinking produced virtuous habits of mind. And it had already found a certain quasi-governmental acceptance in the National Academy of Sciences. While the Academy was the gov-

[5] "Report of National Academy . . .," in Allison Commission, *Testimony*, *1–*10.
[6] *Ibid.*, pp. 178, 381.

ernment's official scientific adviser, it was also a private, self-perpetuating organization. As such, its leaders insisted, it was safely insulated from political influence; then, too, the professional merit of the membership assured to the government advice that was not only expert but scientifically disinterested. Still, the Academy could exercise only advisory functions. In their politically elitist fashion, Powell and his colleagues were arguing for something more—that the direction of federal research itself should turn on similarly scientific, not "political," criteria.

Secretary of the Navy William B. Chandler understood the issue raised by the proposals for the reorganization of federal science. In the early hearings of the Allison Commission, Chandler, a seasoned politico, testified that to erect a department of science would be "a most anomalous proceeding" for the United States government. While the secretary hardly held a limited view of federal power when it came to aiding business, he now insisted upon an administratively strict construction of the Constitution, which did not specifically provide for scientific research. The executive departments were created to help the President carry out the government's explicit constitutional duties. They were "political," Chandler claimed, and incorporated scientific agencies only so that they might better perform their political functions. To relocate the research bureaus would impair the operations of the existing departments. To make them autonomous would improperly free them from political purposes. To create a department of science would be "unjustifiable" and "set the scientists to grasping [the reins of] Government."[7]

Besides, however passionately scientists believed in the necessity of keeping their bureaus free from lay interference, no agency of a democratic government, *The New York Times* insisted, could be permitted to "set itself above examination." Scientists might consider themselves a disinterested elite; if all men were angels, James Madison once noted, government would be unnecessary. Youmans' *Monthly* was sufficiently disgusted over the Coast Survey scandal to permit itself the opinion that scientists were anything but angelic. "Like other men they are self-seeking, ambitious, and have their personal ends to gain. Can we assume that morally they are any better than their neighbors; or that, if they get possession of place and power, they will not use and pervert them to the promotion of their selfish objects? It is to be hoped that in the future science will become so developed as to react upon character and give us men morally as well as intellectually superior; but we are far from any such happy result as yet."[8]

Whatever scientists were, the debate revealed that the growth in federal patronage of science had raised a major question of public policy: How was governmental science best and properly controlled—through

[7] *Ibid.*, pp. 66–67.
[8] *New York Times*, Sept. 27, 1885, p. 8; *Popular Science Monthly*, 27 (October 1885), 846.

a democratically political or politically elitist mechanism? It was a question that would force itself increasingly upon Americans in the future.

In the 1880s the issue of control was overshadowed by the more pressing question: Just how much general "theorizing" was required, or even allowable, in federal research for the pursuit of "practical" purposes?

Naval critics of the Coast Survey, pointing to the elaborate studies it pursued to measure the curvature of the earth, stressed that sailors did not need to know every outcropping and landmark. Before his dismissal, Hilgard allowed that the paleontology sponsored by Powell, rather than satisfying "a public want," was "strictly scientific." One of Hazen's own military subordinates—he had a number of antagonists among his regular officers—said that the Study Room could be disbanded, since more meteorological research was unlikely to add much to the accuracy of weather forecasts. Most of the scientific witnesses countered by stressing with Cleveland Abbe that their research was practical, not "vain theorizing."[9] But they failed to persuade Congressman Hilary Abner Herbert of Alabama, a member of the Allison Commission who had decided to wage war on federal research.

At the start of the hearings, Herbert had known very little about federal science; by the end of 1885 he knew a good deal and did not like what he had learned. A planter's son and a former officer in the Confederate Army, a lawyer from Montgomery, Herbert was a Bourbon Democrat, a late-nineteenth-century liberal who considered himself "radically democratic." ("I believe in as little government as possible—that Government should keep hands off and allow the individual fair play.") In the opinion of Herbert, a member of the House Naval Affairs Committee, the civilian-managed Coast Survey had strayed into the "boundless" field of physical research. Naval officers would stick to practical matters and would not be tempted by "the prospect of gaining *éclat* among scientists."[1] In the suspicion of Herbert the radical democrat, John Wesley Powell had extended the Geological Survey into too many scientific areas. Did the Survey really have to do topographical mapping and go into paleontology? the puzzled Mr. Herbert asked Alexander Agassiz.[2]

Though Agassiz had been an early supporter of the Geological Survey, he disliked the way it had expanded under Powell. He was a staunch defender of the Coast Survey, whose facilities he had used for zoological research, and now he suspected Powell of maneuvering to help destroy the Coast Survey by swallowing up its topographical work. In Agassiz's

[9] Allison Commission, *Testimony*, pp. 894, 54, 988–89, 259.

[1] Quoted in Manning, *Government in Science*, p. 133; U.S. Congress, Joint Commission . . ., *Report*, 49th Cong., 1 Sess., June 8, 1886, Sen. Report 1285, Series 2361, pp. 70, 79.

[2] Herbert to Agassiz, Nov. 27, 1885, in Allison Commission, *Testimony*, p. 1014.

opinion, Washington science was "getting its purseful" and centralizing research to a dangerous extent. American geologists were becoming mere "satellites" of Powell.[3] Agassiz, the son of the famed Louis Agassiz, was no one's satellite. He had earned a fortune in part-time copper mining; he published a good deal of his own research, richly endowed the Museum of Comparative Zoology at Harvard, preferred to be known for his zoology rather than to win "cheap notoriety as an American millionaire."[4] Responding to Herbert, Agassiz fractured the alliance between academic and federal scientists in a letter that was as highly principled as an independently wealthy university scientist could possibly write.

Science, he insisted, deserved no more federal assistance than any other branch of knowledge or the fine arts. The government's publication of research was "wasteful and extravagant"; its paleontological work, unnecessary. Private individuals could study fossils just as ably and more cheaply. Agassiz admitted that without a good topographical map, a good geological map of the country would be impossible. But, "if the States are not willing to go to that expense, it seems plain that they do not wish the Government to go to that expense for them."[5]

The prestigious Agassiz's assault against the Geological Survey was tailor-made for Herbert. Here were no uninformed charges of corruption; here was an expert's sharp critique, phrased in the language of states' rights and private initiative. Toward the end of January 1886, Herbert made his correspondence with Agassiz public. Toward the end of April, he introduced a bill in the House that called for radical surgery on the Geological Survey. Its major point: No money could be spent for paleontology, the publication of research, or "the general discussion of geological theories."[6]

A thorough geological survey might be useful, Herbert explained to the House, but the possibilities of research were limitless; a prudent government had to count the cost. Economy aside, no government, Herbert argued, following Agassiz's lead, should subsidize a field to the extent of establishing a monopoly over it. A federal monopoly would not only make private research impossible but deliver science into political "bondage." Some sixty-nine consultants and more than twice as many regular employees worked for Powell. All of them, in Herbert's opinion, tended "to think and act as one man, to defend the Survey against all criticism and to maintain its infallibility against all comers."[7]

[3] Agassiz to Simon Newcomb, June 10, 1884, SN, Box 14; Agassiz to T. H. Huxley, Oct. 16, 1886, G. R. Agassiz, ed., *Letters and Recollections of Alexander Agassiz* (Boston, 1913), p. 231.

[4] Agassiz, ed., *Letters and Recollections*, pp. 399–400.

[5] Agassiz to Herbert, Nov. 27, 1885, in Allison Commission, *Testimony*, pp. 1013–14.

[6] See Herbert's bill in Joint Commission, *Report*, p. 96.

[7] *Ibid.*, pp. 95, 124.

To a degree, Herbert was behaving like the ambitious politico he was. If as a member of the House Naval Affairs Committee he favored the transfer of responsibility for shoreline mapping from the Coast Survey to the navy, as a good Democrat he endorsed the removal of the Weather Service from the Signal Corps, meaning Hazen, to a civilian agency. But on the matter of the Geological Survey, even Southerners, a Louisiana journalist reported, seemed unable to divine the reason for Herbert's assault; mineral-rich Alabama was, after all, "among the greatest beneficiaries of the Survey's work."[8] Agassiz's charge of "geological dictatorship" seems to have touched Herbert's sense of principle. No yahoo, the congressman quarreled with the political implications of Powell's system. Herbert set down his essential concern by calling in the judgment of the English historian Henry Buckle upon the fawning court of Louis XIV: "They, who dispense the patronage, will, of course, receive the homage; and if on the one hand government is always ready to reward literature, so on the other hand will literature be always ready to succumb to government."[9]

Later, once the federal government moved massively into the support of academic research, many Americans would worry about Herbert's political point. In the 1880s Herbert's critique was largely ignored, and his bill appalled Powell and his friends. For Charles Sanders Peirce, enraged at this "abominable and scandalous attack," pendulums were for the moment beside the point. "Let the Congressmen hear of science, no longer as merely giving reasons, but as an *interest*, saying *We want so and so*."[1]

John Wesley Powell let them hear. Did Agassiz and Herbert claim that Powell had created a "geological dictatorship"? Anyone, Powell declared, who tried "to coerce or improperly control scientific opinion" would "utterly destroy" his effectiveness. Did Agassiz consider the Geological Survey extravagant? Not even "a hundred millionaires," Powell argued, could support the current research of the federal agencies. Did the progress of American civilization have to "wait until the contagion of [Agassiz's] example shall inspire a hundred millionaires to engage in like good works?" Only "narrow and dilettante" scientists, Powell added, held themselves above utility. Only "wiseacres" boasted of their devo-

[8] Quoted in Manning, *Government in Science*, pp. 142–43.

[9] Joint Commission, *Report*, p. 96. Herbert declared that it was not his intention to show that Powell was "personally dictatorial. . . . The objection was to the system." *Ibid.*, p. 124. At a meeting with Simon Newcomb and Powell, Herbert, "perfectly frank and courteous," expressed his "decided opinion that the Geological Survey had gained influence by employing professors in various parts of the country. . . ." Newcomb saw "no reason" to doubt Herbert's "entire conscientiousness in what he has said and done." Newcomb to O. C. Marsh, n.d. [May 1886], SN, Box 6.

[1] Quoted in Manning, *Government in Science*, pp. 138–39.

tion to "pure science." Surely the national government should support and publish whatever science advanced the welfare of the people.[2]

In April 1886 Simon Newcomb, who thought Powell had defended himself with great skill, happily remarked of the fate of the scientific bureaus: "I think we are coming out all right."[3] If Agassiz had broken the normally unified front of science, his was the sole rupture in the alliance of academic and federal research. The Treasury Department and even the punctilious *Nation* had decided that the charges of corruption in the Coast and Geological surveys were unfounded. While the secretary of war abolished Hazen's school at Fort Myer, a committee of the House had given the general's management of the Weather Service a clean bill of health.[4]

In the final report of the Allison Commission, June 8, 1886, all three Republicans and Congressman Robert Lowry, a highly respected judge and Democrat from Indiana, flatly recommended against meddling with either the Coast or the Geological Survey. The sole restriction recommended for either agency, expressed in the form of a bill, merely required them to itemize their annual printing estimates. On the Weather Service, Allison, Republican Senator Eugene Hale of Maine, and Congressman Lowry acknowledged that weather prediction was not "in any sense military work"; apparently loyal to Hazen, they nevertheless endorsed the retention of the Service in the Signal Corps. They did criticize spending money for "purely speculative" investigations and recommended the abolition of the Study Room. But they applauded Hazen's employment of "highly skilled meteorologists."[5]

Herbert, together with his fellow Alabamian Senator John T. Morgan, filed his expected minority reports on each agency, and late in June he carried his dissents to the floor of the Congress. Neither the House nor the Senate seriously considered a transfer of the Weather Service (Congress did move with dispatch to relocate the Service in the civilian Department of Agriculture after Hazen's death in 1887). Northerners of both parties vigorously defended the surveys. The Democratic House cut the Coast Survey's budget—"to correct some abuses . . . alleged to exist," the friendly chairman of the Appropriations Committee explained; the cuts were restored in the Republican Senate.[6] Even the Commission's bill to impose closer control over the publications of the Geological Survey failed. Herbert was sectionally isolated and, probably for the most part, isolated within his own section. By late July 1886, after two years,

[2] Joint Commission, *Report*, pp. 124–25; Powell to Allison, Feb. 2, 1886, in Allison Commission, *Testimony*, pp. 1078, 1075.

[3] Newcomb to J. N. Lockyer April 15, 1886, SN, Box 6.

[4] "The Condition of the Coast Survey," *The Nation*, March 11, 1886, p. 208; *New York Times*, March 5, 1886, p. 1; April 28, 1886, p. 3; May 3, 1886, p. 4.

[5] Joint Commission, *Report*, pp. 52, 4, 18, 19.

[6] U.S. Congress, House, *Congressional Record*, 49th Cong., 1st Sess., June 25, 1886, XVII, pt. 6, p. 6137.

1,084 pages of intermittent testimony, and the broadsword attack of a southern congressman, the scientific bureaus emerged essentially unscathed.

Yet the Allison Commission did agree with Herbert's contention that there was "no reason why a scientist . . . should be more competent than a Congressman to say how much topography a sailor would want on his chart. If it were simply a question as to the best method of obtaining absolute exactness, the judgment of a pure scientist would outweigh the opinion of the man of affairs; but not so when the question is as to the adaptability of the work accomplished to the purpose of the law."[7] The Commission reported unanimously against the creation of either a department of science or any other scheme for reorganizing governmental research. Federal scientific agencies were to remain in Chandler's political departments, their activities subject to democratically defined purposes and direct political control.

The Allison Commission amounted to a direct political assessment of the post-Civil War expansion of federal science. Perhaps no man was more responsible for the outcome than Allison himself. Not many years later, Othniel Marsh told him that he deserved the "thanks of every scientific man in the country" for his continuing support of federal research.[8] Wherever the main responsibility lay, the Congress had clearly reaffirmed the contemporary scope and purpose of governmental science.

In the course of the Allison Commission's proceedings, physicists and astrophysicists had urged that the government should create a bureau of electrical standards and a "physical observatory" for spectral research. Both warhorse politicos like Allison and liberals like Herbert agreed with *The Nation:* "The Government should not keep a school for original research, and it is not advisable to see established at Washington laboratories of chemistry, physics, biology, etc., etc., intended only for such work as can be done elsewhere."[9] In the late nineteenth century the scope of federal science was to remain limited to fields pertinent to the use of the land, which did not include physics as such. And however forceful Rowland's plea for pure science, the veteran Washington hand Simon Newcomb understood that federal research had to stay within practical bounds. Congress was always ready to spend money for science in a "lavish way," Newcomb reflected—so long as it "received satisfactory assurance that scientific results are not the main object."[1]

[7] Joint Commission, *Report*, p. 68.
[8] Marsh to Allison, July 9, 1892, OCM.
[9] Report of the Academy Committee, Allison Commission, *Testimony*, p. 8*; "The National Government and Science," *The Nation*, Dec. 24, 1885, p. 526.
[1] Newcomb to Lewis Boss, Dec. 16, 1886, SN, Box 6.

V

Research and Reform

More homes and offices shone with electric lights, more factories whirred with electric dynamos, more trolleys clanged along on electric drives. The growing demand for electric power helped stimulate a shift from the commonly used direct to the new alternating current, or AC, which could be transmitted over great distances at low cost. With the advent of alternating current, the analysis of electrical circuits became decidedly more complex. As early as 1886, *Electrical World* fittingly declared: "The day of rule-of-thumb methods has gone by." A few years later, an assistant asked Edison for some technical advice, and the inventor brusquely told him to consult one of the mathematically adept members of the staff. "He knows far more about [electricity] than I do. In fact, I've come to the conclusion that I never did know anything about it."[1] Edison, who had opposed the introduction of alternating current, sold his interests to J. P. Morgan's new combine, the General Electric Company, and left the light and power business for good. General Electric relied for technical advice on a German immigrant with advanced training in physics, Charles Proteus Steinmetz.

Steinmetz, the hunchbacked son of a burgher in Breslau, Germany, had gone off to university, become a socialist, and fled the country when threatened with arrest. While working for the electrical firm of Rudolph Eickmeyer in Yonkers, New York, Steinmetz confronted the difficulty of determining the energy dissipated in iron as its magnetic field changed under

[1] *Electrical World*, Dec. 11, 1886, p. 279; Edison is quoted in Matthew Josephson, *Edison* (New York, 1959), p. 362.

the influence of alternating current. He developed a mathematical formula for this energy loss, and in 1892, when he reported the full results of his study before the American Institute of Electrical Engineers, he won professional fame. That year GE bought the Eickmeyer firm, and Steinmetz went with it. Still a socialist, a modest master of the complexities of alternating current equipment, he happily taught GE engineers how to design the new machines. In 1893 the twenty-nine-year-old immigrant became chief of the company's electrical calculating department in Schenectady.

In the early 1890s engineers like Steinmetz were beginning to set the pace not only in the electrical industry but in chemicals, petroleum, pharmaceuticals, and steel. Andrew Carnegie himself warned: "The trained mechanic of the past . . . is now to meet a rival in the scientifically educated youth, who will push him hard—very hard indeed."[2]

Appropriately, across the land, as an observer remarked, the idea was dawning that the colleges were "not institutions for a fortunate few . . ., but that they hold a very close relation to a sound national life."[3] Sound businessmen were eagerly hiring college graduates, sound farmers deciding that agricultural schools might after all be teaching something useful. The percentage of Americans taking degrees started to rise in the late 1880s. Engineering seemed to be getting the lion's share of the new matriculants, but whatever the subject, the dramatic climb in enrollments, administrators agreed, was astounding.

Most students spent their four years on the campus and left; some headed for the professoriat. Between 1887 and 1892 the fraction of the population in graduate school doubled, and science won its share of the doctoral candidates.[4] Students who came to college to study engineering often found themselves attracted to the intellectual adventure of physics or chemistry, geology or biology. Besides, the academic gown opened the doors of the cultivated town. (The *grandes dames* of New Haven patronized the Yale professoriat, Henry Seidel Canby recalled, "soothing vanities with a kind word, snubbing with a vacant look the strange uncouth creatures that science was bringing into the university, but not too emphatically since one never knew nowadays who might become famous."[5]) Above all, the increasing enrollments in engineering were creating a demand for teachers, particularly in physics and chemistry.

In all fields, the expansion of the professoriat provoked considerably more competition for place in the college hierarchy, and younger faculty especially grew eager to change the prevailing standard of judging aca-

[2] Quoted in Irving G. Wyllie, *The Self-Made Man in America* (New Brunswick, N.J., 1954), p. 112.

[3] Hamilton W. Mabie, "The American College President," *Outlook*, Aug. 19, 1893, p. 340.

[4] Daniel J. Kevles, "The Study of Physics in America, 1865–1916" (unpublished Ph.D. dissertation, History, Princeton University, 1964), Appendix II.

[5] Henry Seidel Canby, *Alma Mater* (New York, 1936), p. 10.

demic merit. If the older faculty members were content to submit to a presidential assessment of their merits as teachers, most of them knew the president personally. With the enlargement of the staff, few instructors or assistant professors could expect to dine at the presidential table, and pedagogical quality could all too easily depend on hearsay or popularity among students. But one's professional peers could directly judge one's merits as a scholar. For the younger faculty, who, unlike the majority of their seniors, had earned Ph.D.'s, promotion through the college hierarchy ought to depend less upon the quality of one's teaching and more upon the merits of one's research.

The new academics were given dramatic institutional encouragement by the founding in 1891 of the University of Chicago, whose president knew all about getting ahead swiftly in academia. William Rainey Harper, the thirty-six-year-old son of an Ohio storekeeper, was out of high school at ten, a Ph.D. in theology at eighteen, a Yale professor at twenty-nine. A Baptist, he evangelized for biblical criticism on the Chautauqua circuit and won notoriety among the modernist leaders of his church. Church elders had been trying to pry money for an educational institution from good Baptist John D. Rockefeller. They were talking of a college, but Harper, who knew from Yale how slowly colleges moved on to bigger things, wanted to create a great university overnight. Rockefeller promised $200,000 for theology and $800,000 for graduate work. Chicago, the young president insisted, would "make the work of investigation primary, the work of giving instruction secondary."[6]

Harper may have sounded like an echo of Daniel Coit Gilman at Johns Hopkins but, unlike Gilman, he was much less interested in advancing a particular set of cultural goals than in building a great educational institution. Harper threw enormous energy—an associate once called him a "steam engine in pants"—into the realization of the project.[7] Funded with an additional $1,000,000 from the businessmen of Chicago, he planned great libraries for the campus, which was rising, like the ebullient symbol of academic progress it was, opposite the grounds of the planned Columbian Exposition. Harper scoured the country for first-rate faculty, offered some of his departmental chairmen the unprecedented salary of $7,000 a year; he managed to attract the ablest staff members from many institutions, including the new but impoverished Clark University, which lost Albert A. Michelson. Harper received nearly one thousand applications for faculty positions and the university opened with an enrollment of more than two hundred graduate students.[8] The

[6] Quoted in Frederick Rudolph, *The American College and University: A History* (New York, 1962), p. 352.

[7] Quoted in [University of Chicago], *The First Fifty Years* (Chicago, 1941), p. 10.

[8] Thomas Wakefield Goodspeed, *William Rainey Harper* (Chicago, 1928), p. 123; Rudolph, *American College and University*, p. 351.

response bespoke not only the American professoriat's deepening commitment to the advancement of knowledge but the new academic's identification of research with opportunity in an expanding institutional milieu.

When the University of Chicago opened in 1892, federal scientists were warily watching Capitol Hill, where a two-thirds majority of Democrats—many of them economy-minded conservatives, including Hilary Herbert—dominated the House. The Congress also contained agrarians of both parties, who, sensitive to the mounting populist revolt on the cotton fields, plains, and prairies, wondered about funding scientific frills. Representative Henry Clay Snodgrass of rurally poor Sparta, in White County, Tennessee, attacked an appropriation for the new Zoological Park of the Smithsonian Institution: "I do not believe the American people, hundreds and thousands of whom are today without homes, ought to be taxed to afford shelter and erect homes for snakes, raccoons, opossums, bears, and all the creeping and slimy things of the earth."[9] The agrarians and the pro-economy Democrats, each for their own reasons, formed a coalition that threatened expenditures for superfluous science.

In the House, the coalition applied the economy ax to both the Geological and the Coast and Geodetic Survey. Congressman William S. Holman of Indiana, the wry old chairman of the Appropriations Committee who liked to poke at the House for overtaxing poor people, told the spenders that the practical value of geodesy was "very remote" and the government could not indulge in purely scientific work when the Treasury was "cramped." Herbert added his dissent from the extravagance of measuring the shape of the earth and weighing it "exactly to a pound," then attacked the Geological Survey by hauling out his minority report from 1886, reminding the House of Agassiz's authoritative opinions and insisting that the Survey was going far beyond the proper bounds of a frugal and limited government.[1] Yet in 1892 Herbert was more the Democrat than radically democratic. Hardly any Westerners wanted to do away with Powell's topographical work, since it was essential for irrigation. Leaving topography alone, Herbert urged the House to eliminate the $37,000 appropriated for paleontology—he assured the Westerners of its irrelevance to irrigation—and to prohibit the Survey from engaging in it forevermore. Herbert's amendment carried easily.[2] Not even the opposition of Holman, whose son was working for Powell, could block the coalition against scientific frills.

The chiefs of the scientific bureaus, hardly any of which escaped the House's display of thriftiness, had to trust their fortunes to the Re-

[9] U.S. Congress, House, *Congressional Record*, 52d Cong., 1st Sess., May 12, 1892, XXIII, pt. 5, 4238.
[1] *Ibid.*, May 24, 1892, pp. 4632, 4634.
[2] *Ibid.*, May 10, 1892, pp. 4283–84, 4389–96.

publican Senate and William B. Allison, still chairman of the Appropriations Committee. But in the upper chamber, where there were thirty-nine Democrats and two populist Midwesterners, the GOP commanded a majority of only six. Moreover, Powell was in deep trouble with far western members of the GOP, against whom he had lost a bitter fight in 1890 over his efforts to withdraw irrecoverably arid lands from traditional homestead settlement. Equally important, both eastern and western Republicans were worrying about the coming elections. Charges of extravagance had helped transform the House in 1890; now the Senate chamber was echoing with testimonials to economy from the Republican side of the aisle, too.

Allison's Appropriations Committee reported out a partial restoration of the Coast Survey budget and the resurrection of paleontology in the Geological Survey. The Senate dutifully gave the Coast Survey its money, but Senator Joseph M. Carey of Wyoming, who had not forgiven Powell for withdrawing from settlement "every acre of ground . . . west of the one hundredth meridian," proposed a massive but shrewdly selective slash—ample funds would remain for topography—of some 45 percent in the Geological Survey budget. The major cuts were aimed at the scientific staff, general geology, and paleontology. "We in the West want more of substantial benefit and less of ornament," Carey argued.[3] His amendment carried by a narrow margin. Powell's friends hurried about the chamber trying for a reversal. A few minutes later the Senate sealed the reduction by a vote of twenty-eight to twenty-five. In the House the Democratic majority ratified the cut and, in addition, refused to concur in the partial restoration of the Coast Survey funds.

The Republicans from the Far West had inaugurated the senatorial move against Powell, but by the final tally the anti-Powell phalanx included twenty Republicans, some from the Northeast. The rest of the margin was provided by six Democrats and the two Populists, who had voted against Powell and paleontology on every test. An astounded reporter wrote: "For the first time in the memory of man, Major Powell . . . has encountered opposition which he was unable to sweep aside."[4] In 1892 every major scientific bureau emerged from the congressional session with a reduced appropriation. The reporter might have explained that not even John Wesley Powell could beat back a coalition of economy-minded conservatives and midwestern agrarians, joined now—as reformers and conservatives would unite more than once in the future—to cut the fat out of federal science.

In 1893 Grover Cleveland returned to the Presidency, with a Democratic House and Senate. For the first time in memory, the new President

---

[3] *Ibid.*, July 14, 1892, pt. 6, 6155.
[4] *New York Times*, July 15, 1892, p. 5. The key shift of eastern Republicans into the anti-Powell column is explicable perhaps by the strong link Powell's enemies made between his survey and extravagance.

opened his term with a White House reception for the government's scientists, but, not least because of the depression, the order of the day in the White House and in Congress was retrenchment. Secretary of Agriculture Levi P. Morton soon announced that he would drop all "useless scientists" from the Weather Bureau. "What the people most want is the knowledge beforehand of what is to happen . . ., rather than a scientific diagnosis . . . after it is all over." The secretary of the treasury kept an irritatingly close watch on Thomas Corwin Mendenhall, now superintendent of the Coast Survey and a staunch Republican.[5] Hilary Herbert, now secretary of the navy, appointed to the superintendency of the Naval Observatory an officer whom Benjamin Gould considered nothing less than a "self-seeking and self-styled ignoramus."[6] His convictions of 1886 still vigorous, Herbert also tried to engineer the abolition of the Coast Survey—his congressional henchmen exhumed the indictment of Charles Sanders Peirce for the exemplary frivolousness of his work with pendulums—and succeeded in reducing its staff.[7]

Cleveland Abbe, harassed by the secretary of agriculture and bitter over the administration's "animus" toward science, considered resigning.[8] He and Newcomb hunched up and waited for better times. Powell and Mendenhall left. Scores of lower-ranking scientists had been forced out of the government by the cuts of 1892, and now Cleveland's hardnosed cabinet was firing scores more, among them Carl Barus.

Fresh from winning his Ph.D. in physics, Barus had joined the Geological Survey at its founding and, through the years under Powell, had won an international reputation for his pyrometric studies of rocks. Literate and knowledgeable, he enjoyed the good talk of Bell's salon; his wife, a reform-minded Vassar graduate, liked having the capital's luminaries in to dinner. Barus had just been elected to the National Academy when in 1892 the ax fell on the Geological Survey. He managed to find a post in the Weather Bureau, but a year later the new administration summarily dubbed him a useless scientist. An unhappy fifteen months under an irascible boss at the Smithsonian, a few more months of worrisome unemployment, and an old friend finally helped Dr. Barus secure a professorship at Brown.[9]

Barus was lucky to get his professorship. When the depression struck, the fraction of Americans taking undergraduate degrees stopped rising, and so did the demand for more teachers. If only because corporate jobs were difficult to obtain, graduate enrollments increased, enlarging the pool of available faculty. Academic posts in science were

[5] Morton is quoted in *The New York Times*, July 13, 1893, p. 10; Mendenhall, "Autobiographical Notes," TCM, VI, 147.

[6] Gould to Simon Newcomb, Feb. 4, 1896, SN, Box 24.

[7] U.S. Congress, House, *Congressional Record*, 53d Cong., 2d Sess., March 15, 1894, XXVI, pt. 3, 3006–7.

[8] Abbe to Charles W. Eliot, July 15, 1893, CWE, Box 134.

[9] Carl Barus, "Autobiography," CB, pp. 65–70, 111, 116–17, 120–27, 129, 161–62.

at a premium everywhere. Meetings of the AAAS were sparsely attended, and membership in the Association dipped back to the level of the early 1880s. In 1894 Bell's magazine folded, a victim of sharply declining receipts. In the middle of it all, the nation's scientists paid slight attention to enlarging their professional opportunities. Like millions of other Americans, they were simply holding on.

By 1897 prosperity had returned to the country and the friends of science were back in power at the White House and on both sides of Capitol Hill. As the century drew to a close, appropriations for the Geological and Coast surveys and for their sister agencies in the Department of Agriculture swung sharply upward. Yet the fiscal turnabout was not to yield just more of the same. The existing scientific bureaus were all children of the sustained thrust to master the continent and pry wealth from the land. Now the frontier was closed, the nation swiftly urbanizing. Despite the sharp rise in agricultural prices, at the turn of the century manufacturing was yearly supplying some 30 percent more of the national income than mining and agriculture combined. Books and articles extolled America's inventive past and painted marvelous portraits of the technological future. Equally important, the Spanish-American War had given the United States an overseas empire. By the turn of the century, exports had passed the billion-dollar mark and exceeded imports for the first time in American history, which sensitized businessmen more than ever to competition for foreign markets.

The trend made more timely the long-term advocacy by American physicists and chemists of greater federal investment in laboratories devoted to their disciplines. The governments of Germany and England had recently established handsomely equipped laboratories for the determination of standard physical and chemical units. The lack of such a laboratory in the United States was said to be humiliating—and commercially costly. The increasing technical complexity of industry made more pressing the need for uniform standards not only of weight but of chemical and electrical quantities. Most measuring devices were imported, and domestic manufacturers were disinclined to use the products of America's nascent instrument industry unless they bore the seal of a European testing laboratory.

The physicist Samuel Wesley Stratton aimed to do something about the situation. The ambitious son of an Illinois farmer, Stratton had gone to Illinois Industrial University. He boarded with the president, acted as his private secretary, and learned the social amenities along with something about the niceties of administration. Appointed to the faculty, he completely revised the physics curriculum, then moved on to the University of Chicago, where he oversaw the construction of the physics laboratory. In 1899, appointed head of the government's meager effort in weights and measures, Stratton set out to create a major agency to deal with standards across the entire spectrum of science and engineering.

He marshaled testimonials from scientists and engineers, professional societies and manufacturing associations, executives in steel, railroads, and the electrical industry.[1] Witnesses at the hearings in 1900 included Henry Rowland, who testified that the nation's reliance on European standards made impossible the enforcement of the laws governing weights and measures, and Secretary of the Treasury Lyman G. Gage, who averred that in applied science and trade, "in all the great things of life," the country was in competition with the older and more thoroughly established nations of the world.[2] On March 3, 1901, Stratton's bill to create a National Bureau of Standards passed in Congress by a large margin.

It was a eupeptic event for America's physicists and chemists. The law authorized $250,000 for a laboratory. It also enabled the agency to engage in whatever research might be necessary to establish standards, which opened the door to the entire realm of physics and chemistry. Equally important, since civil service rules would govern employment, the new Bureau appeared well protected against some of the sharper political vicissitudes that had marked federal science in the late nineteenth century. The organic act provided for a "visiting committee" of private citizens in relevant fields to advise upon the agency's needs and activities. The secretary of the treasury, an advocate of the new Bureau remarked, would "scarcely" fire a director whom the committee considered able or hire one whom it counted incompetent.[3] The administration awarded the inaugural directorship to the decidedly competent Samuel Wesley Stratton, physicist.

At the opening of the twentieth century, a qualified scientist headed every major research bureau. Most jobs, by way of a legacy from Cleveland's second administration, were securely under the civil service. Above all, however high appropriations had been in the heyday of John Wesley Powell, economic nationalism was now driving them to far more opulent levels. All of federal science surged with new opportunities.

In 1901 Andrew Carnegie was astir with a new eleemosynary idea. Years before, earning some $50,000 annually, the son of desperately poor Scottish immigrants had instructed himself: Quit business soon, enroll at Oxford, meet "literary men," buy a newspaper, aim at the "education and improvement of the poorer classes." "Amassing of wealth is one of the worst species of idolatry."[4] He may have spent the next three decades building an empire of steel, but he consorted with cultivated men on

[1] See the correspondence in SWS, Boxes 1 and 5.

[2] U.S. Congress, Senate, Subcommittee of the Committee on Commerce, *Hearings upon the Bill to Establish a National Standardizing Bureau*, Sen. Doc. 70, 56th Cong., 2d Sess., Dec. 28, 1900, pp. 1–2, 9.

[3] Henry S. Pritchett, "The Story of the Establishment of the National Bureau of Standards," *Science*, 15 (Feb. 21, 1902), 284.

[4] Quoted in Joseph Frazier Wall, *Andrew Carnegie* (New York, 1970), p. 225.

both sides of the Atlantic. In the late 1890s, already a giver of libraries, Carnegie set himself up at Skibo Castle, where a tartan-clad piper would skirl eminent guests to his breakfast court. Early in 1901, having sold his sprawling company for a large share in the new U.S. Steel Corporation, America's first billion-dollar combine, Carnegie proceeded to devote himself to good works, especially to good cultural works that might embellish the name of Andrew Carnegie.

During a visit to Carnegie's Fifth Avenue mansion in mid-November 1901, Daniel Coit Gilman outlined an intriguing philanthropic idea. Just retired from Johns Hopkins, Gilman was now president of the new Washington Memorial Institution, a private foundation that aimed to provide qualified people with greater opportunities for research, particularly by helping to open the government's scientific facilities to private citizens. The foundation had already persuaded Congress to legalize private use of federal research bureaus.[5] Now Gilman apparently urged Carnegie to found in the capital an organization to facilitate the research of private investigators in federal laboratories and elsewhere through scholarships, fellowships, and jobs.

On November 28, in a letter to President Theodore Roosevelt, Carnegie proposed to establish just such an organization, in the manner of the Smithsonian Institution, under the unofficial auspices of the federal government. If the federal government would designate the land, Carnegie would pay for the buildings and supply an endowment of $10,000,000. But James Smithson's bequest had arrived in gold; Carnegie offered his gift to the government in bonds of U.S. Steel. And, according to the newspapers, he had attached a proviso that the government would have to hold the bonds for a period of years. Many Americans expected the government to prosecute the giant steel combine for violating the Sherman Act. To accept the bonds, dissident congressmen protested, would be tantamount to stamping the billion-dollar corporation with a seal of official approval. The embarrassed President, it was reported, might consider taking the gift if it were in cash.[6]

The newspapers were wrong; Carnegie had attached no explicit provisos about the bonds in his offer to Roosevelt. He had only implied that he expected the government to keep the bonds until they matured fifty years hence. "Please remember," Carnegie plaintively told Roosevelt after the storm broke, "the Carnegie Steel Co. was no Trust, but a

---

[5] Congressman Joseph Cannon opposed the request from the backers of the Memorial Institution to open federal laboratories: "It looks to me very much as if somebody were getting ready to establish a national university—left-handed, if you please—with its various departments and rooms and ramifications. . . ." U.S. Congress, House, *Congressional Record*, 56th Cong., 2d Sess., Feb. 21, 1901, XXXIV, pt. 3, 2773.

[6] Carnegie to Roosevelt, Nov. 28, 1901, AC, vol. 85; *New York Times*, Dec. 10, 1901, p. 1; Dec. 11, 1901, p. 8; Dec. 13, 1901, p. 7; Dec. 15, 1901, p. 1; Chicago *Daily Tribune*, Dec. 13, 1901, p. 2; San Francisco *Chronicle*, Dec. 12, 1901, p. 1.

Limited Partnership. No stock for sale—I never bought or sold stock on the Exchange. My money was all made in making iron and steel, no gambling."[7]

In New York on December 14, newsmen queried Carnegie—he had been incommunicado over the previous few days—while he was out laying a high-school cornerstone. Would he convert the bonds?

The bonds were easily negotiable, Carnegie pointed out. If the government could not properly sell them, he could. In any case, the matter would turn out all right. Carnegie refused to discuss the matter further.[8]

Four days later, Carnegie lunched at the White House. Once again he refused comment to the press, save to declare that everything would turn out all right. He might have said that he would not go ahead with his effort to emulate James Smithson. Carnegie and Roosevelt had evidently agreed that, even if converted into cash, his prospective gift to the government was already surrounded with too much controversy.[9] Carnegie would now use the bonds to create a new, wholly private institution of research in the capital.

The Carnegie Institution of Washington, established in 1902, was a spectacular monument to Andrew Carnegie's cultural beneficence. His $10,000,000 gift equaled Harvard's entire endowment, and it amounted to far more than the total endowment specifically for research in all American universities combined. The Institution's target: to "encourage investigation, research, and discovery." Its principal methods: to discover and give opportunities to the "exceptional man"; to fund university facilities and research; to underwrite the costs of publication. Carnegie, sure of "our national poverty in science," aimed to do nothing less than "change our position among the nations," and newspaper editorials from New York to San Francisco applauded his gift and his purpose.[1]

Industrialists—Rockefeller, Vanderbilt, Stanford—had been dispensing their vast largess to higher education for years. Through the funding of graduate schools, laboratories, and salaries, their gifts had indirectly provided some opportunities for research. Now, at the turn of the century, Andrew Carnegie had spotlighted a new kind of good work for wealthy Americans, particularly for the newly rich with a yearning for

---

[7] Carnegie to Roosevelt, Dec. 2, 1901, AC, vol. 85. In the original proposal to Roosevelt, Carnegie merely said: "These bonds do not mature for fifty years, when redeemed, [sic] it will rest with Congress to decide what rate of interest will be paid upon the principal." Carnegie to Roosevelt, Nov. 28, 1901, AC, vol. 85.

[8] New York *Herald*, Dec. 15, 1901, p. 6; Chicago *Daily Tribune*, Dec. 15, 1901, p. 8.

[9] *New York Times*, Dec. 19, 1901, p. 2; David Madsen, "Daniel Coit Gilman at the Carnegie Institution of Washington," *History of Education Quarterly*, IX (Summer 1969), 158.

[1] *Carnegie Institution of Washington Yearbook, 1902*, pp. xiii–xiv; Carnegie to Simon Newcomb, Jan. 3, 1902, SN, Box 18; *New York Times*, Jan. 10, 1902, p. 8; St. Louis *Post-Dispatch*, Dec. 10, 1901, p. 6; San Francisco *Chronicle*, Jan. 31, 1902, p. 6.

cultivated respectability. Charles William Eliot noted: "The endowment of research is becoming an attractive object for private benevolence."[2]

At the turn of the century, too, Americans of all sorts, including many scientists, were surging together behind a progressive program of social and economic reform. Amid the spreading respect for the colleges, many progressives wanted the program shaped by the judicious use of disinterested expertise. "There was never a time," the dean of pure science at Columbia remarked, "when talent, energy and enterprise in young men were so much in demand . . . men who can study aright the mighty questions of industrial and social economy now confronting us."[3]

In 1903, as the progressive movement was catching hold, Charles R. Van Hise, a farmer's son and a geologist who had been influenced by John Wesley Powell's reformism as a member of his Survey, assumed the presidency of the University of Wisconsin. In his inaugural, President Van Hise pledged that Wisconsin graduates, schooled to science-minded disinterestedness, would "fight corruption and misrule." Her professors would provide their expertise to the government. But for Van Hise, who was himself a Ph.D., the professor's "greatest service" lay in "his own creative work and the production of new scholars." History surely showed that the advancement of knowledge ultimately added to the wealth and well-being of society. To serve the state properly, Van Hise's university would insist that scholarship and research, both practical and pure, "must be sustained."[4]

In state universities across the country, new presidents, many of them Ph.D.'s, most of them progressives and products of the pro-science educational movement of the late nineteenth century, were taking office and making the promotion of research the order of these days of reform. So were the presidents of the leading private universities, though with a difference. Whatever their political preferences, they were responsible to boards of trustees usually dominated by businessmen rather than to regents appointed by progressive state governments. They stressed scholarship, but usually not so much in tough phrases of reform as in the bland language of moral elevation and public service. Woodrow Wilson, in 1902 a friend of corporations, delivered an entirely neutral call for research when he said in his Princeton inaugural: "Science and scholarship carry the truth forward from generation to generation and give the certain touch of knowledge to the processes of life."[5]

Whatever his praise of science, Wilson, not atypically among the leading private university presidents, was intellectually the heir of James

[2] *Report of the President, Harvard University, 1900/1901*, p. 34.

[3] Robert S. Woodward, "Education and the World's Work of Today," *Science*, 18 (Aug. 17, 1903), 68.

[4] "Inaugural . . . of Charles Richard Van Hise," *Science*, 20 (Aug. 12, 1904), 201, 202, 203, 204.

[5] Woodrow Wilson, "Princeton for the Nation's Service," *Princeton Alumni Weekly*, Nov. 1, 1902, p. 97.

McCosh ("a little archaic," Eliot said of Wilson's educational theories). If Van Hise beamed over the infusion into every subject of the "scientific spirit," Wilson once attacked the noxious atmosphere of the laboratory for making "the legislator confident that he can create, and the philosopher confident that God cannot."[6] While Wisconsin welcomed engineering, Princeton—and Yale—preferred to avoid what Wilson deemed mere vocational endeavors. For Wilson, science was properly a modest element of general culture and only as such a fitting associate of the liberal arts.

Wilson and Van Hise may have differed, yet all the new university presidents departed in highly important ways from the educational reformers of the post-Civil War. They insisted that professors were there to teach students not only how to live right but how to earn a living. They used the phrase "pure science" simply for descriptive purposes, stripped of Rowland's haughty connotations. Instead of advocating pure research because it was impractical, they were saying with President Nicholas Murray Butler of Columbia: "Scholarship has shown the world that knowledge is convertible into comfort, prosperity, and success.[7] In 1900 thirteen major institutions formed the American Association of Universities, to "raise the opinion abroad," as the letter of invitation read, "of our own Doctor's degree."[8] Now new graduate schools were created and old ones injected with vitality. Emphasis went to funding graduate fellowships and higher faculty salaries. The invigoration of graduate work and the increasing availability of fellowships made costly study in Europe less necessary. In the forty largest universities, professors averaged $2,500 a year, just some $300 less than railroad executives and more than five times the wage of coal miners and telephone operators.[9] Like Gilman a quarter century earlier, the new presidents argued that the best teachers were those who did research; unlike their predecessors, they made the Ph.D. an increasingly commonplace requirement for faculty posts, accomplishment in research more of a hurdle for advancement. They also responded to increasingly insistent demands from their faculties, including scientists, for more power over appointments and promotions, a move

[6] Eliot to D. C. Gilman, Jan. 17, 1901, DCG, Eliot file; "Inaugural . . . Van Hise," *Science*, 20 (Aug. 12, 1904), 200; Wilson is quoted in Laurence Russ Veysey, "The Academic Mind of Woodrow Wilson," *Mississippi Valley Historical Review*, 49 (March 1963), 620–21.

[7] Nicholas Murray Butler, "Scholarship and Service," *Educational Review*, 24 (June 1902), 3.

[8] Copy of the letter is in CWE, Box 109, file 133.

[9] See the table of salaries in *Science*, 28 (July 24, 1908), 104–5; Beardsley Ruml and Sidney G. Tickton, *Teaching Salaries Then and Now* (New York: The Fund for the Advancement of Education, Bulletin #1, 1955), pp. 32–33. Noting that professors were probably better paid than "the great mass of their fellow citizens," James McKeen Cattell, the editor of *Science*, pronounced academics "the least privileged members of the privileged classes." *Popular Science Monthly*, 76 (June 1910), 615.

that, the psychologist Joseph Jastrow said, was needed to help assure "men of promise in the oncoming generation the utmost encouragement to rise rapidly in their profession."[1]

In the universities a new day, and in print a new spokesman: As science in the age of Grantism had found an evangel in Youmans, science in the progressive era had its advocate in James McKeen Cattell. A professor of psychology at Columbia University and the well-to-do son of a college president, Cattell bought Bell's defunct *Science* magazine in 1895; in 1900 he moved into the editor's chair of *Popular Science Monthly*. Like Youmans, Cattell believed that a bridge between scientists and laymen was worth retaining. He filled both journals with accolades to science and the social beneficence of its method (Tennyson's were great poems "for their own epoch" but "not like science universal").[2] At times defiantly outspoken, he once editorialized against McKinley's "flabby character" (and provoked Cleveland Abbe to warn that a journal hostile to the administration would "ruin its usefulness as a promoter" of scientific research in the government[3]).

But Cattell differed from Youmans in crucial respects. He may have celebrated the culture of science; before and after 1900, when the AAAS made the resurrected *Science* its official organ, complete with subsidy, his principal loyalty lay with the scientific professions. Cattell veered from Youmans in his recognition that, while talent and character might dispose a young man toward a scientific career, an encouraging environment made his pursuit of it more possible. Unlike Youmans, Cattell heartily approved increasing state and federal spending for research. Along with the university presidents whose addresses appeared in his journals, Cattell was a progressive. Research might stimulate economic development, but Cattell worried not only about the creation but about the distribution of wealth. If disinterested habits of mind helped men think right, he backed governmental efforts to make them act right—and so did many other scientists of the day.

Reformist scientists ᴜke Cattell made their influence felt in local, state, and federal agencies, including the new Bureau of Standards. Officials there knew that petty larcenous shopkeepers were selling short-weight pounds of butter, short-volume pints of milk, short-measure quantities of myriad other items. In 1902 a Bureau staff member reported back from

[1] Jastrow, "The Academic Career as Affected by Administration," *Science*, 23 (April 13, 1906), 567; James McKeen Cattell, *University Control* (New York: The Science Press, 1913), pp. 20, 24; Maurice Caullery, *Universities and the Scientific Life in the United States*, trans. James Haughton Woods and Emmet Russell (Cambridge, Mass., 1922), p. 52; Merle Curti and Vernon Carstensen, *The University of Wisconsin: A History, 1848–1925* (2 vols.; Madison, 1949), II, 49.

[2] *Popular Science Monthly*, 75 (Sept. 1909), 308.

[3] Abbe to Simon Newcomb, Feb. 14, 1898, SN, Box 14, which also quotes the editorial.

a discouraging trip to New York: Most of the state's inspectors of weights and measures were doing a loose job, some no job at all.[4] Then, in 1904, the Bureau discovered that shoddy production procedures, if not deliberate cheating, were costing the U.S. government, which purchased some one million light bulbs a year, a good deal of money. Stratton's investigators tested a shipment and certified three quarters of it inadequate. The Bureau of Standards, a child of economic nationalism, started marshaling its expertise in the service of the reform of consumer standards.

The Bureau made itself a house of materials analysis for the nation's largest consumer, the government. Its scrutinizing activities ran the gamut from an elevator cable for the Washington Monument to paper and ink for the Printing Office. By 1907 it was doing thousands of tests a year, two thirds of them for federal agencies, and impartially rejecting hundreds of shipments. Soon manufacturers were clamoring at the Bureau's door for assistance in standards control. In 1905 the Bureau, which had no regulatory power, called a conference of state weights and measures officials. Local laws, the sparsely attended gathering concluded, were "exceedingly lax . . . with nothing obligatory."[5] Two years later the conference, now an annual affair with seventeen states represented, drafted a model law. Most legislatures ignored it, so Stratton decided to prod with a little muckraking. In 1909, granted a small special appropriation, he sent his men out to investigate the weights and measures situation.

Reporters trailed the sharp-eyed little army as, over a period of two years, its physicists and chemists visited every state and investigated more than three thousand stores and shops. Magazine articles publicized what a contributor summarized for readers of *Cosmopolitan* as the shocking extent of "cheating through false measures, crooked scales and the sale of disguised goods."[6] In 1911 the model law or multiamended variations of it began to sweep through state legislatures (some forty-one states would tighten their weights and measures regulations by 1914). Cities and towns asked the Bureau for help, and Stratton's experts drafted a model municipal ordinance.

The testing for the government and the crusade for consumers imposed heavy demands on the Bureau. Year after year Stratton asked Congress for more physicists, more chemists, more facilities. Year after year he got them. Another laboratory was added to the agency's complex, and by 1910 its staff authorization had multiplied almost tenfold, to 135 men.

The progressive emphasis produced a wide-ranging enlargement of

[4] Louis A. Fischer, "Recent Developments in Weights and Measures in the United States," *Popular Science Monthly*, 84 (April 1914), 345–46.

[5] Quoted in Rexmond G. Cochrane, *Measures for Progress: A History of the National Bureau of Standards* (Washington, D.C.: National Bureau of Standards, U.S. Department of Commerce, 1966), p. 87.

[6] Sloane Gordon, "Is the Housewife Guilty?" *Cosmopolitan*, 50 (Dec. 1910), 75.

federal science. The thrust to conserve and develop the land gave birth to the Reclamation Service under Frederick H. Newell, an irrigation-minded veteran of the Geological Survey. The effort to manage the woods created the Forestry Service under the indomitable Gifford Pinchot. Progressivism's effort to prevent Americans from poisoning themselves found support from the Department of Agriculture's small Bureau of Chemistry, whose chief, Harvey W. Wiley, had been campaigning against adulterated foods for years. Once the Pure Food and Drug Act had passed, Dr. Wiley's apothecary shop needed chemists by the scores to analyze the myriad patent nostrums and products that Americans ate and drank.

In the middle of it all a progressive economist expounded in Cattell's *Science*: "Democracy has not until lately joined itself with the educated classes for the promotion of scholarship, because . . . scholars and scholarship . . . have been the allies of aristocracy." But now, when science for reform was combining with science for economic development, when government was at once aiding industry and regulating its bewilderingly complex products, science was becoming an ally of democratic purpose—and careers in research, as the progressive economist said, were opening "to all who have natural talent and capacity."[7]

[7] David Kinley, "Democracy and Scholarship," *Science*, 28 (Oct. 16, 1908), 501, 508.

VI

Joining the Revolution

At the end of 1895, Wilhelm Conrad Röntgen, a professor of physics at the University of Würzburg in Germany, reported a startling experimental result: Under certain conditions an electrical discharge produced an extraordinary type of invisible radiation—Röntgen dubbed it X rays—which could pass through matter and even render a photograph of the bones encased in human flesh. Accounts of Röntgen's discovery swept through the lay and professional journals of both Europe and America. In the interest of common modesty, a New Jersey state assemblyman introduced a bill "prohibiting the use of X rays in opera glasses."[1] Serious scientists spoke of the promise of X rays for medicine, wondered about the nature of the mysterious radiation—and soon found themselves wondering about a good deal more.

In 1896 Henri Becquerel announced to the French Academy of Sciences that uranium spontaneously emitted invisible radiations which would blacken a photographic plate yet which seemed different from X rays. In 1898 Pierre and Marie Curie detected two new elements that also emitted radiations. Madame Curie named one polonium, after her native country. The Curies called the other radium and christened the phenomenon of spontaneous emissions "radioactivity."

In the year between the work of Becquerel and the Curies, Joseph John Thomson, the Cavendish Professor of Experimental Physics at Cambridge, found that both magnetic and electric fields deflected the

---

[1] Quoted in Otto Glasser, *Wilhelm Conrad Roentgen and the Early History of Roentgen Rays* (Springfield, Ill.: Charles C Thomas, 1934), p. 45.

rays of electricity, or "cathode rays," flowing from the cathode in a partially evacuated tube. The deflection confirmed Thomson's suspicion that these rays consisted of charged particles. He also found that the ratio of the charge on the particles to their mass was always the same, regardless of the gas in the tube. Thomson's bold conclusion: The charged particles in cathode rays were "matter in a new state," and from this state of matter "all the chemical elements are built up."[2] By 1900 evidence from investigations in quite different fields, including radio-activity and spectral analysis, had convinced most physicists that Thomson's charged particles, or "electrons," to use their eventual name, were indeed fundamental building blocks of the atom.

At the opening of the twentieth century, physics was suddenly alive with new and revolutionary questions. What was the nature of X rays? How account for radioactivity? How reconcile the apparent endlessness of its emanations—radioactive processes yielded far more energy than any known chemical reaction—with the conservation of energy? How incorporate the electron into a general theory of electricity and magnetism? And how was the electronic atom constructed?

Amid the turmoil and excitement, an American observer assessed the outlook for young men in physics: The recent advances had "vastly multiplied the opportunities for new discoveries."[3]

For physicists in the United States, the professional high spot of the 1890s had been the founding of *The Physical Review*. Edward L. Nichols of Cornell, who had inaugurated the journal in 1893 with a subsidy from the university's trustees, hoped to make it a worthy forum "with the support of the younger physicists."[4] Amid the hard times, so few young men joined the profession that by the turn of the century it was not much larger than in the late 1880s. Cornell physicists published more than a quarter of the articles in the *Review*. The usual fact-gathering type of research filled most of its pages. The year of Thomson's new matter a knowledgeable scientist sighed of American physics: "Not a great deal of first class work [is] done."[5]

America's "painfully small" contributions to the science annoyed Professor Arthur Gordon Webster of Clark University, who was a Harvard graduate and a Helmholtz Ph.D.[6] His isolation at marauded Clark evidently bothered him. He could obtain no inspiration from the get-togethers of the National Academy, to which, in his somewhat jealous

[2] Joseph John Thomson, "The Cathode Rays," *Philosophical Magazine*, 5th series, 44 (October 1897), 312.

[3] Henry Crew, "Outlook for Young Men in Physics," *Science*, 27 (June 5, 1908), 879.

[4] Nichols to Carl Barus, Feb. 16, 1893, CB.

[5] Edward S. Dana to Henry Rowland, Dec. 2, 1897, HAR.

[6] Webster, Presidential Address to the American Physical Society, *Physical Review*, 18 (April 1904), 306.

opinion, "few of us can hope to belong, and which we might not enjoy if we did." Webster decided to create a nonexclusive forum where scientists like himself could hear papers by the leading members of the profession. Early in 1899 he started organizing an American Physical Society.[7]

Rowland, Michelson, Webster, Nichols, Abbe—the only representative of the government—and thirty-three others assembled to launch the Society at Columbia University (Gibbs, still chary of organizations, declined to participate). The constitution they wrote opened membership to virtually anyone with a reasonable interest in physics and, to alleviate any geographical obstacles, provided for local sections. As though to symbolize their intent, the founders elected as their first leader the man who had pleaded for pure science. Rowland was now like a patriarch of American physics, awed at the coming age of the atom but, mortally ill with diabetes, destined never to see it. In his presidential address some months later, he blessed the new society with a final testament: "Much of the intellect of our country is still wasted in the pursuit of so-called practical science. . . . But your presence here gives evidence that such a condition is not to last forever."[8]

A few years more and the new day in science was giving Rowland's prediction the sanction of new institutional possibilities. In Washington Samuel Wesley Stratton realized that standards for light bulbs ramified into electromagnetism; for pyrometry, into infrared radiation; for chemical compounds, into spectral analysis. Achieving reliable measures for the diverse products of technology marched the Bureau of Standards into the major fields of physics and, as Stratton remarked, necessarily included "pure research."[9] At the same time, the electrification of homes and factories was accelerating, and electrical engineers were in enormous demand. Since the Steinmetzes now dominated the technical arena, four years of training on the campus had become a professional necessity. Electrical engineering enrollments multiplied yearly. Industry-bound students spilled into physics courses, and departmental chairmen found themselves confronted with overcrowded laboratories and overtaxed staffs.

The chairmen were a breed of builders more administratively inclined than their nineteenth-century predecessors. All were Ph.D.'s and eager to promote research. They realized that the need for more staff and more facilities for undergraduates promised dividends for investigations in physics. At Princeton two University of Berlin Ph.D.'s, William F. Magie, the chairman of physics, and Henry B. Fine, the head of mathe-

---

[7] Webster to Carl Barus, May 1, 1899, CB. Webster was not elected to the Academy until 1903.

[8] Rowland, "The Highest Aim of the Physicist," *Science*, 10 (Dec. 8, 1899), 826.

[9] "Report of the Director of the Bureau of Standards," in *Report of the Department of Commerce and Labor, 1908*, p. 677.

matics and President Wilson's trusted dean of the faculty, burned to put McCosh's college on the scientific map. Both men knew Wilson from their undergraduate days, and they persuaded him that physics qualified as a humane study and as a gentlemanly way of preparing for electrical engineering.

In 1905 Fine and Magie brought to Princeton James Jeans, one of Britain's leading young mathematical physicists; a year later, Owen W. Richardson, one of J. J. Thomson's most promising students. Richardson arrived to find the space allotted for his research a shocking "kind of dark basement ventilated by a hole in the wall . . . and inhabited by an impressive colony of hoptoads."[1] The university erected a well-equipped laboratory with $200,000 that Fine managed to get from two alumni. Soon Richardson was pursuing the research on electron emission that would win him the Nobel Prize. "It is a remarkable and interesting fact," Cattell noted, "that Princeton is becoming . . . a great scientific center almost in spite of those in control."[2]

Elsewhere, at Wisconsin, California, Michigan, and Illinois, at Pennsylvania, Harvard, Columbia, and Cornell, physics department chairmen won appropriate dispensations for new laboratories as well as for more professors, and the institutional enrichment combined with the new questions in physics to create major opportunities in the profession. The quality of high-school physics teaching had improved considerably. And with the increasing engineering enrollments, a growing number of students not only took physics courses but became aware of the burgeoning career possibilities in the discipline. Through the 1890s American universities had graduated a total of fifty-four physics Ph.D.'s; they turned out twenty-five in 1909 alone. That year membership in the American Physical Society reached 495, a fivefold increase since 1901. The profession fully matched the swift pace at which all of science in the United States was growing (between 1900 and 1910 the rolls of the AAAS jumped by some 400 percent). A physicist at the Carnegie Institution marveled at the "wonderful spirit of research that has literally seized us in this country."[3]

Many of the new physicists, like their peers in other fields, came from the homes of small farmers and businessmen. Products of midwestern towns instead of the urban East, a sizable percentage did their undergraduate work at state universities, technical schools, and colleges rather than at institutions like Harvard and Yale. Most, avoiding the expense of

---

[1] Quoted in Alan Shenstone, "Princeton and Physics," *Princeton Alumni Weekly*, Feb. 24, 1961, p. 6.

[2] *Popular Science Monthly*, 76 (June 1910), 519.

[3] Daniel J. Kevles, "The Study of Physics in America, 1865–1916" (unpublished Ph.D. dissertation, History, Princeton University, 1964), Appendices I, III, IV; Lewis P. Bauer, "A Plea for Terrestrial and Cosmical Physics," *Science*, 29 (April 9, 1909), 569.

graduate work abroad, took their Ph.D.'s in the United States.[4] The upper-class America of the Rowlands continued to contribute to the ranks of the profession, which remained overwhelmingly white, Anglo-Saxon, and Protestant. But now, amid its wealth of opportunity, physics was drawing at an increasing rate from the middle class as well.

Some of the new physicists joined the Bureau of Standards; most entered the academic world. They usually lacked the independent means of their elders, and, while they earned comfortable salaries, they faced some difficulty in maintaining the standard of living expected of men who moved in cultivated circles. (The president and trustees want us to mix in town society, the professor's wife remarked. "Who can afford the evening dress to go?"[5]) The new physicists were aware that the key to rank and salary was increasingly research, and, along with their colleagues at the Bureau of Standards, they produced. The number of papers delivered before meetings of the American Physical Society rose steadily after the turn of the century; so did the number of articles published in *The Physical Review*.

As though to symbolize the vitality of American physics, in 1907 Michelson won his Nobel Prize. His colleagues cheered, and Cattell announced that Rowland's plea for pure science had by now been handsomely fulfilled. In a way it had, because so much more support was available for pure research. More important, by reason of the disparate yet reinforcing trends within both itself and higher education, the physics profession was proceeding toward the institutional fulfillment of Rowland's best-science program. Centers of graduate training were emerging; the great majority of young physicists took their Ph.D.'s at only five schools—Hopkins, Cornell, Yale, Harvard, and Chicago.[6] The establishment of *The Physical Review* and the American Physical Society also gave the profession institutional instruments necessary to set standards of quality in research on a national scale. But despite the increasing thickness of the *Review*, despite the extent to which Rowland's best-science program was being fulfilled, American physics still operated deep in Europe's shadow.

If the profession had acquired some of the institutions essential to Rowland's best-science program, to set high standards it also needed an appropriate best-science elite. As in Rowland's day, it had a productive elite—one eighth of the practitioners published more than half of the research—but most members of this productive elite still did not merit inclusion among the world's pace-setting best scientists. The advances of

[4] Kevles, "The Study of Physics in America," Appendix VI.

[5] Anon., "A College Professor's Wife," *The Independent*, 59 (Nov. 30, 1905), 1282.

[6] These data are based on an analysis of the entries in M. Lois Marckworth, comp., *Dissertations in Physics: An Indexed Bibliography of All Doctoral Theses Accepted by American Universities, 1861–1959* (Stanford, 1961).

X rays, radioactivity, and the electron had been made entirely without their help. In the decade following these discoveries, they sponsored few American doctoral dissertations along the lines of inquiry opened by Röntgen, Becquerel, the Curies, or Thomson, and four out of five of the articles they published in the *Review* concerned pre-X-ray physics.[7]

Still, in this respect, American physicists hardly differed from Europeans, who in the post-1900 decade also did a small fraction of their research in X rays, radioactivity, and electrons.[8] The trouble with the work of practitioners in the United States was less in its subject than in its quality; old or new physics, a good deal of it remained in the fact-gathering tradition. Compounding that weakness, the productive physicists tended to play a prominent role in the institutions of their discipline. Their influence, combined with the special circumstances of the institutions themselves, made for a prevalence of low professional standards in the graduate schools, in the new professional organizations of physics, and in both the old and new national agencies of science.

At each of the five most heavily attended graduate schools, doctoral work brought some students into significant types of research. Johns Hopkins, which produced almost 30 percent of the physics Ph.D.'s in this period, was notably fecund in its output of spectroscopists. At Cornell, which graduated almost 20 percent of the Ph.D.'s, Edward L. Nichols trained a number of young physicists in fluorescence and other important aspects of radiation; at Yale, a few students did their doctoral research under Bertram Boltwood, one of a handful of American experts on radioactivity. But at Chicago, Michelson paid almost no attention to graduate students, and Harvard had no one comparable to Rowland, Nichols, or Boltwood. Neither did Hopkins after Rowland's death in 1901, which marked the beginning of a precipitous decline in the quality of the department. On the whole, in the early twentieth century the leading graduate schools tended to foster a good deal of pedestrian doctoral research, and Carl Barus pointed out in 1903 that "Most of our physicists are . . . young men still under the influence of their thesis work or of ideas derived from somebody else."[9]

If there were graduate centers emerging, there were no centers of productive faculty research. Institutionally, the productive elite of physicists was at least as widely scattered as in Rowland's day, which meant that its members still tended to lack the local stimulation of able colleagues. Nationally, Nichols and his editorial colleagues at Cornell con-

---

[7] *Ibid.*, and an analysis of articles in the *Physical Review*; Daniel J. Kevles and Carolyn Harding, "The Physics, Mathematics, and Chemical Communities in the United States, 1870 to 1915: A Statistical Survey," California Institute of Technology, Social Science Working Paper No. 136, March 1977, Tables 2, 4.

[8] Tetu Hirosige, "Rise and Fall of Various Fields of Physics at the Turn of the Century," *Japanese Studies in the History of Science*, 7 (1968), 93–113.

[9] Statistical data based on Marckworth, comp., *Dissertations in Physics*; Barus to James McKeen Cattell, June 7, 1903, JCM, Box 3.

ducted *The Physical Review* with a seeming Baconian eagerness to publish the report of any experiment, and a disproportionately large number of articles in the *Review* continued to be published by Cornell Ph.D.'s and faculty.[1] Lacking prestige, the *Review* lost much of the good work of the younger men to foreign journals or, in the case of spectral studies, to the newly founded *Astrophysical Journal* in the United States. Many of the papers published in the *Review* were first presented before the American Physical Society, whose gatherings attracted fewer than fifty physicists early in the century. Virtually any paper could get a place on the Society's program, even after participation in the meetings increased and the secretary no longer had to scramble to find enough papers to make a program of decent size.[2]

The National Academy, whose membership was now overwhelmingly academic, wielded even less influence on physics than in Powell's day. With bureaus of physics and chemistry proliferating in Washington, the federal government hardly ever awarded the Academy the prestige of an advisory request. Legacies had enlarged its endowment by about a third; the total income added minuscule leverage to its featherweight authority. The Academy's annual get-togethers were still sparsely attended, but when Michelson proposed the suspension of anyone absent for five consecutive years, he was almost hooted out of the meeting.[3] In 1906, at the urging of a few reformers, the Academy doubled the limit on annual elections from five to ten. Despite this response to the swift expansion of the scientific community, on the whole it failed to adapt itself to the quickening tempo of American science.

The National Bureau of Standards, for all its vigor, money, and authority, was in no position to do for American physics what Powell and his Survey had done for American geology. Where Powell had enjoyed a free hand with his appropriations, Congress bound all of Stratton's expenditures to relatively specific purposes. Stratton hired no consultants. He did put graduate students on the payroll—a few universities agreed to accept research done in Washington for the doctorate—and he often hosted meetings of the American Physical Society. But while Stratton encouraged pure research, routine testing still occupied the bulk of the Bureau's time. Although it issued numerous publications, articles on the intricacies of light bulb filaments or the specific gravity of cement could hardly inspire the profession's young men, and the establishment of technical standards usually drew upon the old physics, not the new.

Enormous potential for shaping the course of contemporary Amer-

[1] About 35 percent of the articles were by Cornell Ph.D.'s and faculty, who accounted for 21 percent of all authors. Kevles and Harding, "The Physics, Mathematics, and Chemical Communities," Tables 12 and 13.

[2] *Ibid.*, Table 9 and Appendix II; Ernest Merritt, "Early Days of the Physical Society," *Review of Scientific Instruments*, 5 (April 1934), 146–47.

[3] Michelson to T. C. Mendenhall, Oct. 28, 1909, TCM, Box 6.

ican physics rested in the Carnegie Institution, whose endowment was ninety times as great as the National Academy's. After its creation, the trustees asked scientists in various disciplines to advise how this considerable resource might best be used for the advancement of knowledge in the United States. Like Carl Barus, the advisory physicists urged the Institution to aid the exceptional man by distributing the lion's share of its income through fellowships and grants primarily to exceptional academics. But a quite different view was proposed by a scientific member of the board of trustees, Charles D. Walcott. No Ph.D., Walcott had learned to be a scientist through an apprenticeship on the staff of the U.S. Geological Survey, where he eventually won both distinction as a paleontologist and Powell's confidence as a right-hand man. Walcott succeeded Powell in the directorship and rehabilitated the Survey by reemphasizing mining studies and staying clear of controversy. If Walcott had none of Powell's reformist zeal, he fully shared his commitment to the cooperative way of scientific research. As a Carnegie trustee, Walcott, arguing that science needed "strong effective direction from a central office," specifically proposed that the Institution emphasize the funding of large, well-organized research projects.[4]

To sponsor individual scientists or large, cooperative projects—the choice before the Carnegie trustees signaled the emergence of a major issue of research policy in American science. Would it be wise and appropriate to extend the large-project approach in a substantial way from federal into the rest of American science, where the bulk of pure research was conducted? The advantages of large, cooperative projects included the efficient use of resources, the concentration of energies, and the likely achievement of well-defined goals. Among the disadvantages was the risk of less intellectual adventurousness, less freedom for the cooperating scientist. Old Alexander Agassiz warned that a Carnegie commitment to cooperative research would amount to another dangerous step in the direction of centralized scientific dictatorship. Grants for small projects to exceptional men allowed the individual scientist his freedom, including the freedom to fail. Still, as Theodore Richards, the Harvard chemist who would win the nation's first Nobel Prize in that discipline, put the chief justification for the small-grant approach: "The making of a great original discovery is not unlike the writing of a great poem or the painting of a great picture. The thought and its execution must be hammered out by genius alone."[5]

[4] Barus, "Autobiography," p. 150. CB. Walcott is quoted in Howard S. Miller, "A Bounty for Research: The Philanthropic Support of Scientific Investigation in America" (unpublished Ph.D. dissertation, History, University of Wisconsin, 1964), p. 273.

[5] Agassiz to Simon Newcomb, Oct. 14, 1903, Oct. 22, 1903, SN, Box 14. Richards is quoted in Nathan Reingold, "National Science Policy in a Private Foundation: The Carnegie Institution of Washington, 1903–1920" (unpublished manuscript, 1975), p. 10.

But Walcott had the ear of his fellow trustees, many of whom, drawn from the world of big business, no doubt felt at home with his organizational approach to research. The Carnegie Institution would make some small grants; large projects would prevail. Walcott himself may have written the statement of policy: The Institution would aim, wherever appropriate, to substitute "organized for unorganized effort." "The most effective way to discover and develop the exceptional man is to put promising men . . . under proper guidance and supervision."[6] Whoever wrote the statement, Daniel Coit Gilman, the first president of the Institution, accepted the policy; Robert S. Woodward, who succeeded him in 1904, cemented it into practice.

A veteran of a decade in the Coast and Geological surveys, Woodward, like Walcott, had been trained in federal science. After a dozen years as dean of pure science at Columbia University, he also was accustomed to coping with the expanding demands of the academic community. Woodward preferred the large-project approach not least because, he kept saying, the dispensation of small grants was an unrewarding administrative burden. The Carnegie Institution's charter might stipulate aid to educational activities; in Woodward's opinion, giving substantial money to the academic world would merely relieve universities of the duty to fund their faculties' research.[7] By 1910, the Institution was pouring about six times the money going to small grants into some ten large projects, all of them under its direct control.

Three of these projects related to physics. There was the Solar Observatory located atop Mount Wilson in Pasadena, California. It had one of the world's largest reflecting telescopes and a first-rate staff managed by an able director. The operation was administratively efficient, its work distinguished—and also pleasing to the founder, who in 1911 enlarged the endowment with another $10,000,000 in bonds of U.S. Steel. There were also the Geophysical Laboratory and the Department of Terrestrial Magnetism. At the Observatory the staff organized around the telescope; in the other two divisions, around the subject matter (problems in terrestrial magnetism and geophysics could be cut up as easily as the map of the earth). But experiments with electrons, X rays, or radioactivity did not lend themselves to large-scale cooperative attacks. Mount Wilson might obliquely contribute to the study of atomic structure; none of the three divisions dealt with the challenge directly. The Carnegie Institution established enviable models of quality for Americans in astrophysics, geophysics, and geology. On the whole, its large-project activities gave young men in the new physics nothing to emulate.

In principle, the Institution could have set standards in the new physics through the discriminating allocation of its small grants. But these

[6] *Carnegie Institution of Washington Yearbook, 1902*, p. xxxviii.
[7] *Carnegie . . . Yearbook, 1905*, pp. 28–30.

dispensations were spread across the entire spectrum of science, and physics won few of the awards. Moreover, Woodward made a point of administering the program so as to deny funds to "the amateur, the dilettante, and the tyro." If the Institution aimed at helping the exceptional man, he must, in Woodward's interpretation, have proved himself exceptional before he merited consideration.[8] By awarding its largess only to established men, the Institution in the main ignored its opportunity to shape the work of the young physicists, who were the promise of the profession.

Like Joseph Lovering some thirty years before, Arthur Gordon Webster argued that Americans had unearthed "few capital discoveries in physics" because of their generally "insufficient equipment . . . in mathematics."[9] But higher mathematics had been unnecessary for the detection of X rays, radioactivity, and the electron, and dissenters from Webster's view could always point to unconventional Robert W. Wood of Johns Hopkins, who had been born of his wealthy father's middle age, raised amid the iconoclastic circles of Emerson's Concord, admitted to Harvard after twice flunking out of the Roxbury Latin School, and indulged through five years of party life while doing graduate work at home and abroad. Wood never bothered to apply for his Ph.D., but his research was prolific—and usually to the point. He could pry reams of data significant for atomic theory from the simplest spectral apparatus. An intuitive master of optical technique—"his fingers knew what to do," a friend once remarked—he accomplished it all with something like the mathematical competence of an Edison.[1]

All the same, American physics did continue to lack theorists—no one had replaced Gibbs, dead in 1903—and the deficiency was more crippling because theoretical issues were occupying an increasingly central position in the discourse of the discipline. The issues were rooted in the existence of fundamental dissymmetries between the mechanics of Newton and the electrodynamics of Maxwell, between the physics of discrete particles on the one hand and the continuities of the ether on the other. The issues were made evident partly by the failure to detect any motion of the earth through the ether. They found additional expression in the difficulty of accounting theoretically for certain kinds of interactions between discrete matter and continuous radiation. By the turn of the century, largely as the result of the profound attempts of the Dutch physicist Hendrik A. Lorentz to deal with these problems, many young physicists on the Continent, especially in Germany, were rejecting mechanical explanations as the ultimate program for their science. Instead,

[8] *Carnegie . . . Yearbook, 1906*, p. 35; *Carnegie . . . Yearbook, 1914*, p. 17.
[9] Webster, Presidential Address, *Physical Review*, 18 (April 1904), 306–7.
[1] Interview with the late James Franck.

they proposed to construct a universal physics by reducing all phenomena, including mechanical ones, to an electromagnetic basis.[2]

Not so Max Planck or Albert Einstein—Planck, in his forties, a professor of theoretical physics at Berlin, his manner stiff, his dress somber, his reverence for church and Kaiser strong; Einstein, in his early twenties, an unknown employee of the patent office in Berne, Switzerland, carelessly clothed, dreamy, something of the cafe bohemian, a Jew by birth who had made the universe his church and its laws his theology. Yet the two men shared a sensitivity to the prevailing difficulties in physical theory, a recognition of their profundity, and a deep faith in the dependability of thermodynamics—Einstein respectfully called it a "theory of principle"—as a guide to understanding.[3] Einstein was especially bothered by the dissymmetry between the physics of Newton and of Maxwell, and he insisted, with aesthetic conviction, that a fundamental harmony must underlie the two systems.

At the turn of the century, Planck drew upon the statistical mechanical form of thermodynamics to deal with the interaction of radiation and matter in a so-called black body, a closed cavity in thermal equilibrium. The energy in the radiation was experimentally known to be distributed unevenly across the spectral frequencies, but the distribution depended in no way upon the material of the body, only upon its temperature. Planck managed to account for this distribution by constructing an *ad hoc* mathematical formula. He then penetrated to the meaning of the formula itself by violating a cardinal point of theoretical physics, that radiators emitted energy in a continuous, infinitely divisible fashion. In Planck's interpretation, the radiators in a black body emitted energy in discrete, indivisible units mathematically equal to the frequency of the radiation times a constant. Planck dubbed the constant h and called it the "quantum of action."

In 1905 Einstein, one of the few physicists aware of Planck's bold quantum theory, used statistical mechanics to analyze not the radiators in a black body but the electromagnetic radiation. He calculated the entropy of the radiation and emerged with an expression equivalent to the entropy for an ideal gas. Einstein concluded that the radiation behaved as though, like a gas, it consisted of discrete particles: in his phrase, "independent energy quanta."[4] Einstein's interpretation of elec-

[2] Russell McCormmach, "H. A. Lorentz and the Electromagnetic View of Nature," *Isis*, 61 (1970), 459–63, 472–77, 483–91, 495–96. In 1910, about one in five physics papers published in the world was mainly theoretical; in 1930, more than two in five. George Magyar, "Typology of Research in Physics," *Social Studies of Science*, 5 (1975), 81.

[3] Quoted in Martin J. Klein, "Thermodynamics in Einstein's Thought," *Science*, 157 (Aug. 4, 1967), 510.

[4] Einstein, "Concerning a Heuristic Point of View about the Creation and Transformation of Light," reprinted in Henry A. Boorse and Lloyd Motz, eds., *The World of the Atom* (2 vols.; New York, 1966), I, 553.

tromagnetic radiation as a stream of particles flatly challenged the wave theory of light, which rested on a century of accumulated experimental evidence. But Einstein showed that the quantum theory of radiation could explain, among other things, the photoelectric effect, the phenomenon that cathodes exposed to ultraviolet rays emitted electrons in a way that the wave theory of light failed to predict.

In the same year, 1905, Einstein addressed himself to the fundamental theoretical difficulties in mechanics, electrodynamics, and the ether and produced the special theory of relativity. Einstein had come to believe that the assumption of a stationary ether implied the existence of an absolute frame of reference, a unique benchmark in space with respect to which all motion in the universe could be measured. Einstein had concluded that no such preferred reference frame could exist; the failure to detect the motion of the earth in the ether simply confirmed that fact. Since there was no absolute reference frame, Einstein made it his task to relate phenomena occurring in one frame to the way they appeared in a second, which was moving with respect to the first at a constant velocity. He postulated that if Newton's laws held good in the two frames, so must the laws of electrodynamics. He also postulated that the speed of light was constant no matter in what frame of reference it was measured.

Building upon these two postulates, Einstein was led to results which he himself considered peculiar and remarkable, including the relativity of length; the increase of mass with velocity; the impossibility of material bodies achieving the speed of light; and the equivalence of mass and energy. Most striking was the relativity of time, which, as Planck remarked, made the "most serious demands upon the [physicist's] capacity of abstraction."[5] In the opening of the paper Einstein announced that, as a result of the principle of relativity, any reliance upon "a 'luminiferous ether' will prove to be superfluous."[6] And so it did. By denying the possibility of an absolute frame of reference, Einstein had eliminated the ether from physics.

Over the next few years the theory of relativity provoked little interest in France, a good deal of debate in Germany, and emphatic opposition in England, where physicists remained tenaciously attached to the ether. J. J. Thomson typically insisted: "The ether is not a fantastic creation of the speculative philosopher; it is as essential to us as the air we breathe."[7] Knowledgeable British physicists summarily dismissed the quantum theory as an absurdity, in part because the pure thermodynamic

[5] Max Planck, *Eight Lectures on Theoretical Physics*, trans. A. P. Wills (New York, 1915), p. 120.

[6] Einstein, "On the Electrodynamics of Moving Bodies," in *The Principle of Relativity: A Collection of Original Memoirs on the Special and General Theory of Relativity*, trans. W. Perrett and G. B. Jeffrey (New York, 1923), p. 38.

[7] Thomson, "Address of the President of the British Association for the Advancement of Science," *Science*, 30 (Aug. 27, 1909), 267.

reasoning of Planck and Einstein was not of a kind to commend itself to the model-prone school of English physics, in part because it contradicted the well-established wave theory of light. The contradiction also disturbed the Continental physicists—they were few in number and mostly German—who took quanta seriously. Planck himself shied away from accepting Einstein's idea of light and, as late as 1914, when proposing the theorist of relativity for membership in the Royal Prussian Academy of Sciences, he urged that Einstein be excused for his wild extension of the quantum hypothesis.[8] Einstein repeatedly tried to resolve the contradiction between the wave and quantum theories of light. Repeatedly, he failed.

In the United States, few physicists knew about quanta, and, for much the same reasons as their English brethren, those who did dismissed the idea out of hand. Relativity won more attention but not too many adherents, certainly not among the older generation, most of whom remained emphatically loyal to the ether. At the turn of the century the older men had done little, if anything, to sensitize their graduate students to the theoretical difficulties that so troubled European physicists. In the first decade they raised few questions about relativity, quanta, or, evidently, even the program of reducing all physical phenomena to electrodynamics.

The editors of *The Physical Review* made no special effort to call attention to the physics of Einstein and Planck. Neither did the officers of the American Physical Society. The Society was democratic in form and included physicists of all ages, but it operated hierarchically in practice. In the usual way of scientific associations, it had a ruling Council which nominated candidates for both its own at-large and the Society's executive posts. In science, elders of standing could usually count on the deference of the profession, and the membership apparently always rubber-stamped the Council's choices. As a result, the Society, like the profession on the whole, operated under an oligarchy of eminence. Its eminence won in the nineteenth century, the current oligarchy was unequipped by training, intellectual conviction, or interest to stimulate young physicists into confronting the increasingly critical theoretical issues of the twentieth.

Still, the oligarchy had no more power nationally than the power of example. Locally, in the universities, its members were only a small fraction of the growing physics faculties. Unlike some of their counterparts in Europe, especially Germany, they did not control all the means for research in their departments; in the United States equal access to those means was generally considered, and usually allowed, as the right of every

[8] Barbara Cline, *The Questioners: Physicists and the Quantum Theory* (New York, 1965), p. 120. See also Russell McCormmach, "Henri Poincaré and the Quantum Theory," *Isis*, 58 (Spring 1967), 52–53.

professor.[9] Withal, the younger American physicists enjoyed reasonable opportunities to pursue their choice of research topics. They could easily learn about the new physics from European journals, which were now readily available. Some sensed that, in the wake of X rays, radioactivity, and electrons, not to mention quanta and relativity, physics was marching off in sharply different, exciting new directions. Some, including Robert A. Millikan, also recognized the garlands to be won in contributing to the new departure.

In Maquoketa, Iowa, the Reverend and Mrs. Millikan raised their son to fear God, respect hard work, and improve his mind. By the end of his sophomore year at Oberlin, science had done nothing to hone his amiable intelligence. But he had shone in Greek, and the professor asked him to teach the physics course in the college's preparatory department. Millikan protested ignorance. The professor replied: "Anyone who can do well in my Greek can teach physics."[1] In need of money, he took the job. Millikan plunged into the subject, liked it, and decided to make it his career.

In 1893 Millikan arrived at Columbia, a self-taught M.A. Two years later he graduated into the depression, an unemployed Ph.D. A friendly professor urged the stopgap of advanced study in Berlin and obligingly lent him $300 at 7 percent to make the trip. The news of X rays and radioactivity broke during Millikan's *Wanderjahr*. In Germany he heard a good deal of debate over whether cathode rays were particles or processes in the ether. In 1896 he returned home to take a job at the University of Chicago, where he had spent a summer working under Michelson. In Chicago Millikan met Greta Blanchard, whose father, a successful manufacturer and elder in his church, considered his daughter's suitor "somewhat hazardous," as Millikan remarked, "because . . . I was not a man of property and had little prospect of ever being such."[2] Millikan successfully bargained for Mr. Blanchard's approval of the match at such time as he could support Greta properly. Enormously energetic—six hours' sleep was enough, and a round of golf at dawn a tonic—he developed the teaching side of the department, turned out a complement of first-rate physics texts, and soon earned enough for father Blanchard to bless the marriage. Largely on the basis of his pedagogical accomplishments, Millikan rose to an associate professorship, but in 1906, with two children and a new mortgage, he decided that he had to do a good deal better.

Knowing that at Chicago the major rewards went for scholarship,

[9] Joseph Ben-David, "The Universities and the Growth of Science in Germany and the United States," *Minerva*, 7 (Autumn 1968), 6, 8.

[1] Robert A. Millikan, *The Autobiography of Robert A. Millikan* (New York, 1950), pp. 14–15.

[2] Millikan to William Rainey Harper, April 12, 1898, UCH, Millikan file.

Millikan had already spent a good deal of time in the laboratory; most of his work had led to a dead end. But he read the journals habitually and had been stimulated by attendance in 1900 at a conference of physicists in Paris to a sharp awareness of the significant controversies swirling through his science. Now determined to concentrate on research, Millikan started probing after one of the most elusive and critically important quantities of the new physics: the charge of the electron, the fundamental atomic building block found by J. J. Thomson.

By now the ratio of the charge of the electron to its mass was both reliably known and one of the fundamental physical constants. But various ongoing attempts to measure the charge by itself were yielding only approximately consistent results. Aside from the intrinsic need for accuracy, the determination of the charge ramified directly into the theory of radiation as well as general atomic physics. Some physicists argued that, because all electrons displayed the same ratio of charge to mass, it did not follow that each necessarily had the same charge. None of the experiments done so far to measure the charge by itself had proved indisputably that electrons were all identical particles.

Millikan launched his search for the precise charge of the electron with a technique first developed in Thomson's laboratory. By measuring, first, the rate at which a charged cloud of water vapor fell under the influence of gravity, then, the modified rate under the counterforce of an electric field, one could in theory evaluate the charge. But, while proceeding with this approach, Millikan unintentionally applied too much electrical power to the field and found the cloud replaced by a few water drops. He concluded that each must carry just enough charge for the electric field to hold it in suspension against gravity. Moreover, he found that, in contrast to the diffuse and volatile cloud, he could measure the rate at which each droplet fell in the absence of the field with high precision. Millikan spent the spring of 1909 squinting at water drops in Chicago's Ryerson Laboratory. The measured charge almost always came out to a whole number—2, 3, or 4, etc.—times the same irreducible figure. In Millikan's conclusion, the figure was the value of the charge on a single electron, to an estimated accuracy of 2 percent.[3] Late in August, he went up to Winnipeg, Canada, to report his findings to a meeting of decidedly interested British scientists.

On the train home an epiphany struck: a way to infuse the measurement with greater precision. The water drops themselves inevitably evaporated, which limited the time of observation and, in turn, the accuracy of the data. Why not use a much less volatile liquid—oil? By the end of January 1910, Millikan was successfully balancing oil drops

[3] Millikan, "A New Modification of the Cloud Method of Determining the Electric Charge and the Most Probable Value of That Charge," *Philosophical Magazine*, 19 (Feb. 1910), 216–18, 227.

and confidently, though overoptimistically, predicting results accurate to about one-quarter of 1 percent.[4]

That year he reported the experiment in the scientific journals of America and Europe. He had been able to catch oil drops at will, toy with them for as long as four-and-a-half hours, see them jump at the touch of a stray charge from the air. The feat was tantamount to manipulating individual electrons. As with the water drops, the charge on the oil drops almost always measured out to an exact multiple of the same irreducible quantity.[5] Millikan had not only measured the electronic charge with considerable accuracy; he had also proved conclusively that electrons were unique, identical particles. Accolades came from physicists on both sides of the Atlantic—and from the University of Chicago, a full professorship.

By 1910 a small but growing number of the younger American physicists, combining ambition with intellectual excitement, were turning to the problems of the new physics. They studied the subjects of Röntgen, Becquerel, and Thomson. Some started teaching courses, as an MIT catalogue read, in the "Constitution of Matter in the Light of Recent Discovery" and guiding their graduate students into doctoral research on radioactivity, X rays, and especially electrons.[6] In 1908, at the Christmas meetings of the American Physical Society, Gilbert N. Lewis, a thirty-three-year-old professor of physical chemistry at MIT, and his student Richard C. Tolman, delivered an authoritative paper on the theory of relativity and opened major debate on the subject in the United States. Their paper soon appeared in *The Physical Review*, which was now publishing an increasing percentage of articles in the new physics.

The physics of the younger men was not only the new but a higher-quality physics. The shift away from mere fact-gathering occurred in part because the new physics was decidedly rich in significant problems of research. It also occurred because, with the substantial increase in the number of practitioners, American physics included, if only by statistical chance, a larger number of capable people. In the United States of the early twentieth century, in short, physics profited from a fortunate simultaneity between the advent of a new intellectual era in the discipline at large and the opening of its American branch to the talents. Now, given the greater opportunities for research in the universities, the younger, capable practitioners were winning reputations for themselves and nudging their profession into the revolution of modern physics.

[4] Millikan to Carl Barus, Jan. 31, 1910, CB-AIP.

[5] Millikan, "The Isolation of an Ion, a Precision Measurement of Its Charge, and the Correction of Stokes's Law," *Science*, 32 (Sept. 30, 1910), 436–38.

[6] *Massachusetts Institute of Technology Catalogue, 1907*, p. 251. Between 1896 and 1905 the number of physics doctoral theses on subjects related to electrons was five; between 1906 and 1910, fourteen. Data from Marckworth, comp., *Dissertations in Physics*.

VII

A Need for
New Patrons

The pace of the new physics was accelerating, so quickly that professors, John Trowbridge of Harvard marveled in 1909, had to rewrite their lectures on "energy and radiation from month to month."[1] A few more years and Niels Bohr, of Copenhagen, Denmark, compelled the professors to scrap even their most recent treatments of the atom.

Bohr spoke in ponderous sentences but commanded a formidable theoretical talent. His insight was acute, his sense of physical processes almost tactile. When in 1911 he arrived in England for a year of postdoctoral study, he already had the habits of probing tenaciously for the core of theoretical conundrums, of advancing ingenious ideas with quiet modesty and implacable insistence. After a stint at Cambridge with Thomson, Bohr went to the University of Manchester and the laboratory of Ernest Rutherford, who had won the Nobel Prize for his elucidation of radioactive transformations. There, near the end of his postdoctoral year, Bohr turned to the problem of atomic structure. He doggedly pursued it back in Copenhagen over the next nine months and in 1913 published the results in an epoch-making series of papers.

What made Bohr's theory so remarkable was his quantum method of explaining the emission of atomic spectra. Regular numerical relationships had been found among the frequencies of the various spectral lines emitted by hydrogen, but these numerical formulas could not be

[1] John Trowbridge, "Physical Science of Today," *Atlantic Monthly*, 103 (March 1909), 322.

physically accounted for by assuming that atomic electrons radiated out the lines in accord with the laws of ordinary electrodynamics. To produce such a complicated spectral array, a scientist once said, the atom must be so complicated as to make "a grand piano seem exceedingly simple by comparison."[2]

Bohr's theory proceeded from a simple planetary model of the atom then recently developed by Rutherford, in which all of the positive atomic charges were concentrated in a central core, the nucleus, and surrounded at a distance by an equal number of negatively charged electrons. Bohr argued that the electrical force of attraction from the positive charges kept the electrons in nonradiating stationary orbits around the nucleus, much as the sun bound the planets in the solar system. Bohr interpreted the characteristic spectral lines of the atom as the radiation emitted when an electron jumped from one of these orbits, or stationary states, to another of lower energy. The energy lost in the transfer appeared as light whose frequency was exactly determined by Planck's quantum theory of radiation.

Bohr's mixture of Planck's quantum with Newtonian mechanics appeared to rest on a kind of legerdemain. Rutherford wondered in his penetratingly simple way: "How does an electron decide what frequency it is going to vibrate at when it passes from one stationary state to the other? It seems to me that you would have to assume that the electron knows beforehand where it is going to stop."[3] Equally perplexing, the quantum-jump process of radiation confounded ordinary physical sensibilities, since the atomic electrons were undefined in space and time while they crossed between the allowed orbits. But Sir James Jeans wryly remarked: "The only justification at present put forward for these assumptions is the very weighty one of success."[4] Bohr manipulated his theory to predict the known hydrogen spectrum. He also argued that certain spectral lines assumed to belong to hydrogen were actually fathered by helium, and a test in Rutherford's laboratory quickly confirmed his analysis.

Bohr's theory did run into trouble on various points, including the audacious X-ray results of young Henry Gwynn Jeffreys Moseley, the most gifted experimental physicist of his generation in England. The heavier elements in the periodic table emitted characteristic X rays, or spectral lines of very high frequencies. Late in 1913, in a few feverish months of work, Moseley found that, for each element, these character-

[2] Quoted in Clifford L. Maier, "The Role of Spectroscopy in the Acceptance of an Internally Structured Atom, 1860–1920" (unpublished Ph.D. dissertation, History of Science, University of Wisconsin, 1964), p. 130.

[3] Quoted in Ruth Moore, *Niels Bohr: The Man, His Science, and the World They Changed* (New York, 1966), p. 59.

[4] Quoted in Léon Rosenfeld and Erik Rüdinger, "The Decisive Years, 1911–1918," in Stefan Rozental, ed., *Niels Bohr: His Life and Work as Seen by His Friends and Colleagues* (New York, 1967), p. 58.

istic frequencies were mathematically related by a simple formula to a "fundamental quantity," which he inferred to be "the charge on the central positive nucleus."[5] Traditionally, the identification of an element required a measurement of its atomic weight along with laborious analysis of its chemistry and its complex visible spectrum. With Moseley's mathematical relationship, or law, one could confidently accomplish the identification—and the placement of the element in the periodic table—simply by measuring its characteristic X-ray line.

Pushing this stunning achievement further, Moseley noticed that the formula relating the frequencies of the characteristic lines to the nuclear charge closely resembled Bohr's equation for the visible spectrum of hydrogen. Moseley announced that his survey of X-ray spectra confirmed Bohr's quantum theory of atomic radiation. Actually, the two equations differed slightly, and neither Bohr nor Moseley could resolve the troublesome discrepancy.[6] But few physicists attempted to enlarge the difficulty into an assault against Bohr's theory. By 1914, on both sides of the English Channel, the theory was becoming the principal tool for the analysis of atomic structure, and that year a British physicist remarked: "The quantum hypothesis has spread like fire during a drought."[7]

The hypothesis also spread through the ranks of American physics. Membership in the Physical Society passed seven hundred by 1914, and the vast majority were younger men. Some imbibed the heady atmosphere of quanta during sabbaticals abroad. At home or abroad, most were confounded by the remarkable difficulties that quanta forced upon physics, but young physicists eager for professional advancement shared the realization of an enthusiast of Bohr's ideas at Iowa: "What a limitless field for research seems ahead!"[8]

The older men continued to hold most of the profession's academic chairmanships, but they were delegating a good deal of responsibility to the acolytes of the new physics. Everywhere the younger men were introducing courses in electrons and matter, X rays, the interpretation of atomic spectra, and quantum theory. Everywhere they were making research in these subjects, especially photoelectricity and various electron phenomena, the order of the day. At Princeton in 1912, Owen W. Richardson and his student, the twenty-five-year-old Karl T. Compton, managed the first tentative experimental confirmation of Einstein's quantum equation of the photoelectric effect. At Chicago a few years later, Millikan not only proved the validity of the equation conclusively

[5] Moseley, "The High-Frequency Spectra of the Elements," in Henry A. Boorse and Lloyd Motz, eds., *The World of the Atom* (2 vols.; New York, 1966), II, 879.

[6] John L. Heilbron, *H. G. J. Moseley: The Life and Letters of an English Physicist, 1887–1915* (Berkeley, 1974), pp. 102–10.

[7] A. S. Eve, "Modern Views on the Constitution of the Atom," *Science*, 40 (July 24, 1914), 117.

[8] G. W. Stewart, "The Content and Structure of the Atom," *Science*, 40 (Nov. 6, 1914), 662.

but obtained an independent measure of Planck's constant. At Harvard, Theodore Lyman detected the ultraviolet lines in the hydrogen spectrum predicted by Bohr's theory but never actually observed before. The new physics occupied a growing fraction of American doctoral dissertations and of papers published in *The Physical Review* or presented before the American Physical Society.

All the while, the new physicists started to win offices in the American Physical Society and to play a generally more influential role in its affairs. The Society's 1913 meeting was highlighted by a symposium on quantum theory, its 1914 meeting by one on atomic structure. In 1913 the Society also assumed responsibility for *The Physical Review*, whose founding editors had sensibly argued for the shift to increase the prestige of the journal and to attract the many able articles then being sent abroad for publication.[9] As the younger physicists moved into the administrative leadership of the profession, its graduate departments, Society, and journal were turning into stimulants for research of high quality in the field of the quantum-governed atom.

American physicists were acutely aware that in atomic structure their European brethren had done just about all the important work, and the younger men in particular wished to relieve the embarrassment of the situation. But while the amount of funds available for research was unprecedented, so, in all fields, was the number of scientists. By 1914 membership in the American Association for the Advancement of Science had pushed past eight thousand, a fourfold increase since the turn of the century. Robert S. Woodward, always the tight-fisted guardian of the Carnegie's wealth, remarked: "The income of no single research institution . . . can come anywhere near meeting the wants of the great army of competent investigators now pressing for financial assistance."[1]

The endowment of the Carnegie Institution still towered over the pecuniary landscape of research; Woodward kept siphoning the bulk of its funds into large projects. In the academic world proper, state university administrators promoted scholarship; state legislatures appropriated very little money for it. If Andrew Carnegie had dramatized research as a specifically worthy object of private benevolence, few wealthy Americans had followed his example. They did continue to settle large general endowments upon the nation's universities, but with undergraduate enrollments rising at a geometric rate, nine times as much academic income went to teaching as to research.[2]

Physics probably fared neither worse nor better than most other sciences. Fundamental to many fields of engineering, its introductory

[9] *Physical Review*, 1 (Jan. 1913), 63; Leonard Loeb, "Autobiography," AIP, p. 20.

[1] Woodward, "The Needs of Research," *Science*, 40 (Aug. 14, 1914), 224.

[2] Report of the Committee of 100, *Science*, 41 (Feb. 26, 1915), 316.

courses bulged with ever more students, and because of the mushrooming enrollments, departmental chairmen continued to obtain more staff and, in some places, more laboratory facilities. Still, like Lyman at Harvard, they repeatedly pointed to the inadequacy of their research budgets and blamed the "considerable" amount of money spent on teaching. In the tradition of Agassiz, Lyman, a patrician and independently wealthy Bostonian, eased the difficulty with funds from his own pocket.[3] But given the profession's increasingly middle-class complexion, few physicists could afford such generosity. Given its growth, benefactors like Lyman could hardly meet more than a minuscule fraction of the fiscal need.

Actually, Europeans envied the means for research available in the United States, but American scientists considered themselves caught in a fiscal squeeze tight enough to require extraordinary action. A circular of the American Association for the Advancement of Science would soon advise that, if scientists wanted money, they had to organize into associations for the "promotion of their interests." Only the strong voice of numbers could impress the "supreme importance" of research upon the public, the press, and the government.[4] Quantum-minded physicists could easily agree. Yet, however vigorous their appeals, by 1914 they could expect no relief from the general public, the Congress, or even cultivated America.

The public at large knew little if anything about the intricacies of the atom, or even much about the Millikans. The effectiveness of John Wesley Powell's campaign for geology had rested to no small degree on his fame as conqueror of the Colorado. No physicist enjoyed his prestige —or his press. Many newspapermen of the day, as an editor recalled, regarded scientists in every field as subjects for the "staff humorist." Reporters often attended scientific gatherings only to collect tongue-twisters from paper titles or, in the recollection of a newsman, to comment upon the "length and luxuriousness of the beards worn by the assembled savants." Scientists responded by holding the press in contempt.[5]

Not so Samuel Wesley Stratton, who cultivated newsmen and congressmen alike with compelling reports about his agency's ever-widening contributions to progressive reform and economic development. By 1914 the Bureau of Standards was deeply enmeshed in the development of equitable standards for gas and electric utilities. It was also about to

---

[3] *Report of the President of Harvard University 1912/1913*, pp. 193–94. The Harvard physics department depended heavily on a capital gift for research, made in 1901, which yielded $2,500 annually. *Ibid., 1900/1901*, p. 35. A. Lawrence Lowell to Lyman, April 23, 1914, ALL. Typically, the appropriation for research to the Princeton physics department in 1916 was $1,600, UPP, p. 75. In 1915/16 the research endowment for all departments in the Ogden School of Science at Chicago was $14,531, UCH, National Research Council file.

[4] Membership Circular, AAAS, n.d. [1917], copy in SI, AAAS folder.

[5] David Dietz, "Science and the American Press," *Science*, 85 (Jan. 29, 1937), 108.

pay the public a dividend by incorporating the results of its testing program for federal purchases into a bulletin for consumers. While the bulletin scrupulously refrained from mentioning brand names, the simplicity of the discussion as well as the wide publicity given the document in the *Ladies' Home Journal* often left little doubt about the identity of "product X." On the economic side, industry benefited from the Bureau's forays into the physics and chemistry of myriad compounds, raw materials, and gadgets like that astonishing new device, the wireless. Between 1910 and 1914 an appreciative Congress more than doubled Stratton's appropriation to almost $700,000 a year.

Still, physics for reform and economic development was not generally the physics of atomic structure but the physics that appeared practically relevant. In 1914 the Bureau's spectroscopy aimed principally at identifying rather than understanding the chemical elements. Its research in X rays and radioactivity proceeded under the aura of medical purpose. When, a few years later, Stratton allowed two of his men time—but no money—for a type of atomic research, he reported their work as "investigations in electronics," a practical subject in the opening days of radio.[6] To protect his staff, Stratton had to keep the fifty-odd physicists at the Bureau within bounds that he could justify. After all, in the America of 1914 a good many congressmen doubtless wondered with the member of an appropriations subcommittee: "What is a physicist? I was asked on the floor of the House what in the name of common sense a physicist is, and I could not answer."[7]

Physicists were not well known even among the high-status Americans who once—though much less so now—called themselves "cultivated." The word was going out of fashion, no doubt in part because it had acquired a connotation of preciousness but, more important, because it implied an easy familiarity with a catholic core of culture now made impossible by the professional advance of science. *The Nation* lamented the passing of "that real popularization of science which had its period of efflorescence thirty or forty years ago. . . . To listen to a course of lectures on light by Tyndall, to hear Huxley tell of the nature of protoplasm . . ., to have the meaning of conservation of energy expounded by Helmholtz, was to get, not indeed mastery, but at least some kind of comprehension of what science is and a great deal of real sympathy with its aims and methods."[8] In 1915 the publisher of *Popular Science Monthly*, which was now losing some $10,000 a year, sold the name to a firm that intended to put out an illustrated technological gazette. Cattell

---

[6] Quoted in Rexmond G. Cochrane, *Measures for Progress: A History of the National Bureau of Standards* (Washington, D.C.: National Bureau of Standards, U.S. Department of Commerce, 1966), p. 249.

[7] Quoted *ibid.*, p. 148.

[8] *The Nation*, 92 (May 4, 1911), 441.

replaced the journal with his new *Scientific Monthly*, whose content was socially less purposeful than that of Youmans' old magazine. Youmans, Cattell explained, had aimed "to publish articles reviewing scientific progress and advocating scientific, educational and social reforms. The objects are both important, but as science grows in complexity it becomes increasingly difficult to unite them in the same journal."[9]

Progressivism, cementing the identification of research with opportunity, had intensified the scientist's natural attachment to the advancement of knowledge over its popularization. The emergence of a much larger and geographically more widespread scientific community had diminished the importance of lectures delivered before local groups in calculations of professional recognition. Besides, in the progressive era, a knowledgeable observer noted, theories of social reconstruction had assumed the place held by science in the days of Huxley.[1] Instead of a Tyndall, the educated public read John Dewey, Edward A. Ross, Charles A. Beard, and dozens of other reform writers. If the post-Civil War popularizers had taught science to propagate the social beneficence of its method, progressive writers employed the method and ignored the science.

Educated America's receptivity to science depended to a significant extent upon the stimulus of the colleges, but as Nicholas Murray Butler remarked, the attempt to make science an instrument of general education had to be counted a failure.[2] In the spirit of the nineteenth-century university presidents, most critics attributed the failure to the emphasis on research over teaching. Promoting professors on the basis of their scholarship, the critics argued, encouraged them to fasten on particulars and, worse, to concentrate on the subject rather than on the student. But the critics, heirs of the nineteenth century's determination to block the fissioning of culture, were fighting a nearly impossible rear-guard action in the twentieth. As the faculty won more control over appointments and promotions from the president, scientists had more opportunity to run their academic affairs as they wished. So could professors of liberal arts and the social sciences, but professors of science spoke a more highly technical and cumulative language. Under the elective system physics professors tended to attract a smaller band of acolytes than did their colleagues across the campus. As a result, for most of them teaching increasingly meant preparing students for careers in science and engineering: to stress the particular over the general, to train rather than educate.

Popularization or education—in the end, the success of both rested

[9] *Popular Science Monthly*, 87 (Sept. 1915), 307.

[1] "Influence of the War on Science," *Scientific American*, 113 (Nov. 27, 1915), 462.

[2] Butler, Speech at the AAAS, *Science*, 25 (Jan. 1907), 51.

on the intellectual accessibility of science to the layman. The year after Einstein published the theory of relativity, *The Nation* had complained: "Today, science has withdrawn into realms that are hardly [intelligible]. . . . Physics has outgrown the old formulas of gravity, magnetism, and pressure; has discarded the molecule and atom for the ion, and may in its recent generalizations be followed only by an expert in the higher, not to say the transcendental, mathematics. . . . In short, one may say not that the average cultivated man has given up science, but that science has deserted him."[3]

To industrialists like Frank B. Jewett, a Chicago Ph.D. and student of Michelson's, the complex advances of science appeared ripe with economic potential. Michelson had frowned upon his industrial ambitions, but Jewett, the son of a California businessman, had no qualms about the pursuit of science for profit. Early in the century he joined the engineering department of Western Electric, the manufacturing subsidiary of AT&T. Soon he was occupied with one of the Bell system's major technical problems: the development of long-distance service.

At the time, New Yorkers could call no farther than Chicago. Telephone men needed what they called a repeater, a means of amplifying weakened signals by other than mechanical means. For Jewett, the key lay in the three-element vacuum tube that Lee De Forest, an imaginative Yale Ph.D., had patented in 1907. Since a small voltage on one element would control a heavy flow of electrons between the other two, the tube could in principle faithfully amplify the weakest signal. But De Forest's device was too primitive to be relied upon for the distortionless transmission of the human voice over hundreds of miles of wire. Jewett, the best man at Millikan's wedding, was knowledgeable in the physics of electrons and he confidently believed that an organized research effort could solve the repeater problem at a relatively low cost. In 1910 he urged Western Electric to mount just such an attack by hiring "skilled physicists" familiar with the most recent advances in their science.[4]

Jewett's recommendation decidedly interested Chief Engineer John J. Carty, who appreciated the competitive importance of patents. For years AT&T had been coasting on the Bell patents, acquiring rights in the relevant inventions of outsiders, and using the small engineering department at Western Electric only to perfect equipment. Now, Carty looked ahead to the advantages that AT&T might enjoy if it developed a reliable vacuum tube and concluded: "whoever can supply and control the necessary telephone repeater, will exert a dominating influence in the [coming] art of wireless telephony." To prevent "outsiders" from beating

---

[3] "Exit the Amateur Scientist," *The Nation*, 83 (Aug. 23, 1906), 160.
[4] Quoted in Oliver E. Buckley, "Frank B. Jewett," *Biographical Memoirs of the National Academy of Sciences*, XXVII (1952), 246.

AT&T to the technical punch, the firm had to adopt "vigorous [research] measures from now on."[5] Carty obtained the company's approval of Jewett's plan and appointed him chief of research on the transcontinental line. By 1911 Jewett had gathered a small group of talented physicists at the Western Electric headquarters on West Street in downtown New York. The staff was to range beyond immediately practical problems of engineering and, with the aid of Millikan himself as a consultant, to explore some of the fundamental characteristics of electrons. One of the nation's first industrial research laboratories, the enterprise, Carty was confident, would "unquestionably pay liberally for whatever investigations may be made."[6]

For much the same reasons, Du Pont, Standard Oil of Indiana, Westinghouse, and General Electric were also hiring physicists and chemists and letting them plunge into ostensibly impractical research. At General Electric some suspicious stockholders scrutinized the accounts of Willis R. Whitney, the head of the company's new research laboratory, and complained that his outlays were frivolous. Whitney, an MIT graduate and a Ph.D. in chemistry from Leipzig, was sure that "good men put on [pure] research in any field could not fail to show returns in time."[7] He allowed considerable latitude in choice of research topics to some of the more able members of his staff, especially Irving Langmuir, who had been raised in circumstances comfortable enough to breed respect for the Rowland-like admonition of his brother: "You will betray your true self if you devote your life selfishly to private enterprises and personal acquisition. And the minute you allow yourself to deviate from the path of pure science, you will lose something in character."[8]

Langmuir had done his doctoral research at Göttingen on the behavior of gases around the highly heated filament of an incandescent lamp. When he joined General Electric in 1909, Whitney's scientists were wrestling with a fault of the newly developed tungsten bulb: its tendency, after brief use, to blacken and suffer the crumbling of the tungsten filament. Langmuir suspected that the difficulty might be the result of the absorption of gas by the heated tungsten. He soon discovered that, as the lamp heated, its glass walls slowly gave off water vapor which then reacted with the tungsten to produce hydrogen. Going back to his doctoral subject, Langmuir started experimenting with the behavior of gases in the neighborhood of heated filaments, assured by Whitney that

[5] Quoted in N. R. Danielian, *AT&T: The Story of Industrial Conquest* (New York, 1939), pp. 104–5.

[6] Quoted *ibid.*

[7] Quoted in Kendall Birr, *Pioneering in Industrial Research: The Story of the General Electric Research Laboratory* (Washington, D.C., 1957), p. 66.

[8] Quoted in Albert Rosenfeld, "The Quintessence of Irving Langmuir," in G. Guy Suits and Harold E. Way, eds., *The Collected Works of Irving Langmuir* (12 vols.; New York, 1960–62), XII, 63.

the laboratory could stand to know as much as possible about what went on in light bulbs.

Three years of such pure research brought Langmuir to a highly practical result. He found that the two atoms of the ordinary hydrogen molecule would dissociate in the presence of the heated tungsten filament. The atomic hydrogen would then react with the tungsten to create a compound which deposited itself on the walls of the lamp. Here was the process that blackened the bulb while disintegrating the filament. Langmuir discovered, too, that nitrogen would not dissociate under conditions of incandescence and would greatly retard the rate at which the filament evaporated. In the opinion of most engineers, the key to longer lamp life was to get a better vacuum in the bulb. Langmuir's results pointed instead to filling the bulb with nitrogen. The nitrogen lamp, for which GE filed patents in 1913, eventually turned an enormous profit for the company.

By 1914, though only some of the larger, more abundantly capitalized manufacturers had gone into industrial research, discussions of its dollars-and-cents value were sweeping through the technical press. *Electrical World* remarked that the precedent of industrial laboratories had become "firmly established."[9]

Jewett and Whitney continued to encourage their physicists and chemists to probe on the frontiers of science. Langmuir extended his research into the properties of surfaces and filaments, others explored the electronic characteristics of matter, still others the X-ray analysis of crystals. But even in the better industrial research establishments, not even a Langmuir could do whatever he pleased. Whitney kept the better scientists within bounds by indirection; committee discussions took care of the more average men. As Jewett once remarked: "The performance of industrial laboratories must be money-making. . . . For this reason they cannot assemble a staff of investigators to each of whom is given a perfectly free hand. . . ."[1]

The overwhelming majority of the new physicists preferred the unfettered environment of the academic world. There the ceiling on funds and facilities limited the latitude of their research, but industry's developing appreciation of pure science did seem to promise academics relief from their fiscal difficulties. General Electric was helping to support work in X-ray spectra at Harvard. AT&T announced the donation of an endowment for research at MIT, which prompted *Electrical World* to applaud the "growing feeling on the part of large commercial interests that it is

[9] *Electrical World*, Nov. 1, 1913, p. 877.

[1] Birr, *Pioneering in Industrial Research*, p. 73; Jewett, "Industrial Research," *Reprint and Circular Series of the National Research Council*, #4 (Washington, D.C.: National Research Council, 1919), p. 7. See also Willis R. Whitney, "Organization of Industrial Research," *Journal of the American Chemical Society*, 32 (Jan. 1910), 74–75.

good judgment to invest in scientific researches of broad scope"—not only in their own laboratories but in those of the universities.[2]

In 1914 many scientists still insisted upon the cultural importance of their disciplines and the social value of their disinterested method. But the more American physicists committed themselves in research to meeting standards of productivity and merit that were internal to their profession, the more they disaffected the high-status, ex-cultivated Americans who had once responded to their claims of cultural and social leadership. The more the physicists marched toward the fulfillment of Rowland's best-science program, the more they had to find their support among groups with a natural stake in the best science. "The appeal to American interest is utility," a geologist had remarked early in the century.[3] Now, as in Joseph Henry's day, the appeal was once again becoming utility for the emerging leaders of American physics. Along with proponents of the best science in other fields, they increasingly stressed the ultimate technological return of pure research and, with Robert S. Woodward, increasingly hoped to find patronage for it among the "entire class" of the nation's industrialists.[4]

[2] Theodore Lyman, "The Work of the Jefferson Physical Laboratory," *Science*, 43 (May 19, 1916), 707; *Electrical World*, March 29, 1913, p. 677.

[3] John M. Coulter, "Public Interest in Research," *Popular Science Monthly*, 67 (Aug. 1905), 311.

[4] Woodward, "The Needs of Research," *Science*, 40 (Aug. 14, 1914), 224–25.

VIII

"War Should Mean Research"

Late one April afternoon in 1915 thick yellow clouds of chlorine gas rolled toward the French line on the Belgian front at Ypres, and soon hundreds of men were choking and vomiting and dying. A few weeks later a German U-boat sank the *Lusitania*. The Great War, John J. Carty observed, combined the "dreams of Jules Verne with the horrors of Armageddon."[1]

For years thoughtful Americans had assumed that the new forces of science were tying the world more closely together, that the advance of destructive technology was rendering war unthinkable. These assumptions had been especially marked in educated circles. Now Roy K. Hack, a Rhodes scholar and Harvard classicist, charged that the new forces of science had not only failed to prevent war; they had enabled men better to "destroy each other on land . . . assassinate under the sea, and . . . defile the air with Zeppelins." The *New Republic* dismissed Hack's indictment as "hysterical pedantry" and insisted that the war was a failure of man, not of science.[2]

Whatever the case, the Armageddon abroad created down-to-earth economic difficulties at home. The British blockade cut the United States

---

[1] Carty, "Science and the Industries," *Reprint and Circular Series of the National Research Council*, #8 (Washington, D.C.: National Research Council, 1920), p. 1.

[2] Hack, "Drift," *Atlantic Monthly*, 118 (Sept. 1916), 352; "Science as Scapegoat," *New Republic*, 8 (Oct. 7, 1916), 238.

off from its accustomed imports of German dyes, scientific instruments, and optical glass; Britain's massive demand for raw nitrates, which were essential to the manufacture of fertilizers as well as munitions, sharply reduced the supply available from English-dominated companies in Chile, the principal prewar source. U.S. corporations possessed little of the know-how required to produce or synthesize these materials. *Scientific American* issued timely advice: For a "comparatively small outlay," research scientists could develop new processes and make many American businesses "absolutely independent of Europe."[3] Elaborating on that theme, trade journal editors devoted a mushrooming number of pages to the advantages of industrial research, including the profitable opportunities in the present and the challenge of the peacetime future. After the conflict of arms, the predictions went, the United States would face a trade war with a Europe whose industry was honed to superb efficiency. Research, Willis R. Whitney of General Electric proclaimed, was now a "necessity to any people who are ever to become a leading nation or world power."[4]

The American military scarcely measured up to Whitney's worldpower criterion. The army and navy technical bureaus all had some sort of laboratories and could also detail research to the Bureau of Standards. But congressional isolationism had kept the military's technical establishment poorly funded; the bureaus, jealously determined to maintain their respective sovereignties, had persistently refused to pool their limited resources in any servicewide undertaking. While both services had their technical specialists, tradition, law, and policy discouraged ambitious officers from technical careers.[5] The military tended to rely for new weapons upon the innovations of civilian inventors and industrial firms, like the Sperry Gyroscope Company, which, by exploiting U.S. Navy experimental facilities and the remarkable brain of Elmer Sperry, was rapidly developing effective devices for the stabilization of ships.[6] On the whole, armed service laboratories tended to devote themselves to the simple testing of materials and devices, to the cut-and-try improvement of guns, cannons, engines, and gadgetry.

[3] "The War, Industrial Research, and the American Manufacturer," *Scientific American*, 111 (Dec. 26, 1914), 518.

[4] Whitney, "Research as a National Duty," *Science*, 43 (May 5, 1916), 629.

[5] In the army, captains and lieutenants not required for duty in the cavalry and infantry could be detailed to the technical services, but only for a maximum of four years. Since ship's captains had to know their vessels from the engine room to the bridge, naval officers had to rotate their assignments or suffer a delay in promotion. See *U.S. Statutes at Large* 31, pt. 1, 748, 750 (1901); testimony of Admiral Robert Griffin, U.S. Congress, House Committee on Naval Affairs, *Hearings, Estimates Submitted by the Secretary of the Navy, 1916*, 64th Cong., 1st Sess. (3 vols.; 1916), I, 761, 763, 773.

[6] Thomas P. Hughes, *Elmer Sperry: Inventor and Engineer* (Baltimore, 1971), pp. 123, 251, 201.

In the newest field of military technology, aviation, both services had a small group of able enthusiasts. Many were self-made specialists like Henry H. ("Hap") Arnold, engineering officer of the army's aviation section—it was located in the Signal Corps—who had gone to West Point when the horse cavalry was deemed "the last romantic thing left on earth" and who had learned some of his aeronautical principles while listening to the Wrights and their friends swap theories in an old shed in Dayton, Ohio.[7] Air officers had some test facilities available at service flying schools and the Bureau of Standards, and the Washington Navy Yard boasted a model basin and wind tunnel. Still, when in 1912 Arnold wanted to measure the power of an engine, he could not find a dynamometer rated at more than 100 horsepower. In both services, the pro-aviation efforts of men like Arnold often ran into opposition from superior officers who saw no significant military future for airplanes. Between 1909, when the army's first plane was delivered, and 1915, appropriations for army aeronautics totaled $600,000, and when war broke out in Europe, the United States Army and Navy had fewer than two dozen planes between them, almost every one a technical anachronism.[8]

Military aviation officers recognized, an official said, that "to place the development of mechanical flight on a correct engineering basis, the cut-and-try methods of the pioneer must give way to both theoretical and practical investigation of the laws of aerodynamics. . . ."[9] Foreign governments were fostering such investigations. In the opinion of a small group of military men and civilians in Washington, D.C., the U.S. government should be doing the same.

The central figure in the group was Charles D. Walcott, now secretary of the Smithsonian Institution, a skilled scientist-politico who was wont to breakfast with a congressman, a senator, or a President. Walcott's predecessor as secretary, Samuel Pierpont Langley, had doggedly pushed forward experiments with his flying machine, funded by the army and his friend Alexander Graham Bell; Langley's work had interested the Wright brothers as they struggled toward their stunning achievement. In the prewar years Walcott and his group tried unsuccessfully to obtain a governmentally funded aeronautical commission and laboratory. Then, early in 1915, spearheaded by Walcott, the Smithsonian Regents memorialized the Congress to create a National Advisory Committee for Aeronautics to consist of both governmental representatives

[7] Henry H. Arnold, *Global Mission* (New York, 1949), pp. 7–8, 26.

[8] Archibald Turnbull and Clifford L. Lord, *History of United States Naval Aviation* (New Haven, 1949), p. 55; Hiram Bingham, *An Explorer in the Air Service* (New Haven, 1920), p. 23; Carroll V. Glines, Jr., *The Compact History of the United States Air Force* (New York, 1963), p. 71; I. B. Holley, Jr., *Ideas and Weapons: Exploitation of the Aerial Weapon by the United States During World War I* (New Haven, 1953), pp. 28–29.

[9] Quoted in Arthur Sweetser, *The American Air Service* (New York, 1919), pp. 18–19.

and nongovernmental experts "acquainted with the needs of aeronautical science."[1]

Walcott warned the Congress that Americans had fallen far behind Europeans in aeronautical development. Every "first-class nation in the world" except the United States had an advisory committee for aeronautics, and to remain without one would be to trust the nation's aviation "to luck" in an era of airpower's obvious martial importance. Defense aside, the flying machine would surely pay as great a commercial dividend as would the automobile.[2] A provision establishing the National Advisory Committee, or NACA, slipped through the Congress without debate as a rider to the Naval Appropriations Act of March 1915.

A committee might have been born; funds for aeronautics, like funds for all the emerging technological needs of national defense, were still severely limited. The chief signal officer, anxious for the planes he could not get, hoped that the NACA, with all its "force of finality and authority," would sanction the budgetary needs of army aviators.[3] But Congress had given the Committee itself an annual authorization of only $5,000. In its first year of operation, the NACA could afford to let only a single contract. It was not for a major attack on the unknowns of aeronautical science but merely for a special report on carburetor design.

In late May 1915, shortly after the sinking of the *Lusitania*, Thomas Edison assured the country in the pages of *The New York Times* that Americans, as "clever at mechanics . . . as any people in the world," could overcome any "engine of destruction," including U-boats. Edison, who said his emblem was the "dove," argued for preparedness without provocation: not a giant military establishment but the mobilization of ingenuity.[4]

Josephus Daniels, the shrewd North Carolina newspaper editor who was now the secretary of the navy, read the interview with considerable interest. For months he had been weathering the attacks of the pro-preparedness press and the discontent of admirals anxious about the submarine threat. The day after the interview appeared, Daniels drafted a request for the mobilization of the ingenious Edison himself. He sent it out early in July when President Woodrow Wilson authorized the

---

[1] U.S. Congress, House, *National Advisory Committee for Aeronautics, Letter from the Board of Regents . . .*, February 1, 1915, 63d Cong., 3 Sess., House Doc. 1549, 1915; U.S. Congress, Senate, *Advisory Board for Aeronautics, Letter from the Secretary of the Smithsonian Institution . . .*, February 1, 1915, 63d Cong., 3 Sess., Senate Doc. 797, 1915; A. Hunter Dupree, *Science in the Federal Government: A History of Policies and Activities to 1940* (Cambridge, Mass., 1957), pp. 284–87.

[2] *Letter from the Secretary of the Smithsonian . . .*, February 1, 1915, pp. 5, 6, 9.

[3] General George P. Scriven to the [National] Advisory Committee for Aeronautics, April 16, 1915, NACA, Organization and Rules, 1915–1921 file.

[4] Quoted in Edward A. Marshall, "Edison's Plan for Preparedness," *New York Times Magazine*, May 30, 1915, p. 6; Hughes, *Sperry*, p. 246.

military to follow a policy of preparedness, "to be prepared, not for war," as he later told the country, "but only for defense."[5]

The secretary of the navy, the letter declared, intended to establish a group to consider the practicality of devices submitted by civilians or the military, and if Edison were to advise it, the chances of public support would be "enormously increased." Daniels could promise nothing in return save the gratitude of the nation, but it was "imperative" for the navy to use "the natural inventive genius of Americans to meet the new conditions of warfare. . . ."[6] Edison came down to Washington to confer with Daniels on what was to be called the Naval Consulting Board. In September the secretary announced the membership, which included such able men as Willis R. Whitney of General Electric; Leo H. Baekeland, the inventor of Bakelite plastic; Frank J. Sprague, the developer of electric trolleys; and Elmer Sperry of Sperry Gyroscope. With such a board a few years earlier, Daniels confidently told the country, "we would today have control of the submarine and the aeroplane." The Board earned widespread praise, including *Engineering Magazine*'s typical assessment of Edison: "French in his brilliance, more than German in his thoroughness, he is totally American in the application of his genius to practical ends."[7]

A few days after the announcement of the membership, Daniels told the press that, since the navy had no adequate research department for experimentation on a scale demanded by new conditions, he was asking the Board to recommend a plan for a naval laboratory. The Board's plan ambitiously called for an appropriation of $5,000,000 to build the laboratory and $3,000,000 annually to operate it. The laboratory would be run by naval officers but have a staff of civilian physicists and chemists. When the House Naval Affairs Committee heard testimony on the proposal in March 1916, high-ranking navy officers endorsed the scheme and Daniels, in turn, promised that "of course" the navy's bureaus would control the main direction of the work.[8] No witness was more enthusiastic than the sixty-nine-year-old Edison, his unkempt hair now white, his convictions extravagantly Edisonian.

Edison later privately advised Daniels that he did not think "scientific research" would be necessary "to any great extent"; for that the navy could rely on the Bureau of Standards and industrial laboratories. Besides, it would be "useless to go on piling up more data at great expense and delay while we are free to use this Ocean of Facts."[9] At the hearings

[5] Daniels to Edison, draft, May 31, 1915; Daniels to Edison, July 7, 1915, JD, Box 39. Wilson is quoted in Arthur A. Ekirch, Jr., *The Civilian and the Military* (New York, 1956), pp. 163–64.

[6] Daniels to Edison, July 7, 1915, JD, Box 39.

[7] *New York Times*, July 16, 1915, p. 1; *Engineering Magazine*, 50 (Nov. 1915), 199.

[8] House Committee on Naval Affairs, *Hearings Estimates . . ., 1916*, III, 3574.

[9] Edison to Daniels, Dec. 15, 1916, JD, Box 39.

Edison explained that the laboratory would improvise rough ideas into models ready to manufacture. In a laboratory like this "we have no system; we have no rules, but we have a big scrap heap." With the right approach you could put together a submarine in fifteen days.

Congressman Edwin E. Roberts of Nevada was skeptical. Could you turn out a submarine that fast and still have proper inspection along the way?

> *Edison:* Why not? All you have to do is to walk around and see that it is done right.
>
> *Roberts:* The present method of Government inspection is rather closer than that. . . .
>
> *Edison:* Yes; and the Government inspection, by using this high mathematical business and unnecessary fineness, delays the thing two or three times longer. I have seen it in brass works, working to the ten-thousandths of an inch, which is absurd.[1]

The Naval Affairs Committee gave Edison a standing ovation, but, aware that in the modern technological era merely working with big scrap heaps would not quite do, the members endorsed a laboratory to deal with "the practical working out of scientific deductions." Congress appropriated $1,500,000 for an experimental and research laboratory, whose staff would include "scientific civilian assistants."[2]

In the month that Edison testified, March 1916, President Wilson appointed a new secretary of war, Newton D. Baker, the former progressive mayor of Cleveland. Aware of the need to prepare the army better on the technical side, Baker was especially sure that "all the ingenuity and brains of . . . Government experts and . . . scientists ought to be aiding in the development of the aeroplane." Baker urged Congress to authorize a research corps of civilian experts and create a research laboratory to keep the army in the "very forefront of engineering science."[3] He also reorganized the aviation section of the Signal Corps under the remarkable Colonel George O. Squier.

Stationed at Fort McHenry in Baltimore after his graduation from West Point, Squier decided to take a doctorate in physics at Johns Hopkins. Morning parade conflicted with classes, but Squier had an idea. As highly eligible escorts in the social circles of Washington and Baltimore, most young lieutenants disliked weekend duty. Squier, who would

[1] House Committee on Naval Affairs, *Hearings Estimates . . ., 1916*, III, 3345, 3350.

[2] U.S. Congress, House Committee on Naval Affairs, *Report, Naval Appropriations Bill, 1916*, 64th Cong., 1 Sess., 1916, Report No. 743, p. 26; U.S. Navy Department, *Annual Report, 1920*, p. 64.

[3] U.S. Congress, House Committee on Military Affairs, *Hearings, Army Appropriations Bill, 1917*, 64th Cong., 1 Sess., 1916, p. 844; U.S. War Department, *Annual Report, 1916*, I, 57.

never marry, traded duty on Saturdays and Sundays for morning classes at Hopkins and in 1893 became perhaps the only Ph.D. in the U.S. Army. He eventually won an admirable reputation in the international electrical world and numerous patents in telegraphy. A scientific interest in planes also turned Squier into a devotee of aviation. Assigned to London as military attaché in 1913, he made a special effort to learn as much about European military aviation as he could. British officialdom, quick to recognize his scientific knowledgeability, gave him access to the front lines. Ambassador Walter Hines Page found Squier, the son of a Michigan farmer, "a thoroughbred if ever I knew one," and commended him to the President.[4]

When in the spring of 1916 Squier returned home to head up army aviation, the navy was beginning to pay more attention to its own air power advocates, and the wind tunnel at the Washington Navy Yard was busy. Congress raised the appropriation of NACA seventeenfold, to $85,000. In the National Defense Act of June 1916 it also created an Engineers Reserve Corps and called upon the President to promote nitrate research for the production of munitions and fertilizers (that section had the support of the National Grange, the Farmers' Union, the Patrons of Husbandry, and the Farmers' Educational and Cooperative Union of America).[5] In all, the nation's military preparedness program seemed to be ratifying Willis Whitney's science-minded definition of world-power status. But Frank Sprague unintentionally delineated the meaning of it all when he enthused that the creation of the Naval Consulting Board bespoke a recognition of the "engineering profession"—not, he might have added, of the profession of science.[6] With the exception of the National Advisory Committee for Aeronautics, the case was the same across the entire gamut of technical preparedness.

The Naval Consulting Board typically busied itself with organizing state committees of engineers to consider problems, not of science, but of manufacturing and standardization. Its membership was drawn overwhelmingly from the country's major engineering organizations and included only two scientists: Robert S. Woodward and Arthur Gordon Webster, both as representatives of the American Mathematical Society. In the summer of 1915, Webster pressed Daniels to appoint members from the American Physical Society and the National Academy. The secretary refused.[7] Early in October, at the Board's first meeting, Webster de-

[4] Quoted in Burton J. Hendrick, *The Life and Letters of Walter Hines Page* (3 vols.; Garden City, N.Y., 1926), III, 209.

[5] See the petition from these agricultural organizations in U.S. Congress, House Committee on Military Affairs, *Hearings, To Increase the Efficiency of the Military Establishment of the United States*, 64th Cong., 1 Sess., 1916, p. 26.

[6] Frank Sprague, Memorandum, Oct. 1915, BuS, Entry 88, file 15828.

[7] Arthur Gordon Webster to Ernest Merritt, Oct. 10, 1915, EM, Personal file.

nounced whoever had compiled the original membership list. Told that Edison had done it, he demanded to know why the inventor had omitted the American Physical Society.

M. R. Hutchison, who was Edison's chief engineer, explained: "Because it was his desire to have this Board composed of *practical* men who are accustomed to *doing* things, and not *talking* about it."

Why was the Mathematical Society included? Webster asked.

"Because Mr. Edison realizes," Hutchison replied impatiently, "that *very few really practical* men are . . . expert mathematicians, and thought it advisable to have one or two men on the Board who could figure to the 'nth' power, if required."

A few moments of talk on another subject, and Webster returned to his theme. Why hadn't the National Academy been represented on the Board?

"Possibly," the inventor Peter Cooper Hewitt speculated, provoking considerable laughter, "because they have not been sufficiently active to impress their existence upon Mr. Edison's mind."[8]

In Pasadena, California, in October 1915, George Ellery Hale, the distinguished astronomer, editor of the *Astrophysical Journal*, director of the Mount Wilson Observatory, and foreign secretary of the National Academy of Sciences, grumbled at the membership of the Naval Consulting Board. Depressed by the omission of his National Academy, Hale wondered: Could that body's reputation not be raised to the point where it might "penetrate to the sanctum of the Secretary of the Navy"?[9]

"Make no small plans," Hale liked to say.[1] The son of a wealthy Chicago businessman and a graduate of MIT, Hale had his own small observatory, invented the spectroheliograph, and was elected to the Royal Astronomical Society by age twenty-four. He relished English literature—Keats was a pleasant companion for long nights at the telescope—as much as his early editions of Newton. The background, the drive, the reputation, the cultivation all helped Hale as he turned increasingly from the doing to the promotion of research. Philanthropists like Andrew Carnegie—it was chiefly Hale who persuaded Carnegie to fund Mount Wilson—found him a bundle of animated enthusiasm, difficult to refuse. In the prewar decade Hale's entrepreneurial vision extended beyond astrophysics to the promotion of every field of science in the

[8] Hutchison to Josephus Daniels, Nov. 6, 1915, JD, Box 400–1. Hutchison's report of the conversation was based on notes that he took at the meeting.

[9] Hale to Edwin Grant Conklin, Oct. 12, 1915, GEH, Box 11. Part of the material on Hale, the National Research Council, and the International Research Council in chapters X–XII previously appeared in *Isis*, copyright © 1968, 1971 by *Isis*; and it is reprinted here in different form with the permission of the editor.

[1] Frederick H. Seares, "George Ellery Hale: The Scientist Afield," *Isis*, 30 (May 1939), 244.

United States, and in 1915 he brought his ideas together in a slim programmatic volume, *National Academies and the Progress of Research.*

Like his professional peers, Hale recognized the need for greater public appreciation and financial support of research. He argued for educating the public better to the intellectual adventure of science—its treatment in the daily press was "synonymous with rank sensationalism"—and urged that, if scientists were to stress to manufacturers how Faraday's work had laid the foundation for electrical engineering, they could "multiply the friends of pure science and receive new and larger endowments."[2] Much more than most of his peers, Hale realized that American research, particularly in the physical sciences, had to be shifted away from simple fact-gathering, from its tendency to focus on, in his phrase, "the details and the merely technical elements of investigation."[3]

Attributing this tendency to overly narrow specialization, Hale argued that interdisciplinary awareness would encourage even the most concentrated specialist to confront the "large relationships."[4] In his own discipline, astrophysics, students of stellar evolution profitably drew upon physics, chemistry, and mathematics (for much the same reasons that John Wesley Powell had found it advantageous to promote a cooperative, multidisciplinary analysis of the evolution of the land). As director of the Mount Wilson Observatory, Hale had actually inherited, by way of the policies of Walcott and Woodward, the Geological Survey's institutional emphasis on cooperation, an emphasis that was bolstered by the natural tendency of the staff to organize its research around the telescope. Hale might have cited the vitality of the Observatory, along with the texts of Powell and Woodward, when he pointed to the double advantage of group undertakings: They stood a better chance of financial support than did isolated appeals from individuals, and they tended to encourage the concentration of many minds upon a few significant subjects. In Hale's fundamental analysis, American scientists could best get at the large relationships by, in effect, adopting the large-project approach in every discipline and on a national scale.

Impressed by the influential standing of scientific academies in Europe, Hale took the appropriate executor for his multifaceted program to be the National Academy. He recognized the drawbacks, to be sure. He might aim at the patronage of practical manufacturers, but the Academy had repeatedly refused to admit engineers, even Edison himself. In the unhappy judgment of the noted Princeton biologist Edwin Grant Conklin, his fellow members generally preferred to continue as a mere 'blue ribbon society." Some nonmembers, Hale himself understood, re-

[2] George Ellery Hale, *National Academies and the Progress of Research* (Lancaster, Pa.: New Era Printing Co., 1915), pp. 115, 130–31.
[3] Hale to Simon Newcomb, March 21, 1906, SN, Box 25.
[4] Hale, *National Academies*, pp. 100–101. See also Hale to Newcomb, Jan. 5, 1899; March 10, 1906, SN, Box 25.

garded the Academy's best-science elitism as "a menace to true democracy."[5]

While enough of a best-science elitist to stand by the idea of an Academy, Hale was willing to endorse an increase in the limit on annual elections to fifteen. He called for the Academy to cooperate with local and national societies, a venture that would mitigate both the traditional geographical difficulty and the exclusionist objection. He also urged that the Academy, which was something of a gerontocracy, enlarge its influence by dispensing money for research, particularly to promising young men. And the Academy, he insisted, should publish a proceedings, not least to keep the nation's scientists abreast of progress outside their own specialties, and should acquire a building, a permanent meeting home with lectures and exhibits open to the public.

But research grants, like a building and a proceedings, required money. In 1915 the Carnegie Corporation of New York turned down Hale's request for a substantial endowment, in part because the Academy was doing so little to fulfill its chartered purpose of advising the government. The rejection only ratified what Hale had already learned from his study of science in Europe: "to accomplish great results," academies had to "enjoy the active cooperation of the leaders of the state."[6]

Hale shared the members' traditional wariness of thrusting the Academy upon the government, but surely in a national emergency they could offer its services to the administration without appearing like mere political supplicants. In June 1915, after the sinking of the *Lusitania*, Hale —who supported the President's firm demands, rejoiced at the resignation of pacifist Secretary of State William Jennings Bryan, enthused for the Navy League, and sided with pro-Allied interventionists—supposed that President Woodrow Wilson might welcome an offer of assistance in the event of war. Conklin, a fellow Academy activist who knew Wilson from his Princeton days, advised against the idea as premature. Hale agreed to drop it for the time being. Ten months later, April 19, 1916, the day after Wilson issued an ultimatum to Germany over the sinking of the *Sussex*, Hale rose at the Academy's annual meeting in Washington to offer a resolution: In the event of a break in diplomatic relations, the Academy would place itself at the disposal of the government. While blocks away the President rallied the Congress, Hale's unprecedented resolution was endorsed unanimously.

A week later William Henry Welch, the Johns Hopkins physician and president of the Academy, led a delegation that included Hale,

[5] Conklin to Hale, March 28, 1913, GEH, Box 11; Hale, *National Academies*, p. 22. When Edison was nominated for membership in the Academy in 1911, he received only three votes. R. S. Woodward to T. C. Mendenhall, Nov. 5, 1911, TCM, Box 6.

[6] Henry S. Pritchett to Hale, Feb. 3, 1913, copy in EGC, National Academy of Sciences file; Hale, *National Academies*, p. 53.

Conklin, Walcott, and Woodward to the White House. After Welch rehearsed the Academy's special relationship to the government, Hale went on to stress to President Wilson the importance of research for defense and argued that the Academy could plan an arsenal of science for the country.

After a few questions Wilson approved the undertaking. Would the Academy form a committee? Would it also keep the matter confidential? German-American relations were in a delicate stage and the public might misinterpret an open request for military help from the President to his official scientific adviser.[7]

Whatever the caveats, the undertaking, which was to proceed war or no war, promised to put the Academy within reach of that governmental cooperation so fundamental to Hale's ambitions. Hale was jubilant: "I really believe this is the greatest chance we ever had to advance research in America."[8]

In June the Academy responded to Wilson's official request by quietly forming a National Research Council. The NRC's objective: to encourage both pure and applied research for the ultimate end of "the national security and welfare." Its strategy: to promote "cooperation" among all the research institutions of the country. Its planned composition: leading scientists and engineers from universities, industry, and the government, including the military.[9] Broad cooperation, industrial contacts, federal representation—the NRC gave institutional life to the principal points Hale had made in his *National Academies and the Progress of Research.*

Some scientists grumbled at the Council's ambitions, complaining that it was trying to impose a dangerous centralization on American research. Others agreed with James McKeen Cattell, a pacifist, who indicted the Council as "militaristic."[1] But Hale laid down the reassuring policy that, while attempting to organize American research along cooperative lines, the Council would safeguard "individual initiative."[2] In any case, faced with the threat of war, few scientists were especially sensitive on the traditional dangers of overcentralization.

Compared to the bulk of the population, a disproportionate fraction of the research community had traveled abroad, and its members were more concerned than most Americans with the troubled affairs of the

[7] "Memorandum of a Conference with the President . . .," n.d. [April 1916], GEH, Box 60.

[8] Quoted in Helen Wright, *Explorer of the Universe: A Biography of George Ellery Hale* (New York, 1966), p. 288.

[9] "Report of National Research Council to President of the National Academy," attached to Hale to Welch, Oct. 23, 1916, copy in BuS, Entry 88, file 13158.

[1] See Edwin Grant Conklin to Hale, Dec. 21, 1916, EGC, Hale file; Cattell to Conklin, Oct. 30, 1916, EGC, Cattell file; Cattell to E. C. Pickering, Jan. 10, 1917, EP, Harvard Observatory Director's papers, Cattell file.

[2] "Report of National Research Council . . .," attached to Hale to Welch, Oct. 23, 1916, BuS, Entry 88, file 13158.

Old World. In addition, every leading scientist enjoyed close personal and professional ties to individual Europeans, Britons in particular. Before the war these bonds had of course extended to Germans, but American scientists were sufficiently attached to the equation of *science* with *progress* to consider the Central Powers' employment of submarine and poison gas perverted uses of the study of nature. In the East and even in the anti-interventionist Midwest many scientists were talking preparedness, some outright involvement.

Whatever the attitudes toward Germany, the burning of the library at Louvain, Belgium, by the Kaiser's troops symbolized the extinction of the lamps of learning in Europe. In August 1915, hardly a year after his brilliant work on X rays, twenty-seven-year-old Henry Moseley, Signaling Officer, 38th Brigade, went down at Gallipoli with a bullet through his head (the snuffing out of this young life alone, Millikan declared, would make the war "one of the most hideous and most irreparable crimes in history").[3] No matter what the outcome in Europe, scientists in America, it was widely argued, were obligated to ensure the continued advancement of knowledge.

Pro- or anti-involvement, few of the country's scientists could fault the National Academy's position that "true preparedness" meant the encouragement not only of applied but of pure research. Few could disagree with what Hale told President Wilson: "We must not prepare poisonous gases or debase science through similar misuse; but we should give our soldiers and sailors every legitimate aid and every means of protection."[4] Many scientific societies, research foundations, and universities quickly pledged the NRC their cooperation.

The new intertwining of science and business guaranteed the Council some powerful entrees into the corporate world: Whitney, Baekeland, and Carty agreed to serve. The Academy, no doubt prodded by Hale, also enhanced the bid for industry's friendship by finally creating a section for men who had contributed "to the science or art of engineering."[5] Michael I. Pupin, a Columbia physicist, member of the Academy, and wealthy inventor of a device that had started AT&T on the road to long-distance service, helped the NRC obtain New York office space from the Engineering Foundation, as well as the Foundation's income for the year. In the government, the civilian agencies seemed enthusiastic, particularly Samuel Wesley Stratton at the Bureau of Standards. But in the defense-minded atmosphere of Washington, the cooperation of the secretaries of war and the navy would be crucial, and while Baker seemed guardedly friendly, Daniels had his Naval Consulting Board.

In Hale's judgment, the real power in Edison's advisory group was

[3] Millikan, "Radiation and Atomic Structure," *Science*, 45 (April 6, 1917), 322.
[4] National Academy of Sciences, *Proceedings*, 2 (1916), 507–8; Hale to the President, July 25, 1916, GEH, Box 60.
[5] National Academy of Sciences, *Annual Report, 1916*, p. 20.

William L. Saunders, a director of Ingersoll Rand, leading figure in the Wilson Business Men's National League, and devoted Democrat who liked Edison's cut-and-try way and was tending to the Naval Board's affairs in Washington. Hale, eager to stake out the NRC's territory, arranged a meeting. Saunders, his manner domineering, emphasized that the Consulting Board could take care of submarine defense better than any group of scientists. He did suggest that the Research Council might possibly work with Edison's group in a subordinate capacity.[6]

Hale did not like the idea. If the Research Council were to get anywhere, it had to cooperate with the top of the government directly, not through some intermediary. The question of the Consulting Board, as Hale saw it, was "bound to affect the whole future of the Academy."[7] Early in July, Hale decided that to ease the NRC's way in the labyrinth of the Washington bureaucracy and to neutralize the threat of the Consulting Board, the Council had to have Wilson's *public* approval. It also needed an explicit assurance of federal cooperation: in Hale's definition, the presidential appointment to the Council of official representatives from the government.

Wilson had to be prodded. The Republican Party had nominated Charles Evans Hughes, an unalloyed advocate of preparedness, to oppose Wilson in November. Hale told James Garfield, the GOP campaign manager, about the Research Council. Would Hughes give it his support? On July 20 Garfield telephoned: Hughes was decidedly interested in the NRC and wanted a statement about it for his letter accepting the nomination.[8]

Hale had already discussed the Research Council with Wilson's confidant, Colonel Edward M. House, and now he informed House of Garfield's call, adding that "endorsement from every quarter will help our work." Hale was preparing a press release and wished he could say that the President approved the NRC's activities. He hated to push things. But the Council, he continued, had progressed to the point "where the appointment of representatives of the Government becomes essential." Hale would of course defer the publication of his own press release as long as possible, but since it had to appear before Hughes's letter, to no later than July 27.[9]

House quickly advised the President: Send Welch a letter and telegraph its contents to Hale. "I think one of the best things that has been done in the way of preparedness is the work of these men and I would hate to see the Republicans get any benefit from it."[1]

[6] Hale to Evelina Hale, April 21, 1916, GEH, Box 80.

[7] *Ibid.*

[8] Hale to Gano P. Dunn, July 21, 1916, GEH, Box 14. Hale to Edward M. House, July 21, 1916, attached to House to the President, July 22, 1916, WW, Series 2, Box 149.

[9] Hale to House, July 21, 1916, WW, Series 2, Box 149.

[1] House to the President, July 22, 1916, WW, Series 2, Box 149.

The letter and telegram went out from the White House on July 24. The contents included everything that Hale wanted: the approval of the NRC, the assurance of government cooperation, the promise of presidential appointments from the departments. July 29, three days before Hughes's letter was to appear, the White House announced the Research Council on its own: "Preparedness, to be sound and complete, must be solidly based on science."[2]

Hale elatedly told Conklin that the NRC was out of danger in Washington. To Hughes, "in the interest of candor," he wrote that the President's letter would be included in the Council's press release, but he hoped that the Republican nominee would still come out strongly for scientific preparedness. Hughes's acceptance of the nomination did not mention the Council explicitly; it did insist that preparedness had to include "promoting research and utilizing . . . science."[3] Hale's NRC would have the cooperation of the government no matter who won the election.

Churning ahead, Hale went to Europe with Welch to study the Allied scientific effort. When he returned to the United States in mid-September 1916, the NRC had its governmental representatives, including Stratton, Squier, Walcott, who was soon to replace Welch as president of the Academy, and Rear Admiral David W. Taylor, an air enthusiast who was presiding over the wind tunnel experiments at the Washington Navy Yard. But Congress had created a Council of National Defense, a group of cabinet officers who were to have, in the words of the legislative act, an Advisory Commission empowered to develop "special knowledge" through "investigation, research, and inquiry." The members of the Commission included Bernard Baruch, Julius Rosenwald, Samuel Gompers, and Hollis Godfrey, the president of the Drexel Institute, a technical school in Philadelphia. Secretary of War Baker expected that Godfrey's appointment would "cause great happiness in many quarters, as it is a recognition of pure science." By profession Godfrey was an engineer, author, and eccentric devotee of time-saving through scientific management—scarcely a fitting representative of science to Hale and his colleagues.[4] And Charles Walcott predicted that unless the NRC, which had no Washington headquarters, put a forceful executive in the capital and got down to defense work, it would "disappear" behind Godfrey's group.[5]

Robert Millikan was with Hale in Pasadena, California, when on

[2] Wilson to Hale, July 24, 1916, GEH, Box 60; copy of the White House press release is in WW, Case File 206.

[3] Hale to Conklin, July 27, 1916, GEH, Box 11; Hale to Charles Evans Hughes, July 25, 1916, GEH, Box 22: *New York Times*, Aug. 1, 1916, p. 6.

[4] *U.S. Statutes at Large*, 39, pt. 1, 650 (1916); Baker to the President, Oct. 14, 1916, WW, File VI, Case 206.

[5] Walcott to Hale, Dec. 8, 1916, SI, National Research Council file. See also the minutes of the meeting of the NRC Executive Committee, Jan. 6, 1917, NRCM.

February 3 the United States severed diplomatic relations with Germany. Two days earlier, immediately upon the resumption of unrestricted submarine warfare, an alarmed Admiral Taylor had assigned the NRC a critical task—to find a way to detect prowling U-boats. Hale wired the President: intensive research on the subject would begin at once.[6]

Millikan, appointed chairman of a Research Council committee against submarines, left for Washington, where he proposed to Admiral Robert S. Griffin, chief of the Bureau of Steam Engineering, the commencement of submarine detection experiments at the New York City laboratories of Western Electric. Griffin informally approved the venture and promised some funds from Steam Engineering. A few weeks later, the project now authorized officially, Millikan got down to full-time work at Jewett's facilities on West Street.

On February 17 the NRC Military Committee met at the Smithsonian, and in midmorning Walcott announced that the secretaries of the cabinet would join the group for lunch to discuss the relationship of Hale's Council to Godfrey's Advisory Commission. Godfrey was accommodating. By the end of the afternoon, everyone seemed to agree that the NRC would handle all scientific matters for the Council of National Defense.[7] But even with that bureaucratic problem out of the way, the academically flavored NRC still had to contend with the Naval Consulting Board.

The Board also was now trying to develop devices for the detection of submarines, an effort entirely independent of Millikan's and the object of greater naval interest. Edison himself was reportedly working eighteen hours a day on inventions for defense. *The New York Times*, normally friendly to scientists, said of the Consulting Board's illustrious chairman: "Superior persons are somewhat too eager to assure us how much pure science overshadows the application of it, the pure scientist the 'mere' inventor. Well, this man has invented a new world."[8]

Hale left for Washington early in March, eager to get the NRC's affairs fully "out of the academic state." Preparedness had brought the Council a long way. Now he knew: "War should mean research . . ., and unless we get it started, some other agency will do so."[9]

[6] U.S. Navy Bureau of Construction and Repair, Memorandum for Military Committee, Feb. 1, 1917, NRCM; Hale to Cary T. Hutchinson, Feb. 6, 1917, copy in Hutchinson to Walcott, Feb. 9, 1917, SI, NRC file.

[7] Minutes of meeting of the NRC Executive Committee, Feb. 17, 1917, AAF, Series 87, Aircraft Board files, folder 685; Conklin to Hale, Feb. 19, 1917, GEH, Box 11.

[8] *New York Times*, Feb. 11, 1917, Section VIII, p. 2.

[9] Hale to Cary T. Hutchinson, March 5, 1917, Feb. 10, 1917, GEH, Box 23.

IX

The War Work of the Physicists

ale opened offices for the Research Council just a few blocks
from the White House in the Munsey Building, where both
the Council of National Defense and the National Advisory
Committee for Aeronautics had their headquarters. Millikan
was to preside over technical affairs as the third vice-chairman, director
of research, and executive officer. "If the science men of the country are
going to be of any use to her," he was sure, "it is now or never."[1]

Hale realized the importance of now or never, but he had made
trouble for the Research Council by constructing it in the scientific
community's tradition of political elitism. Expecting the NRC to operate
"more by moral pressure than by force of [legal] authority," he intended
that the Council, a private body like its parent the National Academy,
perform a public function, the mobilization of science for defense, without
governmental oversight.[2] To keep the Council safe from federal intervention, Hale funded the NRC with gifts from sympathetic businessmen
and private groups. Administratively, when the NRC was asked to serve
as the scientific department of the Council of National Defense, he took
"great precautions" to have the NRC merely "act" as such a department,
without becoming a part of the agency and "without surrendering any
authority to it." Though the government could nominate representatives
to the NRC, Hale kept the actual power of appointment in the hands of

---

[1] Robert Millikan to Greta Millikan, April 1, 1917, RAM, Box 49.
[2] George Ellery Hale to John C. Merriam, Feb. 5, 1918, NRCM, Hale file.

the National Academy, which eliminated "all questions of political preferment."[3]

Being privately funded, however, and lacking any legal authority, the National Research Council was in no position to inaugurate on its own any major defense research programs. To participate in defense research, it had to wangle its way into the good graces of the federal, especially the military, bureaucracy. Fittingly, it put a premium on staff adept at operating, as an associate of Hale's remarked, by "tact" and "influence."[4] Through the spring of 1917, Millikan, though not always a paragon of tact, politicked and maneuvered to obtain a role for the NRC in wartime science, aided by his cheery self-confidence and boundless energy —and by the deplorable state of technological preparedness.

At the declaration of war in April 1917 the supply of optical glass, essential for rangefinders and gunsights, was entirely inadequate for the military's needs. The army had neither a gas shell in its arsenal nor a gas mask in its standard equipage, and its meager air force, which totaled fewer than 250 planes, was unequipped with the machine guns or the flight instruments essential to combat operations. No manufacturer had yet mastered the techniques for producing high-grade optical glass or poison gases in quantity. In all of 1916 nine different factories had managed to fill less than 20 percent of a military order for 366 aircraft.[5] The U.S. Navy's best submarine detection apparatus could not even sense the presence, let alone determine the location, of a submerged U-boat cruising more than two hundred yards away. Worse, after three months of experiments, the likelihood of improvement, in Admiral Griffin's recollection, appeared "rather remote." The British were desperate enough to toy with training sea lions as submarine hounds.[6]

German U-boats had sunk 1,300,000 tons of Allied and neutral

[3] Hale to James R. Angell, Aug. 13, 1919, NRCM, Hale file; Hale to Arthur Schuster, April 18, 1918, GEH, Box 47.

[4] Arthur A. Noyes to Gano Dunn, July 22, 1918, NRCM, Research Information Committee file.

[5] Benedict Crowell, *America's Munitions, 1917–1918* (Washington, D.C., 1919), pp. 139–40; "Report of the Director of the Chemical Warfare Service," in U.S. War Department, *Annual Report, 1919*, I, 4381; G. W. Mixter and H. H. Emmons, *United States Army Aircraft Production Facts* (Washington, D.C., 1919), p. 5; Alfred Goldberg, ed., *A History of the United States Air Force, 1907–1957* (Princeton, 1958), p. 2.

[6] Josephus Daniels to Naval Consulting Board, April 5, 1917; Office of Naval Intelligence, "Detection of Under-Water Sounds [with Sea Lions]," Report No. 321, June 11, 1917, SNV, Declassified file; [Robert S. Griffin], *History of the Bureau of Engineering, Navy Department, During the World War* (Office of Naval Records and Library, Historical Section, Publication No. 5; Washington, D.C., 1922), p. 47. The Chief of Naval Operations asked the NRC whether seals might be trained to detect submarines. Not least because of the British experience with sea lions, the NRC did not consider this approach "at all promising." Millikan to Admiral Benson, Aug. 21, 1917, SNV, Declassified file 28754.

merchantmen in the first quarter of the year; in April alone they were expected to cost the Allies close to 1,000,000 tons more, an intolerable rate of loss. On April 9 Rear Admiral William S. Sims arrived in England as chief of the United States naval mission, and in London the next day he learned the seriousness of the submarine situation from John R. Jellicoe, an old friend and now First Sea Lord of the Admiralty.

"Is there no solution for the problem?" Sims asked.

"Absolutely none that we can see now," Jellicoe replied.[7]

Sims pushed through the brilliant tactical solution of the convoy, in which destroyers escorted large groups of merchant vessels through the danger zone around the British Isles. During May merchant losses dropped almost one third below the ominous April high. At home, army and navy officers, grimly aware how unprepared their services were on the technical side of defense, hurried to consult the scientific experts in the Munsey Building.

There the National Advisory Committee for Aeronautics was helping, in the commendation of the Boston *Transcript*, to spur ahead the development of a "tremendous air fleet"—ten thousand planes within a year and an "All American engine" to power them.[8] Two floors below, the NRC joined with army and navy officers to inaugurate research and production for high-quality optical glass, drawing heavily upon the substantial expertise of scientists at the Geophysical Laboratory of the Carnegie Institution. Under the auspices of the NRC, the Bureau of Mines launched a program of research in chemical weapons—the distinction between offensive and defensive chemical warfare disappeared once American soldiers were committed to fight in France—and by May the Bureau had farmed out problems to twenty-one university laboratories. And all the while, Millikan and John J. Carty of AT&T wrestled with approaches to the detection of submarines, with a considerable sense of urgency—and no small disapproval of the way that the Naval Consulting Board was going at the problem.

The Board had no laboratory of its own (the naval research facility funded in 1916 had not yet been constructed because of a dispute over its purpose, control, and location).[9] While the dispute continued, the Board had authorized the Submarine Signal Company of Boston, which had a rudimentary detection device, to establish an official experimental station on the coast of Massachusetts at Nahant. General Electric was to share the facility, and the station opened the day after the declaration of war, with Irving Langmuir on its small staff and the development of the

[7] Quoted in Elting E. Morison, *Admiral Sims and the Modern American Navy* (Boston, 1942), p. 342.

[8] Boston *Transcript*, June 29, 1917, clipping in NACA Scrapbook, NACA.

[9] Thomas P. Hughes, *Elmer Sperry: Inventor and Engineer* (Baltimore, 1971), pp. 252–54.

Submarine Signal device its major purpose. But the device, an array of transmitters and receivers installed below the water line, had been designed originally to transmit Morse code through the sea by sound, not to detect submarines.[1] And though engineers from AT&T were soon added to the Nahant group, academic physicists were excluded, because, Admiral Griffin explained, their presence would complicate the patent situation. Millikan emphatically agreed with Professor Ernest Merritt of Cornell, who had resented the exclusion of the American Physical Society from Edison's Board: "I should be glad if we could give a demonstration that the physicists ought to have been included."[2]

The leaders of the NRC hoped to open submarine detection research to academic physicists with the help of an Allied scientific mission. Mainly French, the mission was armed with full knowledge about the latest military applications of science and included, from Britain, Ernest Rutherford, who had been experimenting with submarine detection devices in the basement of his Manchester laboratory. In lengthy conferences on June 1–2, the mission talked mainly about submarines to a full panoply of military bureau chiefs and Research Council scientists. A fine meeting, Rutherford thought. So it was for the NRC, not least because the French and British revealed various devices for submarine detection of considerably greater and more immediate promise than the apparatus under development at Nahant.[3] A week later, pointing to the lessons of the Anglo-French mission, Hale and Millikan urged the navy to establish another experimental station at New London, Connecticut, the site of an existing submarine base; and to bring in ten of the "most competent . . . scientists of the country" to develop the Allied detection devices.[4]

Secretary Daniels remained a partisan of Edison, but most members of Edison's Consulting Board had college degrees and many understood the advantages of the new trend in industrial research, including the usefulness of putting scientists onto technological problems. One member of the Board generously admitted to Millikan: "The subject of sub-

[1] The device, invented by Reginald Fessenden, is discussed in Cleveland Moffett, "A New Defense against the Submarine," *The American Magazine*, 79 (April 1915), 11 ff., and in Harvey C. Hayes, "Detection of Submarines," *Proceedings of the American Philosophical Society*, 69 (March 1920), 22.

[2] Admiral Benson to the Secretary of the Navy, May 8, 1917, SNV, Entry 19, file 8011-175; Millikan, "The Beginnings of the New London Experimental Station as Recorded in the Personal Notes of Lieutenant Colonel Robert A. Millikan," p. 9, attached to Millikan to Commander Calhoun, July 10, 1919, BuS, Entry 988, Bureau of Steam Engineering [hereafter, BSE] file 797. Edward House to Josephus Daniels, May 8, 1917, JD, Box 40; Merritt to Millikan, March 28, 1917, EM, New London file.

[3] Rutherford to his wife, June 1, 1917, quoted in Arthur S. Eve, *Rutherford* (New York, 1939), pp. 257–58; [Griffin], *History of the Bureau of Engineering*, pp. 50–51.

[4] "Recommendations of the National Research Council as to the Organization of Research for the Detection of Submarines," June [8], 1917, BuS, Entry 988, BSE file 797, tray 3793.

marines is big enough for all of us."[5] Even Daniels had to agree to that when in mid-June the NRC sponsored a conference in its offices—the participants included members of the Allied mission and Edison's Board, U.S. and British naval officers, and fifty academic physicists—which ranged over methods of submarine detection for a full three days. After the meeting, Willis R. Whitney wrote Daniels how "aggrieved" Millikan and his colleagues were. Surely the Research Council could be supplied with experimental facilities "without great cost." "If a little of the energy which, I fear, may otherwise develop into hard feeling, could be utilized in the submarine work, time would be saved."[6]

Daniels turned the matter over to a small group of naval officers—his new Special Board on Antisubmarine Devices. Composed mainly of younger career officers eager to make a success of their task, the Special Board sensibly wanted to use all available talent, and after more than a month of experiments at Nahant, it judged the Submarine Signal apparatus entirely inadequate. In the closing days of June, the Special Board neutralized the squabbling between the Research Council and Edison's group by attaching to itself as informal advisers an engineer from Submarine Signal, and Whitney, Jewett, and Millikan. It also authorized the NRC to start research at New London with its physics professors.[7]

GE, AT&T, and Submarine Signal would continue to work at Nahant, but now the academics would have their chance, and Millikan was optimistic. The detection of submarines, he put it, was only a "problem of physics pure and simple."[8] The Research Council appointed a staff of six to work at New London, and on July 3 Millikan joined the group at the Mohican Hotel to discuss a submarine detection device brought over by the Allied mission.

French in origin and called the Walzer apparatus, this device consisted of two blisterlike metal surfaces, each bulging into the water on opposite sides of the ship (they were multifaceted and, from the outside, looked like the eyes of a fly). Each operated as a sound lens and would focus the noise of a distant submarine somewhere on the center beam of the listening vessel. A line drawn from this focus through the middle of the lens would in theory run straight to the U-boat. The great advantage of the Walzer apparatus was that in principle it would not pick up confusing noises, like piscatorial swishes or the throb of the patrol vessel's own motors, coming from other directions. But in practice,

[5] B. G. Lamme to Millikan, June 13, 1917, NRCM, IR: Inter-Allied Exchange of Information: French Scientific Mission to U.S.: General file.

[6] Whitney to Josephus Daniels, June 16, 1917, NRCM, Ex Com: Committee on Physics: Subcommittee on Submarine Detection: General file.

[7] Special Board to Navy Department, June 14, 1917; Daniels to Bureaus, June 23, 1917, NR, Subject file OD; Millikan to Ernest Merritt, June 25, 1917, EM, New London file; Hale to Evelina Hale, June 20, 1917, GEH, Box 80; Robert A. Millikan, *The Autobiography of Robert A. Millikan* (New York, 1950), pp. 154, 161–62.

[8] Chairman, Submarine Committee, NRC to anon., June 11, 1917, NCB, Box 3.

the focal point of the apparatus was indistinct, which prevented directional precision, and it brought so little of the incident sound energy to the focal point that its range was quite limited.[9]

At the conference in New London these difficulties intrigued Professor Max Mason of the University of Wisconsin. Over the years Mason's seemingly effortless talents had gone more into chess, golf, bridge, the violin, and billiards than into research, but as a Ph.D. student at Göttingen he had astonished his professor by returning an elegant solution to a difficult thesis problem just ten days after it had been assigned. Now, with similarly swift insight, Mason suggested replacing the blisterlike Walzer apparatus with a horizontal row of short capped pipes, or Broca tubes, connected by more piping to an earphone. Transmitted through the pipes, a sound from the water would lose very little energy, and simply by varying the pipe lengths until the sound was loudest, a listener could determine with considerable accuracy the direction from which the noise was coming.[1]

The next day Mason returned to Wisconsin, where he built a crude model and experimented with it on the Madison lakes. On July 10, back in New London, Millikan found him dabbling off a dock with an ungainly, trombonelike affair of pipes and tubes. Mason exclaimed: "By Jove, Rob, it works."[2] Mason had more good news: The awkward trombone slides could be advantageously replaced with a compact circular compensator. Instead of manipulating a cumbersome array of pipes, one would need only to turn a dial until the sound reached a maximum, then read off the direction of the submarine. In a test at sea later in the summer, this modified version sensed the presence of a submarine over the noise of a listening ship making four knots.[3] By the early fall the New London group was well on the way to engineering Mason's device into a practical, seaworthy detector. The MB listener was the first in what the Navy designated the "M"—for multiple tube—series of submarine sensors, all of which would be forged from the principle of focusing sound by compensation.

In Nahant, Massachusetts, all the while, General Electric physicists had been going at the U-boat detection problem with a British device, a pair of capped Broca tubes deployed as a pair of rotatable underwater ears. This apparatus detected submarines on the so-called binaural principle, which makes one hear a noise loudest when facing it directly. In the British version, the Broca tubes were capped with metal. Irving Langmuir's group at Nahant found that the simple substitution of rubber

[9] Hayes, "Detection of Submarines," p. 21.

[1] Millikan, "Beginnings of the New London Experimental Station . . .," pp. 15–16.

[2] Quoted in Millikan, *Autobiography*, p. 164. The trombone device is described in Max Mason, "Submarine Detection by Multiple Unit Hydrophones," *The Wisconsin Engineer*, 25 (Feb. 1921), 76–77.

[3] Millikan, "Beginnings of the New London Experimental Station . . .," p. 17; Mason, "Submarine Detection," p. 99.

for metal increased the range to almost three miles.[4] Moreover, two rubber-capped Brocas suspended at the tips of an inverted "T" pipe acted as a remarkably good binaural detector. In mid-August three small boats equipped with the C-tube, as the Navy designated the new Nahant listener, chased a submerged submarine around Boston Harbor, used light bulbs for depth bombs, and scored a symbolic kill.[5]

Both the C- and MB-tubes left a good deal to be desired, but for range and bearing, both detectors, Millikan reported in late September, far outclassed "any which the British have sent over. The situation is most encouraging." At Nahant, Langmuir's group was on the verge of completing a microphonic listening device with electrical amplifiers, the K-tube, whose range exceeded ten miles.[6] By using it to steam within reach of the C-tube, which was directionally more accurate, a patrol vessel could ferret out a submarine cruising anywhere under some three hundred square miles of ocean.

The navy cloaked listening devices with top-secret status, warned the scientists at Nahant and New London against breaches of security, and in August authorized the installation of C-tubes on all antisubmarine vessels, particularly those bound for overseas duty.[7] Sometime in the fall, it seems, similar recognition went to both the K- and the MB-tubes. Earlier, Secretary Daniels had refused to supply money for the physicists at New London; Mason had been forced to draw upon the Research Council's limited funds and a special subvention from the University of Wisconsin. Evidently Daniels soon reversed himself. By October the Special Board had $300,000 in hand for research on detection devices, and that month the navy took over the effort at New London completely by transforming it into an official Experimental Station.[8]

[4] Percy W. Bridgman, "The C-Tube and the K-Tube," Lecture, May 20, 1918, NR, subject file LA; Kendall Birr, *Pioneering in Industrial Research: The Story of the General Electric Research Laboratory* (Washington, D.C., 1957), p. 64. The advantage was in the shift from a resonant to a nonresonant receiver. [Griffin], *History of the Bureau of Engineering*, p. 65.

[5] [Griffin], *History of the Bureau of Engineering*, pp. 55–56.

[6] Millikan, "Monthly Report of the . . . National Research Council . . .," Sept. 25, 1917, NRCM, Reports file; "Notes of Submarine Hunting by Sound," attached to William S. Sims Circular Letter No. 32, May 1, 1918, SNV, Declassified file C-26-99. The various detectors and their comparative advantages are discussed in Millikan, testimony before the General Board, U.S. Navy, Jan. 18, 1918, BuS, Entry 988, BSE file 797.

[7] Hale to William Wallace Campbell, Aug. 2, 1917, GEH, Box 36, San Pedro file; Admiral Griffin to Chief of Naval Operations, Aug. 1, 1917, and Secretary Daniels' indorsement, Aug. 4, 1917, BuS, Entry 988, BSE file 797, tray 3794.

[8] Franklin D. Roosevelt to Senior Member of Special Board, Oct. 20, 1917, BuS, Entry 988, BSE file 797, tray 3795. It is interesting to note that when the French members of the Allied scientific mission called on Daniels to say good-bye, August 7, 1917, they urged that the U.S. Navy, Daniels told his diary, "not depend too much upon scientific discoveries but more on fighting ships and guns." Some days later, August 18, Daniels reported the advice to his Naval Consulting Board, stressing the warning against too great a reliance upon scientists. E. David Cronon, ed., *The Cabinet Diaries of Josephus Daniels, 1913–1921* (Lincoln, Neb., 1963), pp. 188, 192.

In November 1917, Griffin sent an assistant abroad to demonstrate the latest detectors to Admiral Sims, who had been disappointed by the inaccuracy of the Submarine Signal apparatus and the failure of the Allies to overcome the inadequacies of their own detectors. When Griffin's assistant arrived, the British and American naval commands, Sims recalled, were "all more or less doubtful." But during tests in the English Channel in the early days of January 1918, crews operating with the C- and K-tubes—Mason's, which had developed a leak, was out of action—managed to locate an enemy submarine. Depth bombs were hurled into the water and a great slick of oil spread over the surface of the sea. The detectors, Sims recalled, "astonished everybody who was let into the secret."[9] The British immediately ordered large quantities of the American listening devices, and Sims, along with creating his own anti-submarine section in Europe, threw the full weight of his prestige behind the research and development effort in the United States.

New London, which steadily acquired more equipment and more physicists, was by the spring of 1918 virtually absorbing Nahant, and the Research Council overshadowing the Consulting Board. Early in May, Daniels paid New London an inspection visit in the company of Edison. Out at sea on a mock U-boat hunt, Daniels, ignoring the business at hand, cornered Mason by the wheelhouse to talk politics. Edison keenly scrutinized the techniques of the chase. Suddenly, the submarine surfaced just where science said it should be. Edison hauled Daniels to the rail, and, while clouting him on the back, shouted: "There he is, the ———— ———— stingarino!" A few days later, Edison told Daniels: "My opinion is that if you back up these young officers and scientific men, give them the kind of boats they want, and do it quick, . . . the submarine will be reduced from a serious menace to a minor annoyance."[1]

The best submarine hunters were destroyers, but few could be spared from convoy duty. Besides, destroyers required a submarine detector which could be used while under way, and in the spring of 1918 an appropriate device was not yet available. Eager to exploit the development of listening devices, the navy committed coastal patrol vessels to the submarine war in Europe. Merely 110 feet long, made of wood, lightly armed, and powered by gasoline engines, the boats had a cruising range of only nine hundred miles and a maximum speed five knots slower

[9] William S. Sims, with Burton J. Hendrick, *The Victory at Sea* (London, 1920), pp. 172–73; Frank Parker Stockbridge, *Yankee Ingenuity in the War* (New York, 1920), p. 224; R. H. Leigh to Force Commander, Jan. 8, 1918, NR, subject file OD. The British Admiralty later classified the sinking of the submarine as doubtful. Anon., untitled memorandum, on subchaser operations, n.d., NR, subject file OD.

[1] Daniels' visit to New London is related in Harvey C. Hayes, "World War I: Submarine Detection," *Sound: Its Uses and Control*, 1 (Sept.–Oct. 1962), 48; Edison to Daniels, May 9, 1918, JD, Box 39.

than the Germans' best U-boats. All the same, C-, K-, and MB-tubes were installed. In March 1918 hastily trained crews, most of them college undergraduates drawn from the Naval Reserve, crowded into the frail ships, together with depth bombs, listening gear, and radio. They sailed the vessels abroad, refueling in mid-ocean, and, somehow, by mid-July managed to bring 104 chasers safely across the Atlantic.[2]

Admiral Sims assigned the majority of them to the English Channel, the rest to the Straits of Otranto at the mouth of the Adriatic, where the British had established a barrage of trawlers and drift nets to bottle up U-boats working out of bases in Austria-Hungary. In the interest of listening, all other boats in their area would, by general order, stop engines five minutes every hour (changing the timing each day kept the U-boats off guard). The ships worked in groups of three, each separated from its sisters by about a quarter mile. On signal, all would stop, listen, take bearings on a suspicious sound, which might prove to be the rumble of a distant explosion or the moan of a wreck tossing on the bottom. Then all would speed to the spot where the bearing intersected. Again they would stop and listen, again race to the new intersection.[3] Since in shallow water a U-boat might lie motionless on the bottom, the hunt could become a battle of sheer waiting and persistence. If the hunters were quick enough and blessed with no little luck, they would ultimately find themselves within a few yards of their evasive target. After depth bombs were dropped, the listening would resume.

At Otranto one midnight, shortly after the chasers arrived in July, eight submarines were detected heading for the Straits. The hunters raced up the Adriatic, stopping, listening, bombing, bombing, bombing. The rain of depth charges broke up the enemy formation. Some of the U-boats, trying to dodge the hail, crunched into the submarine nets. None got through; four never returned to their bases. Within weeks after the chasers started patrolling the Otranto Barrage, no Austrian crew, in the postwar report of villagers on the coast, could be compelled to venture a submarine through the Straits.[4] Though the English Channel squadrons never scored any verified kills, they constantly harassed German

[2] Julius A. Furer, "The 110-Foot Submarine Chasers," *U.S. Naval Institute Proceedings*, 45 (May 1919), 760, 763; Ray Millholland, *The Splinter Fleet of the Otranto Barrage* (New York, 1936), pp. 63–64; Dudley W. Knox, *A History of the United States Navy* (New York, 1936), p. 414. A summary of the monthly arrivals of the subchasers in Europe is in NR, subject file OD.

[3] "Operation of U.S. Submarine Chasers on Otranto Barrage," NR, subject file 1911–1927; Furer, "The 110-Foot Submarine Chasers," pp. 761–62. Admiral Griffin remarked: "The procedures of conducting a successful chase and attack . . . involved details of ship handling, navigational plotting, and communication between ships with a degree of speed unheard of before in any kind of naval craft." [Griffin], *History of the Bureau of Engineering*, p. 58.

[4] Millholland, *The Splinter Fleet*, pp. 119–28; Anon., untitled memorandum on subchaser operations, n.d., NR, subject file OD.

submarines, demoralized their crews, and prevented them from marauding with their accustomed impunity.

As the war drew to a close, Admiral Sims was crediting victory over the U-boat not only to the convoy system but to listening devices. By the armistice the New London physicists had developed Mason's detector to the point where it could locate a U-boat from a destroyer slashing through the sea at fourteen knots; it was by far the best detector in the arsenal of the Allied navies.[5] In Sims's opinion, had more destroyers been available, Mason's device would have helped write a victorious chapter in the battle against the U-boat. As it was, in 1918 the subchasers and Allied ships equipped with the American detectors together sank at least six submarines with depth charges.[6]

At the armistice, the New London Experimental Station, born only of Millikan's dogged insistence upon a role in submarine research for academic physicists, included thirty-two professors, a large plant of laboratories and test facilities, three submarine chasers, three yachts, a precious destroyer, and more than seven hundred enlisted men.

Back in the spring of 1917, shortly before the Allied scientific mission sailed for the United States, the first soldiers of the American Expeditionary Force arrived in France. They learned that many of the enemy's most threatening artillery were hidden in an area from one to five miles beyond the trenches. Distance and hills protected the location of these guns from ordinary ground observers, and cloud, camouflage, and dummy emplacements often shielded them from observation by pilots. To overcome these difficulties, the French and British developed apparatus that spotted German batteries by the sound and flash they emitted when firing. By the time the United States entered the war, Allied army intelligence considered their flash- and sound-ranging services highly advantageous, and in late June 1917 General John J. Pershing cabled home to the War Department: He would need physicists to staff a similar service for the AEF in France.[7]

The National Research Council knew just the physicist to command the service: Augustus Trowbridge, a member of a wealthy New York family who was a product of tutorial sojourns and doctoral years abroad, a connoisseur of French wines, cuisine, and architecture, urbane, hardheaded, and capable. A forty-nine-year-old professor at Princeton, Trowbridge was known for his mastery of intricate apparatus. Over the summer,

---

[5] "Remarks by Vice Admiral Sims . . .," Sept. 6, 1918, copy in GEH, Box 38; Sims to the Secretary of the Navy, Nov. 13, 1918, NR, subject file OD.

[6] Sims, *Victory at Sea*, p. 168; [Griffin], *History of the Bureau of Engineering*, p. 73.

[7] Pershing's cable, June 30, 1917, is quoted in "Preliminary [Historical] Report," 29th Engineers, p. 147, AEF, Entry 915. The material on flash and sound originally appeared in a somewhat different form in *Military Affairs*, copyright © 1969 by the American Military Institute, and is reprinted here with the permission of the editor.

commissioned as a major in the Signal Officer Reserve Corps, he investigated the Allied flash- and sound-ranging systems before choosing the standard equipment for the new American ranging service.

In flash ranging, observers with high-powered telescopes were stationed at five or six widely separated dugouts on an arc close to the German lines. Wires connected each to a central station, where an operator, by using a system of synchronizing signals, would coordinate the observers so that at least two, and usually more, simultaneously sighted their instruments on the same muzzle flash. Once the operator had received the bearings by telephone, he would lay them out on a map board, and the intersection of the plots would mark the location of the gun.[8]

Sound ranging exploited the fact that if two microphones were set down about a mile apart, the boom of a distant battery would arrive at one slightly later than at the other. In accord with a few simple laws of physics and analytic geometry, the time differential determined the curve of a hyperbola on the map, somewhere along which lay the gun. To spot the battery, several pairs of microphones were used, all of them connected to a central station that housed apparatus to measure the various arrival times. Analysts would plot the resulting family of hyperbolas and read off the location of the gun from the intersection.[9] But sound-ranging apparatus had to distinguish between two quite different sound waves: the crack of the shell as it passed overhead, and the boom at the muzzle, which was what revealed the location of the gun. In the principal French sound-ranging device—called the TM, for *Télégraphe Militaire*—an oscillograph located at the central station transformed the microphone signals into a picture that allowed the observer to distinguish the arrival times of the two waves. In the British Bull-Tucker, the field microphones ignored the weak shell wave, but upon receipt of the muzzle wave would send sharp pulses of current to an oscillograph in the central station for the determination of the arrival times.[1]

Trowbridge had no trouble choosing for his troops a flash-ranging system that was technologically straightforward. In comparative tests of sound-ranging systems, the Belgian Dufour, a version of the French TM, edged out the other devices, principally because of its superior sensitivity and ability to distinguish between the shell and gun waves.[2]

[8] *Observation Section, Flash Ranging*, pp. 1–3, AEF, Entry 155, G–2 Report file, Appendix XXVI.

[9] "Sound Ranging," AEF, Entry 210.

[1] Service Géographique de l'Armée, *Description et Mode d'Emploi du Matériel T.M. 1916* (Paris, 1917), pp. 1–3, copy in WDGS, War College file 9652-6; "Sound Ranging," pp. 7–12, 18–19; William Lawrence Bragg, "How Sound Ranging Was Done in Theory and Practice," in John Innes, comp., *Flash Spotters and Sound Rangers: How They Lived, Worked and Fought in the Great War* (London, 1935), pp. 135–36.

[2] Trowbridge, "Final Report on Flash and Sound Ranging," Dec. 22, 1918, pp. 4–5, AEF, Entry 155, G–2 Report file, Appendix XXVI.

But before initiating full-scale production, Trowbridge decided to wait until he had investigated firsthand in France the reliability of the Dufour under the taxing conditions of battle. On September 9, 1917, he sailed for Europe, together with Theodore Lyman of Harvard, a veteran of Platts-burgh and now a captain in the Signal Corps Reserve assigned to duty with flash and sound. The army physicists headed for Paris, the location of the Signal Corps command, then, unexpectedly, to the central nervous system of the AEF, Pershing's headquarters at Chaumont. Flash and sound had been transferred to the Corps of Engineers, and Trowbridge was to work under Colonel Roger G. Alexander, head of G–2–C, the Topographical Section of Army Intelligence.

Second in his West Point class, briefly a professor of military engineering at the Academy but otherwise the product of the Army Engineers' routine tours of duty, the thirty-four-year-old Alexander shouldered considerable responsibilities, including the production for the entire AEF of maps reliably marked with the location of enemy artillery. The young colonel and onetime professor welcomed Trowbridge and Lyman with open arms. He had already sent First Lieutenant Charles B. Bazzoni, a young American physicist who had volunteered for the AEF while on a research fellowship in London, around the front to assess the relative merits of the various sound-ranging systems. In Bazzoni's conclusion, the best apparatus was not the Belgian Dufour but the British Bull-Tucker. After their own visit to the front, Trowbridge and Lyman agreed that, while not as sensitive as the Dufour, the Bull-Tucker was certainly more robust, and they were prepared to sacrifice scientific neatness to battlefield practicality.[3]

Early in November the AEF adopted the Bull-Tucker as its official sound-ranging equipment. The British promised two sets by the end of the month, ten by April 1, and a Western Electric representative in Paris assured Trowbridge that his company could supply up to fifty more sets by the beginning of March, when the AEF would be moving into the trenches in force.[4] Trowbridge and Lyman, aware that the surrender of Russia would make the German spring offensive a juggernaut, rushed to prepare the first full ranging company for the field. They took some of their officers from the Ambulance Service, their troops from the stock of Army Engineers accumulating in France. Early in January 1918, Lyman led the novices to Ft. de St. Menge, a medieval fortress some forty-two kilometers south of Chaumont. There, in a leaky basement, the Harvard physicist opened the army's own ranging school.

Only war could have put together so improbable an academic mix. There was Lyman with his "funny New England old maidishness,"

[3] Bazzoni, "Preliminary Report on Bull-Tucker System," Oct. 18, 1917, AEF, Entry 210; "Report of the Work of G–2–C . . .," 1919, p. 16, AEF, Entry 207.

[4] Chief of Topographical Section to Chief of Intelligence Section, "Report on Visit to British Front," Nov. 1, 1917, AEF, Entry 210.

Trowbridge mused, commanding "some college boys . . ., a really good technical officer or two, a few rough-necked enlisted men, and a few splendid specimens . . . like the big trees of their native Oregon."[5] There was Bazzoni, teaching time differentials and hyperbolas to troops who had no more than a smattering of science. Lyman, grumbling and growling, doctoring and disciplining, won respect and affection. Bazzoni managed to get across the principles of flash and sound. By February, when a second company of rangers was due to arrive from the States, the school was operating smoothly, and the first company was out on field maneuvers shaping into a competent outfit.

At Chaumont, where he and Lyman established their own mess with a good French cook, Trowbridge preferred to end the bleak winter days with a shot of whiskey and a warm bed, instead of stumping "around the dark and cold streets to call on men . . . with whom one has very little in common."[6] Less aloof professionally, he set up ranging equipment in his office and explained the system to curious generals, whom he gradually warmed to. He also earned their respect, and early in March permission came to get the first sound-ranging section into the line. Since the Bull-Tucker sets had not yet arrived from Western Electric in the United States, the section would have to operate with a British-made ranging device, along with French wire and electrical batteries salvaged from a fallen German Zeppelin. Trowbridge, aware that "it is *so* very important for me to deliver the goods to our artillery," felt that he had "to gamble a bit."[7]

Bazzoni's Sound Ranging Section No. 1 headed down to St. Mihiel and stationed itself at Mandres, a small shell-torn village on the south side of the German salient. To the north, on the flat, marshy plain of the Rupt de Mad, lay the enemy's trenches and patches of wood laced with gun emplacements. Beyond these stood the Germans' principal advantage, Mont Sec, rising to a height of some five hundred feet and bristling with watchtowers and batteries. SRS No. 1, as the first sound-ranging section was designated, was in operation only a few weeks when the French pulled two of their own ranging sections from the salient. Trowbridge replaced them with SRS Nos. 2 and 3. Together, the three American sections were responsible for more than twenty-three kilometers of the front, which was subject to constant artillery fire. On the map, the wires running back from the microphones to the central stations looked like a giant tentacular array groping toward the German guns hidden in the patches of woods.

Through the spring the three American sections did "a land office

<hr />

[5] Trowbridge to Sarah Trowbridge, Feb. [5], 1918, "The War Letters of Augustus Trowbridge," *Bulletin of the New York Public Library*, 43 (1939), 728. (Hereafter cited as *NYPL* with volume number.)

[6] Trowbridge to Cornelia H. Fulton, Dec. 17, 1917, *ibid.*, 656.

[7] Trowbridge to Horatio B. Williams, March 25, 1918, AEF, Entry 210.

business."[8] Fog and cloud paralyzed air reconnaissance a good deal of the time, and the Rupt de Mad made for good sound ranging. In the wrecked houses where the centrals were located, the operators tended their oscillographs and plotted their maps amid the fall of gas shells and high explosives. The linesmen had to maintain the thousands of yards of wire, every inch of it vulnerable to shrapnel (in the battle of Seicheprey the lines were severed forty times in a single night). Bazzoni's men did their jobs courageously and effectively, even managing to pinpoint guns that French sections had been unable to locate.[9]

But St. Mihiel was a stable front: Could the rangers cope with a war of movement? Trowbridge had his doubts. The task of stripping down, then reestablishing the Bull-Tucker system took up to a full week, and it was no simple matter to transport the microphone wire, since the lines for a complete outfit weighed eight tons. At Princeton University a research group was trying to develop a mobile sound-ranging set by modifying the Belgian Dufour. In the meantime, Trowbridge's only hope for a war of movement lay with the flash rangers, whose lines were less bulky. Still, relocating a flash section took two full days.[1]

The end of the spring found the AEF's Second Division blocking the Germans at the riverbanks just thirty-seven miles from Paris, and on June 10 Pershing asked Alexander to send flash and sound troops up to the fiercely contested town of Château-Thierry. Trowbridge threw in SRS No. 2 and Flash Ranging Section No. 1, both under the command of Lyman. FRS No. 1, veteran of only a few weeks on the salient, was still green when it arrived, and the two sections shared a common central in a wood no more than a few kilometers from the German line. Shells rained down with unremitting fury and more than once blew telescopes and microphones to bits. Lyman, who had hunted big game before the war, was cool, sure, always bolstering morale. Under his leadership, FRS No. 1 developed into a reliable locator of enemy batteries; it even outpaced the sound section, which had to cope with contrary wind conditions.[2]

At 4:35 A.M. on July 18, the doughboys went over the top and counterattacked through Belleau Wood, and the war of movement began. FRS No. 1, too far in the rear by July 21, packed up and moved forward; another two days and it pushed on again; a move in retreat, then forward once more. On August 4 the section settled in at Cherry-Chartreuse, a

---

[8] Trowbridge to Sarah Trowbridge, April 19 and 21, 1918, "War Letters," *NYPL*, 43 (1939), 907.

[9] Jesse R. Hinman, *Ranging in France with Flash and Sound* (Portland, Ore.: Dunham Printing Co., 1919), pp. 46–47, 53–64; *Procedure Followed by American S. R. S. in the Field*, p. 16, AEF, Entry 155, G-2 Report file, Appendix XXVI.

[1] *Procedure Followed by American S. R. S. in the Field*, p. 5; Trowbridge to Sarah Trowbridge, June 10, 1918, "War Letters," *NYPL*, 44 (1940), 23.

[2] Lyman to Colonel Alexander, July 14, 1918, AEF, Entry 210; Trowbridge to Sarah Trowbridge, July 3, 1918, "War Letters," *NYPL*, 44 (1940), 31.

village on the Vesle some forty kilometers to the northeast of Château-Thierry. There, despite severe shell fire, the loss of seven men to gas, exhaustion, and wounds, the flash rangers managed to locate thirty-four enemy batteries before they withdrew a week later. SRS No. 2 had been unable to keep up with the swift advance, but FRS No. 1 had proved that it could meet the taxing demands of a war of movement and it was commended for its "excellent" work by the AEF chief of artillery.[3]

In September both the flash and the sound rangers won numerous accolades during Pershing's lightning reduction of the salient at St. Mihiel. Over the last six weeks of the war, in the area west of Metz misty weather made sound ranging indispensable to the spotting of artillery; in the Argonne Forest, a crack flash section kept up with the infantry's tumultuous advance. When the guns fell silent, five flash and five sound sections were spread over the front from a point south of Verdun all the way to the Moselle.[4]

Trowbridge, now a lieutenant colonel, had shaped 83 officers and 1,068 enlisted men into an efficient technical service. Elsewhere in the AEF—in the Gas Service, the Signal Corps, the Air Service, and the Corps of Engineers—other physicists and chemists made their wartime mark, by able work at the level of operations or by staffing field laboratories to deal with the urgent technical problems that inevitably rose on the front.

If physicists and chemists like Trowbridge joined the army to fight with the AEF, even at home it was common for scientists interested in defense research to enlist in the armed services. Civilian scientists usually disliked working under military control; the hierarchy of command and the constraints of military regulations did not seem conducive to the freewheeling professional conditions thought to be required for good research. Yet industrial research laboratories and federal civilian bureaus provided only limited alternatives to the armed services, and the National Research Council lacked the resources and authority to prosecute military research independently of the military. The army and navy could have funded research projects administered by the NRC, but that procedure would have violated a fundamental tenet of the professional armed service outlook: Military research had to be conducted under military control.

If civilians inaugurated research programs like the experimental station at New London, the armed services tended to swallow them up once they

---

[3] Hinman, *Ranging in France*, pp. 87–93; Trowbridge, "Final Report on Flash and Sound Ranging," Dec. 27, 1918, pp. 28–29; the chief of artillery is quoted in "Report of the Work of G–2–C, GHQ," 1919.

[4] Lyman to Chief, G–2–C, First Army, Oct. 9, 1918, AEF, Entry 210; Trowbridge, "Report of the Work of G–2–C, GHQ," 1919, pp. 14–15; "Final Report on Flash and Sound Ranging," p. 15.

became important. By the spring of 1918 the Bureau of Mines was supervising the work of some seven hundred chemists and, funded with a substantial amount of army money, had become the center of the nation's research program in chemical weapons. But months before, a General Staff colonel had concluded: "As . . . the magnitude of the work increases, it becomes more and more apparent that a civilian organization financed by the War Department and over which the War Department has no direct control . . . is a mistake."[5] Now Major General William L. Sibert, the strong-willed career officer who had helped construct the Panama Canal, demanded the consolidation of the entire poison gas program under himself in a new Chemical Warfare Service. The Bureau of Mines chemists adamantly opposed the transfer to the army. Sibert, who admitted that the Bureau had done an outstanding job, insisted on the necessity, in principle, of army control over so important an army program —and won the day.[6]

Whatever their penchant for sheer bureaucratic aggrandizement, military officers worried about the maintenance of secrecy in civilian-controlled laboratories. They suspected that, left to themselves, civilian scientists might simply "ride [their projects] as a hobby," to cite the warning of a General Staff officer, "for the purpose of obtaining data for research work and the future benefit of the human race."[7] Above all, working independently, civilian scientists could all too easily produce devices and weapons that, while scientifically neat, might be impractical on the battlefield. Military control, Admiral Griffin said of the navy's taking over New London, kept paramount "the question of practicability" from a military standpoint.[8]

The professional military counted its own expertise superior to any civilian's in all matters of military technology, including, Millikan learned, the strategic exploitation of new devices of warfare. In October 1917 Millikan and Max Mason promoted the idea of building an attack ship designed to take advantage of the advances in submarine detection devices. The navy's own Special Board on Antisubmarine Devices recommended the construction of a thousand Eagle submarine chasers, enough to bottle up the North Sea passage between the German U-boat bases and the

---

[5] Charles L. Parsons, "The American Chemist in Warfare," *Science*, 48 (Oct. 1918), 380; Colonel P. D. Lochridge to Chief of Staff, Sept. 21, 1917, WDGS, War College file 9967–8.

[6] Van H. Manning, "Memorandum Regarding Conference Held in the Office of the Secretary of War . . ., May 25, 1918, Regarding Proposed Transfer of the War Gas Investigations of the Bureau of Mines to the War Department . . ."; Manning, "Memorandum on Bureau of Mines Gas War Work," June 3, 1918, BuM, War Gas Investigations file.

[7] Colonel R. J. Burt to Chief of Staff, June 18, 1918, p. 5, WDGS, War College Division file 10195–25.

[8] [Griffin], *History of the Bureau of Engineering*, p. 56. See the remarks of the chiefs of the Bureaus of Ordnance and of Construction and Repair in U.S. Navy Department, *Annual Report*, 1917, p. 216, and 1918, p. 512.

Channel.[9] Henry Ford offered to build the U-boat hunters without profit —and without diverting a single worker from the crucial destroyer construction program. But the navy ordered only one hundred Eagle boats, one tenth the number recommended by the Special Board, and the contract called for delivery by December 1918.[1]

The navy refused either to enlarge the order or to authorize an accelerated pace of construction; by the armistice only seven Eagle boats had left the factory. In the assessment of a member of Edison's consulting group, the senior admirals appeared to consider the demand of their junior officers on the Special Board for a new boat "somewhat as a confession of professional weakness," a professional weakness brought on possibly because they had gotten too closely associated with the civilian scientists at New London.[2]

Whether it was an issue of strategy or of research and development, joining up for the duration gave the "damn professors," to use the military's phrase for civilian scientists, no necessary advantage in dealing with the armed services. They usually enlisted as reserve officers. In the AEF, by order of General Pershing, no major technical decision could be made by a reserve officer without the sanction of a regular.[3] And at home or abroad, reserve officer scientists usually required the support of the regulars to hold their own in the inevitable battles, jurisdictional and otherwise, that erupted in the civilian and military research bureaucracies.

Millikan headed for such a battle when in June 1917 General George O. Squier, now chief of the Signal Corps, asked him to join the army to take charge of all Signal Corps research. Millikan was reluctant, not least because he was worried about getting locked into the military chain of command. Squier was all insistence and assurances, and Hale sided with the general ("a great opportunity," Hale called Squier's offer, which will "really place all this scientific work in our hands").[4] Soon Millikan became Major Millikan, Army Reserve, a swivel-chair officer stationed in the Munsey Building, and chief of the Signal Corps Science and Research Division.

[9] Millikan to Greta Millikan, Oct. 24, 1917, RAM, Box 49; "Minutes of Meeting [of the Special Board] . . .," Oct. 27, 1917, Nov. 2, 1917. Captain S. S. Robison, Special Board, to the Navy Department (Operations), Oct. 28, 1917, BuS, Entry 88, file 16555.

[1] Millikan, *Autobiography*, pp. 173–76; [Griffin], *History of the Bureau of Engineering*, p. 39.

[2] Thomas Robins to the Secretary of the Navy, March 18, 1918, JD, Box 400–401, Naval Consulting Board file. Millikan reported that Admiral Griffin had chastised him for having "overstepped our authority in pushing the boat problem so far," and Admiral Taylor was "quite indignant" about the matter. Millikan to Greta Millikan, Nov. 10, 1917, Nov. 11, 1917, RAM, Box 49.

[3] Millikan, *Autobiography*, p. 195; Frederick Palmer, *Newton D. Baker: America at War* (2 vols.; New York, 1931), II, 9.

[4] Millikan, *Autobiography*, pp. 149–50; Hale to Evelina Hale, June 25, 1917, GEH, Box 80.

Early in 1918, assigned responsibility for the development of aeronautical instruments, the Division opened a small laboratory on the marshy flats of Langley Field, Virginia. National Research Council scientists donned khaki to staff the enterprise on the ground. The commanding officer was Charles Mendenhall of the University of Wisconsin physics faculty, their test pilot David Webster, a physicist from Stanford University. When a letter arrived at Langley directing officers to tell their men when to wear overshoes, Mendenhall drew himself to attention and said to Webster: "I hereby direct you to wear your overshoes just when you damn please." Whatever the lack of spit and polish, Science and Research enjoyed the emphatic support of Squier the regular, and Millikan counted his relations with the Signal Corps more "intimate" than with any other military department.[5]

In spring 1918 turmoil enveloped army aviation—the production of aircraft was so far behind schedule as to provoke charges of scandal and corruption—and President Wilson reorganized the entire aircraft program.[6] Army aviation was removed from the jurisdiction of Squier and the Signal Corps. It was divided into a civilian-controlled Bureau of Aircraft Production and a military-controlled Division of Military Aeronautics. Administratively, Millikan's Science and Research Division landed in the Bureau of Aircraft Production, whose headquarters were in Washington. It was supposed to work for both the Bureau and the Division of Military Aeronautics, which had its headquarters and its own research facilities out in Dayton, Ohio. But Lieutenant Colonel Thurman H. Bane, the chief technical officer in Dayton, insisted that the Division of Military Aeronautics, not the civilian Bureau of Aircraft Production, "must control the . . . design of the equipment which it is to operate." To Bane, Millikan's men kept going off "on interesting and fascinating research to the total neglect of the immediate problems at hand."[7] In contrast, the Bureau of Standards was eager to work in close accord with Bane's expert conception of military needs. Prompted by Samuel Wesley Stratton himself, Bane supposed that Science and Research might just as well be "busted up," with part of its work turned over to the Bureau of Standards and the rest to Dayton.[8]

Science and Research survived, but the laboratory at Langley Field was always overcrowded, the supply of planes inadequate for airborne tests. From the reorganization to the armistice, the Division endured repeated attempts, many of which originated in Dayton, to reduce its

[5] David Webster interview, AIP, p. 39; "Monthly Report of the . . . National Research Council . . .," Sept. 25, 1917, NRCM, Reports file.

[6] See New York Times, March 19, 1918, p. 3, April 28, 1918, p. 4, May 7, 1918, pp. 1, 4; Department of Justice, Report of Aircraft Inquiry (Washington, D.C., 1918).

[7] Bane, Memo for the Director of the Division of Military Aeronautics, June 5, 1918, AAF, Series 166, file 321.9A; Bane to Director of the Air Service, Aug. 21, 1919, Series 167, file 334.8.

[8] Bane to H. H. Arnold, May 13, 1918, AAF, Series 166, file 321.9D.

technical responsibilities.[9] On the technical side of post-Squier aviation the pivotal regular was Bane, who continued to regard Millikan's men as a claque of impractical academics. And Stratton, Millikan lamented to his wife, simply did not wish "to see any scientific organization grow up inside the Bureau of Aircraft Production which could act in any way as an intermediary between the Bureau of Standards and the military forces."[1]

All the while, Augustus Trowbridge battled with the Bureau of Standards over sound-ranging equipment. Edward B. Rosa, a high-ranking physicist at the Bureau, had instructed his staff to continue to work on the development of the French TM even after Trowbridge decided in favor of the British Bull-Tucker. By mid-January 1918 Rosa was so confident that the Bureau TM would ultimately supersede the British apparatus that he got Trowbridge's formal requisition to Western Electric cut—from thirty Bull-Tuckers back to ten.[2] Adding to Trowbridge's difficulties, as late as June 1918 none of the ten remaining Bull-Tuckers had yet arrived in France. Apparently with the support of the National Research Council, Western Electric had taken the liberty of experimenting with the original designs for the sake of perfecting the apparatus.[3]

Trowbridge smoldered at Rosa for having "the nerve to get my requisition cut to one third just because *he thinks* what he has been working on is just as good ! ! !"[4] Trowbridge knew that Rosa had failed to overcome the advantage of the British device, which was its superior ability to distinguish the locations of different enemy guns when all were bombarding the line at once. In mid-June 1918 a cable went out from Pershing's headquarters to Washington insisting that experimentation no longer be allowed to interfere with the fulfillment of Trowbridge's requisition.[5] In August, with Rosa still pushing the Bureau TM, Trowbridge returned to Washington and met Rosa together with the Bureau staff in Stratton's office. He found the "whole bunch . . . sore." Later, the Bureau would blame the inadequacy of its TM on Trowbridge's failure to maintain adequate technical liaison. But only six weeks before the meeting, Stratton himself had urged the AEF to use the Bureau's device on the front; the lack of information was no explanation of the agency's persistence. Trowbridge discerned at the meeting itself that

[9] Lt. Col. R. A. Millikan to Col. L. S. Horner, Oct. 24, 1918; Horner to C. W. Nash, Oct. 30, 1918; Horner to S. D. Waldon, Oct. 30, 1918; Horner to M. W. Kellogg, Oct. 19, 1918, AAF, Series 22, file 360.05, Langley Field.

[1] Millikan to Greta Millikan, Nov. 18, 1918, RAM, Box 49.

[2] "Report of Sound Ranging Investigation of the Bureau of Standards," January 1918, OCE, Entry 504; Horatio B. Williams to Trowbridge, Jan. 27, 1918, AEF, Entry 210.

[3] Trowbridge to Sarah Trowbridge, May 14, 1918, "War Letters," *NYPL*, 43 (1939), 913.

[4] Trowbridge to Williams, May 25, 1918, AEF, Entry 210.

[5] Trowbridge to Williams, March 25, 1918, AEF, Entry 210; cable for Chief of Engineers, June 23, 1918, in "Preliminary [Historical] Report," 29th Engineers, p. 143.

Rosa's group had received thousands of dollars from the War Department and would be embarrassed if its TM did not get into action.[6]

Whatever Rosa's reasons, Trowbridge told him that, since the Bureau TM still could not match the Bull-Tucker, he would not replace the British apparatus on a stable front. If Rosa really wanted to cooperate, he would put a simple microphone into the Bureau TM and save face by trying to come up with a set appropriate for mobile warfare. More important, Rosa was not to let the work delay the shipment of even a single Bull-Tucker.[7]

Rosa agreed. So did the engineers at Western Electric when Trowbridge insisted that they scrap their modified designs of the Bull-Tucker (the company's model was so "complicated," Trowbridge remembered, that he would have needed "a corps of trained physicists to operate it").[8] Trowbridge was not only the scientific equal of the stateside experts; right down to that sartorial signet of Pershing's staff, the Sam Browne belt strapped across his chest, he was Major Trowbridge from Chaumont.

Trowbridge had grown close to the regulars at Chaumont, had even given up his gourmet's table to become president of the officers' mess. Particularly important, he had developed a warm friendship with his young superior, Colonel Alexander. On long auto trips around the front, the cultivated eastern academic would discourse on architecture; the West Pointer from Paris, Missouri, on military intricacies. Their conversation, Trowbridge once mused, covered "cook stoves and clerestory, triforium and tri-nitrotoluene, gargoyles and guns." In France, Major Trowbridge assumed that he was "managing a service of the Army and not running a laboratory at the front," and he was determined to guarantee his men a device usable under "the nervous strain of field warfare."[9]

Trowbridge may have been a mere reserve officer, but, in his quite accurate analysis of his victory over Rosa, "None of the Bureau or Research [Council] people dare question my authority and the Engineer Dep[artment] people back me up to the limit."[1]

George Ellery Hale raised a quite different matter of military policy when in June 1917 he se..t Secretary Baker a thoughtful memorandum by Ernest Rutherford. The document's main point: Scientists were already

[6] Trowbridge to Colonel Alexander, Sept. 5, 1918, AEF, Entry 207, Trowbridge file; [Rosa?], "Description of Improved Sound Ranging Apparatus of the Bureau of Standards," c. Dec. 1918, OCE, Entry 504; Stratton to Major W. D. Young, July 13, 1918, AEF, Entry 207, Trowbridge file. The records of the Bureau of Standards in the National Archives (RG167) are unaccountably mute on the Bureau's side of the sound-ranging story.

[7] Trowbridge to Alexander, Sept. 5, 1918, AEF, Entry 207.

[8] Trowbridge, Introduction to "War Letters," NYPL, 43 (1939), 604–5; Trowbridge to Chief of Engineers, Aug. 30, 1918, AEF, Entry 207, Trowbridge file.

[9] Trowbridge to Sarah Trowbridge, May 8, 1918, May 17, 1918, "War Letters," NYPL, 44 (1940), 9, 10; Trowbridge, Introduction to "War Letters," NYPL, 43 (1939), 603.

[1] Trowbridge to Alexander, Sept. 5, 1918, AEF, Entry 207, Trowbridge file.

in short supply, and nothing symbolized the blind waste of sending such critically important talent into the trenches more than Moseley's death at Gallipoli.[2] But Baker emphatically opposed draft exemptions for any groups as a class, even, he told Hale, "a group of such importance to our national welfare as the men of science." Everyone would have to apply to his local board for deferment on the basis of his individual role in the civilian war effort. In any case, Hale was assured, the army certainly planned to use scientifically qualified draftees to best advantage.[3]

At first, the Selective Service system operated so that obtaining draft-eligible scientists, as Samuel Wesley Stratton complained, was "uncertain, indeterminate, and subject to accidental personal contact," but by early 1918 the War Department was officially detailing drafted scientists from the infantry into appropriate military or civilian technical bureaus. Requests had to be made by name rather than discipline; a colonel warned that asking the "Adjutant General of the Army for the transfer of ten chemists is as likely to produce drug clerks as real chemists." A few months more and, to ease the situation in the factories, the army was explicitly urging local boards to defer technical experts.[4]

Chemists and chemical engineers dominated the ranks of the experts, and observers then and since have, with considerable justification, called World War I a chemist's war. At home, by the armistice optical glass, nitrates, and poison gases were pouring out of the factories in quantities ample enough for the military's vast needs; abroad, some of the nation's best organic chemists were in the AEF. But besides the highly important submarine detection devices, physicists helped develop a variety of aeronautical instruments as well as airborne photographic devices. Robert W. Wood drew upon his optical virtuosity to provide the infantry with signaling lamps of both unprecedented range and, since some blinked with ultraviolet light, of reassuring invisibility to the enemy. Physicists and electrical engineers at Western Electric whirled ahead the wireless transmission of voice, in some estimates by a decade.

[2] Ernest Rutherford, "Memorandum on the Utilization of . . . Scientific Men During the Period of the War," June 8, 1917; Hale to the Secretary of War, June 11, 1917, AGO, Central File 334.8, Box 774.

[3] Secretary of War to Hale, June 16, 1917; Adjutant General J. T. Dean to Chairman of the NRC, July 6, 1917, *ibid.* The army's lawyers considered class exemptions illegal, many high-ranking officers thought them unwise, and the President was reported to believe that the public would regard a class exemption for college men as favoritism for the sons of the rich. See Colonel D. W. Ketcham to the Chief of Staff, Jan. 24, 1918, WDGS, War College file 9860–22; Ray Lyman Wilbur, *The Memoirs of Ray Lyman Wilbur, 1875–1949*, ed. Edgar E. Robinson and Paul C. Edwards (Stanford, 1960), p. 247.

[4] Stratton to the Secretary of Commerce, Nov. 14, 1917, and attached memoranda of conversations with the Provost General, the Judge Advocate General, and the Director of the Bureau of Mines; S. W. Stratton to G. O. Squier, Dec. 20, 1917, NBS, Entry 51, file AG; Col. D. W. Ketcham to the Chief of Staff, Jan. 24, 1918, WDGS, War College file 9860–22; Brig. Gen. Henry Jervey to the Adjutant General, Sept. 21, 1918, WDGS, War College file 9860–41.

Chemical or physical, the war—and the Millikans—had "forced science to the front," as Hale aptly said, from the technical troops in the AEF to the governmental and industrial laboratories at home. Army and navy projects even found their way into the laboratories of some forty of the nation's colleges, and the campuses operated under the constraint of tight security regulations for the first time in American history.[5] The professors continued to profess, but, like Bazzoni at Ft. de St. Menge, many now faced classes of green technical troops bound for the AEF. In war research or teaching, so many academics were diverted from their normal pursuits that a Yale physicist remarked: "Whatever services science may render to war, it is plain that a state of war is not favorable to the [intellectual] progress of science."[6]

This particular war had not been favorable to Edison. Provided with money and vigorous administrative support from "Friend Daniels," as the inventor liked to call the secretary, Edison fashioned some forty-five devices for the military. They were "all perfectly good," he said, but the navy ignored them all.[7] Edison's Consulting Board received some 100,000 technical suggestions from the public; the National Advisory Committee for Aeronautics, more than 16,000 in addition. Few were promising. The two agencies had to remind the public to consult standard scientific texts or competent scientific authorities before burdening governmental officials with the examination of new devices or plans ("there is no known method of 'charging the sea with electricity,'" the Consulting Board typically pointed out).[8] If some of the suggestions were technically silly, many, as a knowledgeable naval officer said of Edison's ideas, were based on the "wrong premises" or covered "old ground."[9]

For thoughtful military observers the meaning of it all—of Edison's failure and, whatever the conflicts of authority, of Millikan's success—was clear: The advance of defense technology required the organized efforts of scientists and engineers whose first steps often had to be, as Admiral Griffin said of submarine detection, "in a sense, backwards into the unexplored regions where fundamental physical truths and engineering data were concealed."[1]

[5] Hale, "The Purpose of the National Research Council," in *National Research Council Bulletin*, I (1919–21), 2; Parke R. Kolbe, *The Colleges in War Time and After* (New York, 1919), p. 93.

[6] Henry A. Bumstead, "Present Tendencies in Theoretical Physics," *Science*, 47 (Jan. 18, 1918), 52.

[7] Quoted in Matthew Josephson, *Edison* (New York, 1959), p. 454. Edison enjoyed Daniels' support to the end. See Daniels to Edison, Oct. 14, 1918, JD, Box 39.

[8] *The Submarine and Kindred Problems* (Bulletin No. 1; Naval Consulting Board, July 14, 1917); Hughes, *Sperry*, pp. 254–55; L. C. Stearns, "Report on Inventions Handled by the National Advisory Committee for Aeronautics," April 11, 1918, SI, NACA file.

[9] William S. Smith to Daniels, n.d., JD, Box 400–401, Naval Consulting Board file.

[1] [Griffin], *History of the Bureau of Engineering*, p. 52.

X

Cold War in Science

At the opening of 1918, a short while before Trowbridge's rangers settled in on the St. Mihiel salient and not long after the demonstration of the submarine listening devices in the English Channel, Robert Millikan, now tirelessly attending meetings in New London, overseeing Signal Corps research at Langley Field, and often catching his sleep in railroad berths, reported that he no longer felt "the necessity of giving scientific men a chance to do their part."[1] The National Research Council, its activities burgeoning, had just moved into new quarters, a twenty-two-room building on 16th Street a few blocks north of the White House. More academic institutions, more scientific societies, and more individual scientists had volunteered their services than the NRC had tasks to assign. The Council's contacts with industrial managers, spreading far beyond the small, enthusiastic circle of Carty, Whitney, and Jewett, included many of the dollar-a-year men who flooded into Washington, their private railroad cars crowding the Union Station, their names making hotel registers read like a corporate Who's Who. Working together with the Millikans in numerous agencies and committees, industrialists had learned that academics could be hard-headed administrators and could treat technical matters with a practicality worthy of an Edison.

Whatever the jurisdictional disputes with the federal bureaucracy, the NRC was the sole agency in Washington that straddled every scien-

[1] Millikan to Greta Millikan, Jan. 15, 1918, RAM, Box 49.

tific constituency, and federal bureau chieftains used the Council to get in touch with the right scientists and, most important, to keep informed of projects and programs outside their own bailiwicks. Impressed by the usefulness of the NRC, the Carnegie Corporation had reopened the possibility of an endowment for the Academy. John J. Carty, George Ellery Hale noted, had also predicted that the "large-minded" leaders of industry would be giving "great support" to science in the future, particularly to physics and chemistry. By Hale's report, the close contact of the NRC with federal and industrial officials was winning "many new friends for pure science," and as a "coalition of widely different interests," the Council itself was "certain to be beneficial to research in all of its aspects."[2] In March 1918 Hale moved to make the Council and its governmental connections permanent by enlisting the helpful Colonel House to obtain an executive order for the purpose from President Woodrow Wilson.

Hale himself wrote the draft of the order. It provided for the government to continue the appointment of federal representatives to the Council, and for the Council to continue the fostering of cooperative efforts in research, both governmental and otherwise. Hale assured Wilson that the NRC had "no desire for *authority*" over any governmental agency. It merely wished "such official recognition as will give it the influence . . . to secure cooperation." But Wilson, who doubted his "right to give any outside body" the power to coordinate the scientific work of the government, asked Secretary of Agriculture David Houston to revise the order in an appropriate manner.[3]

Houston, a respected political scientist and former university president, disliked the draft order. Politically elitist in key respects, it would permit the NRC to promote research in the name of the government yet to remain a private organization beyond governmental control. It would establish permanently an agency independent of the political process with responsibilities that the Constitution made political and, as William B. Chandler had said in the days of the Allison Commission, properly so. Like Chandler, Houston considered politics an instrument of democratic control of government. All of the secretary's revisions, from the finest changes in language to the slashing away of complete paragraphs, bespoke a determination to keep federal scientific policy in the hands of elected officials.[4]

[2] Hale to Arthur Schuster, April 18, 1918, Sept. 22, 1917, GEH, Box 47.

[3] Copy of the draft order is with David Houston to Wilson, April 30, 1918; Hale to the President, April 22, 1918; Wilson to Hale, April 19, 1918, WW, File VI, Case 206.

[4] Houston denied the NRC the responsibility either to coordinate the federal scientific bureaus or even to promote cooperation among them. And he went further. Hale would have had the Council secure the establishment of research fellowships and help shape the work of industrial laboratories. To Houston, it would be "scarcely appropriate" for the President officially to suggest that a "semi-govern-

All the same, the modified order, which Wilson signed on May 11, 1918, left intact the most fundamental points for Hale's promotional purposes. It requested the Academy to perpetuate the National Research Council, and the NRC to stimulate pure and applied research through Hale's multifaceted strategy of cooperation. It also explicitly provided for future U.S. Presidents to appoint federal representatives nominated by the Academy, which assured the Council the continuing participation of the leaders of the state, without their interference. Hale, who was principally concerned with what could be done for science outside the federal establishment, elatedly declared, "We now have precisely the connection with the government that we need," and he intended to make good use of it at home—and also abroad.[5]

Like most scientists of his day, Hale grew up to a proud tradition—that scientists everywhere belonged to an international fraternity above politics, where national differences made no difference. By the turn of the century the tradition had found considerable institutional expression in the establishment of numerous organizations, including the International Association of Academies, which had a membership of twenty-one academies drawn from fourteen European nations and the United States. After 1914 knowledgeable observers liked to repeat the remark of the British chemist Humphry Davy, who in 1807 accepted a prize for his research from Napoleon: "If the two countries or governments are at war, the men of science are not—that would indeed be a civil war of the worst sort." The scientific community, Davy added, should "soften the asperity" of armed conflict.[6]

But in October 1914 ninety-three German professors, among them Wilhelm Röntgen, Max Planck, and thirteen other scientists of equally high repute, issued a manifesto justifying the burning of the library at Louvain.[7] Lashing back, angry fellows of the Royal Society demanded the removal of all Germans and Austrians from the list of foreign members, and the French Academy dropped the signers of the manifesto.[8] In mid-1917 the eminent French mathematician Emile Picard, a former

---

mental agency" create research fellowships. Similarly, the President might ultimately find it "embarrassing" to approve, even tacitly, industrial research activities "by an agency over which the Government has very slight control. Such activities might take a course quite at variance with the policy of the administration." Houston to Wilson, April 30, 1918, and accompanying draft of the order with penciled changes, *ibid.*

[5] Hale to James R. Garfield, May 16, 1918, NRCM, Hale file.

[6] Brigitte Schröder, "Caractéristiques des relations scientifiques internationales, 1870–1914," *Cahiers d'Histoire Mondiale*, 13 (1966), 170–74; Davy is quoted in T. F. Crane, "Scholars and the War—Then and Now," *The Nation*, Aug. 3, 1916, pp. 107–8.

[7] A full text of the manifesto and list of signers is in Georg F. Nicolai, *The Biology of the War* (New York, 1918), pp. xi–xiii.

[8] Hale to Arthur Schuster, June 9, 1915, GEH, Box 37; *Nature*, 94 (Oct. 29, 1914), 236.

president of the French Academy of Sciences, told Hale that "personal" relations of any kind would be "impossible" with German scientists even after the war. They had to be ostracized from the structure of international research. Picard's specific points: Dissolve the International Association of Academies and replace it with a new organization—without the neutrals at least for the duration, without the Central Powers indefinitely.[9]

Hale had dreamed of transforming the International Association of Academies, which was mainly ceremonial, into "a really effective nucleus" for general international scientific cooperation.[1] But after the United States entered the war, Hale came to hold German professors responsible for the acts of the Kaiser's government and pronounced himself unwilling to conjoin with German scientists after the war. Stimulated by Picard, Hale proposed to reorganize international science completely, both to exclude the Central Powers and to achieve a more robust multinational cooperation.[2]

In England, Arthur Schuster, the secretary of the Royal Society and Hale's old friend and fellow spirit in international scientific cooperation, worried that if the neutrals chose to remain in the Association of Academies, then Allied, not German, scientists would be ostracized.[3] At home, most of the leaders of Hale's own National Academy agreed with Picard about future relations with German scientists, yet because of the problem of the neutrals, a sizable fraction of these exclusionists doubted the wisdom of reorganization before the end of the war. Another—and much smaller—group among the Academy leaders dissented from wartime reorganization on the grounds, tellingly summarized by Arthur L. Day, the Home Secretary of the Academy, that Hale was calling for reorganization to wage a war in science after the war on the battlefield. Hale, Day warned, would be building on "quicksand."[4]

Hale sensed that to reconstruct international science before the peace he would have to find his way out of certain difficult dilemmas. How to reorganize research under the nations of the Entente without losing the neutrals to the Germans? How to persuade the more cautious and moderate members of his own Academy to back a reorganization in the present that would exclude the Central Powers in the future? And Hale believed that, just as the domestic activities of the National Academy required the cooperation of the federal government, so, to even a greater

[9] Picard to Hale, July 22, 1917, GEH, Box 47.

[1] Hale to Henricus Bakhuyzen, Sept. 16, 1915; to Gaston Lacroix, May 2, 1917, GEH, Box 3, Box 47.

[2] Hale to Edwin Grant Conklin, Oct. 8, 1917; to H. H. Turner, July 25, 1917, GEH, Box 47, Box 41.

[3] Schuster to Hale, Aug. 27, 1917, GEH, Box 47.

[4] Day to Hale, Oct. 1, 1917, GEH, Box 47. For the division between exclusionists and dissenters, see the replies from the members of the National Academy Council to Hale, Sept. and Oct. 1917, *ibid.*

extent, would any foreign venture. Yet President Wilson was insisting that America's quarrel was with the rulers, not the people, of Germany. How to obtain the blessing of the Wilson administration, Hale wondered, for an international scheme that excluded enemy scientists principally on grounds of war guilt?[5]

By the spring of 1918, Hale had glimpsed a way out of the dilemmas in the international extension of the Research Council that resulted from the appointment of American scientific attachés in London and Paris. The two attachés were funneling home a large volume of reports, and some fifty to seventy physicists, engineers, and military officers were convening every Thursday evening at the NRC offices to discuss the week's communications from abroad. Encouraged by the usefulness of this process, Hale proposed that each of Germany's enemies ought to form its own NRC; all would federate in an Inter-Allied Research Council to which the Allied and American governments would send army and navy officers, scientists, and industrialists for the consideration of military problems.[6]

By creating the new agency for war purposes, the Allies could naturally omit the Central Powers without openly committing themselves to a recriminatory policy. As a result, the exclusionist innovation would neither alienate the neutrals, who could be invited to join later, nor give the moderates in Hale's own academy any grounds for opposition. Most important, Hale supposed that Wilson just might approve an agency which would strengthen the war effort while leaving the issue of German scientists to the future.

The leadership of the National Academy united behind Hale's scheme. Arthur Schuster, ill, weary, his son at the front, doubted the acceptability of Hale's plan, but he at least agreed that Hale ought to have a chance to argue for his proposal at a conference on the future of international science scheduled for London in October. Schuster also expected that Hale might win the support of both Britain's scientists and Britain's government if his scheme were crowned with the approval of the highly esteemed American President.

Hale tried twice for Wilson's endorsement—surely the President would wish "to pass upon any international arrangements" to be proposed by "representatives of the United States"—but while Wilson entered no objection to the plan, neither did he approve it.[7] Hale, disappointed, was also worried over reports from Europe that the French, Belgians, and even some British scientists intended to force the London conference to

[5] Hale to Schuster, Sept. 22, 1917, *ibid*.

[6] Hale to Schuster, April 18, 1918; Hale, "Suggestions for the International Organization of Science and Research," *ibid*.

[7] Hale to the President, April 29, 1918; Wilson to Hale, May 8, 1918; Hale to Wilson, May 10, 1918; Wilson to Hale, May 13, 1918, WW, File VI, Case 206, Case 206a. Hale to Wilson, Aug. 3, 1918; Wilson to Hale, Aug. 14, 1918, GEH, Box 47. See also, Hale to the Secretary of State, May 1, 1918, DST, File 763.72/9887.

issue sweeping denunciations of Germany. He told the President about that possibility when on September 10, his departure for London imminent, he went to the White House for a face-to-face talk.

The President, Hale summarized the gist of this conversation, was *"very emphatically opposed* to any resolution directed against German men of science." Such recriminations would be the "best possible way to play Germany's game," since they would add credibility to "the claim that Germany is surrounded by vindictive enemies."

Hale countered with how organizing for war would sidestep just that pitfall, but the President objected to conjoining with the Allies in any more organizations of a formal kind. Hale recorded Wilson's reasoning: "France and other European nations have felt the war much more than we have, and are therefore likely to take drastic action that might bind us to do things contrary to our natural intent." Hale jockeyed. On the point of a new formal organization, Wilson—apparently not the Wilson who went to Versailles entirely naïve about the British and French—refused to budge. He would acquiesce only in inter-Allied scientific conferences of a strictly informal nature.[8]

London was cold and damp when Hale arrived in early October, but amid the prospect of imminent victory and by no means dreary in spirit. Under the circumstances, Hale apparently decided, the Inter-Allied Council could be created without violating Wilson's policy: The President had objected only to conjoining formally with the Allies while the war lasted, and Hale's proposed reorganization was unlikely to take real shape until some time after the surrender of the Central Powers. At the urging of Hale and his fellow American delegates, the Anglo-American group came to the conference armed with a declaration of principles in line with Woodrow Wilson's position that a reformed Germany could enter the League of Nations.[9] The conference unanimously adopted the declaration as the political preamble to a series of organizational resolutions: The nations at war with the Central Powers were to form new scientific unions for particular disciplines and these unions were to be tied together through a new international council, which was both to advance pure and applied science in general and to deal with matters related to national defense.[1]

In mid-July 1919 the new International Research Council held its constitutive assembly in the splendid Palais des Académies, Brussels, the inaugural session graced by the king of Belgium, the deliberations, in the proud remark of one of the 224 delegates, those of a "little peace

[8] Hale, "Memorandum of Interview with President, Sept. 10, 1918," GEH, Box 60.

[9] Hale to Wilson, Oct. 15, 1918, WW, File VI, Case 206c. A copy of the declaration of principles is in GEH, Box 47.

[1] "Preliminary Report of Inter-Allied Conference on International Scientific Organizations, Held at the Royal Society on Oct. 9–11, 1918," GEH, Box 48.

conference."[2] Along with adopting statutes, the assembly unanimously invited thirteen additional nations to membership in the Council and its unions. Soon fifteen countries belonged to Hale's new organization of international science, including, though most did reserve the right to deal with the Central Powers independently, almost every neutral nation invited to join.

In 1919 the National Research Council reorganized on a peacetime basis and appointed to its new Government Division nominees from the technical bureaus in the federal establishment, military as well as civilian. Thoughtful officers in the armed services, made aware by the war of the impact civilian scientists might have upon the advance of defense technology, did not want the military to retreat from its historically unparalleled involvement with civilian science.

Both reflecting and advancing the thinking of such officers in the navy was Commander Clyde S. McDowell, an Annapolis graduate and a veteran of New London for whom the wartime experience had flatly contradicted the happy notion that "we simply have to express a need for some new device or apparatus and the American genius will arise with the answer." In McDowell's opinion the war had been "won by *research*." He urged the navy to incorporate scientists into its peacetime organization, to go full steam ahead not only with applied but with pure research, even to become "the greatest patron of science in this country."[3] Few senior navy regulars were quite so ambitious as McDowell, but Secretary Daniels and his ranking officers all expected the bureaus to maintain programs in what Admiral Sims and others were now calling "research and development."[4]

In the army, wartime expansion had transformed what was now called the Air Service into a powerful military arm, with as much right to control its technical affairs as a Thurman Bane—who intended to make Dayton into an experimental facility of gleaming capabilities—could want. In the foot soldier's army, the technical bureaus, serving the diverse needs of the infantry, worked without a common technological purpose and at different levels of technological sophistication. Not even Squier, who emphasized how "military supremacy must [now] be looked for primarily in the weapons and agencies provided by scientists and engineers," called for what would have been the equivalent of McDowell's aim, a

[2] Arthur Schuster, ed., *International Research Council, Constitutive Assembly Held at Brussels, July 18th to July 28th, 1919, Reports of Proceedings* (London, 1920), p. 10.

[3] McDowell, "Naval Research," *U.S. Naval Institute Proceedings*, 45 (June 1919), 898–99, 908. It is interesting to note that McDowell's ideas impressed Louis Howe, who called them to the attention of Assistant Secretary of the Navy Franklin D. Roosevelt. Howe to Roosevelt, March 10, 1919, JAG, Entry 108, McDowell file.

[4] Force Commander to the Secretary of the Navy, Nov. 13, 1918, NR, Subject File OD. See also U.S. Navy Department, *Annual Report, 1920*, p. 716.

concerted program of "army research."[5] In the postwar army, the conversion to science varied from bureau to bureau, the degree of commitment running from mere gestures in the Corps of Engineers to McDowell-like fervor in the Chemical Warfare Service.

Critics might charge that poison gas was a barbaric weapon, but General Sibert, the chief of the Service, liked to point out that while some 24 percent of casualties from rifles and artillery had died, fewer than 2 percent of those from gas had not come back. "Gas," Sibert concluded, "is twelve times as humane as bullets and high explosives."[6] But President Wilson himself remarked that the United States would never initiate chemical warfare, and so vast a majority of Americans apparently agreed with him that in the months following the armistice the issue was not awarded the dignity of a major public debate.

Strategic reliance on poison gases enjoyed little enthusiastic support even within the highest echelons of the regular military itself, certainly not from Chief of Staff Peyton C. March, a regular's regular, his respect for the army's best traditions as sharp as his pointed goatee. To March, the war had not been won "by some new and terrible weapon of modern science"; like "every other war in history," it had been won by "men, munitions, and morale."[7] He found chemical warfare reprehensible, not because of the way poison gases affected combatants but because, by capriciously following the wind, they threatened civilians. Eager to see the distinction between combatants and noncombatants preserved, March wanted the nations of the world to abolish gas warfare "altogether."[8]

By the spring of 1919, March and Secretary Newton Baker, who agreed with the Chief of Staff's objection to poison gas, had broken up the Chemical Warfare Service. But Brigadier General Amos A. Fries, a West Pointer who had gone to France to build roads and had quickly shifted over to the work of Chief of Gas Service, AEF, lobbied to have the CWS restored. In 1919 he took his case to the Senate Military Affairs Committee hearings on army reorganization, where Senator Duncan U. Fletcher of Florida had been wondering: If you "kept on with these scientific means of destruction," would you not run the danger of "wiping out the whole human race?"[9]

As a weapon that placed noncombatants in jeopardy, Fries said, gas was not unique. Some two hundred people had been killed when the Germans managed to reach Paris with a long-range gun. Hundreds more had died when German bombers attacked French cities and even distant

[5] Squier, "Aeronautics in the United States . . .," Nov. 1918, p. 79, WDGS, Army War College file.

[6] "Report of the Director, Chemical Warfare Service," in U.S. War Department, *Annual Report*, 1919, I, 4391.

[7] "Report of the Chief of Staff," *ibid.*, I, 473.

[8] U.S. Congress, Senate, Subcommittee of the Committee on Military Affairs, *Hearings, Reorganization of the Army*, 66th Cong., 1st and 2d Sess., 1919, I, 93.

[9] *Ibid.*, I, 178.

London. For Fries, the issue was not some new form of warfare but the historic one of war itself. The more "deadly" the weapons, Fries argued, "the sooner . . . we will quit all fighting." If war could be made "so terrible" that it would not last "more than five or ten minutes," then war would never begin.[1]

Congress granted the Chemical Warfare Service permanence, and Fries replaced Sibert as chief. The nation's developing chemical arsenal, Fries declared in his first annual report, would go a long way toward "deterring" other countries from starting wars.[2]

Yet if Fries's reasoning adumbrated an element of some significance in a later generation's drive to embrace military research and development, in 1919 the nation was fast retreating into isolationism and demanding cutbacks in military expenditures. In the 1920s, every military technical bureau had to reduce activities drastically below what their chiefs considered desirable peacetime levels; so, too, did the National Advisory Committee for Aeronautics, which the army and navy wanted to push forward basic aeronautical research. And among the casualties of the budgetary difficulties was the military's new partnership with the civilian scientific community, particularly with the National Research Council.

Fiscal problems aside, Millikan and his colleagues, whose lobbying had been crucial to the NRC's active relations with the military during the war, were gone from Washington. The Council's scientific attachés came home, since armies and navies on both sides of the Atlantic halted the exchange of technical information. The U.S. armed services now had their own mechanisms for learning about new technical developments and each could hire consultants, appoint scientific advisory groups, or let contracts to outside laboratories directly. More important, once released from the pressure of war, the army and navy preferred to do military research in military laboratories. That way, the armed services could keep the work targeted on projects of martial practicality. That way, too, they could best prevent violations of security, as always a cardinal point of military thinking. Even McDowell cautioned against relying on civilian scientists who were not "familiar with service . . . conditions."[3]

Now the navy had a research laboratory on the banks of the Potomac to help nurture a naval-minded nucleus of scientifically literate personnel. The National Defense Act of 1920 authorized the army to send up to

[1] *Ibid.*, I, 364–65.

[2] "Report of the Chief of Chemical Warfare Service," in U.S. War Department, *Annual Report, 1920*, I, 1887–88.

[3] Director, Naval Research Laboratory, to the Chief of Bureau of Construction and Repair, Sept. 14, 1922, BuS, Entry 88, file 19123; McDowell, "Naval Research," *U.S. Naval Institute Proceedings*, 45 (June 1919), 899–900, 902. Admiral W. Strother Smith told the Secretary of the Navy: "Experience obtained by both branches of the service has shown that all research work for war purposes must be under military control." Smith to the Secretary, [July 1922], BuS, Entry 88, file 19123. See also Thurman Bane to the Director of Air Service, July 8, 1920, AAF, Series 166, Organization, Air Service: file 321.9B.

2 percent of the officers corps each year for advanced study in colleges and universities and gave technical careers the same legal status as careers in the line.[4] The army invited academics to join the reserve corps, but it refused the request of the NRC to excuse such recruits from basic training. Without that indoctrination the professors, in a regular's distrustful phrase, would be inclined to "ride their particular hobbies."[5] In degrees that varied from bureau to bureau, army and navy technical chiefs expected academics not so much to consult from the campuses as to spend summers or sabbaticals in military laboratories, where they could familiarize themselves sufficiently with service requirements.

General Sibert explained that, by working in a military environment, scientists would get "so imbued with the necessity of secrecy" as to stop wanting to "publish their results in the scientific books and periodicals"[6] But while civilian scientists were willing to put up with secrecy in wartime, in peacetime they refused to submit to such regimentation. In all, the military's wartime partnership with the civilian-dominated NRC broke down for reasons more profound than peacetime budgetary difficulties. Chief among them was the disjunction between, on the one hand, the freedom demanded by civilian scientists and, on the other, the insistence of the armed services on controlling their own research.[7]

If not science for the military, then science for industry—"never again," Edwin B. Rosa read the lasting lesson of the war, "would anybody question the . . . economic value of scientific investigation."[8] So it seemed, with industry boosting research as a weapon of competition; with the American Federation of Labor urging science as a stimulant to a higher standard of living through higher productivity; with Stratton managing to rescue the Bureau of Standards from a severe appropriations cut by emphasizing its services to industry; and with Elihu Root, the respected corporation lawyer, former secretary of war, trustee of the Carnegie Corporation, and Hale's friend and adviser, pointing to the efficiency with which Germany had mobilized research. "The prizes of industrial and commercial leadership," Root predicted, "will fall to the nation which organizes its scientific forces most effectively." At Hale's instigation, Cleveland H. Dodge, George Eastman, Andrew W. Mellon, Ambrose Swasey, and Pierre S. du Pont all publicly declared, to use Root's phras-

[4] *U.S. Statutes at Large*, 41, pt. 1, 760–62, 786 (1920).

[5] The regular was the acting Chief of Engineers, Col. Frederick W. Abbot. Abbot to War Plans Division, Aug. 20, 1919, WDGS, War Plans, Division file 796.

[6] U.S. Congress, House Committee on Military Affairs, *Hearings, Army Reorganization*, 66th Cong., 1st sess. 2 vols.; 1919, I, 544–45.

[7] See the summary analysis of Frank Jewett, in U.S. Congress, House Committee on Military Affairs, *Hearings, Research and Development*, 79th Cong., 1st Sess., May 1945, p. 58.

[8] Rosa, "Expenditures and Revenues of the Federal Government," American Academy of Political and Social Science, *Annals*, 95 (May 1921), 33.

ing, that without pure science the "whole system" of industrial progress would "dry up."[9]

The future of medical progress concerned Simon Flexner, the Director of Laboratories at the Rockefeller Institute of Medical Research, an intimate of the Rockefeller Foundation trustees and enthusiastic member of the National Research Council. His father an itinerant German-Jewish peddler, Flexner had scrimped his way to a medical degree in Louisville, Kentucky, then gone to study at Johns Hopkins under William Henry Welch. For Welch, medical research at its best meant the application of physics and chemistry to the problems of physiology. Flexner, who after a lecture demonstration would come to fish the specimen out of the discard barrel and continue the experiment, became a devotee of Welch's kind of medicine and, after honing himself on postgraduate work in Europe, won a reputation as the uncle—Welch was the father—of modern medical education in the United States.

By World War I, Flexner was eager to bolster American physics and chemistry, for the benefit of medical research and for their own sake. By early 1918, his fellow Rockefeller officials were also pondering the importance of the physical sciences to the economy. Now, at Flexner's urging, George E. Vincent, the head of the Foundation, who had formerly presided over the University of Minnesota in the progressive vein of Van Hise, sent a letter of inquiry to a small group of leading scientists, including Millikan. "The industrial competition which will follow the war will test the scientific resources of the nation," Vincent observed; the United States had to keep filled its reservoirs of fundamental knowledge. What would Millikan and his colleagues think of the Rockefeller Foundation's endowing a new, independent research institution? It would be like the Rockefeller Institute for Medical Research, but devoted wholly to the advancement of physics and chemistry.[1]

While generally enthusiastic, Millikan's colleagues in the NRC could not agree exactly how to establish such an institution, but they all endorsed the funding of a research fellowship program for promising young men.[2] In April 1919 the Rockefeller Foundation awarded the NRC

[9] See the AFL resolution, *Science*, 50 (July 4, 1919), 15, and comment, *ibid.*, 50 (Sept. 5, 1919), 221; Root, "Industrial Research and National Welfare," *Science*, 47 (May 29, 1918), 533. Copies of similar statements by Dodge et al. are in GEH, Box 52.

[1] The letter also went to Julius Steiglitz, Alexander Smith, Albert A. Michelson, and John Zeleny. Vincent to Steiglitz et al., Feb. 5, 1918, RF, Series 200, Box 37.

[2] On the one side was Flexner, who insisted on a single, independent research institute. On the other were Hale and Millikan, their ambitions centered in the pluralist university system, who argued both for several institutes and for attaching them to selected universities across the country. Hale to Arthur Fleming, March 4, 1918; "First Draft of a Plan for the Promotion of Research . . .," [1918]; "Proposal for the Endowment of Research in Physics and Chemistry," [1918], GEH, Box 16, Box 60; Hale to John C. Merriam, Nov. 11, 1918, JCM, Box 227; Nathan Reingold, "World War I: The Case of the Disappearing Laboratory" (unpublished manuscript, 1975), pp. 10–13, 25–39.

$500,000 to be spent over five years for postdoctoral fellowships in physics and chemistry.[3] It was the first program of its kind in the United States, and it was ample enough to operate on a national scale.

In 1919, too, the Carnegie Corporation, responsive to Root, who helpfully argued Hale's case, and to the economic utility of science, granted the Academy an endowment of $5,000,000; about a third of it was to construct a building for the Academy and the NRC, the rest to fund NRC operations. All the same, Willis R. Whitney, worried about the competitive position of American industry, had inveighed to Hale against letting American science remain "subject to the whims of philanthropists."[4] The governments of Europe, following Germany's example, were now subsidizing pure and applied research, and Hale wondered whether the NRC ought not to push for federal support of scientific research in the universities.

College deans, presidents, scientists, and engineers had been attempting to agree on an appropriate federal support program since 1916, when Senator Francis G. Newlands of Nevada introduced a bill to provide $15,000 annually to each state and territory for an engineering experiment station at its land-grant college. In twenty of the forty-eight states, mainly in the South and West, the state university and the land-grant college were separate institutions. Spokesmen for these state universities vigorously protested Newlands' bill and maneuvered to amend it; land-grant college spokesmen objected to any deviation from the precedent of the Morrill Act. Neither faction budged, and the legislation fell victim to deadlock.

In Washington, the chief proponent of the state universities was Phineas V. Stephens, a consulting engineer on retainer at the Georgia School (later, Institute) of Technology. Georgia Tech was starting an engineering research laboratory and hoped to benefit from the Newlands bill, but it was not a land-grant college. Soon after the new Congress convened in 1917, Stephens, hoping to prevent the land-grant colleges from monopolizing the prospective money, wrote a new bill which provided for the legislature in each state to locate its engineering station at the school with the best engineering department. For the sake of the national interest, it also vested the power to approve the choice and coordinate the work of all the stations in the National Bureau of Standards. Since Senator Newlands had died, Stephens found two "strong" men to introduce his bill: Senator Hoke Smith and Congressman William

[3] Hale to Vincent, Feb. 24, 1919; Rockefeller Foundation, "Minutes . . . Executive Committee . . . April 9, 1919," RF, Series 200 E, Box 168.
[4] Whitney to Hale, June 11, 1918, GEH, Box 43.

Schley Howard, both advocates of aid to education and, more important, both of Georgia.[5]

When the Smith-Howard bill came to hearings in June 1918, some land-grant college administrators went on record in its favor, pointing out that, because of local circumstances, a number of the Morrill Act institutions would get the money allocated to their states. But the leaders of the National Research Council, having come to know the darker side of bureaucratic Washington, adamantly opposed settling the power to coordinate the work of the proposed experiment stations in the hands of Stratton's Bureau of Standards. In the fall of 1918 the Council rewrote the bill. Of course some official body in Washington would coordinate the work of the stations, but the new draft included the clause: "There shall not be vested in this agency control over the grants, nor shall it prescribe the researches undertaken nor the mode in which they shall be conducted."[6] Here was a political elitism which went beyond that of the generation of John Wesley Powell. Hale and his colleagues proposed to deny politically responsible officials not only the power to determine the execution of the federal research program but even the power to decide the substance of the program itself.

Going beyond political elitism, the Council objected to the distribution of the research funds on an institutional basis, whether to land-grant colleges or state universities. Like the proponents of the National Academy, the advocates of the Council tended to equate the national interest with the interests of the elite who did the best science. In the past, this scientific elitism had been proposed only for such private bodies as the Academy itself. Now, joining best-science to political elitism as a recommendation of public policy, the Council insisted that the Smith-Howard bill promote first-rate scientific and engineering research—specifically, that each state grant its funds not to any single institution but to meritorious research projects in whatever college they might arise.[7]

The Council's best-science proposal sharpened the public policy problem raised by Stephens' original change in Newlands' bill. On the one hand, awarding funds to the most meritorious projects was the most efficient way to foster productive research. On the other, funding the most meritorious projects meant concentrating the grants in the rela-

[5] Stephens to Samuel W. Stratton, June 3, July 6, July 20, 1916; Stratton to Stephens, July 12, 1916, NBS, Entry 51, AG file; Stephens to Stratton, Jan. 9, 1917, *ibid.*, IG file. The text of the bill is in U.S. Congress, House, Committee on Education, *Hearings, Engineering Experiment Stations for War Service*, 65th Cong., 2d Sess., June 11, 1918, pp. 32–33. This legislative discussion first appeared in different form in *Technology and Culture*, copyright © 1971 by the Society for the History of Technology, and is republished here with the Society's permission.

[6] "Action of the National Research Council with Respect to the Proposals of the Smith-Howard Bill," p. 7, RAM, Box 7; Hale to Gano Dunn, Dec. 24, 1918, NRCM, Organization file.

[7] "Action of the National Research Council . . .," RAM, Box 7.

tively small number of well-equipped institutions where the better scientists and engineers were located. A best-science policy thus promised to add, in each state, to the quality and resources of the leading scientific colleges and universities at the expense of their less capable brethren. Not surprisingly, scientists in the private, and some of the better public, institutions rallied around the Council's draft of the bill; more than a few state university officials were cool, and the land-grant colleges rejected it out of hand. In January 1919, when the contesting parties gathered in an attempt to resolve their differences, the meeting turned stormy, a summary report suggested, with "much human nature . . . injected into the proceedings."[8] The wrangling continued into deadlock again, and, like the Newlands measure, the Smith-Howard bill died in committee.

The NRC did not reopen the issue of federal funding. Elihu Root, a decidedly conservative Republican, bolstered Hale's own conservative Republican conviction that organizing or financing American science through the federal government would open the door to distasteful supervision if not tyranny. To Hale the cooperative, voluntarist NRC was certainly preferable as an organizational instrument to any "scheme in which a very small group of men, appointed as a branch of the Government, attempt to dominate and control the research of the country."[9] Hale would "not be disappointed if we have to rely on private funds for a long period in the future."[1]

James McKeen Cattell, an advocate of public rather than private funding, attacked Hale for staking the future of American science on "aristocracy and patronage."[2] A thoroughgoing internationalist, Cattell also sided with the distinguished Dutch astronomer Jacobus C. Kapteyn, a close friend of Hale's, who circulated an eloquent public letter—it gathered almost three hundred signatories from six neutral nations—protesting the exclusion of the Central Powers from the International Research Council. Cattell printed excerpts of the letter in both his magazines,

[8] Memo on conference, Jan. 7, 1919, NRCM, Ex Bd: Projects: Consideration of Congressional Bills for Engineering Experiment Stations file.

[9] Hale to James R. Angell, Jan. 19, 1918, Aug. 13, 1919, NRCM, Legal Matters: Opinion re NRC-Council of National Defense Relationship: Attorney General file.

[1] Hale to Willis R. Whitney, June 19, 1918, GEH, Box 43. Hale had concluded on the basis of a conversation with Root: "It is almost never possible to obtain any Congressional assistance for scientific projects outside of the Government Departments, and then only under a Republican administration. . . . [According to Root, in one case a few years ago] the Democratic Senators spoke eloquently against all such organizations [as the American Academy in Rome] for research and advanced study, and . . . demanded that attention be concentrated on the needs of 'the Little Red School House on the hill,' standing for light and leading to the lowly of the land! . . . Although it contains some notable exceptions, the Democratic party as a whole represents the most depressing elements in American life. The National Academy would never dream of going to a Democratic Congress for funds, as it would be sure to lose rather than gain by doing so." Hale to H. H. Turner, March 6, 1916, GEH, Box 41.

[2] Cattell to Hale, May 29, 1920, GEH, Box 10.

but only a handful of American scientists joined him in openly deploring what Kapteyn called the division of science, "for the first time and for an indefinite period, into hostile political camps."[3] Few scientists evidently cared about the issue, and those who did could scarcely expect to influence, let alone change, the foreign—or the domestic—policy of the NRC. The politically elitist Council was not bound to respond to the members of the scientific community at large as citizens—or even as scientists. For all the ecumenical character of the NRC, its members were not responsible to the various scientific societies that nominated them to the Council but to the self-perpetuating National Academy, which officially appointed them to it.[4]

Situated beyond the control of the government or of the scientific community, the Academy and Council were, Hale of course believed, free to follow policies that were wholly apolitical. But by advocating a best-science policy in the contest over the Smith-Howard bill, the leaders of the NRC did cast a political vote in favor of the better state and private universities. And by endorsing the exclusion of the Central Powers from international science, the NRC made a political decision. On the whole, the freedom of the Academy and Council from "political interference" simply left both free to follow what amounted to political predilections of their own.

The Council failed to prevail in major areas of public policy not least because it operated under the same constraints as any other political interest group. The NRC lost out in the Smith-Howard controversy since it represented a political faction too weak at the end of World War I to win the day for best-science elitism. In foreign policy, just as Secretary Houston had counseled against awarding official duties to the NRC at home, the State Department legal staff advised that the government owed the international dealings of the NRC no more than "informal" cooperation. More important, Wilsonians in the State Department refused to sanction any new scientific organization with the anti-Wilsonian policy of squeezing out Germany.[5] Hale never did manage to pry an official blessing for the International Research Council out of the

[3] Kapteyn et al., "Aux membres des académies des nations alliées et des Etats-Unis d'amérique," attached to Kapteyn to Hale, Aug. 7, 1919, GEH, Box 47.

[4] At a meeting in late December 1916 Cattell argued for making the representatives to the NRC responsible to the societies that nominated them; he was defeated. William H. Welch to Hale, Dec. 30, 1916, GEH, Box 43; Edwin Grant Conklin to Hale, Jan. 3, 1917, GEH, Box 11. Cattell agreed with Franz Boas—who also opposed the exclusion of the Central Powers from the IRC—that "international relations in science should not be controlled by our obsolete academies but by our active scientific societies. . . ." Cattell to Boas, Dec. 4, 1919, FB. According to an opinion of the acting attorney general, C. B. Ames, the NRC was an "agency" of the Academy. The opinion is quoted in National Academy of Sciences, *Annual Report, 1920*, p. 23.

[5] Memorandum, Office of the Solicitor, Sept. 17, 1920, DST, Entry 196, Doc. 592 D 1/6.

Wilson administration because the international policy of his NRC was closer to that of the French and Belgians than to that of the United States.

However much personal animosity lay behind the fierce French and Belgian determination to read the Central Powers out of international science, they were also energized by Picard's political conviction that German scientists especially would scheme for "revenge" in the peace. Back in 1917 the Frenchman had cited the commonplace theme that the war would continue with undiminished ruthlessness in the "economic and intellectual" arenas, and for Germany, he had warned, science was an instrument of "domination." The apolitical tradition of international science may have been viable in Humphry Davy's day, but now, Picard warned, it was "dangerous" to separate science from "national" questions.[6] In much the same way that Georges Clemenceau viewed the League of Nations, Picard and his allies saw the International Research Council virtually as an organization of mutual security in science.

So did Hale, who advocated American entry into the League of Nations, which he regarded as a peacetime coalition against the threat of revanche. Whatever his differences with Picard, he respected the Frenchman's political calculus and understood that the war had given economic nationalism more than equal weight with disinterestedness in the public arguments for science. The United States, he had told President Wilson, could not hope to "compete successfully" with Germany in war or peace without mobilizing science to the maximum.[7] To that end, Hale had done all he could to commit American science through the private, elitist NRC to a virtual cold war abroad and an alliance with the major industries and philanthropic foundations at home.

[6] Picard to Hale, July 22, 1917, GEH, Box 47.
[7] Hale to the President, March 26, 1918, WW, File VI, Case 206.

XI

The Impact of
Quantum Mechanics

In 1919 Dean Frederick J. E. Woodbridge of Columbia remarked: "Those who shared in the consciousness of the University's power and resourcefulness can never be fully content to return to the old routine of the days before the war."[1]

Robert Millikan hurried back to the University of Chicago and dunned the administration for funds in support of a program of atomic studies, including money to build a new laboratory and to operate a special research institute where physicists and chemists could jointly study the structure of matter. The estimated costs required a capitalization of well over $1,000,000, as much for the physical sciences alone as John D. Rockefeller had first promised as endowment for the entire university.[2] The Chicago administration was cool to the proposal, in part because of the considerable money involved, in part because Millikan wanted the university to share him with the new scientific school that George Hale was fathering in California.

For years Hale, the advocate of interdisciplinary research, had wanted to provide the staff of the Mount Wilson Observatory with a nearby cadre of physicists and chemists. During the war, aiming to re-shape a small technical college in Pasadena for that purpose, Hale raised a large endowment, drawing upon a close-knit group of wealthy southern Californians who were alive to the role a center of science might play in the industrial development of the region. To match the endowment with

[1] *Report of the President of Columbia University, 1919*, p. 147.
[2] Millikan to Harry Pratt Judson, June 19, 1920, GEH, Box 29.

a distinguished faculty, Hale persuaded Arthur A. Noyes to head the chemistry division; for physics, he wanted Millikan, full-time if he could get him, part-time if not. Specifically to tempt Millikan, one of Hale's philanthropists pledged the new school his entire estate, amounting to more than $4,000,000. In 1921 Millikan moved to Pasadena as chief executive officer of the new California Institute of Technology, which was, despite its name, mainly an institute of pure science. In Germany, Wilhelm Röntgen was awed. "Just imagine—Millikan is said to have a hundred thousand dollars *a year* for his researches."[3]

Caltech set a standard for emulation among America's more senior physicists, the rising academic administrators. They had grown accustomed in the progressive era to think of departmental expansion as a normal mode of life, and during the war, as the military's attitude toward scientists turned from skepticism or hostility to warm appreciation, they had witnessed spending for research and development burgeon to millions of dollars. Now they were back on the campuses with a fresh sense of professional self-confidence and expansive ambitions for more laboratories, more equipment, and more faculty.

At Harvard, Chairman Theodore Lyman unabashedly called his physics department, now making endless demands upon the financial lifeblood of the university, "a most vociferous daughter of the Horse-leech."[4] But other academic scientists had also entered the peace with their horizons enlarged and with a markedly expanded sense of fiscal possibilities. As a result of the wartime mobilization, at many universities the faculty was organized locally for the promotion of research.[5] Formally constituted or not, lobbies emerged on campuses after the war, calling upon university presidents to provide even greater recognition and a good deal more money for pure science.

The lobbyists stood an excellent chance with the university president of the 1920s. Often a former professor, he agreed in principle with the aims and arguments of his faculty. Whether the head of a private university and faced with a board of trustees dominated by businessmen, or of a public institution and confronted with a state legislature sympathetic to the needs of industry, he could employ to advantage the economic argument for science. And because faculties had won considerable institutional authority by the 1920s, the chief executive officers of academic institutions were becoming increasingly just that: chief executives. No sensible president could afford to ignore the requests of his professors, especially when they were organized on an institutionwide basis.

[3] Röntgen to Mrs. Theodore Boveri, Nov. 13, 1921, quoted in Otto Glasser, *Wilhelm Conrad Roentgen and the Early History of Roentgen Rays* (Springfield, Ill.: Charles C. Thomas, 1934), p. 180.

[4] Lyman to A. Lawrence Lowell, June 13, 1929, ALL, file 491.

[5] Division of Educational Relations of the National Research Council, "Report upon a Questionnaire Concerning the Organization and Facilities for Research . . .," Dec. 15, 1919, NRCM.

Besides, as Charles W. Eliot noted with the authority of an octogenarian proved right, the war had dramatized the "high value" of the college-trained man.[6] Through the decade, enrollments climbed yearly at a steep rate, the undergraduate population almost doubling between 1920 and 1929, the graduate more than tripling. The combined matriculation in science and engineering kept pace with the general trend, creating a sharp need for the laboratories and professors necessary to inculcate basic physics and chemistry. If only as an educational manager, the university president had to provide more scientific facilities and, in order to keep and attract faculty, to upgrade the salaries of teachers, whose purchasing power had been seriously undermined by the wartime inflation.

Withal, the university president emerged more the money raiser than ever, and in the postwar years drives for capital gifts or enlarged legislative appropriations were epidemic. If it had taken Harvard almost three centuries of genteel solicitations to accumulate an endowment of $10,000,000, now some universities hired professional fund-raising firms and most of the major institutions set out to gather millions of dollars by tomorrow. The appeals almost always called for substantial monies explicitly to support and expand research. Academic presidents, citing the lesson of the war, bore down on the ultimate utilitarian benefits of pure science. A few echoed the understated appeal for funds of President John Grier Hibben of Princeton: "We are evidently in an era of still more remarkable discoveries concerning the nature and laws of the physical world."[7]

Among the authors of the remarkable discoveries was Arthur Holly Compton, who in the early 1920s was pioneering in the field of X rays. Young Compton was the product of the bracing household headed by Elias Compton, an ordained minister turned dean and professor of philosophy at the College of Wooster, a small Presbyterian school in northern Ohio. In the Compton home the dinner-table conversation often sparkled with talk of modern science, including the marvels of radioactivity and the electronic atom. The father found especially intriguing the new field of child psychology. He applied some of its tenets to his offspring, blending a no-back-talk discipline with deliberate attempts to cultivate their natural capabilities and imagination.[8]

Apart from the father's tutelage, Compton followed the increasingly typical educational route for the physicists of his day—the undergraduate work at a small college, Wooster, then graduate training at a leading university, Princeton, where he was singled out as the best advanced student

---

[6] Eliot, circular for the Harvard endowment campaign, Aug. 23, 1919, ALL, file 1981.

[7] *Report of the President of Princeton University, 1925*, pp. 8–10. See also Frederick P. Keppel, excerpt from the annual report of the Carnegie Corporation, *Science*, 60 (Dec. 5, 1924), 516.

[8] James R. Blackwood, *The House on College Avenue: The Comptons at Wooster, 1891–1913* (Cambridge, Mass., 1968), pp. 29, 48–49, 56.

in any department. Less typically, Compton settled into a comfortable post at the Westinghouse laboratories, but once World War I ended, industrial research seemed tame compared to the challenges of pure physics. Betty Compton, who had been understanding her gifted, restless husband since their hometown days, urged him to follow his inclinations, and, in 1919, he went off to study under Ernest Rutherford as one of the first National Research Council fellows. Now back from England and a member of the faculty at Washington University in St. Louis, he was probing for an explanation of the peculiar way that X rays scattered upon impact with matter.

In the prevailing theory of the phenomenon, the X rays were a continuous electromagnetic wave which scattered upon interaction with the electrons of the metal. The trouble was that in the prevailing theory the scattered X rays were supposed to be of the same wavelengths as the primary beam. Compton's own painstaking investigations revealed quite the contrary; the wavelengths of the rays from the target metal varied with the angle at which they emerged. Compton fought tenaciously to explain the anomaly within the limits of classical electromagnetic theory. Showing a thorough mastery of the theory itself, he tried to resolve the difficulty by revising the current assumptions about the size and shape of the electron. He almost succeeded—almost.[9] Gradually, his attention shifted to a reassessment of the nature of the incident radiation.

Compton increasingly wondered whether scattering could be understood by considering the primary beam of X rays not as a train of continuous waves, but, in accord with Einstein's hypothesis of light quanta, as a stream of discrete particles. At the time, Einstein's hypothesis remained a heresy for almost all the authorities in physics save Einstein himself. If in 1916 Millikan had conclusively demonstrated the exact validity of Einstein's equation for the photoelectric effect, even Millikan regarded the quantum basis of the equation as inadmissible and, hence, incorrect. In October 1922 Compton rejected his own radical ruminations, reminding himself of the massive evidence in favor of the wave theory of all radiation, including X rays.[1] But Elias Compton's son was stubborn, self-confident, and intellectually bold. If he had fought adamantly to save classical electromagnetic theory, soon with equally strong will—and data—he embraced Einstein's hypothesis completely.

At the 1922 meetings of the American Physical Society, Compton contended that the interaction of matter and X rays had to be pictured as a quantum of radiant energy bouncing off an electron, just as one billiard ball rebounds from another. In justification, Compton offered a

[9] Roger H. Stuewer, *The Compton Effect: Turning Point in Physics* (New York, 1975), pp. 135–58.
[1] Millikan, "A Direct Photoelectric Determination of Planck's 'h,'" *Physical Review*, 7 (March 1916), 355–88; Stuewer, *Compton Effect*, pp. 208–11.

fully quantitative elucidation of all the characteristics of X-ray scattering, including the chief peculiarity, the variation of wavelength with angle.[2]

If Einstein's hypothesis of light quanta had haunted the house of physics for almost two decades, Compton's results forced the disreputable ghost out of the attic, and its presence was decidedly unsettling. How could radiation be simultaneously both a wave and a particle? How could one explain such a well-understood phenomenon as X-ray interference, in which the wave motion of one ray was assumed to reinforce or cancel the wave motion of another? How could one deal with such fundamental interactions of radiation and matter as dispersion, the process by which the misty atmosphere yielded the rainbow? The classical theory of dispersion hinged on the assumption that the electrons in the atom were set to oscillating by an incident electromagnetic wave. The theory had first broken down when Bohr, by locking the atomic electrons into quantum orbits, had denied them the freedom to oscillate in their orbital planes. Compton's results, by validating Einstein's hypothesis, turned the incident electromagnetic wave into a stream of light quanta and, in consequence, seemingly destroyed the remaining legitimacy of the classical approach.[3]

Confounded by the ramifications of the wave-particle paradox of light, physicists on both sides of the Atlantic could only find refuge in the quip of the British Nobel laureate William Lawrence Bragg: God runs electromagnetics on Monday, Wednesday, and Friday by the wave theory, and the devil runs it by quantum theory on Tuesday, Thursday, and Saturday.[4]

At the University of Minnesota in 1925 the young mathematical physicist John Van Vleck reviewed the interpretation of atomic spectra and concluded that modern physics was certainly passing through "contortions in its attempt to explain the simultaneous appearance of quantum and classical phenomena."[5] The basic assumption of Niels Bohr's theory—that atoms radiated a spectral line when an electron hopped from one quantized orbit down to another of lower energy—remained the keystone of the subject. In Germany during the war, Bohr's particular theory of the atom had received important support from Arnold Sommerfeld, the professor of theoretical physics at the University of Munich, who com-

[2] Stuewer, *Compton Effect*, pp. 223–24.

[3] John N. Heilbron, "A History of the Problem of Atomic Structure from the Discovery of the Electron to the Beginning of Quantum Mechanics" (unpublished Ph.D. dissertation, History, University of California, Berkeley, 1964), pp. 413–16; Max Jammer, *The Conceptual Development of Quantum Mechanics* (New York, 1966), pp. 162–65, 181–83.

[4] Quoted in David Webster, interview, AIP, p. 25.

[5] John H. Van Vleck, "Quantum Principles and Line Spectra," *Bulletin of the National Research Council*, 10 (No. 54; March 1926), 287.

manded remarkable powers of exposition, a reputation as one of the best teachers of his generation, and a supple ability to accept radical innovations, then extend them with clarity and elegance.

Applying his talents to Bohr's quantum model of the atom, Sommerfeld assumed that atomic electrons moved in elliptical as well as circular orbits, and he took into account the relativistic increase in the mass of the electron as it sped around the nucleus. Sommerfeld's generalization and refinement of Bohr's theory immediately accounted for the "fine structure" of ionized helium, the separate spectral lines clustered at small wavelength intervals around the main lines of the element. By the end of the war, the Bohr-Sommerfeld theory of the atom yielded further triumphs, especially in the analysis of X-ray spectra and the splitting of hydrogen spectral lines in the presence of an electric field.[6] Sommerfeld called attention to these wartime developments in his *Atomic Structure and Spectral Lines*, first published in 1919, which quickly became the textbook bible of the subject for physicists the world over.

All the while Niels Bohr, who relished paradoxes as a key to deeper understanding, pursued the paradox of the quantized atom. He soon concluded that there existed "a far-reaching *correspondence*" between, on the one hand, certain classical properties of an electron's motion in a given atomic orbit and, on the other, the probability that it might jump to an orbit of different energy.[7] So far as the applicability of this idea to every possible orbit was concerned, Bohr's was less a tight argument than a profound assertion. Still, with the "correspondence principle" Bohr could calculate the nature and intensity of numerous spectral lines and deduce the rules governing the possible quantum jumps. Bohr also successfully developed the principle to reason out the gross arrangement of the electrons in each natural element. He showed how the difference of one electron accounted for the difference in chemical properties of neighboring members of the periodic table; he even predicted the existence of the chemical properties of an unknown element. Two of his collaborators detected the substance, which was given the name hafnium, in 1922; and that year Bohr had the pleasure of announcing the discovery to the world from Stockholm, on the occasion of his Nobel Prize.

In the United States quantum physicists studied their Sommerfeld—despite the exclusion of German physicists from Hale's new international unions, Americans were of no mind to ignore German physics—and they could appreciate the remark in Sommerfeld's preface: "Today, when we listen to the language of spectra, we hear a true atomic music of the spheres."[8] But like Van Vleck, some of them recognized that for all the

[6] John L. Heilbron, "The Kossel-Sommerfeld Theory and the Ring Atom," *Isis*, 58 (Winter 1967), 451–85.

[7] Niels Bohr, *The Theory of Spectra and Atomic Constitution* (Cambridge, England, 1922), p. 24.

[8] Quoted in Heilbron, "The Kossel-Sommerfeld Theory," p. 451.

success of the Bohr-Sommerfeld theory, its marriage of classical and quantum physics left logic in tatters. Typically, the correspondence principle, which Sommerfeld, who did not accept it, snidely called a "magic wand," demanded that classical mechanics both give way before the quantum and, simultaneously, help predict the more intricate characteristics of the quantized atom.[9] Besides, though the Bohr-Sommerfeld theory predicted the spectrum of hydrogen, it failed to explain the optical and much of the X-ray radiations of atoms with more than one outer electron. Such atoms, when subjected to an external magnetic field, emitted complex spectra, many of which Alfred Landé, a facile young German theorist, managed to analyze in terms of a magnetic coupling among the atomic electrons. But if Landé used a magnetic analysis to explain spectral peculiarities of one kind, Sommerfeld of course relied on the theory of relativity to elucidate other types. By 1924, Robert Millikan and his student at Caltech, Ira S. Bowen, had produced a considerable body of fresh spectral data indicating that every odd spectral effect originated in a single, common process. The blunt result: Either a magnetic or a relativistic interpretation had to account for them all, a task that neither could manage.[1]

The situation exasperated many able European physicists, including Wolfgang Pauli, a product of Vienna's brilliant Jewish circle, a theorist of deep physical intuition and rare mathematical power who manifested almost a mystical faith in the order of nature and a ferocious demand for intellectual rigor. Now in his mid-twenties, he might testily tell Bohr, "Shut up, you are being an idiot." However abrasive, he was the most insistent critic in the theoretical enterprise and, in a way, its conscience.[2] In 1924 Pauli partially untangled the problem of complex spectra by arguing that each atomic electron possessed four quantum numbers. He also postulated his profoundly important exclusion principle, that no two electrons could share identical quantum numbers or, put differently, occupy the same quantum state. All the same, while three of the quantum numbers seemed to be physically understandable, Pauli could only attribute the fourth to the electron's having what he called a "classically not describable two-valuedness," a quality as physically mysterious as the phrase was awkward.[3]

[9] Arnold Sommerfeld, *Atomic Structure and Spectral Lines*, trans. Henry L. Bose (3d edition; New York, 1923), p. 275.

[1] Paul Forman, "The Doublet Riddle and Atomic Physics *circa* 1924," *Isis*, 59 (Summer 1968), 167–70; Robert A. Millikan and Ira S. Bowen, "Extreme Ultra-Violet Spectra," *Physical Review*, 23 (Jan. 1924), 1–34; "The Significance of the Discovery of X-Ray Laws in the Field of Optics," *Proceedings of the National Academy of Sciences*, 11 (Feb. 1925), 119–22; Paul Forman, "Alfred Landé and the Anomalous Zeeman Effect, 1919–1921," *Historical Studies in the Physical Sciences*, 2 (1970), 153–261.

[2] Quoted in Barbara Cline, *The Questioners: Physicists and the Quantum Theory* (New York, 1965), p. 136.

[3] Quoted in Jammer, *Quantum Mechanics*, p. 146.

Early in 1925 Pauli declared: "Physics is decidedly confused at the moment; in any event, it is much too difficult for me and I wish I . . . had never heard of it." More optimistically, Max Born, one of Europe's leading theoretical physicists, poured all that he knew about atomic mechanics into a text on the subject and pronounced the work "Volume I." It might be years before Volume II could be written, but, Born contended, the need for a superior mechanics of the atom was unquestionable.[4] Like other thoughtful atomic theorists, Born saw his discipline in a state of mounting intellectual crisis whose resolution would require the invention of an entirely new mechanics, a quantum mechanics.

The path to a quantum mechanics, Born noted, was marked by "only a few hazy indications," most of which were suggested by the sharp divergences between quantum theory and classical physics.[5] Classical physics assumed natural phenomena to be continuous, describable in space and time, causally predictable, and, as such, mathematically expressible in differential equations. Quantum theory could describe atomic behavior only as a series of discontinuous transitions between states, could not locate atomic electrons in space and time when they leaped from one orbit to another, and could not predict when such leaps would occur. To theorists who appreciated these divergences, a genuine quantum mechanics might possibly be a mechanics of discontinuity, mathematically expressed in equations of differences rather than in differential equations. It might have to reject space-time models of atomic behavior and perhaps, at the atomic level, even causality itself.[6]

These radical ideas had highly influential and respectable exponents, especially Niels Bohr and Max Born. Born, raised in a Jewish academic environment, sensitive, intense—he had sometimes found relief from the tensions of World War I in Berlin by playing violin sonatas with Einstein—disdainful of minds less powerful than his own, occupied the key post of professor of theoretical physics at Göttingen University. In Copenhagen, Bohr now headed his own research institute, where physicists from around the world came to study atomic mechanics and where the director pondered the epistemological issues raised by the subject. But the intellectually radical predilections of Bohr and Born had their commanding opponents, including Albert Einstein. However much Einstein had done to advance quantum physics, he refused to concede that physics must ultimately relinquish continuity in space and time. And if strict causality had to be renounced, Einstein grimly confided to Born, instead

[4] Pauli is quoted in R. de L. Kronig, "The Turning Point," in M. Fierz and V. F. Weisskopf, *Theoretical Physics in the Twentieth Century* (New York, 1960), p. 22; Max Born, *The Mechanics of the Atom*, trans. J. W. Fischer, rev. O. R. Hartree (London: G. Bell, 1927), pp. vii–viii.

[5] *Ibid.*, p. vii.

[6] Paul Forman and V. V. Raman, "Why Was It Schroedinger Who Developed de Broglie's Ideas?" *Historical Studies in the Physical Sciences*, 1 (1969), 298–99.

of a physicist he would "rather be a shoemaker or even an employee in a gambling casino."[7]

The stakes here, Einstein realized, were nothing less than the nature of the physical world and what man could know about it. For, as the French physicist Henri Poincaré had rightly said back in 1911, after the first European conference on quanta, a physics of discontinuity would be "the greatest and most radical revolution in natural philosophy since the time of Newton."[8]

The revolution occurred with unexpected and dizzying swiftness, in a brief span during the mid-1920s—J. Robert Oppenheimer later called it "a heroic time"—of astonishing scientific creativity.[9] Scores of scientists from around the world, including the United States, contributed to the reconstruction. All the same, it was accomplished mainly on the continent of Europe by two groups of physicists, one of which took the radical path to quantum mechanics exemplified by the ruminations of Bohr and Born, the other of which fought to cleave to Einstein's more conservative approach.

Among the chief acolytes of the radical approach was Born's assistant, Werner Heisenberg, the son of a classics professor, who had arrived at Göttingen University in 1921 still wearing the khaki shorts of the German youth movement. Keenly aware of the epistemological issues in the crisis of quantum theory, he groped for a theory independent of mechanical, space-time pictures and based strictly on such observables as spectral frequencies. In 1925, one ecstatic night by the sea in Helgoland, where he had gone to relieve his hay fever, Heisenberg employed his enormous resourcefulness and mathematical facility to guess and manipulate his way to the invention of a primitively successful quantum mechanics which was characterized by a strange type of multiplication. Heisenberg showed his work to Max Born, who recognized that this odd multiplication scheme was the well-established mathematical formalism of matrices. Born, Heisenberg, and another assistant, Pascual Jordan, together soon developed a full matrix formulation of quantum mechanics.[1]

On the side of Einstein and continuity stood Louis de Broglie, an

---

[7] Quoted in Martin J. Klein, "The First Phase of the Bohr-Einstein Dialogue," *Historical Studies in the Physical Sciences*, 2 (1970), 32; interview with Werner Heisenberg, SHQP, Feb. 15, 1963, p. 18; Forman and Raman, "Why Was It Schroedinger . . .," pp. 298–99; Paul Forman, "Weimar Culture, Causality, and Quantum Theory, 1918–1927: Adaptation by German Physicists and Mathematicians to a Hostile Intellectual Environment," *Historical Studies in the Physical Sciences*, 3 (1971), 70–71.

[8] Quoted in Jammer, *Quantum Mechanics*, p. 170.

[9] Quoted in William H. Cropper, *The Quantum Physicists* (New York, 1970), p. 36.

[1] Werner Heisenberg, *Physics and Beyond: Encounters and Conversations*, trans. Arnold J. Pomerans (New York, 1971), pp. 60–62; Heisenberg interview, SHQP, Feb. 19, 1963, pp. 6–7; Feb. 22, 1963, pp. 8–9; Max Born, "Memoirs," SHQP, Chapter XIX, pp. 9–14.

introspective French aristocrat with a compelling interest in philosophy, who was one of the few French physicists to worry about the crisis of quanta, even during his six years of duty at the military radiotelegraphic station on the Eiffel Tower. Working alone, guided by an acute sense of symmetry, de Broglie proposed that the wave-particle duality of light might be extended to matter, that by quantizing the energy of mass, matter might be represented as a wave in space and time. "Read it," Einstein instructed Max Born about de Broglie's treatise. "Even though it might look crazy it is absolutely solid." Most physicists rejected de Broglie's theory as preposterous, but not Erwin Schrödinger, professor of physics at Zurich University, who rejected "at the threshold" Bohr's view of the impossibility of a space-time description of atomic phenomena, who was "discouraged, if not repelled," by Heisenberg's "transcendental algebra," not least because it defied "visualization." Schrödinger's attention was drawn to de Broglie's wave theory of matter by a favorable published comment of Einstein's.[2] Boldly generalizing de Broglie's scheme, in 1926 Schrödinger invented the wave mechanics, a quantum mechanics expressed by a wave equation, the solution of which yielded the observable characteristics of atomic behavior, was defined in mathematical space and time, and, as Schrödinger himself soon proved, was mathematically equivalent to the matrix mechanics of Heisenberg.

Young Eugene Wigner read one of Heisenberg's first papers on quantum mechanics and immediately called up his friend Leo Szilard to say that Heisenberg had solved the outstanding problems of physics.[3] Neither Heisenberg, Schrödinger, nor anyone else had solved anywhere near *all* such problems, but with quantum mechanics, physicists found themselves able to resolve numerous longstanding difficulties in the behavior of atoms and radiation with astonishing speed. And what in the old quantum theory had been mere *ad hoc* assumptions reappeared in the new quantum mechanics as logical consequences of the theoretical scheme.

Prior to the work of Schrödinger and Heisenberg, two young Dutch physicists, George Uhlenbeck and Samuel Goudsmit, had proposed that the complex spectra could be explained by postulating that the electrons possessed a positive or negative spin. In classical physics, a point on the surface of an appropriately spinning electron would move with a speed greater than that of light, in violation of the theory of relativity. Soon Paul Adrien Maurice Dirac, an introverted young genius at Cambridge

[2] Einstein is quoted in K. Przibram, ed., *Letters on Wave Mechanics*, trans. Martin J. Klein (New York, 1967), pp. xiii–xiv. Schrödinger wrote in 1926: "Bohr's standpoint, that a spatio-temporal description is impossible, I reject *a limine.*" Quoted in Forman and Raman, "Why Was It Schroedinger . . .," p. 301, n. 36; Cropper, *The Quantum Physicists*, p. 90. Schrödinger had earlier flirted with, and then repudiated, an acausal approach to atomic physics. Forman, "Weimar Culture, Causality, and Quantum Theory," pp. 87–88, 104.

[3] Wigner interview, SHQP, Dec. 4, 1963, p. 1.

University with an aesthetic mathematical mind—he thought it "more important to have beauty in one's equations than to have them fit experiment"—developed a quantum mechanics for the relativistic electron.[4] Dirac's theory yielded the spin as an inherent, nonmechanical property of the electron (precisely the classically nondescribable two-valuedness that Pauli had glimpsed). But even as physicists stood in awe at the power of quantum mechanical formalisms, many of them wondered what ultimate physical meaning lay behind the mathematics.

Einstein welcomed Schrödinger's wave mechanics because at first glance it seemed to reaffirm the physics of continuity in space and time. Schrödinger himself believed that the solution to his wave equation expressed the distribution in the atom of waves of electron matter. With this interpretation, the electrons could be understood to make the transition from one quantum energy level to another, Schrödinger hopefully suggested, by changing from "one form of vibration to another," a process that could occur "continuously in space and time."[5] But Schrödinger's interpretation was recognized to be inconsistent with the implications of his own mathematics. And in Göttingen, Max Born was sure that electrons could hardly be smeared out like a wave if only because in many experiments in James Franck's nearby laboratory they clearly behaved like small, concentrated particles of matter. What the solution of the wave equation represented, Born argued, was not the distribution of a smear of electron matter but the distribution of the statistical likelihood of locating the electron at a particular point in space and time. Schrödinger's wave equation expressed the propagation of a wave of probability.[6]

Bohr embraced Max Born's statistical interpretation enthusiastically, but not Schrödinger, who in September 1926 lectured on wave mechanics at Bohr's institute and stayed on afterward to argue for his matter-wave interpretation. Bohr doggedly attempted to persuade Schrödinger over to the side of statistics and discontinuity, even pursuing the issue from the edge of the sickbed when Schrödinger fell ill. "If one has to stick to this damned quantum jumping," Schrödinger exclaimed, "then I regret having ever been involved in this thing."[7] Schrödinger left unpersuaded, but for months afterward the physical interpretation of quantum mechanics was the chief subject of discussion in Copenhagen, especially between Bohr and Heisenberg, who was living on the top floor of the institute in a small room overlooking the park.

[4] Dirac, "The Evolution of the Physicist's Picture of Nature," *Scientific American*, 208 (May 1963), 47.

[5] Schrödinger, "Quantisation as a Problem of Proper Values [Eigenfunctions], Part I," in Erwin Schrödinger, *Collected Papers on Wave Mechanics*, trans. J. F. Shearer (London, 1928), pp. 10–11; Jammer, *Quantum Mechanics*, pp. 271, 282–83.

[6] L. Rosenfeld, "Men and Ideas in the History of Atomic Theory," *Archive for History of Exact Sciences*, 7 (1971), 84–85; Jammer, *Quantum Mechanics*, pp. 285–86.

[7] Quoted in Jammer, *Quantum Mechanics*, p. 324.

Heisenberg, after considerable struggle, finally found a key to the physical interpretation of quantum mechanics in a glaring contradiction between the results of the quantum mechanical formalism and a crucial assumption of classical physics. The mathematics of quantum mechanics seemed to give a spread in the values of the simultaneous position and velocity of an electron, yet according to classical physics the position and velocity of an electron could always be precisely defined. Heisenberg chose as the critical question: Could the position and velocity of an electron in fact be measured simultaneously with complete exactness? Not at the level of quanta, he decided, not even in principle. At the level of quanta the measurement of certain pairs of variables was limited in accuracy by what he called a—and what became justly celebrated as Heisenberg's—"principle of uncertainty."

To illustrate this principle, Heisenberg considered the observation through a microscope of an electron struck by light of an appropriate frequency. The position of the electron could be determined more precisely by increasing the frequency of the light, but the higher the frequency, the larger the jolt to the electron, and, hence, the greater the indeterminacy in the measurement of the electron's velocity. Conversely, the velocity could be determined more precisely by using light of a lower frequency, but the lower the frequency, the greater the indeterminacy in the measurement of position. On the basis of these considerations, Heisenberg boldly affirmed the acausality of quantum mechanics: For to predict the future, you had to know everything about the present, and according to quantum mechanics, Heisenberg asserted, "We cannot, as a matter of principle, know the present in all its details."[8]

When Pauli saw a preliminary draft of Heisenberg's uncertainty arguments, he sighed with relief that day had finally dawned in quantum mechanics, but Bohr was dissatisfied. Heisenberg insisted to Bohr that he had constructed a consistent mathematical scheme which said everything about what could be observed. Bohr drove Heisenberg to tears with relentless demands for probing beyond mere consistent mathematics, to the conceptual foundations of the physics. To Bohr's mind, the physical meaning of quantum mechanics had to be found by starting with the wave-particle duality of light.[9] Doing just that, Bohr developed a remarkable conceptual interpretation of quantum mechanics through his principle of "complementarity."

According to the principle of complementarity, the description of light required two complementary concepts, waves and particles. The principle proceeded from Bohr's recognition that we must use classical language—position and velocity—to describe the results of experiments.

---

[8] Quoted *ibid.*, p. 330. See also, Heisenberg, *Physics and Beyond*, pp. 72–78.

[9] Heisenberg interview, SHQP, Feb. 25, 1963, pp. 16–18; L. Rosenfeld, "Men and Ideas in the History of Atomic Theory," p. 89; Jammer, *Quantum Mechanics*, pp. 345–47.

But position and velocity were of course mutually exclusive as precisely measurable quantities; they were thus complementary variables. The uncertainty principle came into play, Bohr explained, when physicists tried to measure such complementary variables simultaneously. More generally, causal descriptions of atomic phenomena were complementary to those in space-time; if you modeled them in space and time, you could not simultaneously analyze them in the language of causal continuity. "We meet here in a new light," Bohr summarily concluded, echoing J. B. Stallo's critique of nineteenth-century physics, "the old truth that in our description of nature the purpose is not to disclose the real essence of the phenomena but only to track down, as far as it is possible, relations between the manifold aspects of our experience."[1]

In October 1927, at the Solvay conference in Brussels, the Born-Heisenberg-Bohr interpretation of quantum mechanics was searchingly examined by a group of the world's leading physicists, including Einstein. Discontented with the statistical reading of the wave equation, Einstein had insisted to Born: "An inner voice tells me that this is not the true Jacob. The theory accomplishes a lot, but it does not bring us closer to the secrets of the Old One. In any case, I am convinced that He does not play dice." Einstein remained convinced that it was possible to obtain a model of reality, a theory which represented "things themselves and not merely the probability of their occurrence."[2] Now, at Brussels, Einstein pressed his objections to quantum mechanics. At breakfast he would happily pose an experiment that, he was sure, violated Heisenberg's uncertainty principle, then trundle down the street to the conference, absorbed in discussion with Bohr. By evening, having pondered the experiment all day, Bohr would triumphantly demonstrate to Einstein how the experiment did not violate but reaffirmed the principle of uncertainty. Every morning Einstein would challenge Bohr with a new experiment; every evening Bohr and quantum mechanics would win the day.[3]

"From first to last," Robert Oppenheimer accurately said of the development of quantum mechanics, "the deep creative and critical spirit of Niels Bohr guided, restrained, deepened, and finally transmuted the enterprise."[4] Einstein left the Solvay conference convinced that Bohr's logic was correct; he remained equally unconvinced that quantum mechanics was a final answer. So did some other physicists, but the Solvay conference of 1927 signaled an overwhelming victory for the "Copenhagen interpretation" of quantum mechanics, for complementarity, inde-

---

[1] Niels Bohr, *Atomic Theory and the Description of Nature* (Cambridge, England, 1934), pp. 18, 11–12, 55–56, 90–91.

[2] Quoted in Gerald Holton, "The Roots of Complementarity," *Daedalus*, 99 (Fall 1970), 1020; Cropper, *The Quantum Physicists*, p. 124.

[3] Heisenberg interview, SHQP, Feb. 27, 1963, p. 29, July 5, 1963, p. 10; Heisenberg, *Physics and Beyond*, pp. 79–81; Ruth Moore, *Niels Bohr: The Man, His Science, and the World They Changed* (New York, 1966), pp. 163–69.

[4] Quoted in Cropper, *The Quantum Physicists*, p. 36.

terminism instead of causality, statistical descriptions instead of space-time models of reality—for a revolution in physical thought even more profound in its intellectual depth and its philosophical import than the revolution worked by Isaac Newton more than two centuries before.

In the United States, many senior physicists watched the development of the quantum mechanical revolution with a sense of frustration. For some, the mathematics of the new formalism was simply too difficult. For others, accustomed to thinking of their science as a determinate description of the natural world, the statistical quality of quantum mechanics was, in the phrase of one critic, "like a confession of ignorance." And all who were concerned with the philosophical implications of the new physics—particularly the middle-aged men whose thinking had been formed in a more certain world—were bothered by the seeming subjectivism of Heisenberg's approach. John Zeleny, the fifty-seven-year-old chairman of the department of physics at Yale, told an audience at his alma mater in Minnesota, "I feel that there is a real world corresponding to our sense perceptions. . . . I believe that Minneapolis is a real city and not simply a city of my dreams."[5]

Yet at least some of Zeleny's peers went along with the revolution, recognizing with Zeleny himself: "The tide of scientific progress . . . will engulf any who oppose its flow. The old world is gone, never to return."[6] Quantum mechanics was, after all, stamped with the imprimatur of the highest authorities, from Bohr, if not Einstein, on down. Judged by the magnificent sweep of its results, it was an undeniably powerful construct. Physicists of Zeleny's generation had been practicing their discipline amid a torrential flow of new ideas for a quarter of a century now; they were toughened to the appearance of radical notions. Just about any scientist who had witnessed the vindication of the heresies of Planck, and of Einstein himself, knew it was the better part of wisdom, as Robert Millikan remarked, to put their money on the younger men.[7]

The Zelenys also had to deal with a special administrative necessity. A number of American physicists, like Arthur Compton and Robert Millikan, were doing first-rate experimental work; some, like Compton in his pursuit of the quantized X ray, were conducting high-level discussions and debates about contemporary theoretical issues. But only a few matched John Van Vleck in attempts to resolve theoretically the problems of atomic theory. Early in the decade, as the challenges of atomic studies centered increasingly on theoretical issues, the American profession's relative inadequacy in the special type of creative theoretical physics practiced in Europe had become increasingly disadvantageous. After

[5] Leigh Page, *Science*, 69 (June 14, 1929), 625; Zeleny, "The Place of Physics in the Modern World," *Science*, 68 (Dec. 28, 1928), 634.

[6] Zeleny, "The Place of Physics," 635.

[7] Millikan to F. A. Lindemann, March 8, 1932, RAM, Box 41.

the revolutionary breakthroughs of Heisenberg and Schrödinger, it was intolerable. Many experimentalists needed the help of quantum mechanicians. Moreover, without such theorists on the staff, no department of physics could hope to stay in the first professional rank. With them, high institutional reputation could come quickly.

Millikan and his fellow leaders at the California Institute of Technology, all alive to the profound challenge of the atom, agreed that "great advances" were likely to occur only where active experimentation was coupled with "the most profound theoretical analysis." To proceed with atomic studies, they reasoned, Caltech would have to develop "a center and school of mathematical physics as ably manned as any . . . [other] in the world."[8] Caltech hired Richard C. Tolman, the American theorist of physical chemistry, and imported Paul S. Epstein, a ranking theoretical physicist, from abroad. It also made a point of bringing foreigners over for visiting lectureships, inaugurating the program with the graceful presence of Hendrik A. Lorentz and the historically symbolic one of Charles G. Darwin, a grandson of the great Darwin, and, in his own right, an accomplished mathematical physicist. Within just a few years of its founding the school had won a reputation of enviable distinction where it counted most. From Germany the young American physicist Earle H. Kennard reported in 1927: "Caltech—there is something magnetic about that place [for] . . . Europeans."[9]

From the early 1920s, and especially after the advent of quantum mechanics, the nation's senior physicists made the acquisition of theorists a major part of their expansionist programs. Some cited Caltech as a model for emulation. Impressed with Caltech or not, all agreed with Millikan that in physics the United States faced a "new opportunity in science." If we seize it, Millikan had proclaimed at the end of the war, "in a very few years we shall be in a new place as a scientific nation and shall see men coming from the ends of the earth to catch the inspiration of our leaders and to share in the results [of] our developments."[1]

---

[8] [Millikan, Hale, and Noyes], "Memorandum Relating to the Application of the California Institute of Technology to the Carnegie Corporation of New York for Aid in Support of a Project of Research on the Constitution of Matter . . .," p. 5, attached to Hale to the Carnegie Corporation of New York, Sept. 17, 1921, GEH, Box 67.

[9] Kennard to Ernest Merritt, July 2, 1927, CUP, Kennard file.

[1] Stanley Coben, "The Scientific Establishment and the Transmission of Quantum Mechanics to the United States, 1919–1932," *American Historical Review*, 76 (April 1971), 446; Karl T. Compton, the chairman of the Princeton physics department, typically wrote of theoretical physics: "no field in physics at the present time is of greater importance. . . . In this country we have carried [our] experimental sides to a high degree of achievement, but . . . theoretical developments . . . are coming largely from Germany." Compton to Henry Allen Moe, Dec. 18, 1925, KTC; Millikan, "The New Opportunity in Science," *Science*, 50 (Sept. 26, 1919), 297.

XII

Popularization and Conservatism

I n the mid-1920s the *Saturday Evening Post* noted the public's
familiarity with the works of applied science and, by way of
comparison, remarked: "Pure science is the wallflower, the ugly
duckling, the elder sister who lives secluded and remote, unknown
and unpraised."[1] Eager to fill the coffers of research, physicists and other
scientists decided to dress up their wallflower and present her from Main
Street to Fifth Avenue. The most purposeful in argument were the
scientists clustered at the top of the National Academy and National
Research Council, including Millikan, Hale, Whitney, Jewett, and
Walcott, who now constituted the public leadership of American science.
Leaders or followers, in the 1920s so many of the nation's men of research
went at popularization with gusto that an amused observer commented:
"The High Priests have taken off their false whiskers and have given Mr.
Average Citizen a peep at the ceremonies going on inside the Temples."[2]

Going beyond the popularization of the 1870s, the priests were not
only taking to the hustings as individuals; the priesthood as such was also
organizing what amounted to public relations arms within the church.
Twenty years before, a newsman reflected, the leadership of American
science would have considered such activities probably as "unethical,"
certainly as "undignified."[3] Twenty years before, he might have explained,
the scientific community had not yet been sensitized by the experience of

---

[1] *Saturday Evening Post*, editorial, 198 (June 5, 1926), 38.
[2] Thomas J. Le Blanc, review of *Arrowsmith*, *Science*, 61 (June 19, 1925), 632.
[3] David Dietz, quoted in *Science*, 68 (Aug. 10, 1928), 121.

World War I to the value of concerted efforts at molding public opinion. Now the American Association for the Advancement of Science hired a press officer, sponsored weekly radio talks, and, late in the decade, held a symposium for reporters. Now, too, some of the nation's leading scientists allied themselves with the newspaper magnate Edward W. Scripps for the sake of creating their own authoritative news service.

It was *prima facie* an improbable coalition, this partnership between a group of eminent scientists and Scripps. A farmer's son who managed his newspaper empire from a vast ranch near San Diego, California, Scripps often dressed like a cattle baron, talked like a populist, and regarded scientists as "so blamed wise and so packed full of knowledge . . . that they cannot comprehend why God has made nearly all the rest of mankind so infernally stupid."[4] But Scripps knew William E. Ritter, a California marine zoologist and a fervent evangel of the social beneficence of science and its method. His imagination evidently captured by Ritter's preachments, Scripps endowed the Scripps Institution for Biological Research, vaguely hoping to advance a field of knowledge that might help democracy cope with the orneriness of man. The disaster of the war set Scripps to brooding about the fate of democracy itself. His conclusion: If America was to survive the pitfalls of a scientific age, he had better do something to help the people become, in Ritter's vein, more scientific.

Scripps had always run his papers for "the 95 percent," as he proudly liked to say, and he wanted to found an agency of popularization that would translate the highbrow doings of science into "plain United States that the people can understand."[5] Early in the century, to supply his local dailies with stories of general interest, he had created the Newspaper Enterprise Association; before long the NEA had developed into a highly successful features syndicate, with more sales outside Scripps's own chain than within it. Now, in 1919, mixing philanthropic conviction with a practiced sense of market, Scripps proposed to organize a nonprofit syndicate for distributing general news of science. It would have an immediate major customer in the Newspaper Enterprise Association, and, from Scripps himself, $30,000 a year while it tried to become self-supporting, a $500,000 endowment if and when it got on its feet. By 1920 the syndicate had been launched under the name Science Service, with a board of trustees representing the world of journalism, the American Association for the Advancement of Science, and the National Academy of Sciences–National Research Council.[6]

The trustees chose as editor Edwin E. Slosson, a chemistry Ph.D and the author of the recently published and widely celebrated *Creative*

[4] Quoted in Charles R. McCabe, ed., *Damned Old Crank: A Self-Portrait of E. W. Scripps* (New York, 1951), pp. 231–32.

[5] Quoted in "Edward W. Scripps," *DAB*, XVI, 518, and Negley D. Cochran, *E. W. Scripps* (New York, 1933), p. 114.

[6] Ronald C. Tobey, *The American Ideology of National Science, 1919–1930* (Pittsburgh, 1971), pp. 62–71.

Chemistry. Cloistered at the University of Wyoming in the 1890s, Slosson, a whirring dynamo, had taught his discipline during the day and written articles on a variety of subjects for the *Independent*, a magazine connected with his church, by haunting the library at night. After coming east as literary editor of the journal early in the century, he helped transform it into a widely read secular organ, leavening its pages with sprightly essays on books, religion, science, and politics. In Albany, Kansas, Slosson's father had run a station on the underground railroad; his wife, the first woman to win a doctorate from Cornell, was a director of the YWCA and a ranking suffragette. In New York, Slosson got to know John Dewey and his circle, and a good deal of his political writings added up to an outspoken advocacy of progressive reform.

At the opening of the 1920s Slosson was appalled by the Red Scare's laceration of civil liberties and by the ominous revolt brewing against the teaching of evolution; to his mind, it could scarcely be said that the scientific spirit had yet become "dominant in politics and society."[7] He admired Edward Livingston Youmans' zealous efforts to promote public respect for the disinterested study of nature, and he wanted to do the same. But if Youmans had found his audience among the cultivated public, Slosson, like Scripps, wished to educate "the whole people."[8] Where Youmans had, for the most part, drawn his language from the dictionary of stilted, Anglophilic gentility, Slosson was as richly American a writer as his fellow Kansan and college classmate, William Allen White. In one simple, pointed sentence Slosson could interpret the pragmatism of William James: "There is no difference that does not *make* a difference"; or capsule the chief idea of atomic physics: Among the "vulgar" the quantum theory is known as the "jerk theory."[9] Sharp or breezy, on or just near the literary mark, he could translate the most abstruse subjects into the pungent phrases of plain United States.

Now eager to switch over to the popularization of science full-time, Slosson announced the principal ground rule of Science Service: "It will not indulge in propaganda unless it be propaganda to urge the value of research and the usefulness of science."[1]

In the 1920s, with the availability of the mass-circulation press, propagandists for science could bombard an enormous—and an enormously receptive—audience. Technology had revolutionized warfare, and now technology was transforming American civilization. Radio was bringing

[7] Edwin E. Slosson, "Notes of a Talk to Trustees of Science Service . . . June 17, 1921," p. 13, copy in GEH, Box 75.

[8] *Ibid.*, p. 14.

[9] Quoted in Preston W. Slosson, "Edwin E. Slosson, Pioneer," introduction to Edwin E. Slosson, *A Number of Things* (New York, 1930), p. 17; Edwin E. Slosson, "New Wonders of Science," *The Independent*, 108 (May 13, 1922), 445.

[1] Slosson, "A New Agency for the Popularization of Science," *Science*, 53 (April 8, 1921), 323.

political conventions and prize fights into millions of parlors, and automobiles were adding status and convenience to the lives of as many families. Vacuum cleaners, washing machines, and refrigerators were appearing in the stores, while rayon was inaugurating the day of synthetic fibers; even secretaries and sales clerks could wear dresses that felt as smooth and seemed as fashionable as any made of silk. And now, with the tractor enlarging the output of food crops and accelerating the flight from the farm, more Americans were living in urban than in rural areas—in towns with electric lights and streetcars, in cities with the clangor and jams of traffic, in what was the swiftly emerging technological metropolis.

Scores of articles emphasized that the technological marvels came out of a business establishment where Edisonian practice had given way to organized research. Statistical assessments told how more companies were opening laboratories—by the minimum estimate, the number in operation doubled over the decade—and paying their best scientists up to $20,000 a year. The National Industrial Conference Board announced that the United States was spending some $200,000,000 annually on scientific research and that of the total, American industry was providing twice as much as the federal government. "Science is not a thing apart," one could read in the *Saturday Evening Post*; "it is the bedrock of business."[2]

Near the top of the nation's business-minded government, Secretary of Labor James J. Davis expatiated upon the extent to which the practical applications of science had "freed" the factory hand by shortening his hours, lightening his workload, and, through increased productivity, raising his wages. The U.S. Women's Bureau credited the increasing employment opportunities of its special constituency to the new products and forms of service born of applied research.[3] Administration after administration, presidents and the future president celebrated the study of nature. At the one end of clarity was Herbert Hoover, who, in his widely applauded *American Individualism*, challenged pioneers with the reminder that "the great continent of science is as yet explored only on its borders"; at the other, Warren G. Harding, who, while amiably explaining how the mysterious rays of radium might cure cancer, sublimely added how they might eventually be found to originate in the same source as the warmly beneficent emanations of the "radioactive personality."[4]

The official plaudits, the corporate backing, the technological miracles

[2] "Better Business Through Research," *Literary Digest*, 95 (Dec. 31, 1927), 52; Floyd W. Parsons, "New Trails in Business," *Saturday Evening Post*, 194 (Jan. 7, 1922), 28.

[3] James J. Davis, "Science and the Worker," *Monthly Labor Review*, 26 (Feb. 1928), 241; "Effects of New Inventions upon the Field of Women's Employment," *Monthly Labor Review*, 22 (Jan.–June 1926), 1256.

[4] Herbert Hoover, *American Individualism* (Garden City, N.Y., 1929), p. 64; Harding is quoted in *Science*, 53 (June 3, 1921), 510.

—all combined to give science such enormous prestige that *The Nation* remarked: "A sentence which begins with 'Science says' will generally be found to settle any argument in a social gathering or sell any article from tooth-paste to refrigerators." The burgeoning advertising industry, eager to be up-to-date, responded to the ambiance of the day (buy Palmolive soap, "the mildest cleanser science can produce").[5] So did the flappers hypnotized by Freud, the teachers captivated by intelligence tests, the mothers attached to the latest theories of child development. So, by what they wanted to read, did many Americans who were not necessarily flappers, teachers, or psyche-minded mothers. An editor of the Scripps-Howard newspapers enthused: "Never . . . has the public been so eager for scientific information as it is today."[6]

By the end of the decade Americans had bought over 200,000 copies of Slosson's *Creative Chemistry*, which put it in the same sales league as *Babbitt*. Sensitive to the market, publishers brought out numerous books about science, and at a rate that outpaced the growth of their general lists. Scarcely a single major magazine, of select or mass circulation, seemed to go to press without an article on the doings in the laboratory. Scores of newspapers subscribed to Science Service; some, along with the several national press associations, hired their own science editors.[7] In 1923 Alva Johnston of *The New York Times* won the Pulitzer prize for his coverage of the annual convention of the American Association for the Advancement of Science.

Some of the popularizations extolled the intellectual triumphs of pure science, including pure physics. ("The most interesting and even spectacular achievement of modern science," the editor of *Harper's* said flatly, "has been the laying bare of the structure of the atom.") The bulk of the popularization applauded the benefits of applied science, detailing the wonders at hand, predicting yet more thrilling wonders to come. But even the elucidations upon the technological theme paid pure science rich compliments for having made the existing marvels possible and predicted with Floyd Parsons, a mining engineer, veteran of the government's wartime fuel administration, and an editor of industrial journals who was now writing regularly on business and science for the *Saturday Evening Post*: "The outlook is that from now on science will make life one unceasing revolution."[8]

[5] *The Nation*, 127 (Oct. 17, 1928), 390; the Palmolive soap advertisement is in the *Saturday Evening Post*, 195 (March 31, 1923), 78.

[6] David Dietz, quoted in *Science*, 68 (Aug. 10, 1928), 122.

[7] Preston W. Slosson, "Edwin E. Slosson, Pioneer," p. 28; *Babbitt* sold some 253,000 hardback copies. Mark Schorer, *Sinclair Lewis: An American Life* (New York, 1961), p. 357. On book publishing, see the remarks of Edward M. Crane, the president of D. Van Nostrand, *Science*, 68 (Aug. 3, 1928), 94.

[8] Editor's introductory note to Benjamin Harrow, "The Romance of the Atom," *Harper's*, 149 (July 1924), 251; Parsons, "Science and Everyday Life," *Saturday Evening Post*, 195 (March 10, 1923), 158.

The revolutionists, Edwin Slosson assured the public, were as "clean-shaven, as youthful, and as jazzy as a foregathering of Rotarians. . . . It must be admitted that the scientist of today is fully as much a man of the world as his brother, the businessman."[9] As always, what people were really like meant less, publicly, than what they were likened to. In the 1920s, by associating even the most abstruse theorist with the good works of technology and business, the popularizers helped make the pure scientist as such a highly respected figure of the decade. So, in his own special way, did Albert Einstein.

Midway through the war Einstein had put the finishing touches on his general theory of relativity. This generalization yielded a field theory of gravitation that explained a long-standing puzzle in the motion of the planet Mercury. It also said that the path of starlight would bend and the frequency of spectral lines change in the neighborhood of an enormous mass like the sun. British astronomers measured the deviation during an expedition to observe the solar eclipse of May 1919. In November, before a joint assembly of the Royal Society and Royal Astronomical Society—and in an atmosphere charged with high anticipation—they announced the results: Einstein's prediction had been confirmed with incontestable accuracy. Sir Joseph John Thomson capped this historic occasion by calling the general theory of relativity "the greatest discovery in connection with gravitation since Newton," and within days newspapers made the twentieth-century Newton famous around the globe.[1]

Einstein was widely—and inaccurately—reported to have warned that only twelve men in all the world could comprehend his general theory; in the United States through the 1920s it seemed as though everyone was competing for the honor of thirteenth man. Dozens of books and articles discoursed on relativity. Cranks, as well as physicists still unreconciled to Einstein's theory, advanced their opposition in the press; they merely helped sustain the vogue of interest. It was front-page news when, near the end of the decade, Einstein announced that he would soon publish a unified theory of electromagnetic and gravitational fields. The New York *Tribune* arranged for the transatlantic transmission of the text, including —it was said to be a first for cable communication—the mathematical formulae. When Einstein completed his work, newspapers devoted thousands of words to it and the New York *Post* as well as the *Tribune* printed radiophotofacsimiles of the original pages.

Even old Senator Boies Penrose of Pennsylvania, better known as a GOP warhorse than as a patron of the intellect, got himself a book about relativity. But the senator had to confess on the floor of the Congress:

[9] "Must Scientists Wear Whiskers?" *The Independent*, 115 (Nov. 28, 1925), 601.

[1] Thomson is quoted in *The New York Times*, Nov. 9, 1919, p. 6. The meeting is reported in the London *Times*, Nov. 7, 1919, p. 12.

"I have nearly lost my mental faculties in trying to understand Einstein."[2] The general theory was confounding even in the most lucid popularization. Here was the universe described as a four-dimensional continuum, a closed manifold in space and time. Its local properties were explained to depend upon the amount of mass in the neighborhood, and, somehow, it was supposed to double back on itself at infinity. In the 1870s, an observer recalled, everyone at least felt that he understood the theories of evolution and the conservation of energy; as for relativity in the 1920's: "Nearly everybody has sense enough to know that it is hopeless for him to get more than the vaguest notion of what it is all about."[3]

Many news stories remarked upon Einstein's luminous, childlike eyes; he was the fantastic dreamer incarnate. Past forty, clothes baggy, hair unkempt, his favorite musical instrument the violin—by no stretch of the imagination could Einstein be considered either jazzy, clean-cut, or even youthful. But while Americans likely did not applaud Einstein because they understood his theory, or because he seemed a promising candidate for the Rotary, they could certainly celebrate a theorist who was said to rank with Newton. And this Newton was heralded with two specially appealing postwar twists.

In 1914 Einstein had become a professor at the University of Berlin, the director of the Kaiser Wilhelm Institute for Theoretical Physics, a member of the Prussian Academy of Science, but he had not lent his prestige to the German war effort. Now Mayor Andrew J. Peters of Boston respectfully declared: "Not many of us can follow Prof[essor] Einstein in his discussion of the mathematical properties of space; but all of us can understand his refusal to sign the manifesto of the ninety-three professors." Besides, there was what the editors of *Current History* announced as the "revolutionary" implications of his theory for "practical physics."[4] More than one journalist called attention to the striking revelation of the theory of relativity that, locked in every atom, waiting to be tapped, were enormous, incredible stores of energy.

Atomic energy—once man learned to release it, Floyd Parsons elaborated, he would have available to himself "a source of power inconceivably greater than any possible requirement of the human race." Release atomic energy, and "the world's people will be freed from the need of constant toil," "science will tend to equalize the distribution of wealth and the good things of life," and "statesmen will be glad . . . to compromise their differences without any talk of force, for a power will

[2] U.S. Congress, Senate, *Congressional Record*, 67th Cong., 1st Sess., April 18, 1921, LXI, pt. 1, 404.

[3] Fabian Franklin, "Einstein and America," *Scientific Monthly*, 28 (March 1929), 279.

[4] Mayor Peters is quoted in the Boston *Evening Globe*, May 17, 1921, p. 11; "Professor Einstein's Theory of Relativity and Its Revolutionary Effects in Practical Physics," *Current History*, XII (June 1920), 498.

be available in the world so mighty in its potentialities that no person would dare consider its use except for constructive purposes."[5]

In the 1920s the celebration of Einstein not only helped enlarge the prestige of pure science; it endowed even the abstruse study of the atom with the expectation of wonderful technological promise. But in the middle of it all, popularizers typically reported: "Cocksureness has no place in the science of today, in the science which has wrought the miracles of recent generations."

Here was Einstein, the world's most celebrated scientist, declaring that anyone lucky enough to disclose "something of the beauty of this mysterious universe . . . should not . . . be personally celebrated."[6] Newsmen fastened upon Einstein's unassuming manner, the innocent timidity with which he would greet a drove of reporters, his pipe clutched in one hand, his violin in the other. Merely by being in the same profession as Einstein, every physicist wore a halo of humility. When a journalist described the men to be found in the laboratories as "brilliant, accurate, ingenious, unpretentious, devoting every hour of the day to their labors with little thought of self and career," Einstein's storied self-effacement added no small credibility to the portrait.[7]

If the 1870s had experienced the abrasive iconoclasm of John Tyndall, the 1920s typically knew the soothing, erudite apologia of Alfred North Whitehead, a professor of philosophy at Harvard and coauthor with Bertrand Russell of the monumental *Principia Mathematica*. Whitehead explained in his *Science and the Modern World* how, in the light of recent physics, the mechanistic interpretation of nature had all been a serious mistake.[8] If the 1870s had witnessed the theological controversies of evolution, in the 1920s no one, certainly not Einstein, exploited relativity to assault religious faith. In fact, Einstein's theory found one of its most respected popularizers in Arthur S. Eddington, a British physicist, a member of the eclipse expedition, and a devout Quaker whose widely read *Space, Time and Gravitation* was dominated by a single theme: Relativity insisted that physics could describe only the relationship among measurements, not the "nature of things." Accordingly, scientific knowledge amounted to "an empty shell—a form of symbols," with the

[5] Parsons, "Science and Everyday Life," *Saturday Evening Post*, 195 (March 10, 1923), 157; "Pioneering Beyond the Rim," *ibid.*, 195 (May 5, 1923), 123; "The Stupendous Possibilities of the Atom," *World's Work*, 42 (May 1921), 30.

[6] Dixon Merritt, "The Brotherhood of Scientists," *Outlook*, 139 (Jan. 14, 1925), 57; Einstein is quoted in *The New York Times*, April 27, 1921, p. 21.

[7] C. E. Bechhofer Roberts, "The Almighty Atom," *World's Work*, 50 (Aug. 1925), 416.

[8] "The old foundations of scientific thought are becoming unintelligible. . . . What is the sense of talking about a mechanical explanation when you do not know what you mean by mechanics." Whitehead, *Science and the Modern World* (New York, [1925], 1953), pp. 16, 17.

real world beyond its reach.[9] In the epistemological gap between what physics knew and what it could not, Eddington found a good deal of room for creative mind—and, by emphatic implication, for soul, for spirit, for God.

After the development of quantum mechanics, Eddington promptly told his readers how the new theory meant "a denial of determinism," and one could recognize "a spiritual world alongside the physical."[1] The uncertainty principle in particular, other writers elaborated, made the real world unknowable in its entirety. The concepts of quantum mechanics were of man's, not nature's, making. Its equations were simply operational shorthand forms rather than descriptions of reality, and, as such, they could only predict the results of experiment, tell you what the atom would do, not what it looked like. Bertrand Russell, an occasional visitor to the United States in the 1920s, remarked: "The new philosophy of physics is humble and stammering where the old philosophy was proud and dictatorial."[2] Intentionally or not, by genuflecting before the limits of knowledge, the popular interpreters of quantum mechanics kept the door wide open for religious faith. A good many natural scientists were even humbly saying with G. Raymond Wakeham, professor of chemistry at the University of Colorado: "Science has offered nothing which begins to rival in poignancy, persuasiveness, and general psychological effectiveness the appeal of religion."[3]

Religious fundamentalists agreed so emphatically as to treat science with turbulent intolerance. White, Anglo-Saxon, and Protestant, centered in the dusty towns of rural America, they resented the city, with its heavy concentrations of Catholics and Jews, of speakeasies and flappers, of science and all its intellectually corrosive works. They resented it all the more because urban America was increasingly setting the political and cultural and moral standards of the nation. Powerless to halt the trend, they vehemently, sometimes viciously, lashed back at its symptoms. Rural America revived the Ku Klux Klan, demanded that the country remain desolately dry—and followed William Jennings Bryan in a revolt against the teaching of evolution which climaxed in the spectacle of the Scopes trial.

The hostility of the fundamentalists dismayed the nation's leading scientists, particularly Robert A. Millikan. Through the decade, from

[9] Eddington, *Space, Time and Gravitation: An Outline of the General Relativity Theory* (Cambridge, England, 1920), pp. 200–201.

[1] Eddington, *The Nature of the Physical World* (Cambridge, England, 1928), pp. 228, 288.

[2] See, for example, Percy W. Bridgman, "The New Vision of Science," *Harper's*, 158 (March 1929), 444, 448, 450; Russell, "The Twilight of Science," *The Century Magazine*, 118 (July 1929), 311.

[3] Wakeham, "When Knowledge Is Misused," *The Independent*, 116 (Jan. 16, 1926), 76. See also Vernon Kellogg, who was the secretary of the National Research Council, "Some Things Science Doesn't Know," *World's Work*, 51 (March 1926), 523, 529.

the lecture platform and in the press and a number of books, Millikan campaigned for the reconciliation of science and religion. Regarding the scientific method as crucial to moral development, he reminded the religious world that, historically, the church without science had fostered "dogmatism, bigotry, and persecution." A devotee of Eddington, he announced to his own profession that even more important than science itself was "*a belief in the reality of moral and spiritual values.*" To drive both points home to a wide audience, Millikan drew up a statement in 1923 that testified not only to the harmony of science and religion but to the value of both. When released to the newspapers, it bore the signatures of forty-five prominent Americans, including Herbert Hoover and William Allen White, sixteen Protestant theologians, and some of the most distinguished men in the National Academy. The dozen leading scientists of the nation had attested to their support of a higher being, Millikan asserted, and he had the documents to prove it.[4]

Millikan managed to extract fresh testimony to the Creator's benevolence even from one of his newly developed areas of research. In 1925, in a deft series of experiments, he decisively confirmed the hypothesis of the Austrian physicist Victor Hess that a radiation far more powerful than the most energetic emanations of radium was bombarding the earth from all directions in space. Impressed, *The New York Times* proposed calling the radiation "Millikan rays," to associate them "with a man of such fine and modest personality."[5] Millikan modestly dubbed them "cosmic rays," the name that stuck.

Millikan was sure that cosmic rays must consist of photons, bundles of light quanta, rather than charged particles. From his experimental data, he concluded that these cosmic photons were clustered in three energy bands, each of which, he argued, could result from the energy radiated when various charged particles united to form the nucleus of an atom. Millikan's three bands happened to contain in turn the characteristic energies for helium, oxygen, and silicon, so he theorized that cosmic rays were produced by the creation of photons when four hydrogen atoms fused in interstellar space to form helium, or sixteen to form oxygen, or twenty-eight to form silicon. Since these three elements were among the most abundant in the universe, it seemed reasonable to Millikan that their formation should account for the bulk of the cosmic rays.[6]

[4] Quoted in "Men of Science Also Men of Faith," *Literary Digest*, 78 (July 14, 1923), 30; *Science*, 55 (March 31, 1922), 332. The statement is printed in *Science*, 57 (June 1, 1923), 630–31. Apropos Millikan's efforts to reconcile science and religion, an observer commented: "Science itself nowadays is a form of piety—so we have been told by those who speak for science." Anon., "The Scientist Bends the Knee," *New Republic*, 44 (Aug. 5, 1925), 280. See also the tart editorial, " 'Reconciling' Science and Religion," *New Republic*, 44 (Sept. 9, 1925), 59.

[5] *New York Times*, Nov. 12, 1925, p. 24.

[6] Robert A. Millikan and G. H. Cameron, "The Origin of the Cosmic Rays," *Physical Review*, 32 (Oct. 1928), 533–57.

To many physicists, Millikan's theory seemed a house of dubious hypotheses. His determination that the rays were clustered in three energy bands was specious. Most difficult to accept was the idea of the spontaneous fusion of numerous hydrogen atoms in interstellar space. Such an event was a good deal less probable than the simultaneous collision of sixteen or twenty-eight airplanes, each of which was pursuing an entirely independent course. Aware of the kinetic difficulties, Millikan tried to explain them away. He unhesitatingly popularized his hypothesis, calling cosmic rays the "birth-cries" of atoms.[7] The phrase quickly achieved public notoriety. So did Millikan's religiously flavored inference that the creation of the elements was going on continually throughout the universe, and in the process saving it from the heat death to which, according to various interpreters of the second law of thermodynamics, it was doomed. In Millikan's hands, cosmic rays turned into fresh evidence for the existence of God; they also suggested that He was emphatically not a fundamentalist.

If in the 1920s the iconoclasts of urban America relished Henry L. Mencken's bombardment of the Bryanites, the more solemn public followed the Millikans. To *The New York Times*, what Millikan himself had to say about moral and spiritual values was "even more significant" than what he knew about physics.[8] Protestant and Catholic, the modernist religious press praised the reconcilers and happily cited their piety. The Reverend Harry Emerson Fosdick, an unimpeachable modernist, wryly remarked: "When a prominent scientist comes out strongly for religion, all the churches thank Heaven and take courage as though it were the highest possible compliment to God to have Eddington believe in Him."[9]

Amid all the acclaim for science, many Americans stirred when in 1927 the English bishop of Ripon, a former chaplain to the king, wondered aloud—only jestingly, he later said—whether mankind might benefit from a ten-year moratorium on all scientific research. Calling attention to the idea, the Chicago *Evening Post* suspected that a sizable number of people were troubled by the way "science has been leading us rather a giddy chase."[1]

Unlike the bitterness behind the Scopes trial, this discontent was centered in urban America, and it found voice in a number of highly articulate dissenters from the regime of science. By no means fundamentalists, the critics were descended from Thomas Huxley's educated

---

[7] Millikan, "Available Energy," *Science*, 68 (Sept. 28, 1928), 4.

[8] *New York Times*, Nov. 15, 1923, p. 18.

[9] Quoted in Frederick Lewis Allen, *Only Yesterday* (New York: Bantam, 1957), p. 141.

[1] The bishop and the *Evening Post* are quoted in "Is Scientific Advance Impeding Human Welfare?" *Literary Digest*, 95 (Oct. 1, 1927), 32–33. "My original remark," the bishop later told Millikan, "was made with a broad smile." Bishop of Ripon to Millikan, March 25, 1930, RAM, Box 39.

enemies, and they tended to argue more from a secular than from a religious angle. Professors, writers, and poets, they called themselves "humanists." They charged, frequently with passion and eloquence, that the dominance of science and the machine was throwing civilization into a dangerous imbalance. They received wide attention in the press, and near the end of the decade Charles Beard observed that what not long before had appeared to be "a tempest in a teapot, a quarrel among mere 'literary persons,' has become a topic of major interest among hard-headed men of affairs. A subject mildly discussed in women's clubs has broken into offices, factories, smoking compartments, and political assemblies."[2]

No thoughtful American could help but respond to the humanist indictment of technology. There was the destructiveness of applied science: If it already had much of the carnage of the Great War to its credit, it would make the next war catastrophic, particularly—commentators such as Raymond B. Fosdick, a League of Nations advocate, worriedly pointed out—if man were to discover the secret of atomic energy. There were the dehumanizing effects of the machine: The automobile industry might have created jobs for tens of thousands, but the assembly line had robbed blue-collar workers of the pleasures of craftsmanship, the organization of mass production had turned white-collar employees into faceless cogs in a corporate machine. There was, as Charles Beard summarized it, the blight of urban, technological civilization: "New York City from the elevated railway, huge sections of Pittsburgh and Chicago, shabby and dilapidated water fronts, glorious spots of nature made hideous by factories, endless rows of monotonous dwellings, the shameful disregard of beauty along the highways from Boston to San Francisco, magnificent avenues through forest and valley ruined by billboards and gas-filling shacks, fretful masses rushing from one mechanical show to another, the horrible outpouring of radio nonsense, natural and canned, the unceasing roar and grind of urban life."[3]

Culturally, humanist critics pronounced the worship of Einstein merely another mass craze, the news columns on relativity an overindulgence, the attempts to make science accessible to the layman a failure. *The Nation* sniped that if physics got any more incomprehensible, science was going to find itself "ruling a realm as autonomous and as remote as . . . theology." Schoolmen agreed with the widely read historian James Harvey Robinson, who declared that college scientific courses "neither engender in the student a discriminating and exacting tendency of mind . . . nor do they foster . . . a lively understanding of the workings of nature." They were too professional, too specialized. Worse, almost the entire curriculum was said to have been infected with the disease of scientism. "An army of specialists," an educator lamented, "has routed

[2] Editor's introduction, in Charles Beard, ed., *Toward Civilization* (New York, 1930), p. 1.
[3] *Ibid.*, p. 11.

the humanist from the academic groves, and the old discipline of the liberal arts . . . has been replaced with myriad fragments of knowledge, often without underlying form or pattern. A chorus of youthful critics complains of an education which informs but does not enlighten."[4]

Through all the humanist critique one word kept cropping up: *values*. A Millikan might seem to argue that morality progressed with the progress of research, but the humanists replied that science merely described; it did not speak to timeless questions of right and wrong. While modern man knew more than Socrates, he was surely neither wiser nor more decent. While industry and even nations could write checks against science, science neither checked the rapacious industrialist nor did it make bellicose nationalists any less bellicose. In fact, a number of penetrating critics charged, scientism had so eaten away at traditional values that thoughtful men found themselves burdened with the despair of moral relativism. Where are our great moralists? Henry Seidel Canby wondered. "Our Miltons have gone in for mathematics, biology, and atomic research."[5]

The humanists had alternatives, but these were less stimulating than their critiques. The corrosion of values? More attention needed to be paid the arts and humanities. The widening cultural chasm? Let the colleges adopt broader curricular requirements, with an emphasis upon general rather than specialized courses. The ravages of technology? The humanist critics said nothing about such economic issues as the elimination of jobs by labor-saving machinery or the contribution of the tractor and chemistry to the worrisome weakness of agricultural markets. And they manifested no concern that a disproportionately large share of the new national wealth of the decade went to higher- rather than lower-income groups.

Sometimes the humanists simply sounded as though the United States ought to return to the pastoral days of Jefferson. With good sense, Charles Beard chided the fraternity of letters against such wishful thinking; there was no going back to an agricultural Eden. Instead of thrashing at machine civilization, its critics might better confront the problem of how to control it for desirable social purposes. With that aim in mind, Beard called upon the scientist to exercise a measure of social responsibility, insisting that he could not "escape his obligations by crying out that the upshot is none of his business."[6]

In response to the humanist critique, the nation's scientific leaders counseled the acceptance of social responsibility ("service to man . . .

---

[4] "Einstein's Latest," *The Nation*, 128 (Feb. 13, 1929), 179; James Harvey Robinson, "The Humanizing of Knowledge," *Science*, 56 (July 28, 1922), 99; William E. Wickenden, "Education and the New Age," in Beard, ed., *Toward Civilization*, pp. 261–62.

[5] Canby, "Render Therefore Unto Caesar," *Saturday Review of Literature*, 5 (May 4, 1929), 970.

[6] Introduction, Beard, ed., *Toward Civilization*, p. 16.

is the slogan of the scientist to-day").[7] They admitted that the scientist needed culture as much as culture needed science; some helped found a history of science society, evidently believing with George Sarton, the leading, overoptimistic practitioner of the discipline, that such studies could "bridge the widening gap between the men of science . . . and the men of letters."[8] Various scientists also agreed with James McKeen Cattell, now editing his magazines from his home on the Hudson River, who doubted that the advance of science would by itself lead to an economically more democratic society; while research had most certainly created a great deal of material wealth, Cattell reminded his colleagues, it had not yet "taught us how to distribute and use it properly." But whatever some scientists said or believed, in the public forum the scientific response to the humanist critique was dominated by Robert A. Millikan, who won the Nobel Prize in physics early in the decade, made the cover of *Time* near the end of it—he had the "face of a witty and successful banker," the editors declared—and was the most famous and respected American scientist of the day.[9]

Culturally, Millikan perceived the "most wantonly destructive" threats to civilization not in science but in the "emotional, destructive, over-sexed" content of modern literature and art. Educationally, he thought that no discipline matched the power of the mathematical and physical sciences to teach young men "to see clearly and to think straight." Millikan supposed that Caltech, where these subjects dominated the curriculum, might very well exert "a wholesome, yes, a saving influence" upon all of American education. With high authority, a good deal of support from most physicists, and a special bow to the Creator for building some "foolproof" safeguards into His handiwork, Millikan— who scoffed at the general anxiety over technology—assured his fellow citizens that the energy locked in the atom was destined to stay there.[1] With good sense, he attacked the idea of a moratorium on research as both "impossible and foolish" (impose a moratorium, another commentator explained, and you would get "bootleg science as well as bootleg whiskey"). Less logically, Millikan announced that, whatever the problems of urban congestion, or even of gangsters with getaway cars, the automobile had created "a new race" of responsible, abstemious men.

[7] Charles R. Richards, "Influences Affecting the Advancement of Education and Research," *School and Society*, 16 (Nov. 18, 1922), 564.

[8] *Science*, 59 (Feb. 8, 1924), 138. Ernest Rutherford once observed, "In my general experience, the younger men are blissfully ignorant of the development of the ideas with which they are dealing every day. . . . It is curious that men as a rule take very little interest in scientific history till they reach middle age." Rutherford to Henry Crew, March 1, 1928, HC.

[9] Cattell, "Science, the Declaration of Democracy," *Scientific Monthly*, 24 (March 1927), 203; *Time*, April 25, 1927, p. 16.

[1] Millikan, "Science and Modern Life," *Atlantic*, 141 (April 1928), 490; address accepting the physics laboratory at Caltech, *Science*, 55 (March 31, 1922), 331; "Alleged Sins of Science," *Scribner's*, 87 (Feb. 1930), 121.

One had only to "contrast the clear-eyed, sober, skillful, intelligent-looking taxi driver of today with the red-nosed wreck of a human being who used to be the London cabby of a quarter century ago."[2]

Millikan did not worry about technological dislocations on the farm or in the factory or about the way the growing economic pie was distributed. Quite the contrary, to his mind "no efforts toward social readjustments or toward the redistribution of wealth have one thousandth as large a chance of contributing to human well-being as have the efforts of the physicist, the chemist, and the biologist toward the better understanding and the better control of nature."[3] Millikan's was a trickle-down approach to prosperity. Awed by the businessmen he had met and worked with during the war, he assumed that the public interest would be served by letting private industry, mainly big industry, decide by itself how the benefits of research should be passed on to the public.

Bertrand Russell caught something of the meaning of it all when he remarked that science had become "politically conservative." He might have added that in the era of Harding, Coolidge, and Hoover, the leading exponents of science had also become vigorous advocates of the day's conservative syllogism—that science was good for business, business good for America, and, in consequence, science good for the nation's economic and spiritual well-being.[4]

[2] Millikan, "Science and Modern Life," pp. 491, 495; A. W. Meyer, "That Scientific Holiday," *Scientific Monthly*, 27 (Dec. 1928), 542.

[3] Millikan, "The Significance of Radium," *Science*, 54 (July 1, 1921), 8. Millikan unaccountably revised the odds to one-tenth the chance two years later in "Science and Society," *Science*, 58 (Oct. 19, 1923), 297.

[4] Russell, "Is Science Superstitious?" *The Dial*, 81 (Sept. 1926), 180–81. A liberal observer commented on this conservatism: "When you become a custodian of civilization, the most important thing in the world is civilization, more important even than the science through which you rose to your position of power and esteem." Anon., "The Scientist Bends the Knee," *New Republic*, 44 (Aug. 5, 1925), 281.

XIII

Making the Peaks Higher

In 1925 the *New Republic* noted: "Today [the scientist] sits in the seats of the mighty. He is the president of great universities, the chairman of semi-official governmental councils, the trusted adviser of states and even corporations."[1] Never had the nation's leadership been so well situated to raise substantial funds for research. Never had its public posture been more congenial to the industrialists, financiers, and foundation executives who controlled the purse strings of philanthropy.

The conservative propensities of the Millikans bolstered in particular the big business friendships of the National Academy of Sciences—Cattell contemptuously called it "an exclusive social club for those who have arrived"—and in 1925, at the urging of George Ellery Hale, the Academy launched a bold scheme to raise money for pure science: a National Research Endowment.[2] The endowment was to be capitalized at \$20,000,000, mainly with contributions from industry, and managed by a board of trustees including representatives from the Academy itself as well as Andrew W. Mellon, John J. Carty, Elihu Root, Charles Evans Hughes, Edward M. House, John W. Davis, Owen D. Young, Julius Rosenwald, and Felix Warburg. The chairman of the Endowment was

---

[1] Anon., "The Scientist Bends the Knee," *New Republic*, 44 (Aug. 5, 1925), 281.
[2] Cattell, "The Organization of Scientific Men," *Scientific Monthly*, 14 (June 1922), 572. A discussion of the Endowment first appeared in *Minerva*, copyright © 1974 by *Minerva*, and is reprinted here in different form with the permission of the editor.

Secretary of Commerce Herbert Hoover, the technically proficient mining engineer who in 1919 had been awarded the Academy's Medal for Eminence in the Application of Science to the Public Welfare and who in 1922 had been elected to the Academy. Hoover launched the public campaign for funds himself, in a speech before a key industrial forum, and he took pains to see that his address got nationwide coverage in the press. It was an "appeal from a big man who knows his subject," the *Capital News* in Lansing, Michigan, said of the speech. Scores of newspapers all over the country, big city and small town, Democratic and Republican, agreed. The Galveston *Daily News* went on to predict: "With such distinguished sponsorship, the movement will no doubt . . . raise the required sum without difficulty."[3]

Hoover knew, however, that raising $20,000,000 for pure science would be quite difficult, not least because the trustees aimed to obtain a sizable fraction of the funds out of corporations chartered to make profits, not engage in philanthropy. A British judge had laid down the fundamental doctrine of common law: "Charity has no business to sit at boards of directors *qua* charity." Further, the Internal Revenue Code then contained no provision allowing corporations to deduct the cost of philanthropies as such from taxable income, and attempts to write one into it early in the decade had failed repeatedly in the Congress. By holding deductible only direct operating expenses, the tax code bolstered the philanthropic limits set by the courts. Business corporations could engage in only those eleemosynary activities that promised to yield what lawyers called a "corporate benefit."[4]

To assure that the National Research Endowment qualified for corporate largess, the trustees changed its name to the National Research Fund and proposed to spend, rather than invest, each contribution in the year it was made. That way, a corporate donation would be no mere philanthropy, but a current expense, a payment for new knowledge, and, as such, both legal and tax-deductible. To lay to rest any possible doubts about the legality of corporate participation, Hoover asked Elihu Root to draw up an informal brief on the point. Root studied the precedents, found that the technological lessons of the war had affected the thinking of judges as well as businessmen. In two recent cases, the English and American courts had upheld industrial gifts made for the purpose of enlarging the pool of trained technical talent. Extending the logic of the

---

[3] Lansing *Capital News*, Dec. 8, 1925, p. 4; Galveston *Daily News*, Dec. 22, 1925, p. 4.

[4] Judge Bowen is quoted in "Donations by a Business Corporation as *Intra Vires*," *Columbia Law Review*, 31 (1931), 136. See also, Committee on Corporate Laws, American Bar Foundation, ed., *Model Business Corporation Act Annotated* (American Bar Foundation, 1960), I, 130–42; Henry J. Rudick, "The Legal Aspects of Corporate Giving," in Beardsley Ruml, ed., *The Manual of Corporate Giving* (Washington, D.C.: National Planning Association, 1952), pp. 35–36.

rulings, Root concluded, with the concurrence of John W. Davis and Charles Evans Hughes: Yes, business corporations whose competitive position depended upon the progress of science could contribute, so long as the corporate directors were not "swayed by motives of generosity or by public spirit."[5] By mid-1930 business enterprises and philanthropists had made sufficient commitments to put the Fund into operation. But when the trustees tried to collect in this first year of the depression, one of the donors reneged on a technicality. By 1932, the possibility of re-couping from this defection was clearly gone and the National Research Fund, its promoters had sadly to agree, was dead.[6]

The trustees liked to think that the depression killed the Fund, but the only corporate donors were AT&T, U.S. Steel, and the member companies of the National Electric Light Association and the American Iron and Steel Institute. General Motors had declined to participate; so had the railroads. Even the chemical and oil companies, which could expect a more substantial dividend from further advances in pure science than could the automobile or railroad industries, neglected to join the list of contributors. The trouble was that, when produced in a university laboratory, the results of an investment in pure science were usually published widely; corporations that contributed to the Fund would in effect be helping to supply new knowledge to competitive firms that did not. AT&T could safely contribute because it was a monopoly. So could U.S. Steel, because it dominated the steel industry and because, by prodding a contribution out of the American Iron and Steel Institute, it forced virtually every other major firm in the industry to share the total eleemosynary cost. Similarly, a donation from the National Electric Light Association was tantamount to gifts from all the firms in the lighting and power industry. Besides, Hoover reportedly remarked of the Association: "Don't fool yourself that they care a damn for pure science. What they want is to get into their reports, which will soon be examined by the Federal Trade Commission, that they are giving money for pure science research."[7]

Economic self-interest also governed the type of financial contributions that industry was willing to make to academic science directly.

[5] Hoover, Root, and Hughes to John D. Rockefeller, Jr., Jan. 10, 1926; Davis to Hoover, April 26, 1926, HH; Root, "Opinion and Memorandum Concerning the Power of Corporations to Subscribe to a Fund to be Used in the Prosecution of Research in the Pure Sciences," attached to Harold Phelps Stokes to Vernon Kellogg, April 16, 1926, NRCM, National Research Endowment file.

[6] Robert A. Millikan to John J. Carty, Sept. 21, 1929, GEH, Box 10; Frank B. Jewett, "Report to the Trustees of the National Research Fund . . .," April 19, 1934, RAM, Box 8.

[7] George Ellery Hale to Robert A. Millikan, March 15, 1928, RAM, Box 24. For a more detailed analysis of the failure of the Fund, see Lance E. Davis and Daniel J. Kevles, "The National Research Fund: A Case Study in the Industrial Support of Academic Science," *Minerva*, 12 (April 1974), 213–20.

Corporations and trade associations substantially enlarged the supply of fellowships in science. They also funneled over $1,000,000 into the technical departments of the land-grant colleges, their largess including grants from the National Electric Light Association for research on high-voltage cables, from the National Tanners Association for work on the chemistry of skins, even from H. J. Heinz and the National Pickle Packers Association for studies in the diseases of cucumbers. When these gifts and grants were made through trade associations, every firm in a given industry had to bear the cost. Whatever the type of granting agency, the vast majority of the subventions evidently related directly to the business operations of the contributors; they promised to buy either immediately useful knowledge or employable trained talent.[8] By the end of the decade no corporation or trade association had turned over sizable funds to any university explicitly for pure science.

Industrial corporations generally preferred to make their investments in pure research in their own company laboratories, where technologically promising discoveries could be patented before they were published. At General Electric, Irving Langmuir was proceeding with the work on surface chemistry that would lead to the Nobel Prize. In 1925 AT&T turned the research department that Frank Jewett had created into the Bell Telephone Laboratories. The new organization attracted some highly able physicists, including Clinton J. Davisson, the son of a contract painter in Bloomington, Illinois, who had worked his way through the University of Chicago, taken his Ph.D. at Princeton under Owen W. Richardson, married Richardson's sister, and made a specialty of his brother-in-law's field, the emission of electrons from metals.

Davisson's specialty bore directly upon the problems of vacuum tube development at AT&T. He began to explore how a beam of electrons scattered when it struck a nickel target; then a minor laboratory accident, which changed the physical structure of the target, set Davisson on a particularly intriguing line of investigation. In 1927, after struggling through Schrödinger's papers, he and his assistant Lester H. Germer demonstrated that the electrons were diffracted by the nickel in exactly the pattern of a light wave and with the wavelength predicted by wave mechanics.[9] In England at about the same time, George P. Thomson, J. J. Thomson's son, independently achieved the same results. Together, the two sets of experiments increased the scientific community's confi-

[8] Edward J. Eddy, *Colleges for Our Land and Time* (New York, 1956), p. 174; George D. McLaughlin, "Research and Industry," *Scientific Monthly*, 22 (April 1926), 283–84; "Funds Available in 1920 . . . for the Encouragement of Scientific Research," *Bulletin of the National Research Council* (No. 9; Washington, D.C., 1921); "Funds Available . . .," 2d ed., *Bulletin of the National Research Council* (No. 66; Washington, D.C., 1928); F. Emerson Andrews, *Corporation Giving* (New York: Russell Sage Foundation, 1952), p. 209.

[9] C. J. Davisson and L. H. Germer, "Diffraction of Electrons by a Crystal of Nickel," *Physical Review*, 30 (Dec. 1927), 705–40.

dence in the physical legitimacy of wave mechanics—and a decade later the two experimenters would share the Nobel Prize.

Davisson's work yielded a sizable dividend in prestige for Bell Labs and also made the organization seem like an indulgent academic laboratory. Davisson's published accounts of his experiment emphasized both its serendipitous beginnings and the sustained and sweaty effort before the ultimate triumph. But if in some ways Bell Labs resembled a freewheeling research institute, Robert W. King, the editor of the company's technical journal, pointed out that most of its scientists were essentially team workers and their projects required the approval of the director of the laboratory. At Bell Labs only some ten physicists enjoyed Davisson's liberty, elsewhere fewer or none at all. Reflecting the attitude of most industrial research laboratories, a group of engineering executives complained that American universities were paying too much attention to producing Ph.D.'s qualified for research in atomic structure. "In an industrial laboratory, the physicist or engineer who has the fundamental classical background is of greater value than a physicist who has almost exclusively specialized in the modern . . . developments."[1]

In Washington at the opening of the 1920s, the Bureau of Standards was even less a paradise for pure physics than were industrial laboratories. Salaries were relatively low, the threat of congressional cutbacks in operating funds always present. Stratton ruled the Bureau with an iron hand, delegating little authority, insisting upon close supervision of everyone's activities. Discontented, in 1919 some of the Bureau's staff helped create a scientific and technical branch in the local of the National Federation of Federal Employees, a union of 21,000 members affiliated with the AFL. The new branch hoped to negotiate a more satisfactory administrative environment, higher salaries, and a more stable congressional policy toward research.[2] By the early 1920s the local had managed to get its membership reclassified upward on the civil service scale, and the average wage for scientific workers rose to $3,500 a year, which was higher than the median income in the United States. Otherwise, it accomplished little.

At the Bureau of Standards, the directorship passed from Stratton, who in 1923 left for the presidency of MIT, to Edward B. Rosa, who died in office a year later, to George William Burgess, a physicist who oversaw a more laissez-faire administrative regime. But if the staff now had more control over its own activities, Burgess, like the directors of all executive agencies, was at the mercy of the new Bureau of the Budget.

[1] Robert W. King, "Physics in Industry as a Career," *Scientific Monthly*, 28 (March 1929), 227, 229; the engineering executives are quoted in Albert W. Hull, "Qualifications of a Research Physicist," *Science*, 73 (June 12, 1931), 627; interview with Karl K. Darrow.

[2] R. H. True and P. G. Agnew, "A Union of Scientific Federal Employees," *Science*, 49 (May 23, 1919), 487–89.

However passionate Herbert Hoover's pleas for science, the administration, including the secretary of commerce, was committed to keeping federal expenditures down. Between 1921 and 1928 Burgess' operating budget climbed only about 16 percent, to some $2,500,000 annually. Burgess prudently stressed practical studies. William F. Meggers, one of the leading atomic physicists at the Bureau, complained that his colleagues could get plenty of money for research on "economies in automobile manufacture," but little if any for pure spectroscopy. In 1924 Meggers, already burdened with routine spectrochemical analysis, found his section reduced to two people.[3]

Whatever the constraints on industrial corporations or federal scientific laboratories, the major philanthropic foundations were free, by definition, to ignore the doctrines of corporate benefit or laissez faire. And they possessed a munificent supply of funds increasingly subject to distribution in accord with the judgment of professional philanthropists rather than the social ambitions of industrial captains. The organizational philanthropists may have had their status drives, too; there was a high status dividend to be won from helping university science. Before the war only a few grants from the Carnegie Corporation of New York had gone to the academic world, the bulk of them to the Carnegie Institute of Technology. Now, with Carnegie himself dead and the universities in fiscally straitened circumstances, the Corporation began contributing a good deal more to the nation's colleges, especially to the endowment of science. But however helpful the Carnegie officers, no philanthropist was quite so eager to subvene science as Simon Flexner's colleague at the Rockefeller Foundation, Wickliffe Rose.

Rose grew up in western Tennessee listening to his father's fundamentalist sermons, went off to a teachers' college in Nashville, and soon turned into a secular missionary of learning. As head of the Sanitary Commission, which John D. Rockefeller created to combat hookworm in the South, he encouraged state and local governments to act against the debilitating disease. As director of the International Health Board, the Commission's successor, Rose traveled the globe. All the while he preached a special lesson drawn mainly from the work against hookworm in the South: The battle for public health could not be fought simply by encouraging governmental agencies to act. It had also to be waged by enlarging the number of men trained to deal with the problems of disease scientifically. To that end Rose endowed schools of hygiene in the United States and abroad and created a system of international fellowships so that

[3] William F. Meggers to H. Konen, March 22, 1924; May 12, 1924, WFM. Karl T. Compton observed: "The government is now supporting a few research agencies, but on a very niggardly scale, in very restricted fields, and with such pressure for immediate practical returns as to drive out those scientists who might do big things." Compton, "Specialization and Cooperation in Research," *Science*, 66 (Nov. 11, 1927), 440.

specialists from different countries could migrate to the new centers for advanced study.

In 1923 Rose took over the Rockefeller General Education Board, whose activities were confined to the United States; he also went abroad for five months to explore the international philanthropic possibilities at the scientific centers of Europe. Rose was confident that the world was tending increasingly toward democracy. Sure that the trend would be strengthened by education, he was particularly certain that "all important fields of activity, from the breeding of bees to the administration of an empire, call for an understanding of the spirit and technique of modern science." Rose returned from his trip both excited by the revolutionary momentum of atomic studies and disturbed by the extent to which the war had impoverished research and damaged international scientific collaboration. To restore European science, to advance research, to safeguard the democratic trend, Rose drew upon his philanthropic experience and submitted an imaginative proposal for the consideration of the Rockefeller trustees. Its essential strategy: Find the leading figures in the physical and biological sciences, ease the financial difficulties of their laboratories, and create a new system of international fellowships so that talented young Ph.D.'s could study under these masters in both Europe and the United States.[4]

Some of the Rockefeller officers shared Rose's enthusiasm for the natural sciences. Others, worried like the humanist critics about the runaway pace of technological development, thought that a large fraction of the Rockefeller educational millions ought to be invested in the social sciences and the humanities, in the disciplines which dealt with social problems and questions of values. Abraham Flexner later called Rose's unalloyed faith in science "strangely naïve."[5] Rose, who agreed with Millikan's rebuttal to the humanists, paid only lip service to the humanities, and considered investing in the social sciences a waste of money. To his mind, "the more completely the world of physical, chemical, and biological phenomena can be described and accounted for, the more prestige does the scientific attitude acquire . . . [in] other fields." Promoting a program no less conservative in its essentials than Millikan's, Rose assumed that, by some trickle-down process, the advance of science

---

[4] Rose is quoted in George W. Gray, *Education on an International Scale: A History of the International Education Board, 1923–38* (New York, 1941), p. 10. Raymond B. Fosdick, *The Story of the Rockefeller Foundation* (New York, 1952), p. 148; *Annual Report of the International Education Board, 1924/25*, p. 6; Rose, "Scheme for the Promotion of Education on an International Scale," [1924], RF, series 100, International Education Board, Box 4.

[5] Abraham Flexner, *Funds and Foundations* (New York, 1952), p. 82; "Meeting of the Board of Trustees of the Rockefeller Foundation," Oct. 29, 1930, pp. 88–89, RF, Series 900, Box 22; interview with the late Raymond B. Fosdick, who recalled that a number of Rockefeller Foundation officers were reading Frederick B. Soddy's *Science and Life* (London, 1920), in which, while praising science, Soddy worried that science by itself could neither save nor destroy mankind.

would advance the intellectual and, hence, general welfare of mankind.[6] Whatever his naïveté, Rose could argue with passion. In 1924 the Rockefeller trustees appointed him the director of the new International Education Board, with an endowment that would ultimately reach $28,000,000, and Wickliffe Rose, the fundamentalist minister's son, became virtually a central banker to the world of science.

Rose awarded some forty International Education Board fellowships in the first year, more than ninety in the second, more than five hundred by the end of the decade to men from thirty-five nations, including such physicists from the Central Powers as Werner Heisenberg. Rose began his program of institutional aid with $40,000 for Niels Bohr to build an addition to his institute and he granted subventions to a number of universities of high rank in science, including those at Göttingen, Paris, Leyden, Stockholm, Edinburgh, and Cambridge. The largest single appropriation went to the California Institute of Technology for the giant telescope that would eventually be housed in the Mt. Palomar Observatory. To enrich science at home, Rose relied primarily on the General Education Board, departing from its traditional practice of contributing to the general endowments of colleges and universities and shifting over to an emphasis on raising the quality of advanced training and research, especially in the physical and biological sciences. By 1932 the Board had singlehandedly enlarged the funding of academic science by some $19,000,000—about three times what it had awarded the humanities, and some six times what the total endowment for science in the United States had been at the turn of the century.[7]

"Make the peaks higher," Rose liked to say.[8] Between 1902 and 1925, the General Education Board had contributed to the endowment of 291 institutions; under Rose, who wanted especially to develop the quality of American scientific research, the Board awarded almost two thirds of its total grants for science to just eight institutions—Caltech, Princeton, Cornell, Vanderbilt, Harvard, Stanford, Rochester, and Chicago. The overall funding of scientific research followed a pattern of concentrated distribution. The leading private institutions, with high prestige, did best in tapping science-minded philanthropists; the major state universities, where research was already established, received the most substantial increases in appropriations. George Eastman and T. Coleman Du Pont

[6] Rose, *Annual Report of the General Education Board, 1925/26*, pp. 5–6. Raymond Fosdick recalled that Rose "almost seemed to feel that there was some process of osmosis by which the mood of the fundamental sciences would by diffusion be transferred to the problem of social control." Fosdick, introduction to Gray, *Education on an International Scale*, p. xii.

[7] *Annual Report of the International Education Board, 1925/26*, pp. 11–12; Raymond B. Fosdick, *Adventure in Giving: The Story of the General Education Board* (New York, 1962), pp. 229–34; Gray, *Education on an International Scale*, pp. 22, 50.

[8] Quoted in Fosdick, *Adventure in Giving*, p. 230.

opened the 1920s with a combined gift to MIT of almost $1,000,000 for physics and chemistry, and by 1931 six- and even seven-figure donations for science or research in general had gone to Harvard, Stanford, Princeton, Chicago, Caltech, and Cornell. While legislative appropriations for research in public universities climbed steadily, at Wisconsin, Michigan, and Berkeley they reached a combined total of almost $200,000 by 1932, an income equivalent to the return on new endowment of at least $5,000,000.[9]

Public or private, the institutions favored by the concentrated distribution of the new funds harbored the peak departments of physics. Their administrators won their share of the new money for science, and year after year in the 1920s academic physics grew far better off financially than it had ever been before.[1]

All the while George Ellery Hale remained eager for the National Academy–National Research Council to foster the kind of intimate collaboration among physicists, chemists, and astrophysicists that might move the United States into the front rank in the attack upon the structure of matter. Shortly after the war the Research Council began bringing practitioners of different disciplines together in Washington for conferences and symposia. The conferences and symposia yielded a variety of plans; little ever came of them. Cooperative research on a national scale cost a good deal of money, far more than the NRC could afford. Moreover, with most scientists enmeshed in their own special research interests, few were eager to commit themselves to a program of interdisciplinary cooperation in physics.

Decidedly unwilling, indeed angrily so, was Arthur Gordon Webster, a member of the National Academy since early in the century, but still isolated at Clark University doing generally undistinguished science, increasingly depressed—he would take his own life in 1923—and bitter. Cooperative research might be appropriate in astronomy or geodesy, Webster argued; it was entirely inappropriate in physics. Progress in that, as well as many other, disciplines depended on individual, not group, efforts. The solitary scientist needed financial support—Webster had never received a penny from the large-project-oriented Carnegie Institu-

[9] Ernest V. Hollis, *Philanthropic Foundations and Higher Education* (New York, 1938), pp. 274–75; Merle Curti and Roderick Nash, *Philanthropy in the Shaping of American Higher Education* (New Brunswick, N.J., 1965), p. 222; *Report of the President of Harvard University, 1929/30*, p. 7; *Annual Report of the President, Stanford University, 1927/28*, p. 5; *Catalogue of Princeton University, 1929/30*, pp. 289–90; *Report of the President of the University of Chicago, 1924/25*, p. x; *Report of the President of the Massachusetts Institute of Technology, 1920*, p. 11, *1921*, p. 34; Morris Bishop, *A History of Cornell* (Ithaca, 1962), p. 439; *University of Wisconsin, President's Report, 1918/20*, pp. 12–13; *University of Michigan, Report of the President, 1927/28*, pp. 149–51; *University of California, Berkeley, Annual Register, 1931/32*, p. 27.
[1] Hollis, *Philanthropic Foundations and Higher Education*, pp. 243–44.

tion of Washington—yet here was the cooperative-minded Research Council planning to spend a sizable fraction of its endowment on "a monumental administration building," of its income on a ream of "red tape." Worse, here was the NRC expecting scientists around the country to subordinate themselves to its centralized direction of their work. Webster defiantly announced that he certainly had no intention of abiding by "the wisdom of the authorities of the National Research Council."[2]

Neither did a good many other American scientists, not least because now, as in the days of Alexander Agassiz, the nation's men of research disliked anything that smacked of supervision from Washington. Like the dean of the Graduate School at the University of Wisconsin, they evidently regarded the Council's program and administrative framework as "rather elaborate." Some scientists worried that they would have to spend so much time cooperating as to have none left for research. Others chafed at the memories of their wartime experience with organized cooperation. John M. Clarke, unhappy with what he had seen of the NRC as chairman of its Geology Division, declared: "God save us from any more machinery in science. . . . I have been afraid of machinery ever since I cut my hand in a feed chopper."[3]

To Hale, it was quite "curious how sensitive" some scientists were about the National Research Council. Surely they could not believe that the NRC had any intention of controlling the nation's research. Was it not a representative body, constitutionally committed to democratic procedures? By a phrase Hale himself had written into the executive order, the Council was constrained "in all co-operative undertakings to give encouragement to individual initiative, as fundamentally important to the advancement of science."[4] But the executive order failed to guarantee the protection of individual initiative against an overly zealous administrator. And if the Council was representative in form, its members still were not elected by the scientific community at large. Robert M. Yerkes, a noted psychologist and an official of the NRC, toured the campuses trying to drum up support for its program. He reported that the more he discussed the matter, the more he found that

[2] Webster, "Research and Organization," *The Weekly Review*, 2 (June 30, 1920), 686–87.

[3] Dean of the Graduate School to E. A. Birge, Dec. 14, 1920, UWS, Graduate School General files, Box 1; John M. Clarke to Albert W. Barrows, Oct. 1, 1919, JCM, Box 221. See also William M. Wheeler, "The Organization of Research," *Science*, 53 (Jan. 21, 1921), 62, 66; Francis B. Sumner, "Some Perils Which Confront Us as Scientists," *Scientific Monthly*, 8 (March 1919), 269–70. President Henry S. Pritchett of the Carnegie Foundation for the Advancement of Teaching advised Hale: "Such attacks [as Webster's] always contain a half truth . . ., else they would have no weight." Pritchett to Hale, July 8, 1920, GEH, Box 33.

[4] Hale to James R. Angell, Aug. 13, 1919, GEH, Box 3; the executive order was printed in *Science*, 47 (May 24, 1918), 511–12. For defenses of the NRC, see Hale, "Cooperation in Research," *Science*, (Feb. 13, 1920), 154; James R. Angell, "Organization in Science," *The Weekly Review*, 2 (March 13, 1920), 251–53.

the prevailing feeling around the country was one of "outsidedness."[5]

The feeling of outsidedness—and resentment—applied no less to the Council's internationally related activities. American physicists were naturally eager to learn from German experts in atomic theory, but when in 1920 it was suggested that the NRC sponsor a visit to the United States by Alfred Landé, who would discuss atomic structure, Hale vetoed the proposal. Like a number of American scientists, the leadership of the Academy and its Council still bitterly recalled the wartime manifesto of the ninety-three professors and understood, in the blunt report of a colleague, that German scientists remained "undemocratic, monarchistic, militaristic and entirely unprogressive." But in 1921 Hale even refused to let the National Academy cosponsor a visit to America by Albert Einstein, who could scarcely be considered either monarchist or militarist. The objection to Einstein was that he persistently called for what the French and Belgians especially still adamantly opposed—the immediate resumption of international relations among scientific nationals from both the Allies and the former Central Powers. By sponsoring Einstein, Hale worried, the National Academy would alienate the French and Belgians and thus jeopardize the already uncertain future of his International Research Council.[6]

The International Research Council, which Hale knew was widely regarded as "entirely useless," suffered much the same structural liabilities as its model, the American NRC, but these inherent difficulties were compounded by its politically exclusive character.[7] No nation could join a union without joining the international council, which meant that no union could invite any of the former Central Powers to membership. None could even allow German scientists to be present at its meetings and congresses. Originally resented in principle, the proscription increasingly annoyed many scientists on practical grounds. What was the point of international meetings of mathematicians without the distinguished professors of Göttingen University? What could be the value of an international symposium of physicists or astrophysicists that had to omit Pauli, Heisenberg, Sommerfeld, even Einstein? In 1925, at the International Research Council's general assembly in Brussels, the respected Dutch elder of physics, Hendrik A. Lorentz, pleaded the special concerns of his profession: In certain fields of contemporary research, collaboration with the former Central Powers was simply indispensable.[8]

[5] Yerkes to Hale, Jan. 6, 1921, GEH, Box 46.

[6] F. L. Saunders to Hale, Dec. 1, 1920, Dec. 18, 1920, GEH, Box 60; William F. Meggers to Arnold Sommerfeld, Feb. 24, 1922, WMF; Hale to Simon Flexner, March 4, 1921, GEH, Box 59; the quotation is from Alonzo E. Taylor to Hale, April 14, 1922, GEH, Box 61.

[7] Hale to Arthur Schuster, May 2, 1924, GEH, Box 47.

[8] Lorentz read an open letter from himself and his fellow physicists J. V. Einthoven, H. Kamerlingh Onnes, and P. Zeeman. Sir Arthur Schuster, ed., *International Research Council, Third Assembly, Held at Brussels, July 7th to July 9th, 1925, Reports of Proceedings* (London, 1926), pp. 6–7.

Not until 1926, long after the spirit of Locarno had settled over Europe, did the International Research Council finally remove the ban against the former Central Powers. The Weimar government urged its scientists to adhere to the Council, unsuccessfully. Many German scientists opposed the Weimar policy of rapprochement, and Lorentz' failure in 1925, Einstein told Millikan, had strengthened "the chauvinistic currents among the scholars in Berlin." The leaders of German science conveniently chose to regard the International Research Council as still more political than scientific and pronounced themselves unwilling to join any organization, especially one created by the former enemy, that violated the traditional apolitical internationalism of science.[9] In physics certainly, to recall Arthur L. Day's warning of 1917, Hale had built his structure of international cooperation on quicksand.

Abroad before 1926, physicists sidestepped the ban against German scientists by meeting independently of the international union in their discipline. At home, before and after 1926, American physicists brought over for visits such foreign physicists as the Germans Arnold Sommerfeld, Max Born, James Franck, and Werner Heisenberg, and the Austrian Erwin Schrödinger. In foreign as in domestic matters, American physicists generally ignored Hale's political calculus, his international organizations, and his National Research Council. All these agencies had proved for the most part as superfluous to the practice of physics in America as had their institutional base, the National Academy of Sciences.

Hale expected the Academy to command more respect after the construction of its building. Completed in 1924, the structure resembled a temple of science, with a vast hall, a classical façade, an entrance framed by two marble and bronze lamps to symbolize, in the Academy's explanation, "the enlightenment that will come from within." That year Charles D. Walcott, still president of the Academy, conveyed his sense of the organization's new dignity and importance. Emphasizing how it might at any time be called upon for advice by the government, he argued against the election of an able mathematician, T. H. Gronwall, who was believed to have violated the prohibition laws. The Academy, Walcott insisted, could not risk blotting its official escutcheon by allowing a disreputable scientist to join its ranks.[1]

The Academy might have drawn more attention to itself had it elected Gronwall than if it had not. President Calvin Coolidge dedicated the building and praised the organization for its disinterestedness, but

[9] Einstein to Millikan, Sept. 1, 1925; Henry Lyons to William Bowie, Oct. 19, 1928, RAM, Boxes 37, 9; Paul Forman, "Scientific Internationalism and the Weimar Physicists: The Ideology and Its Manipulation in Germany after World War I," *Isis*, 64 (June 1973), 173.

[1] *Report of the National Academy of Sciences, 1922*, p. xiii, E. B. Wilson to Oswald Veblen, Jan. 12, 1924; George D. Birkhoff to Veblen, April 25, 1923, OV, Boxes 16, 2.

the government asked for its advice almost as infrequently in the 1920s as before the war. Year after year the average age of the members hovered around fifty-nine. "Keeping the Academy very select," a supporter of the rejected mathematician sniped, ". . . also keeps it senile." Through the decade the halls of the Academy echoed with the sound of unrealized plans; James McKeen Cattell called the massive structure out on what became Constitution Avenue a "marble mausoleum."[2] Most of Hale's cooperative schemes were interred there—and with them, moldering away, his aim of seeing the Academy and its Research Council mount a cooperative attack upon the structure of matter.

Despite the failure of Hale's grand plans, the Research Council did have its uses for physics. The Council organized a Physics Division whose officers created a set of committees to consider the major problems in their discipline. In short order the committees began publishing bulletins that reviewed what was known and what was not in the central fields of physics. Given the conceptually uncertain state of the science, such publications as John Van Vleck's *Quantum Principles and Line Spectra*, completed in 1925, provided a timely, authoritative review of the field and also spotlighted the increasingly critical difficulties in the interpretation of spectra.[3] But more important than the bulletins were the postdoctoral fellowship programs administered by the NRC, which in a crucial way moved the physics profession closer to the fulfillment of Henry Rowland's program.

By now the profession had become highly pyramidic in institutional structure. At its base were the numerous universities and small colleges where young Americans could prepare for a career in science; at its midlevel, the twenty graduate schools where more than three out of four physicists took their doctorates and whose capabilities were considerably strengthened in the 1920s by the concentration of the new funding for physics. Here, in these schools, considerable money was spent to construct new or enlarge old laboratories. Here the sums available for research— in some cases as much as $30,000 annually—increased five- to tenfold over the decade. Here were increasingly centered the more able and productive American physicists—they produced more than 75 percent of the papers published in *The Physical Review*—from the generations of Millikan down through those of Compton to Van Vleck. And here, in the main, the foreign physicists came to visit; by 1932 the leading graduate universities had known the presence of some twenty eminent

[2] Oswald Veblen to E. B. Wilson, May 18, 1929, OV, Box 16; Cattell, "The Organization of Scientific Men," *Scientific Monthly*, 14 (June 1922), 576. For an analysis of the age distribution of the Academy membership, see "Memorandum on Academy Elections," OV, Box 25, NAS Executive Committee 1928/29 file.

[3] *Bulletin of the National Research Council*, 10 (No. 54; March 1926).

physicists in Heisenberg's league of merit, including more than a half dozen Nobel Prize winners, present and future.[4]

At the peak of the pyramid were the postdoctoral fellowship programs of the new John Simon Guggenheim Foundation and especially the NRC. In 1923 the Rockefeller trustees extended the original NRC program and agreed to supply additional funds for awards in mathematics; in 1924 the Council was given the power to screen American applicants for Wickliffe Rose's international grants. Young Ph.D.'s naturally coveted the fellowships. Both types of awards were prestigious, meant a year or two free of teaching duties and the opportunity to study at the leading centers for physics in the United States or western Europe. At the same time, the fellowship program was permissive rather than directive; it made awards rather than demands. As such, it faced none of the liabilities inherent in the more ambitious schemes of the National Academy and the Research Council. The distribution of both the Guggenheim and NRC fellowships was determined by boards of outstanding scientists. Drawn from around the country, the NRC board included precisely such physicists as Robert Millikan who were eager to develop the general, and especially the theoretical, strength of their profession and who, through their membership on the board, acquired unprecedented leverage to do just that.[5]

If this pyramidic system had been building since Henry Rowland's day, capped by the postdoctoral programs, it was now genuinely national

[4] Spencer R. Weart, "The Physics Business in America, 1919–1940: A Statistical Reconnaissance" (unpublished paper, 1976), p. 3; Stanley Coben, "The Scientific Establishment and the Transmission of Quantum Mechanics to the United States, 1919–32," *American Historical Review*, 76 (April 1971), 444. At Columbia, the appropriation for physics research typically went from $1,000 in 1920/21 to $7,200 in 1929/30; at Berkeley, to $12,000 in 1928/29. The Princeton department had special supplementary grants of $30,000 annually for three years at the end of the Twenties. I am indebted for the Columbia information to the late Lucy J. Hayner; for Princeton, see Karl T. Compton to A. W. Goodspeed, March 22, 1928, KTC; for Berkeley, *University of California, Report of the President, 1928/30*, pp. 447, 455. Physicists of a later, more opulent era recalled the 1920s as an impoverished period, but they ignored the substantial increases in funding over the prewar years and the extent to which experimentalists of the day relied upon the departmental machine shop, especially the glassblower, rather than upon costly purchases from equipment manufacturers. The high demand for foreign lecturers is evident in Karl T. Compton to K. Herzfeld, March 6, 1928, KTC; Hans Kramers to John Van Vleck, April 28, 1928, SHQP, Van Vleck file.

[5] The original NRC fellowship board also included Hale, Noyes, Henry Bumstead of Yale, and C. E. Mendenhall of Wisconsin. Oswald Veblen's correspondence with the secretary of the board, W. E. Tisdale, reveals the board's emphasis on the development of mathematical physics. See OV, Box 14, Tisdale file. Coben, "The Scientific Establishment . . .," pp. 445–46. Some physicists felt that the Fellowship Board was "a closed corporation" and that "however able and unprejudiced its members may be there is in the nature of the case a considerable permanent influence affecting certain institutions and certain types of research." Karl T. Compton to Simon Flexner, April 8, 1929, KTC. On the Guggenheim fellowships, see Stanley Coben, "Foundation Officials and Fellowships: Innovation in the Patronage of Science," *Minerva*, XIV (Summer 1976), 229, 235–6.

in scope. Pluralist, it also harmonized with the American attachment to regionalism and local independence. Selective, it distilled the pool of talent at each level so as to yield an aristocracy of merit. Structurally democratic in its opportunities, it was nevertheless designed to give its richest opportunities to the best men. It was, in short, a system open to the talents of all yet simultaneously favoring that elite of young Americans most willing and able to confront the leading challenges of atomic studies.

XIV

A New Center
of Physics

In the dozen years after 1920 twice as many young Americans, some
thousand in all, took Ph.D.'s in physics as in the half century
between Appomattox and Sarajevo. About 130 of them won NRC
fellowships and the majority of that group spent their postdoctoral
year or two at a university in the United States. At home there were
first-rate mentors in the chief experimental fields of modern physics—in
X rays, atomic and molecular spectroscopy, in electron scattering and
photoelectric phenomena. For help with quantum mechanics, there was
the steadily increasing number of theoretically *au courant* native physicists and the visiting foreigners (from whom Americans often learned, as
Sommerfeld said at Berkeley: If you want to be a physicist, you must do
three things—first, study mathematics, second, study more mathematics,
and third, do the same).[1] Equally important, there were one's peers, one's
companion postdoctoral fellows. What the NRC fellows could not learn
from their professors they learned from each other. All eagerly kept up
with the latest triumphs, thumbing through the foreign journals so often
as to turn the library's copies dog-eared.

Yet between 1919 and 1932 at least fifty, a significant fraction, of the
most promising young American physicists made their way to Europe

[1] In 1920 American universities granted thirty-one physics doctorates; in 1932,
115. L. R. Harmon, "Physics Ph.D.'s: Whence, Whither, When?" *Physics Today*,
15 (Oct. 1962), 21. Sommerfeld's advice was cited in an interview with Paul H.
Kirkpatrick.

for postdoctoral study on fellowships from the NRC, the International Education Board, or the new Guggenheim Foundation.[2] They were drawn by Göttingen's formidable mathematical physicists; by cosmopolitan Berlin, where one could see seven Nobel Prize winners at a single colloquium; by Munich, where the seminars continued in the local cafes with the equations jotted down on the marble tabletops; above all, by Copenhagen, a visit to which was virtually required, where Bohr held forth in his red-roofed institute, knitting the world of quantum physics together in an untiring and provocative search for meaning. "When Bohr is about," an American postdoctoral fellow marveled, "everything is somehow different. Even the dullest gets a fit of brilliancy."[3]

"Theoretical physics has reached a terrible state," Earle Kennard reported from Göttingen in 1926; "new methods have to be learned every week, almost."[4] Entering the discipline in an era of enormous intellectual upheaval and located at the center of the revolution, many found the going exceedingly difficult. They had to struggle to overcome the inadequacy of their preparation in mathematics and occasional doubts about their ability to cope with both the techniques and the profundities of quantum mechanics. They were measuring themselves against a formidable array of talent, even genius; against not only the highly select group of their fellow young physicists but men who would have been masters in any age. And some of the European authorities were scarcely paragons of tolerance. If Pauli was abrasive, at Göttingen Max Born presided with such brilliant drive and hauteur that more than one American fled to Munich and the comforting arms of genial old Professor Sommerfeld. Even Bohr, often overworked, could be testy. He occasionally refused to accept students, and his inner circle of assistants sometimes rebuffed newcomers with a disconcerting air of exclusiveness.

But while Born was cool and distant, most other luminaries, including Bohr, were more frequently than not hospitable; day-to-day activities were usually informal, and ping-pong games often pitted neophytes against Nobel Prize winners. If the pace of advance was swift and demanding, the Americans soon learned that their brilliant European peers— Enrico Fermi, Edward Teller, John Von Neumann, Hans Bethe, and dozens more—were often just as perplexed as themselves. It was no trouble to meet and work with these able contemporaries from across the Atlantic, since scores were migrating around the Continent on Interna-

---

[2] The NRC and IEB statistics are compiled from "Preliminary Report of the Fellowship Board in Physics, Chemistry, and Mathematics, National Research Council . . .," March 9, 1931, RAM, Box 6; "National Research Council Fellows, 1919–1938, Physical Sciences," AIP; *International Education Board Fellows Voted Prior to June 30, 1928*, RF. The number of Guggenheim fellows is drawn from the lists of physicists starred in successive editions of *American Men of Science*.

[3] I. I. Rabi to George B. Pegram, Jan. 1928, GBP, R-Misc. file.

[4] Kennard to R. C. Gibbs, March 3, 1927, CUP, Kennard file.

tional Education Board Fellowships. Everyone was bound together by high anticipation of what Heisenberg or Schrödinger might publish next, and in Berlin Leo Szilard, a Hungarian of boundless gregariousness, made a practice of introducing all the Americans to everyone else. The friendships formed in these years would significantly affect the course of physics (and of the world's history, when in World War II many of these same men found themselves together again at Los Alamos).

At the time, honing themselves to match the standards set by their foreign colleagues, the American neophytes added a rare seasoning to their training, a fine literacy in quantum mechanics to their repertoire, and, to their inner resources, a self-confidence born of the knowledge that they were surviving a remarkably difficult regimen. By 1932 most had returned home to become active members of a physics profession which—almost three times as large as in 1919—included some twenty-five hundred members, and whose new lights were better trained than their predecessors of any previous generation.

Though some of the new physicists of the day came from upper-income families, even more of them were the products of middle- to lower-middle-income homes, of the smaller colleges and state universities, of the Midwest and now, too, the region along the Pacific Coast. Most of the new physicists were also Anglo-Saxon Protestant males, but not all of them, not in this decade when a woman was permitted to vote, Al Smith nominated for the presidency, and a Jew lionized as the Newton of the century.

"Now that the doors of opportunity have been thrown widely open to women," Simon Flexner told the 1920 graduating class at Bryn Mawr, "one may expect that many more will . . . enter upon the career of science." In the late nineteenth century, parents had discouraged their daughters from going to college, in part because college women were considered unconventional, in part because, as Charles W. Eliot expressed the common assumption, women were supposedly too frail to bear "the mental stress of hard study." Females also found their opportunities for professional training limited. In the empirical assessment of the respected —and not atypical—psychologist Edward L. Thorndike, women were "restricted to the mediocre grades of ability." They were suited for such occupations as teaching, medicine, or architecture, which, Thorndike claimed, required only average intelligence; it was simply a waste of educational resources to prepare them for "administration, statesmanship, philosophy or scientific research." The handful of women who earned Ph.D.'s in science found almost no jobs in industry and not many more in federal scientific agencies. Most women scientists were employed in the academic world, where, like most of their female colleagues, they were consigned to the women's colleges, which meant to inadequate laboratory facilities, limited research funds, virtually no graduate students, and a professionally debilitating isolation.[5]

But now, in the wake of the suffrage movement and the accelerated entrance of women into the work force during World War I, many aspiring young women wanted to go to college and one out of every two undergraduates was female. Women were taking Ph.D.'s in the best graduate schools of science. Industrial corporations, hungry for technical personnel, seemed eager to employ women in general, perhaps even women scientists. The U.S. Civil Service Commission announced in 1921 that it would henceforth administer qualifying examinations for scientific posts without regard to sex. Even if agency heads could still specify "men only" to the Commission, Samuel Wesley Stratton had relaxed his long-time rule against hiring female scientists at the Bureau of Standards.[6]

A committee of the American Association of University Professors surveyed their colleagues and concluded that male academics now generally believed women could teach and bear the stress of professional duties just as well as men. The committee, which included a slim majority of women, also reported that in coeducational universities female academics tended to consider their opportunities equal to men's for scholarship and research. Women's college faculties now employed only about 37 percent of women scientists, as compared to 58 percent early in the century, and almost 20 percent of women scientists were employed on the faculties of state universities, whose administrators were wary of their respective states' new female voters. The trend, in short, was away from the exclusion of females from the mainstream of academic science.[7]

[5] Simon Flexner, "The Scientific Career for Women," *Scientific Monthly*, 13 (Aug. 1921), 103; Eliot is quoted in Helen Wright, *Sweeper in the Sky: The Life of Maria Mitchell, First Woman Astronomer in America* (New York, 1950), 193; Robert W. Smuts, *Women and Work in America* (New York, 1959), pp. 50–51; Edward L. Thorndike, "Sex in Education," *Bookman*, 23 (April 1906), 213; Margaret Rossiter, "Women Scientists in America Before 1920," *American Scientist*, 62 (May–June 1974), 318. Thorndike opposed the automatic consignment of women to domestic roles. According to test data, women displayed a narrower distribution of intelligence than men around the mean. Hence, they were to be discouraged from entering fields that required very high intellectual ability and encouraged into those suited to average abilities, which Thorndike assumed to include teaching, architecture, and medicine.

[6] Patricia Albjerg Graham, "Women in Academe," *Science*, 169 (Sept. 25, 1970), 1284; Martin A. Morrison, "The Employment of Women in Technical and Scientific Positions," *School and Society*, 13 (Feb. 5, 1921), 174–75; Rexmond G. Cochrane, *Measures for Progress: A History of the National Bureau of Standards* (Washington, D.C.: National Bureau of Standards, U.S. Department of Commerce, 1966), pp. 169–70.

[7] "Second Report of Committee W on the Status of Women in College and University Faculties," *Bulletin of the American Association of University Professors*, 10 (Jan. 1924), 68, 70–71; Rossiter, "Women Scientists in America before 1920," p. 318. An earlier report of Committee W included among the reasons for the increased hiring of women on state university faculties: "The success of the woman suffrage movement, which seems to have put the 'fear of God' into the hearts of not a few

In 1925 the academically dominated National Academy of Sciences elected to membership Florence Rena Sabin, a physiologist at the Rockefeller Institute for Medical Research, and the first woman to be so honored in the sixty-two years of the Academy's history.

America's female scientists had a good deal to celebrate when in 1921 Marie Curie paid her first visit to the United States. Here was a wife and mother who had won the Nobel Prize twice, a physical chemist who had helped give mankind radium to cure cancer, an unselfish benefactress who had never patented her discoveries. During a visit to Paris in 1920, Marie Meloney learned that Madame Curie had only a minute supply of precious radium available for her own research. Meloney, the new editor of the *Delineator*, a widely read magazine of women's fashions and affairs, had a propensity for causes—her New York cable address was IDEALISM—and a sharp business sense. She promptly organized a campaign to buy Dr. Curie a gram of radium—it would cost $100,000— and invited her to the United States to accept this gift from the nation's women. By the time Curie's ship docked in mid-May 1921, the fund had been oversubscribed and was still growing.[8]

Curie and her two daughters toured the eastern United States for six weeks. Coeds danced for her on the campus lawns, admirers showered her with floral bouquets, and 3,500 members of the American Association of University Women turned out to fete her at Carnegie Hall. Smith and Wellesley awarded her honorary degrees; so did Northwestern, Pittsburgh, Columbia, Pennsylvania, and Yale. President Harding gave her the radium at the White House itself, before some six hundred senators and congressmen, diplomats and scientists, wives and leading women. In the middle of it all, Robert B. Moore, the chief chemist of the Bureau of Mines, understandably predicted that the American woman's "appreciation of science . . . will be quadrupled by Madame Curie's visit."[9]

Perhaps it was. In 1920, 41 women were awarded Ph.D.'s in science; in 1932, 138. The number of scientific doctorates granted to women over the period totaled almost a thousand. But since men were of course taking scientific doctorates in increasing numbers, too, at the end of the 1920s women represented only a slightly greater fraction, about 5 percent, of the nation's scientific and engineering community than at the beginning. The record was worse in physics, where women earned only 3.4 percent of the doctorates and accounted for less than 3 percent of the entire profession.[1] Curie's achievements may have attested, in the praise of Warren

---

ever-watchful university executives." "Preliminary Report of Committee W . . .," *Bulletin of the American Association of University Professors*, 7 (Oct. 1921), 25.

[8] Robert Reid, *Marie Curie* (New York, 1974), p. 249; *New York Times*, May 11, 1921, p. 8.

[9] Quoted in *The New York Times*, May 18, 1921, p. 17.

[1] Jessie Bernard, *Academic Women* (University Park, Pa.: Pennsylvania State University Press, 1964), p. 71; Phyllis Blanchard and Carolyn Blanchard, *New Girls for Old* (New York, 1930), p. 207. The statistics on physics are taken from M. Lois

Harding, to her sex's "equality in every intellectual and spiritual activity," but she was white-haired, shy, modest, a stooped porcelain figure, her eyesight failing, who everywhere she went had worn black gloves, a black dress, and a black taffeta hat. Her daughter Eve, who was not a scientist, dressed with silk stockings and a gaily colored bonnet; her daughter Irène, a physicist, attired herself in brusque black and cotton. Asked to cite the women of history they admired most, only 3 of 347 girls mentioned the famed radiochemist.[2] Coeds could celebrate Madame Curie; few wanted to emulate her in this day of Clara Bow and the flapper and swish Fitzgeraldian heroines freed from their mothers' Victorian chains.

Zelda Fitzgerald herself revealed the *au courant* attitudes of the day: "I think a woman gets more happiness out of being gay, light-hearted, unconventional mistress of her own fate, than out of a career that calls for hard work, intellectual pessimism, and loneliness. I don't want [my daughter] Pat to be a genius. I want her to be a flapper, because flappers are brave and gay and beautiful." The liberation of women from old notions of sexual and bodily shame was putting a new premium on the traditional commitment to marriage and motherhood. If the old-time feminists had called for a revolt against sexual slavery, the modern woman made a fashion of sexual fulfillment, and even romanticized the bearing of children.[3] Women had always managed to combine careers in art and literature with the satisfactions of home and hearth, but eight out of nine women scientists, including physicists, were unmarried. Scientific careers called up the atmosphere of militantly feminist faculties—some women's colleges would still not hire married women—when to the new generation of the 1920s, the journalist Dorothy Dunbar Bromley remarked, *feminism* was a "term of opprobrium."[4]

If, as Bromley added, the "truly modern" woman hoped for both marriage and a career, at the opening of the decade thousands of Americans were reading A. S. M. Hutchinson's novel *This Freedom*, about a woman whose home had disintegrated because of her outside professional activities. Was Hutchinson's portrait accurate? *Literary*

Marckworth, comp., *Dissertations in Physics: An Indexed Bibliography of All Doctoral Theses Accepted by American Universities, 1861–1959* (Stanford, 1961), and J. McKeen Cattell and Jaques Cattell, eds., *American Men of Science* (5th ed.; New York, 1933), p. 1264.

[2] Harding is quoted in *Science*, 53 (June 3, 1921), 509; Reid, *Marie Curie*, p. 263; William L. O'Neill, *Everyone Was Brave: The Rise and Fall of Feminism in America* (Chicago, 1969), p. 322.

[3] Fitzgerald is quoted in Nancy Milford, *Zelda* (New York, 1970), pp. 125–26; O'Neill, *Everyone Was Brave*, p. 315.

[4] In the late 1920s 615 of 687 women scientists were unmarried, and so, in 1920, were all of the 23 female physicists. Luella Cole Pressey, "The Women Whose Names Appear in 'American Men of Science' for 1927," *School and Society*, 29 (Jan. 19, 1929), 96–100; Rossiter, "Women Scientists in America before 1920," p. 317; Dorothy Dunbar Bromley, "Feminist—New Style," *Harper's*, 155 (Oct. 1927), 552.

Digest asked. No, four out of five professionally accomplished women replied, but they warned that avoiding familial collapse required an emotional balance and talent for efficiency worthy of Lillian Gilbreth. To master physics, or Einstein's theory of relativity for that matter, Walter Pitkin advised young women, was "mere child's play beside raising one child."[5] Most of the increasing fraction of married women who joined the work force, almost 12 percent by 1930, likely did so for the economic necessity of a job. Surveys concluded that most modern young women did not want careers, especially if the careers would interfere with marriage.[6]

No less important, the nation, including the female nation, traditionally expected women to be soft, maternal creatures, emotional and perhaps illogical. In contrast, applied science called up the harsh world of factories and machines, pure physical science, a commitment to ruthless rationality and cerebral dealings with cold, inanimate nature. Women who turned to the physical sciences had always to expect the ambivalent salute that Voltaire rendered his mistress, the brilliant Madame du Châtelet: "A woman who has translated and illuminated Newton . . . [is] in short a very great man." The journalist Anne Shannon Monroe was only reiterating that historic stereotype when she told the readers of *Good Housekeeping* to keep away from motherhood the "walled-in, self-contained" woman and send her to work in "the laboratories, where matter has no feeling, and feeling is not needed to penetrate the mysteries of matter."[7]

Women with a bent for technical subjects might consider themselves engaged in womanly pursuits by turning to schoolteaching, nursing, or the practice of medicine, or perhaps by absorbing themselves in the study of animate, organic nature. Now as in the prewar era, well over half the women scientists in America were in psychology, botany, and zoology, and more than three times as many women took doctorates in the biological and social as in the physical sciences. Even chemistry, by far the most glamorous physical science of the 1920s, proportionately drew far more male scientists than female.[8] If so few women became physicists, it

[5] Bromley, "Feminist—New Style," p. 552; "Can a Woman Run a Home and a Job, Too?" *Literary Digest*, 75 (Nov. 11, 1922), 40 ff.; Walter B. Pitkin, "Can Intellectual Women Live Happily?" *North American Review*, 227 (June 1929), 704.

[6] See the suggestive statistics in *Recent Social Trends in the United States: Report of the President's Research Committee on Social Trends* (2 vols.; New York, 1933), I, 714–15, 723; O'Neill, *Everyone Was Brave*, p. 322.

[7] Voltaire to François Thomas Marie de Baculard d'Arnaud, Oct. 14, 1749, in Theodore Besterman et al., eds., *The Complete Works of Voltaire* (vol. 95: Geneva, 1970), pp. 178–79; Anne Shannon Monroe, "The Woman Who Should Marry," *Good Housekeeping*, 73 (Aug. 1921), 129.

[8] Louella Cole Pressey, "The Women . . . in 'American Men of Science' for 1927," pp. 96–100; Bernard, *Academic Women*, p. 71. Chemists represented 26.3 percent of all, but only 16.3 percent of women, scientists. Cattell and Cattell, eds., *American Men of Science* (5th ed.; 1933), p. 1264.

was not least because to do so required defiance of the prevailing standards of womanhood.

Jane M. Dewey, the youngest daughter of the most celebrated philosopher of the day, nevertheless took her doctorate in physical chemistry from MIT. In 1927, after two years with Niels Bohr in Copenhagen, she won a National Research Council fellowship to Princeton, where the physics faculty, most of them sure that no woman could do physics, grudgingly dubbed her "Magie's folly," after their colleague Dean William F. Magie, who had insisted she be admitted. In 1929 Dr. Dewey sought a job. Her adviser, Karl Compton, wrote all over the country to place her in a professional position. The only sympathetic response came from a physicist at Berkeley, who regretted to report that his colleagues simply refused to have a woman on the staff. Jane Dewey had to spend yet another two years as a research fellow, then join the staff at Bryn Mawr.

A number of American scientists dissented from the celebration of Madame Curie; almost the entire Harvard physics department helped block the award of an honorary degree in Cambridge. Whatever the new appreciation of women in the college and university system, after their extensive survey the committee of the American Association of University Professors soberly concluded that, compared to males of similar quality, female academics were generally paid less, promoted more slowly, and treated with a "considerable degree of discrimination."[9] Like Jane Dewey, most of the thirty-four women who took doctorates in physics between 1920 and 1932 wound up on women's college faculties. A few found posts in the leading coeducational institutions, but only in subordinate research positions, not as members of the regular professional staff.

The American professoriat had never practiced such overt discrimination against Catholics, if only because not many had ever tried to make their way onto the faculties of the major universities. Though the largest religious minority in the United States, by now some 20 percent of the total population, so few Catholics had entered the ranks of research over the previous half century that they accounted for only about 1 percent of the nation's current scientific leadership. Protestant professors often shrugged away the fact by citing the Church's record of intellectual dogmatism. The American priesthood was in fact generally anti-scientific in its attitudes. But if the Church had not yet officially exonerated Galileo, it had removed his books from the *Index*, and the Vatican had repeatedly made clear that, as Karl Herzfeld, a Catholic physicist at Johns Hopkins attested, "no conflict [existed] between the

[9] The Magie quotation is from an interview with Mrs. Karl T. Compton; Leonard Loeb to Karl T. Compton, Feb. 1, 1929, KTC; Reid, *Marie Curie*, p. 255; "Preliminary Report of Committee W . . .," p. 26.

results of science and the *doctrines* of the Church."[1] What had kept American Catholics out of science in the past was not nearly so much the doctrines of the universal Church as the nature of the Church in America.

Herzfeld, who had been born and educated in Austria, knew that part of the difficulty lay in the traditionally low scientific quality of American Catholic schools. Most of the American clergy were drawn from the Irish, the first massive wave of Catholic immigrants to arrive and face the hostility of the Protestant majority. Burdened with the immediate needs of the flock, the hierarchy had neither the money nor the will to create a first-rate educational system, certainly not one with good colleges, any graduate schools, or many well-trained lay faculty. More important, determined to protect the faithful, the clerics made the Church—and its schools—a fortress for the militant defense of the faith.[2]

What latter-day Catholics of Herzfeld's stripe have called the "siege mentality" of the American religious led the hierarchy to call for a parochial school in every parish, to warn parents against public education, to encourage academically talented youngsters away from secular professions into the priesthood. Operating under that mentality, the Catholic colleges presented secular subjects less for their own sake than as a means of confirming the faith or as foils for exposing doctrinal error. In 1887 Georgetown, then the best Catholic university in America, reported of its curriculum: "Whatever is important in natural science is taken into her courses and taught with a philosophical analysis intended to guard the student against confounding mere information with learning, which is the danger of modern education."[3]

Whatever the drawbacks of the colleges, Catholic commentators often argued that the children of most first-generation immigrants could scarcely afford higher education. They failed to recognize what the Notre Dame biochemist Julian Pleasants later observed: "Ours is not an abject

[1] Karl Herzfeld, "Scientific Research and Religion," *Commonweal*, 9 (March 20, 1929), 560–62. Pope Leo XIII declared in 1893: "Whatever [scientists] can really demonstrate to be true of physical nature, let us show to be capable of reconciliation with our scriptures." *New Catholic Encyclopedia* (15 vols.; New York, 1967), V, 684.

[2] James W. Trent, with Jeanette Golds, *Catholics in College: Religious Commitment and the Intellectual Life* (Chicago, 1967), pp. 14, 15, 19, 20. See also Edward J. Power, *A History of Catholic Higher Education in the United States* (Milwaukee: Bruce, 1958), pp. 53, 54, 57, 85, 204, 237.

[3] Quoted in Power, *A History of Catholic Higher Education*, p. 81; Philip Gleason, "Immigration and American Catholic Intellectual Life," *Review of Politics*, 26 (April 1964), 149–54; John Tracy Ellis, "American Catholics and the Intellectual Life," *Thought: Fordham University Quarterly*, 30 (Fall 1955), pp. 354–56, 368, 373. Higher education seemed to reduce the sense of conflict between science and religion for white Protestants but to increase it for Catholics. Gerhard Lenski, *The Religious Factor: A Sociological Study of Religion's Impact on Politics, Economics, and Family Life* (Garden City, N. Y., 1961), pp. 254–55. In the 1920s George N. Shuster found that his fellow American Catholics "even today live in something like an armed camp." Shuster, *The Catholic Spirit in America* (New York, 1927), p. 118.

but a discriminating poverty; it lays bare our scale of values by indicating what we feel we can do without." The modern American Catholic, Pleasants noted, "places a very low value on . . . scientific research."[4] Clannish in the face of the majority's hostility, young Catholics who could afford college preferred to prepare for work among their own group, frequently in the law. Besides, whether they had come from Ireland or the nations of central and southern Europe, American Catholics generally derived from peasant cultures. In the old country, learning as such had been left to the priests; in the new, as an observer who understood his fellow Irish pointed out, "intellectual curiosity . . . was taboo because it was lazy and non-utilitarian. But," he added, summing up the prime occupational aspirations of these new Americans, "a 'good head for business'—ah, that was a gift from heaven, indeed!"[5]

Shortly after World War I the Catholic historian Carlton J. H. Hayes urged the American faithful to contribute more to the nation's intellectual life, remarking: "We are no longer immigrants. We are Americans."[6] If American Catholics were not proportionately as well off as the Protestant majority, those descended from the earlier waves of immigrants had moved to a more comfortable position on the economic ladder. But while Catholic enrollments in the colleges were rising in both Church and non-Church schools, the siege mentality continued to dominate the Catholic academic atmosphere.

The growing number of young Catholics in secular undergraduate and graduate institutions alarmed the hierarchy, who tended to agree with Roy J. Deferrari, dean of the Graduate School at Catholic University, that "Catholic surroundings are almost necessary for the development of a sound research specialist in any field." Eager to reverse the secular trend, the Catholic colleges began to employ a higher percentage of lay faculty, offer a more diverse curriculum, and develop graduate programs in a number of secular fields. All the same, the Catholic colleges harbored no first-rate graduate departments, and most awarded only master's degrees to prospective parochial school teachers, not Ph.D.'s. More important, armed with the humanist assault against science, educationally conservative clerics continually fought the shift to secular concerns and succeeded at least in maintaining an emphasis on the inculcation of faith and morals.[7] In the day of Karl Herzfeld's critique, the Catholic system of higher

---

[4] James J. Walsh, "Catholics and Scientific Research," *Commonweal*, 15 (March 23, 1932), 574; Julian Pleasants, "Catholics and Science," *Commonweal*, 58 (Aug. 28, 1953), 511, 512.

[5] Harry McGuire, Letter to the editor, *Commonweal*, 9 (May 1, 1929), 748; Gleason, "Immigration and American Catholic Intellectual Life," pp. 154–66.

[6] Hayes, "A Call for Intellectual Leaders," in Frank Christ and Gerard Sherry, eds., *American Catholicism and the Intellectual Ideal* (New York, 1961), p. 73.

[7] Roy J. Deferrari, "Catholics and Graduate Study," *Commonweal*, 14 (June 24, 1931), 204; Philip Gleason, "American Catholic Higher Education: A Historical Perspective," in Robert Hassenger, ed., *The Shape of Catholic Higher Education* (Chicago, 1967), pp. 26–28, 42, 46–48.

education still discouraged intellectually talented students away from science and into apologetics.

In the era of normalcy, the resurgent conservatism in the Catholic colleges was no disappointment to their students. The core of the newer curriculum—programs to train for business and the professions—remained intact. Like so many other Americans, more Catholics were going to college not only because they could afford it but because it had become an accepted and advantageous way to get into the most admired and financially rewarding of American occupations. Whether enrolled in parochial or secular institutions, now as in the past, few Catholic students had any taste for careers in pure science and few actually joined their Protestant peers in the profession of atomic physics.[8]

Despite the distinction of Albert Michelson in physics and Simon Flexner in medical research, at the opening of the 1920s Jews were almost as underrepresented in American science as Catholics. This was not surprising, since the majority had come to the United States only recently. But now, like Catholics, many were no longer immigrants, and unlike Catholics, they knew much the same mixture of drives and circumstances in America which had led their brethren in Europe to win so disproportionately large a place in the world of learning.

Why the "intellectual preeminence" of the Jews in modern Europe? Thorstein Veblen asked in 1919. Though some observers of the day considered them an intellectually superior "race," Veblen knew that they were a hybrid people, and, anticipating the conclusions of modern psychologists, he rightly dismissed the assumption that they had "a peculiarly fortunate intellectual endowment, native and hereditary." Intellectually gifted Jews did so well in the "gentile republic of learning," Veblen argued, because they were half-way men, suspended in limbo between their own culture and the majority's. As such, they enjoyed "a degree of exemption from hard and fast preconceptions, a skeptical animus . . . [a] release from the dead hand of conventional finality." Veblen might have added that the Talmud emphasized respect for learning and that in the ghetto rabbis and teachers were honored at least as much, and usually more, than financially successful men. Equally important, though Jews might be robbed of their land, legislated and pogromed out of their trades, or subjected to a crazy-quilt of occupational restrictions, no one could take away their expertise. The refugee physicist Abraham Pais once commented on Veblen's analysis: "The Jew has the tradition of the book first because so it was in the ghetto, but secondly . . . because the contents of the book are inalienable—even if the book itself is not."[9]

[8] Harvey C. Lehman and Paul A. Witty, "Scientific Eminence and Church Membership," *Scientific Monthly*, 33 (Dec. 1931), 546; R. H. Knapp and H. B. Goodrich, *Origins of American Scientists* (Chicago, 1952), pp. 51, 288.

[9] Thorstein Veblen, "The Intellectual Pre-eminence of the Jews in Modern Europe," *Political Science Quarterly*, 34 (March 1919), 35, 38, 39; Pais to Oswald

Most recent Jewish immigrants had come from the ghettoes of eastern Europe to the United States impoverished and uneducated. They settled in the northern cities and eked out their livings as peddlers, shopkeepers, or garment workers. With their strange dress, slum poverty, and heavy accents, the newcomers were scorned by the western European Jews who had preceded them and whose social status they now threatened. Faced in addition with increasingly anti-Semitic practices in the gentile business world, they attached high value to getting ahead in the professions. To become a "professional man," the historian Jerold S. Auerbach has explained this tendency, has always been attractive to individuals from ethnic and religious groups "that suffer social scorn or prejudice. To become a professional is, in familiar American terms, to 'make it,' " not least because the associated prestige "often serves to obliterate the stain that is felt to accompany minority group membership." However poor, many Jewish parents scrimped and saved to send their sons to college, eager to see them enter medicine or law or public school teaching.[1]

Doubtless few encouraged their sons into science. The industrial laboratory was darkened by the anti-Semitism of the large corporations. The academic world, especially its distinguished private sector, was identified with President A. Lawrence Lowell's recommendation of 1922 that Harvard impose a quota on the admission of Jewish undergraduates. "Harvard is not the first American university to attempt to limit the proportion of Jews in its midst," *The Nation* remarked. "It is merely the frankest."[2] Anti-Semitism also operated at every level of the scientific professoriat. There was Hale's warning against the nomination of Simon Flexner for the vice-presidency of the National Academy: Michelson already occupied the presidency, and the membership might well object to having Jews in both offices; Edwin Grant Conklin's comment upon the qualifications of a biologist: "I do not know whether he has any Jewish blood or not but he has some of the characteristics of that gifted race, among which must be listed a large amount of push." Robert Millikan hired Paul Epstein for Caltech "even though [he is] a Jew," then brushed aside another theorist because he doubted that the Institute could have

Veblen, Aug. 19, [195?], OV, Box 10; for the emphasis on learning in the European ghetto, see Mark Zborowski and Elizabeth Herzog, *Life Is with People: The Culture of the Shtetl* (New York, 1962), pp. 73–87, 118–23.

[1] John Higham, "Social Discrimination Against Jews in America, 1830–1930," in Abraham J. Karp, ed., *The Jewish Experience in America* (5 vols.; American Jewish Historical Society; New York: Ktav, 1969), V, 362–63; Nathan Goldberg, "Occupational Patterns of American Jews," *Jewish Review*, 3 (Oct.–Dec. 1945; Jan.–Mar. 1946), 11–12, 16, 22, 162–63, 166; Jerold S. Auerbach, "Legal Profession Project," unpublished manuscript, The American Jewish Committee, William E. Wiener Oral History Library, May 25, 1970, p. 1.

[2] "May Jews Go to College?" *The Nation*, 114 (June 14, 1922), 708. See also Carey McWilliams, *A Mask for Privilege: Anti-Semitism in America* (Boston, 1948), pp. 38–39; Higham, "Social Discrimination Against Jews in America," pp. 369–70.

"more than about one Jew anyway." In 1930 an official of the Rockefeller Foundation queried the advisability of granting postdoctoral fellowships to Jews in view of their "recognized difficulty in . . . getting positions later."[3]

Yet, though Epstein was hired as merely a lecturer, he was so good a theorist that Millikan eventually promoted him to a full professorship. Few administrators of physics were actually willing to let a man's national or religious background stand in the way of improving their staffs. In the decade of the quantum mechanical revolution, pure physics offered Jews opportunity for both professional place and social standing in the gentile world. Both possibilities found symbolic expression in the reception given Albert Einstein when in 1921, together with a small party of Zionists headed by Chaim Weizmann, he visited the United States to help raise money for the Hebrew University to be built atop Mount Scopus in Jerusalem.

Einstein's humility captured the audience at the annual meeting of the National Academy of Sciences, his charm, the faculty in Princeton. Told that an antirelativist had just claimed the detection of an ether drift, Einstein gently quipped: "God is clever, but he is not malicious."[4] When Princeton awarded Einstein an honorary degree, virtually the entire higher educational establishment of the East Coast assembled to do him honor. In New York, overflow crowds turned out to hear him lecture, in German, on the theory of relativity, and politicos voted the modern Newton the keys to the city and the state. Governors sent official committees to welcome his party, hosted them at breakfasts and luncheons. When Einstein and Weizmann addressed a rally of thousands at the 69th Regiment Armory, the politico in the White House wired greetings— "their visit must remind people of the great services that the Jewish race have rendered to humanity"—and down in Washington a few days later, Mr. and Mrs. Harding had Professor and Frau Einstein over to the White House.[5] Amid the picture-taking the President amiably acknowledged that he did not understand the theory of relativity, but like public officials elsewhere, he clearly understood what Einstein meant to the nation's Jewish voters.

Thousands of Jews had greeted Einstein and his party at the Battery; thousands more had lined the streets cheering and waving handkerchiefs

[3] Hale to John C. Merriam, Feb. 24, 1921, GEH, Box 28; Conklin to W. C. Carter, June 13, 1923, EGC, Carter file; Millikan to Hale, July 16, 1921, RAM, Box 29; Robert A. Lambert, "U.S. Fellowships and the NRC," Nov. 10, 1930, p. 2, RF, file 200 E, Box 169. See also George D. Birkhoff to Oswald Veblen, March 26, 192[?]; E. B. Wilson to Veblen, June 13, 1924, OV, Boxes 2, 16; A. G. Worthing to Karl T. Compton, June 10, 1929, KTC; Ludwig Lewisohn, Up Stream: An American Chronicle (New York, 1922), pp. 146–47.

[4] The incident is related in Oswald Veblen to Einstein, April 17, 1930, OV, Box 4.

[5] Quoted in The New York Times, April 13, 1921, p. 5.

as they drove up the Lower East Side. The Jews of Boston met their train with a brass band in the morning, then feted them in the evening with a kosher banquet. The day they arrived in Cleveland, Jewish merchants closed up shop at noon, and what an astonished reporter called "a swirl of fighting, crowding humanity" kept their two-hundred-car motorcade to a slow pace on the way to city hall.[6]

Everywhere Einstein perhaps more than Weizmann was the object of the adulation. For American Jews, Einstein was more than the Newton of the century. One of their own and a lion among gentiles, he was what in the 1920s and through science some of them might hope to be.

In Brooklyn, a few years before Einstein's visit, young Isidor I. Rabi was browsing one day in the Carnegie Library near his family's small grocery store. The Rabis, who had emigrated from Austria at the turn of the century, were orthodox Jews. They saw that their son learned to read Hebrew, and when Rabi returned home from public school to their small rooms in the back of the store his mother often greeted him: "Did you ask any good questions today?" Now, in the Carnegie Library, Rabi stumbled across a book about Copernicus. According to the Judaism he knew, even the rising and setting of the sun were miraculous. How impressive to find here that law and reason governed the workings of nature. Rabi soon decided that he wanted to study the structure of the universe.[7]

Impish, quick with his wit and warming laugh, Rabi organized discussion groups, taught the boys in the neighborhood to play chess, brought some of them together to build a wireless set. He also kept reading voraciously, especially about history, socialism, and science, and the more he read, the more he grew eager to know the larger world beyond his circle in Brooklyn. He enrolled in the predominantly gentile Manual Training High School, a good three-mile walk away; then, with two scholarships, went to Cornell and majored in chemistry. But not even a Jewish fraternity rushed Rabi, who could afford neither good clothes nor parties and was short on glad-handing pleasantries. And chemistry was disappointing, especially when at graduation in 1920 most of his gentile classmates landed decent jobs and Rabi, the top student in his department, landed none.

Warned that a Jew had no chance in the academic world, Rabi found a job in a chemical laboratory analyzing mother's milk and furniture polish; he soon left for a trading firm which promptly set him to computing accounts receivable in a factory. All the while he knocked about with some equally bright and misemployed Jewish friends who maintained their intellectual mettle by arguing science and philosophy, their self-esteem by discussing the rest of the world's stupidity. Rabi spent

[6] Cleveland *Plain Dealer*, May 26, 1921, pp. 1, 2.

[7] Mrs. Rabi is quoted in *Time*, Jan. 2, 1961, p. 42. Unless otherwise noted, this section is based on interviews with I. I. Rabi.

many evenings reading science at the New York Public Library on Fifth Avenue. He attended one of Einstein's lectures at City College; using a mirror and soap in lieu of a blackboard and chalk, he gave his own seminar on the theory of relativity to a neighborhood study group. Finally, in 1923, resolving not to attribute every setback to anti-Semitism, Rabi returned to Cornell for graduate work in physical chemistry. Six months of hard study later, Rabi knew that the part of the subject he really liked was called physics. Rejected for a fellowship in the Cornell department, he transferred to Columbia University, took a full-time job at City College to meet expenses, and plowed ahead with his subject.

Rabi kept warning Helen Newmark that physicists did not make much money. Helen's father was an intellectualish advertising executive; a graduate of the Ethical Culture School, she was delighted that Rabi wanted to be a physicist. The marriage took place in 1926, about the time that Rabi was beginning to delve deeply into the new quantum mechanics. Thrilled by the emerging ideas, he hungrily awaited the foreign journals at school, even complained to the librarian when they arrived late. In 1927 Rabi was awarded his doctorate for a thesis on magnetism in crystals. Though the work was scarcely in the mainstream of quantum physics, he had figured out an ingenious method of doing the experiment, and now he went to Europe, financed by a postdoctoral grant from Columbia.

A disappointing week in Schrödinger's Zurich, five exhilarating ones at Sommerfeld's Institute, two inspiring months in Copenhagen, and by the winter of 1928 Rabi, who could by now even crack jokes in German, had settled down in Hamburg to study theory with Wolfgang Pauli. But in Hamburg there was also Otto Stern, a connoisseur, a bachelor, an eccentric, and an equally discriminating experimentalist and gastronome. With Walter Gerlach in 1921, Stern had demonstrated that a beam of silver atoms split into two distinct parts when it passed through a non-uniform magnetic field. The experiment showed decisively that atoms had magnetic characteristics of a quantum kind, provided physicists with an elegantly simple technique to measure them, and would one day win Stern the Nobel Prize.

Rabi had marveled over the Stern-Gerlach experiment in the United States, and now he saw a way to do it with even greater accuracy and simplicity. Impressed, Stern invited the young American to try the modified technique in his own laboratory. Rabi wanted to be a theorist, but the invitation was too great an honor to turn down.[8] He worked at the experiment painstakingly for almost a year, obtained superb results, and published them in the leading German journal of physics. Awarded an International Education Board fellowship for another year in Europe,

[8] Rabi to George B. Pegram, Sept. 1, 1927; Dec. 14, 1927, GBP, R-Misc. file.

Rabi resumed his theoretical studies, spent time with Heisenberg in Leipzig, then once again with Pauli, who had moved down to Zurich. Meanwhile, Columbia's physics department was looking for someone to teach the new mathematical physics. Heisenberg, in the United States for a lecture tour, recommended Rabi. No Jew had ever been appointed to the Columbia physics faculty, not at least in anyone's memory. Chairman George B. Pegram, a patrician from North Carolina, had already helped Rabi get the International Education Board fellowship; now he offered him a lectureship at $3,000 a year.

Rabi accepted with alacrity. Returning to the United States in 1929, he inaugurated a joint colloquium on quantum mechanics with Gregory Breit, a young theorist from New York University, and some fifty students turned up at the first meeting. Rabi soon discovered that he could teach theory better than do research in it. Shifting back to the laboratory and work with the Stern-Gerlach technique, he was acquiring both a reputation as one of the most adept experimentalists of his generation—he would eventually win the Nobel Prize—and a growing flock of talented graduate students, including Jews of recent immigrant backgrounds like himself, who were now entering the physics profession in steadily increasing numbers.[9]

Like Rabi, most of the new physicists of the 1920s joined the professoriat. The demand for those who could teach quantum mechanics was so strong that departmental chairmen were willing to give masters of the theory special dispensations, including relief from heavy teaching loads at the outset and the promise of swift promotion in the near future. By the late 1920s salaries at the better universities ranged from $2,400 for a fresh Ph.D. to more than $7,000 for a full professor, which even at the lowest rank was comfortably higher than the national median income. And before the war Andrew Carnegie had endowed a pension program for certain groups of college teachers; transformed early in the 1920s into the Teachers Insurance and Annuity Association, it was now open to any professor who cared to participate.[1] Above all, with its commitment to pure science, the academic world was clearly the place to get on with the modern developments.

In the mid-1920s Leonard Loeb, one of the first NRC fellows, told James S. Thompson of McGraw-Hill that American students badly

[9] Pegram to R. de L. Kronig, Sept. 28, 1929, GBP, Columbia, Kronig file; I. I. Rabi and Gregory Breit, "Measurement of Nuclear Spin," *Physical Review*, 38 (Dec. 1, 1931), 2082–83.

[1] George W. Stewart, "Physics as a Career," in *Opportunities for a Career in Scientific Research* (Washington, D.C.: National Research Council, Division of Educational Relations, 1927), p. 6; James R. Angell to W. F. G. Swann, Feb. 22, 1924, WFGS, Angell file; "Physics Department—Appointments and Budget, 1922–1929," UCP; "Budget, 1927/28," in UCBP; Abraham Flexner, *Henry S. Pritchett: A Biography* (New York, 1943), pp. 98–100.

needed good physics textbooks. A liberal arts graduate of the University of Wisconsin and a onetime country newspaperman, Thompson was no expert in science, but he listened to the experts and had an enterprising sense of market. Early in the decade Thompson had originated his firm's distinguished International Chemical Series, and within a few years of Loeb's remark he conceived a sister set, the International Series in Physics. Launched in 1929, most of its volumes dealt with quantum topics, were written by recent postdoctoral fellows, and, quickly becoming standards, revolutionized the advanced physics textbook situation in the United States.[2]

Pushing their discipline ahead in other ways, the new physicists joined enthusiastically in the summer symposia on theory that began in 1928 at the University of Michigan. Physics dominated the long, warm days, from the lectures in the mornings through the family picnics in the afternoons to the parties at night. At the start the lecturers were mainly imported Europeans, but the young American professionals increasingly endowed the proceedings with native leadership. On their own campuses all the while, the new American physicists gradually took over the direction of the curriculum from their elders and emphasized the promulgation of quantum mechanics. Whatever their pedagogical styles, few approached the dissemination of modern theory with quite the dedication of the slender young physicist, his face intense and his hair billowing skyward, who in 1927 wandered into the Caltech physics laboratory and announced: "I'm Oppenheimer."[3]

"J. Rob-ert Opp-en-heim-er," the precisely articulate child used to introduce himself. The son of German Jews, his mother an artist, his father a well-read and wealthy New York importer of men's suit linings, young Robert wrote poetry, collected minerals, continually devoured books, and by the time of his graduation from the Ethical Culture School had mastered five languages as well as a good deal of advanced physics and chemistry. The elder Oppenheimers worried about their prodigy. Shy, dreamy, physically frail and athletically inept, he had trouble making small talk; his acquaintances—he had few real friends—considered him a snob. On Long Island during vacations, his parents had to plead with

[2] Leonard Loeb, "Autobiography," pp. 87–88, LL–AIP. Thompson was guided in the actual formation of the Series by Edward U. Condon and Harold C. Urey, both recent NRC fellows. Curtis G. Benjamin to the author, Feb. 16, 1970. By 1932 the Series included Robert F. Bacher and Samuel Goudsmit's *Atomic Energy States*; Edward U. Condon and Philip Morse's *Quantum Mechanics*; Arthur Hughes and Lee A. DuBridge's *Photoelectric Phenomena*; Goudsmit and Linus Pauling's *The Structure of Line Spectra*; and Arthur Ruark and Harold Urey's *Atoms, Molecules, and Quanta*.

[3] Quoted in "Raymond T. Birge," Transcript of a tape-recorded interview conducted by Edna T. Daniel, Regional Cultural History Project, General Library, University of California, Berkeley, 1960, p. 128.

the other teenagers to include Robert in their outings. Oppenheimer confided to his English teacher: "I'm the loneliest man in the world."[4]

Entering Harvard as a sophomore, Oppenheimer haunted the library, took ten courses a term, delved into poetry and philosophy with Alfred North Whitehead, and probed more deeply into physics under the tutelage of the future Nobel laureate Percy W. Bridgman. At Harvard, he recalled, "I almost came alive."[5] During his first doctoral year at Cambridge University, 1925, Oppenheimer found the quantum mechanics a torturing struggle and seriously doubted his capacity to do theoretical physics. Later, describing the general problems of the young American physicist abroad, Oppenheimer was revealingly self-descriptive: "The difficulties are most acute in men who combine a certain timidity with a quite robust vanity—or, perhaps more accurately, with an urgent desire for excellence. . . . [They fall victim to] the melancholy of the little boy who will not play because he has been snubbed." Stuttering, suffering from incipient tuberculosis, Oppenheimer brooded over his sense of inadequacy. At the time, a psychiatrist was no help. Neither was Niels Bohr, who asked whether his problems with theoretical studies were mathematical or physical.

"I don't know," Oppenheimer admitted.

"That's bad," Bohr said flatly.[6]

A sojourn in Corsica rescued him from his depression, and in 1926 he went to that mecca for mathematical physicists, the University of Göttingen. His cutting tongue, air of the *Wunderkind*, and habit of interrupting the lecturer—the problem can be better done this way, he would say, blithely walking to the board uninvited and taking the chalk—decidedly annoyed his teachers and fellow students. All the same, his theoretical facility increasing and his remarkable erudition evident, he became a favorite of Max Born. The two physicists would often spend the day closeted together, mixing research in quantum mechanics with discussions of literature. His contemporaries sometimes ridiculed him in private, but they agreed with Earle Kennard, who, reporting that "there are three young geniuses in theory here, each less intelligible to me than the others," ranked Oppenheimer with Pascual Jordan and Paul Dirac.[7]

In 1929, already a legendary figure in the world of physics, the author of sixteen papers, confident of his ability to do quantum mechanics and eager to spread its gospel, Oppenheimer returned home to take up a joint appointment at Caltech and Berkeley.

[4] The young Oppenheimer is quoted in Philip Stern, *The Oppenheimer Case* (New York, 1969), pp. 10–11, and in Denise Royal, *The Story of J. Robert Oppenheimer* (New York, 1969), p. 23.

[5] Quoted in Royal, *Oppenheimer*, p. 31.

[6] Quoted in R[ockefeller] F[oundation] *Illustrated*, 2 (Aug. 1974), 2; Oppenheimer interview, Nov. 18, 1963, p. 21, SHQP.

[7] Kennard to R. C. Gibbs, March 3, 1927, CUP, Kennard file.

At Berkeley during his first professorial year, the pace of his presentation left the students dumbfounded. ("It's going too slow . . .," he worried to an older staff member. "I don't know what's wrong.")[8] At Caltech, where his inaugural lectures were jammed, he chain-smoked cigarettes at the blackboard, scribbled equations in a cramped hand, laced his barely audible delivery with obscure references to the classics of literature and philosophy. Though the lectures remained packed, soon only a single registrant was taking the course for credit. Desperately eager to reach his students, his sensitivities sharpened by his own past difficulties, Oppenheimer made it a point to pay as much attention to the troubles of his charges as to the intricacies of his subject. His language evolved into an oddly eloquent mixture of erudite phrases and pithy slang, and he learned to exploit an extraordinary talent for elucidating complex technical matters. Students from around the country flocked to California to study under Oppenheimer. They put in more than forty hours a week on his courses, migrated with him from Berkeley to Caltech and back, imitated his mannerisms, even his walk. By the early 1930s, J. Robert Oppenheimer had turned into the most captivating, commanding, and worshipped teacher of theoretical physics of his generation.

With the advent of Oppenheimer and his quantum-mechanical peers, with the new textbooks and the Michigan symposia, America's neophyte physicists no longer felt the need to study abroad. For the first time in the profession's history, they could get first-rate training at home not only on the experimental but on the theoretical side of the discipline.

Given ample time for research, the new generation published so prolifically that over the decade the number of pages in *The Physical Review* swelled more than two and a half times, to some 3,800 annually. Frequently the appearance of an article lagged behind its submission by some six months, and since fame, even the Nobel Prize, could depend upon priority—Arthur Compton, who was awarded the prize in 1927, had published his quantum explanation of X-ray scattering almost in a dead heat with a European—many physicists disliked the delay. For years foreign journals like *Nature* had accelerated the dissemination of fresh results or ideas by printing brief letters to the editor. In 1929 *The Physical Review* started both doing the same and coming out biweekly. George B. Pegram, who had a hand in the innovation of the letters, happily commented: "I think . . . we shall not hereafter be satisfied to let our European friends be too far ahead of us."[9]

[8] Quoted in Nuel Pharr Davis, *Lawrence and Oppenheimer* (New York, 1968), p. 27.

[9] Pegram to A. L. Loomis, April 23, 1929, GBP, L-Misc. file. See also John T. Tate to K. T. Compton, June 10, 1929, KTC. Arthur Compton, pleading with the managing editor of *The Physical Review* for immediate publication of his article on X-ray scattering, noted that the subject was "exceptionally live" and wondered in general whether there were not "some way to lessen the interval between pen and printed page." A. H. Compton to Gordon S. Fulcher, July 26, 1923, AHC–AIP.

By the opening of the 1930s the new generation had virtually closed the historic gap in quality between European and American research. The work of Davisson and Compton bespoke the profession's increasingly striking accomplishments in experimentation. And if Willard Gibbs had been an anomalous genius, steady contributions to the advancement of physical theory were now coming from such Americans as John Van Vleck; John C. Slater, who collaborated with Bohr and his assistant Hendrik Kramers in a paper of central importance for the leap from the old quantum theory to the new quantum mechanics; Carl Eckart, a Midwesterner of an imaginatively philosophical turn of mind, who demonstrated the equivalence of the wave and matrix mechanics independently of Schrödinger; Oppenheimer, who fruitfully probed the equations of Dirac and broke important ground in the quantization of the electromagnetic field; and Linus Pauling, who established himself as one of the world's leading physical chemists by ingeniously applying the quantum mechanics to the puzzle of the chemical bond.

American physics developed such power and strength in part because of the substantial enlargement of the profession. With the increase in the number of physicists, simple chance made the country likely to have more good ones. Helping chance out were the new theorists and the special intellectual circumstances of the discipline. With mathematical physicists down the hall, experimentalists could easily get expert advice when they needed it to interpret puzzling data or plan a crucial test in the laboratory; with the quantum mechanics at hand, even moderately able theorists could make a respectable mark by working out old problems in new ways. But no less important was the institutional pyramid of Rowland's best-science program, especially the capstone of the pyramid, the postdoctoral fellowships.

The fellowship board of the National Research Council judged applicants by the merit of the research they wished to do and the standing of the institutions where they wished to study. The opportunities of the fellowships were thus given to new Ph.D.'s who were going into the prime fields of physics, and the distinction of the awards spotlighted the kind of work a good young physicist ought to be doing. More important, the NRC designates tended to collect at five centers—Berkeley, Caltech, Princeton, Chicago, and Harvard—where many of the more able, productive physicists were located. The fellowship program thus brought the elite of the new physicists, about one in eight of the period's Ph.D.'s, together with the existing elite, which de facto put into practice the strategy endorsed long before by the trustees of the Carnegie Institution, that the most effective way to discover and develop the exceptional man was to put promising men under proper guidance and supervision. By working with the leaders in their specialties at home or abroad, the NRC fellows not only learned more physics; they also developed what Rabi once termed a "taste" for the significant piece of research or what Hale had gropingly called a sense of "the larger relationships." The fellows themselves added a good deal of intellectual vitality to the departments where they spent their postdoctoral years, and once they settled into teaching posts, they began seeding a new generation of students with the same taste for asking key questions of nature.

The United States may have always had a pool of potentially outstanding physicists. In the 1920s, as a member of the NRC later summarized, "God went on doing about the same as he always had done and we did something different."[1]

Oblivious of their debt to God or man, the new physicists raced ahead, some arrogant about what they understood, most humble before what they did not, all confident that they were teaching, experimenting, and theorizing their profession into the first rank. They knew each other, followed each other's work. They supplied theorists to the editorial board of *The Physical Review* and increasingly dominated meetings of the American Physical Society, where sessions on quantum mechanics drew more than four hundred participants. Gone were the lamentations of Rowland's day, gone the sense of provincial inferiority. And with good reason: However many Americans went abroad to study in the 1920s, one out of every five of the International Education Board's foreign fellows in physics spent his postdoctoral years in the United States. Even established Europeans, John C. Slater proudly recalled, were coming over "to learn as much as to instruct."[2]

And, Slater might have added, to take jobs. Abroad, young physicists often had to spend years teaching high school or assisting a professor before they won a full-fledged university post of their own. Jewish physicists found that anti-Semitism made the limited number of ranking academic places even more limited. But you did not have to worry about getting ahead in the United States, Sommerfeld told his students; there "every young man becomes an assistant professor." By 1931 American universities had hired at least fifteen Continental physicists, including George Uhlenbeck, Samuel Goudsmit, Alfred Landé, and Eugene Wigner. The United States, a British scientist remarked, was "constantly" attracting "some of the ablest men in Europe."[3]

In 1932 Abraham Flexner, the director of the new Institute for Advanced Study in Princeton, New Jersey, suggested that Einstein, who had been gradually spending more and more time in the United States, consider joining the staff. Einstein had endured constant anti-Semitic hostility in the land of his birth; now, amid the mounting turmoil in

[1] Isaiah Bowman, "Report of the Fellowship Program NRC . . . 1934," JCM, Box 220.
[2] Slater, "Quantum Physics in America Between the Wars," *Physics Today*, 21 (Jan. 1968), 43.
[3] Sommerfeld was quoted in an interview with the late Edward U. Condon; J. B. S. Haldane, "Nationality and Research," *Forum*, 75 (May 1926), 720.

Germany, the attacks were growing more virulent. Numerous universities were showering him with offers of appointment, but he was not keen to teach, and the new Institute, amply endowed by the Jewish philanthropists Mrs. Felix Fuld and her brother, Louis Bamberger, would have no regular students. Einstein mulled over the invitation, then, one drizzly evening in the summer of 1932, concluded a long discussion with Flexner: "I am flame and fire for it."[4]

Einstein's migration to America stimulated Professor Paul Langevin of the Collège de France to predict that the United States would now become "the center of the natural sciences."[5] Langevin might have added that by 1932, before the flight of the refugees, the American physics profession was already one of the most capable and vigorous branches of the world physics community.

[4] Quoted in Abraham Flexner, *I Remember: The Autobiography of Abraham Flexner* (New York, 1940), p. 252.
[5] Quoted in Robert Jungk, *Brighter than a Thousand Suns*, trans. James Cleugh (London: Penguin, 1960), p. 51.

XV

Miraculous Year

Shortly before returning to the United States in 1929, I. I. Rabi noted, "The march of theoretical physics . . . isn't the gay rush it was about 18 months ago."[1] Progress in the understanding of cosmic-ray phenomena was slow, and physicists hardly knew how to forge the wave and particle characteristics of radiation into a consistent quantum theory of the electromagnetic field. Then, too, for all the success of Paul Dirac's relativistic theory of quantum mechanics, the solutions of his fundamental equation implied that the electron could fall into a state of negative energy, a seemingly preposterous result. And there were no less perplexing difficulties for relativistic quantum mechanics in the domain of the atomic nucleus.

In the prevailing view, the nucleus consisted of protons—heavy, positively charged particles—and electrons. The protons were believed to account for and equal in number the atomic weight. The electrons seemed necessary to neutralize those protons that contributed to the atomic weight but not to the nuclear charge. All the electrons and protons appeared to be packed into a volume that occupied no more relative space in the atom than did the sun in the solar system. According to the ordinary laws of electromagnetism, at nuclear distances one proton would repel another with a force far greater than their mutual gravitational attraction. Since nuclei obviously did not fly apart, at such small distances a very powerful type of binding force clearly came into play. The nature

[1] Rabi to George B. Pegram, March 15, 1929, GBP, R-Misc. file.

of this force was mysterious but its general magnitude was believed to be at least a million times more powerful than the forces binding the outer atomic electrons to the nucleus itself.

Much of what was then known about the nucleus had come from the laboratory of Ernest Rutherford, the engagingly outspoken New Zealander—"that man is like the Euclidian point," he quipped of a pompous official; "he has position without magnitude"—who was Cavendish Professor of Experimental Physics at Cambridge University. Before coming to Cambridge in 1919, Rutherford was already the world's leading authority on radioactivity, progenitor of the nuclear model of the atom, and a Nobel laureate. Once asked why he always seemed to be on the crest of the wave in physics, he replied, "Well, I made the wave, didn't I?"[2] Rutherford started generating a new wave during World War I when he bombarded the lighter nuclei with the charged alpha particles emitted by radioactive elements. Energetic enough to push their way into the nuclear region, the alphas either scattered from the target nucleus— or actually disintegrated it (the "ultimate importance" of these results, Rutherford remarked, could well be "far greater than that of the war itself").[3] After coming to Cambridge, Rutherford made the Cavendish Laboratory the world's center of nuclear studies, gathering around himself a small group of productive young experimentalists who helped develop scattering and disintegration experiments, among other techniques, into the chief experimental methods of nuclear physics.

But the acquisition of nuclear data depended heavily on the use of radioactive sources and, as the $100,000 for Madame Curie's gram made manifest, radium, the most intense source, was in short supply. The sources in Rutherford's possession emitted such low intensities of alpha, beta, and gamma rays that accurate analysis of the results of nuclear bombardment was decidedly difficult. Moreover, the dependence upon radioactive sources confined nuclear studies to processes that occurred at the energies of the rays from naturally radioactive substances. The energy and intensity limitations especially handicapped disintegration experiments. The fastest radioactive alpha particles possessed energies too small to penetrate the repulsive force field that surrounded the nuclei of the heavier, more highly charged elements; for lighter elements, the small flux of alphas from the available sources made the probability of disintegration extremely small.

[2] Quoted in A. S. Russell, "Lord Rutherford: Manchester, 1907–19 . . .," in J. B. Birks, ed., *Rutherford at Manchester* (London, 1962), p. 90; and in Arthur S. Eve, *Rutherford* (New York, 1939), p. 436. Some of the material in the opening of this chapter first appeared in *The Physics Teacher*, copyright © 1972 by the American Association of Physics Teachers, and is reprinted here with the permission of the editor.

[3] Quoted in Karl T. Compton, "Battle of the Alchemists," *Annual Report of the Smithsonian Institution, 1933*, p. 275.

On the theoretical side, a few physicists had successfully applied quantum mechanics to the analysis of certain nuclear phenomena, but in 1931 Samuel Goudsmit had to say: "Nuclear problems are piling up without any progress in understanding."[4] A dynamic theory of nuclear structure seemed beyond reach without knowing the force law between nuclear particles, and the development of such a force law was nowhere in sight. It also appeared impossible to explain the basic characteristics of nuclear composition or radioactivity without assuming that most nuclei contained electrons, but the assumption of the electron in the nucleus raised confounding theoretical difficulties.

According to classical physics, the electron was much too big even to fit into the nucleus. According to quantum mechanics, if it could somehow be forced into the nuclear space, it would be much too energetic to remain there very long. No less disturbing, the total angular momentum of the nucleus had to equal the total spin-based angular momenta of all its particles. But according to certain types of spectral data, while the nuclear protons seemed to contribute to the total angular momentum, the nuclear electrons did not. Hence the perplexing difficulty for quantum mechanics: Once electrons entered the nucleus, they seemed to lose the spin that Dirac's relativistic theory predicted as an intrinsic property of the particle.

In the face of these anomalies, some physicists speculated that the existing relativistic quantum mechanics might be invalid in the nuclear region. A few, including Niels Bohr, wondered whether the resolution of the nuclear difficulties might not even require still another brand-new mechanics. But in 1931 Rutherford, now knighted and president of the Royal Society, told an audience in Göttingen that the nucleus must be simple in structure. "I'm always a believer in simplicity, being a simple person myself." Time and again Rutherford had expounded to his colleagues at the Cavendish Laboratory how nuclei just could not be built of protons and electrons alone. In his judgment, they had to contain another particle, a combination of the proton and electron. As such, the particle would have zero net charge. Rutherford's staff liked to call it the "neutron."[5]

Among the believers in the neutron was James Chadwick, Rutherford's right-hand man, assistant director of radioactive research at the Cavendish, and a fellow of the Royal Society. One morning near the end of January 1932, Chadwick read with astonishment a report of some radiation experiments just done in Paris by Irène Curie and her husband, Frédéric Joliot. According to the Joliot-Curies, upon bombardment by 5,000,000-volt alpha particles from radium emanation, beryllium emitted

[4] Goudsmit to L. Brillouin, May 6, 1931, SHQP, microfilm roll 65, Sec. 3.

[5] Recording of Rutherford's remarks in Göttingen, AIP. In 1924 James Chadwick told Rutherford how "we shall have to make a real search for the neutron." Quoted in Eve, *Rutherford*, p. 301.

gamma rays that possessed up to 50,000,000 volts of energy and knocked copious numbers of protons out of paraffin. Chadwick was skeptical that 5,000,000-volt alpha particles could provoke beryllium to emit gamma rays ten times as energetic. Within two feverish weeks of experimental activity Chadwick in fact proved otherwise—that the beryllium "gamma" radiation consisted of Rutherford's long-suspected neutrons.

Chadwick's identification of the neutron won swift acceptance in the world of physics. If he claimed to have seen what others had failed to see, he did so with a marshaling of reasoning and evidence that made for one of the most decisive and well-turned papers of modern experimental physics. With the appearance of the neutron the atomic nucleus had to contain only enough protons to account for the nuclear charge; the rest of the nuclear weight could be accounted for by neutrons, instead of protons balanced by electrons. Since the neutron eliminated the principal reason for assuming that the electron was in the nucleus, it removed the main doubts that quantum mechanics could be applied to phenomena in the nuclear region.[6] And the neutron added considerable specific importance to Harold Urey's work on a rare type of hydrogen.

Urey lost his father, a farmer and minister of the Brethren Church in Walkerton, Indiana, at age six, and his faith not many years later. He wandered from schoolteaching to a degree in zoology from Montana State University to wartime chemical work for a firm in Philadelphia. Soured on industrial chemistry, he joined the Montana State staff and, in 1921, at twenty-eight, he went to Berkeley for a Ph.D.

Urey gravitated to the chairman of the Berkeley department, Gilbert N. Lewis. The only physical chemist in America who ranked with Irving Langmuir, Lewis knew all about the theory of relativity, even had his own quantized alternative to the Bohr-Sommerfeld atom. He pressed his ideas with stimulating, sometimes brusque, directness. "Damn it . . .," Lewis once sputtered, waving his cigar at a puzzled graduate student, "if I understood the question I'd know the answer."[7] Some people hated Lewis; under his prodding tutelage, Harold Urey developed a passion for research and an active knowledge of atomic physics. A postdoctoral year in Bohr's institute completed his transformation from an uncertain neophyte into a bantam cock of a physical chemist, and by 1929 he was an associate professor at Columbia University.

Urey became intrigued with the question of the possible existence of an isotope of hydrogen. The isotopes of an element all possess the same number of nuclear charges but differ slightly in atomic weight. No

[6] James Chadwick, "The Existence of a Neutron," *Proceedings of the Royal Society*, 136 (June 1, 1932), 692–708. Of course the neutron did not resolve the issue of the emission of electrons by radioactive nuclei, but that peculiarity seemed to be a feature only of the naturally radioactive elements.

[7] The student was Joseph Mayer, who quoted Lewis in an interview with the author.

one had ever detected an isotope of hydrogen, but a minute discrepancy existed among the atomic weights of hydrogen as determined by different methods. Two American scientists suggested that the tiny difference could be explained if, along with every 4,500 normal atoms of hydrogen, there was an abnormal one, an isotope of mass two. Of course, to believe in the discrepancy—and its interpretation—one had to believe that the relevant data were accurate to better than two parts in 10,000. Urey, who had acquired his own secular brand of faith, decided to look for the isotope.

The experimental problem was obvious: how to isolate enough of the suspected isotope to demonstrate its existence. Doing a few theoretical calculations, Urey learned that if hydrogen was cooled to a liquid state, then permitted to evaporate, its normal atoms of mass one would boil off at a considerably faster rate than those of mass two. The process would accordingly yield a residue rich in the heavier isotope. Urey had F. W. Brickwedde of the Bureau of Standards distill a gallon-size sample of liquid hydrogen until only a few cubic centimeters was left. He then subjected the residues to careful spectral analysis. He found, close to the main spectral lines of normal hydrogen, the faint lines to be expected of a hydrogen atom of mass two.[8] In January 1932, not many weeks before Chadwick announced the neutron to the world, Urey published his proof of the existence of the hydrogen isotope, which was eventually named deuterium.

Deuterium promised to ease a major difficulty in the development of a theory of nuclear structure. The trouble was that all nuclei except those of normal hydrogen contained three or more particles, and in the nucleus, the protons and neutrons were packed so closely together that the effect of one particle on the behavior of all the others could not be neglected. Consequently, nuclear theorists had to analyze the collective interactions of many bodies, all of approximately equal mass. Physicists knew no way to treat with analytical exactness the dynamics of three or more bodies, but deuterium was a nuclear system of just two bodies, a proton and a neutron. As such, its characteristics could be calculated, then compared to the results of experiment. Since what governed the proton and neutron in deuterium presumably governed their interaction in more complex nuclei, deuterium offered an ideal test case for the development of a general theory of nuclear structure.[9]

. . .

[8] F. W. Aston, the British authority on isotopes, said that Urey was "a brave experimenter to attack such a problem with such extraordinarily small differences in which to trust." "Discussion on Heavy Hydrogen," *Proceedings of the Royal Society*, 144 (March 1, 1934), 8.

[9] The physicist Victor Weisskopf later called deuterium "the hydrogen atom of nuclear physics." Charles Weiner, ed., *Exploring the History of Nuclear Physics* (AIP Conference Proceedings, No. 7; New York: American Institute of Physics, 1972), p. 15.

Of course, neither the neutron nor deuterium resolved the key experimental difficulty of nuclear physics: the dependence upon radioactive sources, especially for disintegration experiments. Simple calculation revealed how to get around that obstacle. A machine that accelerated just a minute fraction of an ampere of proton current would spew forth as many projectiles per second as about 180 grams of radium. Since 1919 Rutherford had been calling for such a machine—"a million volts in a soapbox"—and in the later 1920s a few enterprising physicists in Europe and the United States had started trying to fill the order.[1]

But operating above about 300,000 volts, as Merle Tuve knew all too well, ordinary equipment would usually break down electrically. While a graduate student at the University of Minnesota, Tuve had heard Professor W. F. G. Swann, an impressively theatrical-looking British import, talk about the importance of high-voltage accelerators for the study of the nucleus. By 1930, now a staff member at the department of terrestrial magnetism at the Carnegie Institution, Tuve won the year's research prize from the American Association for the Advancement of Science for obtaining 1,000,000 volts with a so-called Tesla coil protected against electrical breakdown by an insulating oil bath. Still, Tuve could not make his machine work reliably enough to be used as a tool for probing the nucleus. After years of virtually obsessive marriage to the Tesla coil, he hated the entire apparatus.[2]

One spring evening in 1929, Merle Tuve's boyhood friend Ernest O. Lawrence sat in the Berkeley library glancing through a Norwegian engineer's report of an intriguing method of accelerating charged particles. Suddenly, Lawrence thought of a way to improve the technique dramatically. Hurrying across the campus the next afternoon, elation written all over his uncomplicated face, Lawrence saw a colleague's wife and yelled: "I'm going to be famous."[3]

Ernest Lawrence had always seemed to be hurrying toward something. His father, Carl, who had come to Canton, South Dakota, to teach in a Lutheran academy, was a vigorous spokesman of progressive-minded education; people called him the La Follette of the state and twice elected him superintendent of public instruction. While an engineering student at the University of South Dakota, young Ernest stayed away from the fraternities and conscientiously attended to his studies; during vacations he earned money to help pay expenses. Hard work and success, Ernest told his younger brother John, were the greatest things in life. In college an able professor persuaded Lawrence that another great thing was

[1] Rutherford is quoted in "Ernest Thomas Sinton Walton," *Current Biography*, 1952, p. 619.

[2] Interview with Tuve. See also Tuve's report, *Carnegie Institution of Washington Yearbook*, 31 (1931–32), 230–31.

[3] Quoted in Herbert Childs, *An American Genius: The Life of Ernest Orlando Lawrence* (New York, 1968), p. 140.

physics. In 1922, graduated with high honors, Lawrence joined Tuve for doctoral work at the University of Minnesota, where he was inspired by W. F. G. Swann. He followed Swann to Yale and gained a Ph.D. as well as an instructorship. Lawrence soon developed a national reputation for his ability to conceive ingenious experimental devices and his uncanny knack for making the devices work.

Yale's physics department seemed on the move with Swann, and New Haven was made all the more attractive for Lawrence by his future wife, Molly Blumer, the daughter of the former dean of the Yale Medical School and a Vassar student who was more than politely interested in science. But senior physicists elsewhere started trying to lure Lawrence away from Yale. At Princeton, the dean proudly told him that the graduate school had only two hundred students. Why, I want that many for myself, Lawrence is said to have replied without a smile.[4] Berkeley, whose physicists had a fierce determination to make their department the equal of any in the East, not to mention of Robert Millikan's Caltech and even Gilbert Lewis' chemistry group, offered Lawrence an associate professorship, as many graduate students as he could handle, and the right to direct research. Lawrence headed for California. "Not the least homesick for Yale (remarkable thing)," he soon noted. "I have a relative importance which I could never have attained at Yale in years."[5]

The importance burgeoned considerably not long after that night in the Berkeley library when Lawrence had his technological epiphany. The Norwegian engineer Rolf Wideröe had built a pipelike acceleration tube broken by two successive gaps and powered by a 25,000-volt alternating electrical signal. When charged particles traveling down the tube entered the first gap, the electrical field speeded them forward with 25,000 volts of push. At this moment, the accelerating field across the second gap pointed in the backward direction. But by the time the particles actually reached the second gap, the field was reversed and the particles got another 25,000-volt push forward. The accumulated result: 50,000-volt particles from only 25,000 volts of electrical potential.

What Lawrence realized was that charged particles could be given such successive pushes indefinitely if they were kept moving in a circle by a magnetic field. Simply apply an alternating voltage signal to a gap cut across the circle's diameter. When the particles crossed the gap on the right-hand side, they would be kicked ahead by the electric field. By the time they traversed their semicircular path to the gap on the left-hand side, the field there would have reversed direction and the particles would be kicked ahead again, to a still higher speed.

Of course, the particles would be moving faster after each kick, but they would also be making ever wider circles. In fact, no matter how

[4] The exchange was related in an interview with the late Edward U. Condon.
[5] Quoted in Childs, *An American Genius*, p. 132.

fast they were moving, they would always take precisely the same time to make their way around from the right-hand to the left-hand side of the gap. That fact made possible applying to the gap a voltage of a particular constant frequency whose cycles would always be in resonance with the cycles of the particle. Here, in this principle of resonance, was the feature that promised to make Lawrence's proposed circular machine workable: By applying to the gap a relatively low voltage signal alternating at the resonance rate, one could in theory accelerate charged particles to energies many times greater than that of the signal itself.

With the help of a graduate student, Lawrence constructed a test model. Only four inches in diameter and made of glass, it seemed to work. In print and on the lecture platform of the National Academy of Sciences itself, Lawrence announced that he saw no difficulty in the way of accelerating particles to 1,000,000 volts of energy, and even far beyond. Berkeley President Robert G. Sproul, showing some of the ambitious spirit of his physics department, overrode considerable opposition from the senior nonscientific faculty and appointed Lawrence to a full professorship at $5,000 a year. He was just twenty-nine.

In the fall of 1930 M. Stanley Livingston, opting to do his Ph.D. thesis on the accelerator, built an improved version of the four-inch model; by the spring of 1931, he had coaxed out of it 80,000-volt particles with a voltage supply of only 2,000 volts. Mysteriously, the particle beam was nowhere near as intense as it should have been. Barging ahead, Lawrence put Livingston to work on an eleven-inch model which was designed to reach the magic goal of 1,000,000 volts. Now the difficulties with the beam grew worse. For some reason, long before the particles reached the maximum voltage level, they got out of focus and smashed into the walls of the acceleration chamber. Lawrence divided his time between fund-raising trips to the East and long nights with Livingston trying to make work what the men around the laboratory were calling his "cyclotron."

By late 1931 Lawrence and Livingston had found two tricks that improved the focus of the beam. One exploited the electric field between the gaps; the other modified the magnetic field that kept the particles moving in circles. Both tricks were forged cut-and-try; the theory came later. Livingston claimed that it was he who had taken the two empirical shots in the dark. Lawrence, and especially Lawrence's friends, always disagreed. Whoever was responsible, the tricks made the cyclotron work. In February 1932 Livingston got the machine going to the point where he could chalk on the blackboard: 1,000,000 volts. Late in the evening Lawrence came in, saw the board, and, Livingston recalled, "literally danced around the room."[6]

[6] Quoted in Nuel Pharr Davis, *Lawrence and Oppenheimer* (New York, 1968), p. 42.

Within weeks of this achievement, news came to Berkeley of the triumph at the Cavendish Laboratory of John D. Cockroft, one of Rutherford's chief assistants, and Ernest T. S. Walton, a recent graduate of Trinity College, Dublin. At home with a screwdriver or a pencil, Cockroft was respected as a general utility man who could build apparatus or solve equations with equal facility. In the late 1920s he had done some quantum mechanical calculations on the accelerator requirements for nuclear disintegration. He found that relatively low-energy protons would provoke disintegrations with a probability comparable to that of much higher energy alpha particles. The implication was striking: Light nuclei could be disintegrated in quantities sufficient for useful experiments by a machine that accelerated protons not a million but to only a few hundred thousand volts.[7]

Cockroft, now joined by Walton, eventually designed a proton accelerator, an ingenious arrangement of electrical condensers and vacuum tubes that multiplied a relatively low input voltage to an output level four times higher. Once constructed, the apparatus occupied a large amount of floor area, with its large glowing vacuum tubes and huge shielded condensers. Thick cables fed the output voltage to the top of an upright accelerator tube which was some ten feet tall and channeled the protons down to the target area near the floor. One day early in 1932, just a few weeks after Chadwick demonstrated the existence of the neutron, Cockroft and Walton inserted lithium at the bottom of the accelerating tube, turned on the beam, and achieved the disintegration of the target—with the machine operating at just 125,000 volts.[8]

In mid-1931 Lawrence's cyclotron had been producing enough of a proton beam to have beaten Cockroft and Walton, but Lawrence apparently had his eye more on reaching 1,000,000 volts than on disintegrating targets.[9] When he learned of Cockroft and Walton's results, Lawrence promptly fired off congratulations and soon achieved his own disintegration with the cyclotron. In the spring of 1932 Sproul turned over an old campus shack for Lawrence's exclusive use; Lawrence dubbed it the Radiation Laboratory and started raising money for a still bigger, more powerful accelerator.

In 1932 Cockroft and Walton's success with their accelerator signaled the end of the confining dependence upon radioactive sources for nuclear

[7] Cockroft was encouraged in his calculations by George Gamow's work on the quantum mechanical probability of the penetration of nuclear potential walls by charged particles. Cockroft interview, May 2, 1963, SHQP, pp. 11, 12; Cockroft, "The Probability of Artificial Disintegration by Protons" [1928], SHQP, Cockroft file.

[8] John D. Cockroft, "Experiments on the Interaction of High-Speed Nucleons with Atomic Nuclei," *Nobel Lectures in Physics* (Amsterdam, 1964), p. 168.

[9] It is interesting to note that Lawrence and his student David H. Sloan seem to have known about a paper of Cockroft and Walton's in which the British physicists discussed the relatively low voltages necessary for nuclear disintegration. See David H. Sloan and Ernest O. Lawrence, "The Production of Heavy High Speed Ions without the Use of High Voltages," *Physical Review*, 38 (Dec. 1931), 2021.

research. In the same year Lawrence's 1,000,000-volt cyclotron opened the door to bombarding nuclei with particle beams of energies and intensities that Rutherford could scarcely have dreamed of reaching.

In Pasadena, Robert Millikan had been pondering anew the Creator's own high-energy radiation, cosmic rays. Still sure that the rays consisted of photons, he also remained convinced that they were the birth cries of atoms forming in interstellar space. But Millikan had developed his atom-building hypothesis by inferring the energy of cosmic rays from the way they were absorbed in matter. To establish his hypothesis on a more solid experimental base, Millikan wanted to measure cosmic ray energies directly, and he thought that the task might be accomplished with a cloud chamber set in a magnetic field.

The cloud chamber was one of the most ingeniously simple and effective devices of modern physics. It was usually constructed in a flat, circular form, contained moist air, and had a glass cover on one side. On the other was a movable piston. Upon the partial withdrawal of the piston, the pressure in the chamber would drop and bring the moist air inside to the point of forming a cloud. If a charged particle now passed through the chamber, it would ionize the gas along its trajectory, water droplets would condense on the ions, and the trail of drops could be photographed through the glass. If the photograph was taken at the right time, a cosmic ray might be caught knocking an electron out of its atomic orbit. Under the influence of Millikan's proposed magnetic field, the path of the electron would bend and the energies involved could be computed from the curvature of the tracks. About 1930, mainly in the interest of confirming his atom-building hypothesis, Millikan asked Carl D. Anderson, who had just gotten his Ph.D. from Caltech, to stay on at the Institute as a postdoctoral fellow and develop this line of cosmic-ray research.[1]

Mrs. Anderson once modestly said of her son: "If he has special ability, I don't know where he got it. . . ."[2] Swedish immigrants from farm families, the elder Andersons had spent the early years of their marriage struggling in New York City, then moved to Los Angeles, where the father scraped out a living managing restaurants. Young Carl, financially too strapped for college out of town, lived at home while he did his undergraduate work at nearby Caltech. He studied hard, accumulated an outstanding academic record, and ultimately got through graduate school on a fellowship.

While in graduate school, after everyone else had dropped out of Oppenheimer's first course in quantum mechanics, Anderson heeded

---

[1] Unless otherwise noted, the material in this section about Anderson and his work is based on interviews with Carl Anderson.

[2] Quoted in William S. Barton, "From Test-Tube to Fame," *Los Angeles Times Sunday Magazine*, Dec. 27, 1936, p. 9.

Oppenheimer's plea to remain registered, even though experimental work was his real métier. In the laboratory, Anderson was equally courageous, and also deft, patient, and a tough-minded judge of evidence. A religious skeptic, he considered Millikan's atom-building hypothesis nothing more than wishful thinking. But when he finished his Ph.D. in 1930, he could not hope to get a position anywhere without Millikan's support. Intrigued by the absorption of high-energy radiation in matter anyway, he agreed to build a cloud chamber and measure the energies of cosmic rays.

Anderson constructed a spacious cylindrical chamber—it was six inches in diameter and two inches thick—and set it in a magnetic field powerful enough to deflect high-energy particles. Through long, often late-night hours of observation, Anderson would create a cloud in the chamber, quickly photograph it, and hope that he had caught a charged particle set in motion by a cosmic ray. By the fall of 1931 he had obtained and studied some thousand photographs. About a dozen of them revealed two, and in one case three, particles careening away from the same point. Since some of them were positively charged, they could not be electrons knocked out of their atomic orbits. To Anderson it seemed reasonable to interpret the double and triple tracks of positively charged particles as those of protons forced out of the nucleus upon cosmic ray impacts. Eager to turn confirmation into proof, Anderson set out to demonstrate conclusively that the positively charged particles were protons.[3]

Anderson took numerous additional photographs, detected many particle tracks—and was thoroughly puzzled. On the one hand, a sizable fraction of the particles seemed to have the same positive charge as the proton. On the other, they seemed to have no more mass than the much lighter electron. Of course, as Anderson realized, his experimental arrangement made for ambiguity in determining whether the particles were positively or negatively charged. The reason was that the curved track of a positive particle moving downward across the magnetic field looked like that of a negative particle moving upward. Since the only charged particles known to physics were the positive proton and the negative electron, Anderson supposed that the odd tracks must have been made by electrons moving upward.

Millikan scoffed. Everyone knows that cosmic rays come from above, he would tell Anderson. The particles must be protons moving downward.[4]

Upward or downward—Anderson decided to settle the argument by placing a thin lead plate horizontally across the middle of the cloud

[3] Anderson to R. A. Millikan, Nov. 3 [1931], RAM, Box 22.

[4] Charles Weiner, interview with Carl Anderson, June 30, 1966, AIP, pp. 27–28; Anderson recalled that his discussions of the point with Millikan were "at times somewhat heated." Anderson, "Early Work on the Positron and Muon," *American Journal of Physics*, 29 (Dec. 1961), 825–26.

chamber. Since a charged particle would lose energy in the plate, its degree of curvature would be different on the far side of its trajectory. Anderson could then tell clearly which way it was moving with respect to the magnetic field and, hence, whether it possessed a positive or a negative charge.

On August 2, 1932, in the normal course of taking photographs, Anderson obtained an extraordinarily unambiguous picture of a *positively* charged particle curving upward. Since cosmic rays did of course move downward, Anderson decided that the particle had somehow been released when a cosmic ray struck the material beneath the cloud chamber. But its ionizing power was too weak to be a proton or an alpha particle. In fact it seemed to have a mass comparable to that of the electron. After taking further photographs and carefully checking his measurements, Anderson boldly concluded that he had found a new particle in nature, one with the mass of the electron and a positive charge. Even Millikan was convinced. In September 1932, not unmindful of the added distinction in store for Caltech, he had Anderson publish a quick preliminary report of the existence of the positive electron, or "positron," to use the contraction Anderson soon adopted.[5]

In later years, the advent of a new elementary particle would scarcely ruffle the intellectual sensibilities of the world's physicists; in 1932, Anderson's announcement of the positron ran into a wall of resistance. If the neutron had resolved many long-standing difficulties of nuclear theory, the positron seemed simply to complicate matters. It is said that Niels Bohr dismissed Anderson's finding out of hand, and when in the fall of 1932 Millikan discussed the positron in a lecture at the Cavendish, various members of the audience coldly suggested that Anderson had doubtless become tangled in some fundamental interpretive error.[6] But not all of Rutherford's physicists were prepared to ignore Anderson's claims, especially not the resident Cavendish expert on cloud chambers, Patrick M. S. Blackett.

Recently, Blackett and a young Italian research fellow, G. P. S. Occhialini, had developed cloud chamber equipment that was specifically designed for the efficient study of cosmic rays; it would go into operation when—and in principle only when—it actually contained a particle track to be photographed. A vast improvement over Anderson's hit-or-miss experimental arrangement, Blackett and Occhialini's apparatus made checking for the positron relatively easy. Within just a few months, by early 1933, they found themselves forced, in their published report, to

[5] Anderson, letter to the editor, *Science*, 76 (Sept. 9, 1932), 239; Anderson, "The Positive Electron," *Physical Review*, 43 (March 15, 1933), 491–94.
[6] Interview with P. M. S. Blackett, Dec. 17, 1962, SHQP, p. 4; interview with John D. Cockroft, May 2, 1963, SHQP, p. 14. In February 1933 Cockroft, then in Berkeley, told Millikan that the Cavendish people had now "capitulated" on the question of the positive electron. Cockroft to Millikan, Feb. [1933], RAM, Box 22.

"the same remarkable conclusion as Anderson" about the existence of the particle.[7] And, after a number of discussions with Paul Dirac, they also knew how to interpret the positron theoretically.

Back in 1931, in the course of pondering his relativistic theory of the electron, Dirac had published an intriguing analysis of its apparent chief difficulty, the negative energy states. Drawing upon the elucidations made by his critics, he argued that these states must be normally occupied and, hence, undetectable. But occasionally, he continued, some of these states could empty out, creating a "hole" in a negative energy sea. Such a "hole, if there were one," he predicted, "would be a new kind of particle, unknown to experimental physics, having the same mass and opposite charge to an electron. We may call such a particle an anti-electron."[8]

The positron, Blackett and Occhialini argued in their paper of early 1933, was precisely Dirac's anti-electron. It was somehow created in the process of collision when a cosmic ray struck a nucleus. Anderson's positron, as forthrightly interpreted by Blackett and Occhialini, dispelled whatever doubts that remained about the validity of Dirac's relativistic quantum mechanics.

Anderson's and Blackett and Occhialini's results have got us "very much stirred up," the Harvard theorist Edwin C. Kemble remarked in March 1933, and as the new year progressed the stirring up continued.[9] The Joliot-Curies found that positrons could be produced without cosmic rays by bombarding aluminum nuclei with alpha particles. Ernest Lawrence obtained some deuterium from Gilbert Lewis, put it in the cyclotron, and observed a remarkable variety of new nuclear reactions. Heisenberg and Eugene Wigner independently considered the neutron and the proton and attempted the first quantitative theories of the forces binding the two particles together. On both sides of the Atlantic theorists and experimentalists were weaving the strands of the 1932 achievements into the tapestry they were now calling the field of nuclear physics.

In October the Solvay Conference for 1933 convened in Brussels, its subject the structure and properties of the nucleus. Here was Cockroft describing the technicalities of accelerating charged particles; Chadwick elaborating upon the neutron; Dirac rigorously probing the theory of the positron; Heisenberg, bold as always, venturing a full-scale theory of nuclear structure. Even Niels Bohr, in a tone that mixed concession with congratulation, acknowledged that the existing relativistic quantum mechanics would do to explore phenomena in the nuclear region. As 1933 came to a close, some physicists were still working primarily with cosmic

[7] P. M. S. Blackett and G. P. S. Occhialini, "Some Photographs of the Tracks of Penetrating Radiation," *Proceedings of the Royal Society of London*, A, 139 (March 1933), 703.

[8] P. A. M. Dirac, "Quantised Singularities in the Electromagnetic Field," *ibid.*, A, 133 (Sept. 1, 1931), 61.

[9] Kemble to Garrett D. Birkhoff, March 27, 1933, SHQP, microfilm roll 53.

rays, others with the quantum theory of the electromagnetic field. But, whatever their special interests, all could agree with Eugene Wigner that the "most urgent question in physics" was the structure of the nucleus.[1]

Wigner might have added that what gave urgency to the question was the extent to which the advances of 1932 had made it seem so much more answerable. Chadwick, Cockroft and Walton, Urey, Lawrence, and Anderson—their achievements would, in themselves or through the lines of research they opened, earn five Nobel Prizes, three of them for work done in the United States, which made 1932 a miraculous year for both nuclear physics and American science.

[1] Institut International de Physique Solvay, *Structure et Propriétés des Noyaux Atomiques: Rapports et Discussions du Septième Conseil de Physique* ... (Paris, 1934), *passim*; Wigner is quoted in *Current Biography*, 1953, pp. 657–59.

XVI

Revolt Against Science

In the United States in 1932, Arthur Compton announced that he would explore nuclear phenomena by studying how high-energy cosmic rays interacted with matter. Most other administrators of American physics wanted to build cyclotrons. Whatever the approach, equipment for nuclear physics cost money—some of it, like cyclotrons, more than the legacy available from the affluent 1920s. But amid the depression, support for research was plummeting everywhere.

In Congress economizers slashed the budgets of all the federal scientific agencies an average of 12.5 percent. The Bureau of Standards, its estimates reduced in committee, then on the floor, where the senatorial leadership drove through cut after cut with hardly a dissent, emerged with an appropriation almost 26 percent below the 1931 level.[1] Around the country, state monies allocated for research fell sharply at such scientific centers as the universities of California, Wisconsin, and Michigan. In Illinois, where the legislature sliced the university budget considerably, the governor reduced it even further. In the private sector, at Stanford and MIT, campaigns for new capital funds collapsed; a sizable part of the Caltech and all the Cornell endowments for research were wiped out, and no private university escaped a decline in its investment income and severe loss in the book value of its portfolio. Washington University

[1] "Reductions in the Appropriations for Scientific Work Under the Federal Government," *Science*, 76 (July 29, 1932), 94; U.S. Congress, Senate, *Congressional Record*, 72d Cong., 1st Sess., 1932, LXXV, pt. 8, pp. 8488–89, 8496–97.

in St. Louis had so little money to equip its new physics building that the third floor of the structure was turned into a children's skating rink.[2] The annual income of the richly endowed Carnegie Institution of Washington fell by some $1,000,000. At Bell Laboratories, General Electric, and numerous other industrial research establishments, retrenchment was the order of the day.

Yet the difficulties of science went beyond the direct economic effects of the crash. The United States had of course suffered previous industrial depressions; none had ever struck after a decade in which science had been so widely hailed for making possible the miracle of modern society. Now, going beyond the earlier humanist critique, thoughtful Americans naturally asked whether science was not responsible, at least in part, for the end of the miracle, for the failure of machine civilization.

To most people the main link between the abstrusities of science and the wonders of normalcy had been technology, tangible machines that spewed forth goods or eased the burden of labor. Many Americans now wondered whether chemistry produced more than consumers could absorb, whether machines destroyed more jobs than they created. The prominent economic critic Stuart Chase told the Women's City Club of New York how the advent of "talkies" had thrown ten thousand movie-house musicians out of work. William Green, president of the AFL, explained how during the 1920s the castoffs of technology could find jobs in new industries and emphasized that there were no new industries to absorb them now. President Hoover's own authoritative Committee on Recent Social Trends, publishing its conclusions at the beginning of 1933, ominously warned: "Unless there is a speeding up of social invention or a slowing down of mechanical invention, grave maladjustments are certain to result."[3]

Exactly, cried Howard Scott, who, preaching his own peculiar brand of social invention, made *Technocracy* the new word of 1932. Disheveled beneath his broad-brimmed felt hat and baggy leather engineer's coat, Scott looked like the technological prophet he claimed to be. In Scott's view, endlessly increasing productivity would lead to permanent unemployment and debt, even to the collapse of the capitalist system— unless the system was drastically revised. The economy, Scott argued, had to be run by a central planning mechanism that would match consumption to output. To make the mechanism work, the value of goods would have to be measured not by price but by the quantity of energy

---

[2] William Knapper, "Arthur L. Hughes," *Washington University Magazine,* August 1962, pp. 18, 20, copy in AIP.

[3] *New York Times,* Nov. 10, 1929, p. 17; Green, "Labor Versus Machines: An Employment Puzzle," *ibid.,* June 1, 1930, III, 5, 8; the Committee was quoted *ibid.,* Jan. 2, 1933, p. 2.

required for their production. To keep the system independent of political machinations, it would have to be run by technicians. Under Technocracy, Scott claimed, Americans could have a standard of living ten times as high as in 1929 and on a mere sixteen-hour workweek.

By fall 1932 Scott had become the sought-after luncheon and dinner guest of industrialists and bankers, editors and future New Dealers; numerous Americans were ready to say with Manchester Boddy, the editor of the *Daily Illustrated News* in southern California: "Out of Technocracy, we believe, will come a movement as important as the great American revolution of 1776." But reporters were unable to verify any of Scott's important biographical details, including his claim to have earned a doctorate at the University of Berlin. Such high authorities as Assistant Secretary of Commerce Julius Klein and Alfred Sloan of General Motors denounced Technocracy for what President Virgil Jordan of the National Industrial Conference Board called capitalizing on "the fears, miseries and uncertainties due to the depression."[4] Liberal reformers had to admit that Scott's ideas scarcely stood the test of close, dispassionate scrutiny. His energy theory of value ignored the plight of agriculture, omitted the problems of service industries, and vastly oversimplified the problem of technological unemployment.

The New York Times jeered at Technocracy's "comic finality," but liberal reformers, anxious over the economic impact of technology, could still agree with Stuart Chase that the question was "not whether Scott has lived in Greenwich Village or Lung Tun Pen, but what his figures show." Whatever the figures actually showed, observers noted that one cause of unemployment cited frequently by Americans was the displacement of men by machines. Dean William E. Wickenden of the Case School of Applied Science caught the meaning in the brief popularity of the Technocratic movement: "John Doe isn't quite so cock-sure as he used to be that all this science is a good thing. This business of getting more bread with less sweat is all right in a way, but when it begins to destroy jobs, to produce more than folks can buy and to make your wife's relatives dependent on you for a living, it is getting a little too thick."[5]

Humanists, drawing fresh energy from the economic disaster of the depression, renewed their critique of science. Quick with I-told-you-so's, the modernist Protestant and Catholic press reemphasized how mistaken man had been to put so much faith in the methods of science. As the liberal minister John Haynes Holmes explained: "We can see, without much difficulty, what is the matter. Science is wonderful, but also terrible

---

[4] Boddy and Jordan are quoted in Harry Elsner, Jr., *The Technocrats: Prophets of Automation* (Syracuse, N.Y., 1967), pp. 37, 8–9.

[5] *New York Times*, Jan. 25, 1933, p. 16. Chase is quoted in Elsner, p. 15; Wickenden, "Science and Every-Day Philosophy," *Science*, 78 (Nov. 24, 1933), 468. See Dixon Wechter, *The Age of the Great Depression* (New York, 1948), pp. 291–92.

—and terrible for the reason that it has no values. . . . Science itself is neutral."[6] Religious writers called upon the nation to turn away from scientism and go back to God, or at least to Christian ethics. Orthodox or modern, secular or religious, various humanists revived the bishop of Ripon's proposal for a moratorium on scientific research; in the early years of the depression the scheme won even more attention from the press and the pulpit than it had in 1927. Arguing for the idea, G. K. Chesterton typically summarized the trouble: "There is nothing wrong with electricity; nothing is wrong except that modern man is not a god who holds the thunderbolts but a savage who is struck by lightning."[7]

Nothing was wrong with scientists either, except that the *New Republic* wanted to know why they should be accepted as "authority on all social questions." Scientists might apply the scientific method with masterful dispassion in their professional work; when it came to emotional issues like economics and war, the noted British popularizer J. W. N. Sullivan told readers of the *Atlantic*, they were "not conspicuous for their detachment and fair-mindedness." By the early 1930s journalists and preachers, not to mention Arthur Holly Compton, soon to become a general chairman of the Layman's Missionary Movement, his public religiosity rapidly approaching Millikan's, were celebrating anew the epistemological agnosticism of physics.[8] If in the preceding decade commentators had applauded the scientist because he professed to know so little, they now seemed to doubt him because, by his recent public acknowledgment, there was so little he could claim to know.

Academic humanists found their guide in the son of a theologian, the new educational leader Robert M. Hutchins, who had returned from the wartime ambulance service to win a Phi Beta Kappa key at Yale, become secretary of the University at twenty-four, dean of the Law School at twenty-eight, and, in 1929, at age thirty, president of the University of Chicago. If the nation was in "despair," he told the university convocation in 1933, it was because "the keys which were to open the gates of heaven have let us into a larger but more oppressive prison house. We think those keys were science and the free intelligence of man. They have failed us."[9] An enemy of narrow, reductionist specialization, his ideas resembling a blend of Thomas Aquinas and James McCosh, Hutchins decided that students ought to be taught a fundamental core of academic knowledge and moral values.

[6] Holmes, "Religion's New War with Science," *Christian Century*, 54 (Oct. 27, 1937), 1323.

[7] Chesterton, "A Plea That Science Now Halt," *New York Times Magazine*, Oct. 5, 1930, p. 2.

[8] "Science—and Other Values," *New Republic*, 68 (Sept. 16, 1931), 114; Sullivan, "Science and the Layman," *Atlantic*, 154 (Sept. 1934), 336. See Charles A. and Mary R. Beard, *America in Midpassage* (2 vols.; New York, 1939), II, 853–54.

[9] Quoted in "The Revolt Against Science," *Christian Century*, 51 (Jan. 24, 1934), 110.

Doubters wondered how Hutchins was going to persuade the Chicago faculty to agree on just what fundamental principles Chicago students ought to learn. But Hutchins possessed remarkable self-confidence— "Gentlemen," he once responded to a glowing introduction, "I have never heard praise . . . so lavishly, so extravagantly expressed—nor so richly deserved"—and at Yale he had managed to revolutionize the teaching of law, leavening the curriculum with courses in economics and psychology.[1] Now he brashly turned the Chicago curriculum upside down, abolishing grades and examinations, substituting "intellectual" mastery for course requirements, prohibiting specialization until the junior year.

Christian Century took enthusiastic note of Hutchins' pronouncements and summarily announced that the depression had fused the long-standing discontent of the humanists into what appeared—and what many scientists certainly took—to be a full-scale "revolt against science."[2]

Despite the revolt, Robert Millikan remained serene about a few things. In particular, his cosmic-ray theories told him that the universe was safe from the heat death predicted by the Second Law of Thermodynamics, that, in short, the Creator was "continually on his job."[3] In December 1930 Millikan aired his confidence from the presidential podium of the American Association for the Advancement of Science. William L. Laurence, the young science reporter from *The New York Times*, was in the audience. An enterprising and intellectual graduate of Harvard, Laurence had already written in the Sunday supplements how Millikan had done battle with the "dreadful Second Law of Thermodynamics." Now Laurence filed what was evidently an equally enticing story. Back in New York the night editors read of Millikan's theology and wired Laurence for more copy. The next morning Laurence's story was page one news, and Millikan's address had been published in full, taking up six columns, an amount of space usually reserved only for speeches by the President of the United States. By New Year's Day 1931 newspaper readers across the country could know that, to use Laurence's slightly inaccurate rendition of Millikan's reassuring phrase, the Creator was "still on the job."[4]

But serious doubt hung over the key postulate of Millikan's

[1] Quoted in *The New York Times*, May 12, 1929, sec. V, p. 9.

[2] *Christian Century*, 51 (Jan. 24, 1934), 110.

[3] Millikan, "Present Status of Theory and Experiment as to Atomic Disintegration and Atomic Synthesis," *Annual Report of the Smithsonian Institution, 1931*, p. 284.

[4] *New York Times*, Sept. 28, 1930, sec. IX, p. 4; Dec. 30, 1930, p. 1; Chicago, *Daily Tribune*, Dec. 30, 1930, p. 6; New York *Herald-Tribune*, Dec. 30, 1930, p. 4. Laurence recalled that the night editors at the *Times* were Catholics. William L. Laurence interview, COH, II, 192–97.

cosmology, that cosmic rays consisted entirely of photons instead of charged particles. If they were charged particles, the magnetic field of the earth would distribute them unevenly around the globe and the intensity of cosmic rays would vary with latitude. Millikan had never found such a latitude effect; hence his inference that cosmic rays were photons. But other physicists had begun noticing a latitude effect, and in 1931, determined to settle the question, Arthur Holly Compton launched a worldwide survey to test conclusively whether the intensity of the radiation varied around the earth.[5]

Compton divided the globe into nine regions and assigned work to some sixty collaborators in half a dozen countries. In 1932 Compton himself traveled about fifty thousand miles in search of the latitude effect, visiting five continents, crossing the equator five times, ranging from as far south as Dunedin, New Zealand, to as far north as the Arctic, exploring down at sea level and high in the mountains. The sheer scope of the effort compelled public interest. Then in September 1932, fresh from the Arctic, Compton announced that cosmic rays were definitely charged particles and that Dr. Millikan was wrong.[6]

Withholding public comment, Millikan went off early in the fall of 1932 on his own expedition to the Arctic Circle. He also sent Henry Victor Neher, a young Caltech Ph.D., on a voyage to South America to look for a variation of intensity in that geographical direction. Late in November, Neher cabled from Mollendo, Peru, that he had found "no change" with latitude. Neither had Millikan himself, who said so to reporters. By now the press was not so much wondering which of the two scientists was right as gloating that one of the two Nobel Prize winners was wrong. *The New York Times* permitted itself the wisecrack: "It is decidedly a comedown, after being invited to contemplate the wondrous glory . . . of the new physics, to be reminded by the creator himself that of course we cannot say with confidence that we really know what we are talking about."[7]

When Millikan and Compton were scheduled to participate in a symposium on cosmic rays at the Christmas 1932 meetings of the American Association for the Advancement of Science in Atlantic City, a knowledgeable young cosmic-ray physicist predicted that it would be "hot . . . at the Compton-Millikan debate, for Millikan has a chip on his shoulder and Compton is . . . [ready] to knock it off." Millikan went at Compton with a vehemence that reminded William L. Laurence of the days when "learned men clashed over the number of angels that could

[5] Compton to John C. Merriam, Dec. 22, 1931, draft in Compton Notebooks, vol. 9, pp. 10–13, AHC.

[6] *New York Times*, Sept. 15, 1932, p. 23.

[7] Millikan to Compton, Nov. 30, 1932, RAM, Box 21; *New York Times*, Dec. 3, 1932, p. 2; Nov. 16, 1932, p. 16.

dance on the point of a needle." Compton, who disliked the "exaggerated emphasis" that the newspapers were giving the dispute, presented a dispassionate analysis of why the bulk of cosmic rays must be charged particles. Bristling at the press for emphasizing the controversial, Millikan fueled the debate even further by adducing all the evidence at his command that cosmic rays were photons. At the end of the session Laurence tried to get the two Nobel laureates to shake hands. Millikan testily refused. Relations between the two physicists, Laurence reported, were marked by "obvious coolness."[8]

Millikan may have been irritated because at Atlantic City he received an unsettling report from Victor Neher. Neher had not detected a latitude effect on his way south because his electroscope had failed to work properly while passing through the area where he would have noticed a dip in intensity. The electroscope had not failed on the way back north. Arriving in Atlantic City, Neher bumped into Richard C. Tolman, professor of physical chemistry at Caltech. "Did you find a latitude effect?" Tolman asked in his soft manner.

"Yes," Neher replied.

"Tch, tch, tch," Tolman chirped, shaking his head.[9]

By February 1933 Millikan had admitted the latitude effect; he soon said no more about the birth cries of atoms and left it up to the clergy to decide whether the Creator was still on the job. When Compton proposed to air a fresh version of the dispute in 1936, Millikan persuaded him against it, not the least because, Millikan warned, "the public . . . will simply look upon the whole performance as a dog-fight between two Nobel prize men, and this doesn't help anybody."[1]

When the dispute of 1932 reached its height, some journalists pronounced it distressing. For decades, the comments went, laymen had been accustomed to think of the scientist as hardheaded, factual, cautious, accurate, and, as such, worth listening to even when he advanced notions about nature and the world contrary to common sense. Now here were two Nobel Prize winners displaying all the passion and fallibility of ordinary men. The Millikan-Compton spectacle may have added another human dimension to the knowledgeable public's idea of scientists, but coincidentally it also reinforced the revolt against science. Paul R. Heyl, a thoughtful staff member at the Bureau of Standards, summarized the reason: The unsettled condition of modern physical theory had "sufficiently penetrated the nonscientific world to produce a mingled

[8] T. H. Johnson to Howard McClenehan, Dec. 13, 1932, WFGS; *New York Times*, Dec. 31, 1932, p. 1; Jan. 1, 1933, p. 16; Compton to Millikan, Dec. 5, 1932, RAM, Box 21; Laurence interview, COH, vol. II, 192–97.

[9] Henry V. Neher, colloquium on cosmic rays, California Institute of Technology, May 1970.

[1] *New York Times*, Feb. 5, 1933, p. 1; Millikan to Compton, Dec. 4, 1936, Box 21.

state of wonder and bewilderment, suggestive of those earlier days when men began to doubt the authority and infallibility of the Church."[2]

Neither doubts nor faith troubled Frederick J. Schlink, a mechanical engineer and former staff member of the Bureau of Standards, who had coauthored a widely read indictment of the food, drug, and cosmetic industries, cofounded Consumers' Research, Inc., and was emerging as one of the country's best-known consumer advocates. The public, Schlink knew, tended to believe that "Science and Research . . . were uncorruptible and unpurchasable, that they must in the nature of things always and everywhere, serve the common weal." The public, he added, was decidedly "mistaken."[3]

Reformers of the day, whether categorical Marxists, pragmatic liberals, or worried humanists, found the trouble in the economic relationships of science, especially its dependence upon private wealth and industry. Schlink himself charged, accurately, that the Bureau of Standards tested numerous products for governmental purchasers but refused to release this valuable data to the public. It even failed to warn consumers against items found to be faulty or injurious. Indeed, Schlink asserted, all the federal scientific agencies were "little more than handy consulting or guidance services to business enterprise."[4] Other critics attacked industrial corporations themselves for treating science as just another weapon in the war for profits. No major industry was exempt from suspicion, and no corporation was more suspect than the giant that controlled some 85 percent of the telephone service in the United States.

By the late 1920s the AT&T applied-research program had given the public the convenience of dial telephones. But the conversion to the new system, which the company accelerated in the early years of the depression, was said to be costing thousands of central station operators their jobs. Moreover, despite the gain in efficiency as well as the general decline in prices, AT&T neither reduced its rates nor failed to pay its regular $9 dividend. Telephone charges, Congressman Joseph Patrick Monaghan of Butte, Montana, declared on the floor of the House, were in many instances "excessive and extortionate." Equally questionable, AT&T was said to be writing off against the telephone business such Bell Laboratory costs as the development of movie sound equipment,

[2] "Cosmic Row," *The Nation*, 136 (Jan. 18, 1933), 54; Heyl, "Romance or Science?" *Annual Report of the Smithsonian Institution, 1933*, p. 284.

[3] Schlink, "Government Bureaus for Private Profit," *The Nation*, 133 (Nov. 11, 1931), 508. The widely read indictment was Arthur Kallett and F. J. Schlink, *100,000,000 Guinea Pigs: Dangers in Everyday Foods, Drugs, and Cosmetics* (New York, 1933).

[4] *Ibid.* See also D. W. McConnell, "The Bureau of Standards and the Ultimate Consumer," *Annals of the American Academy of Political and Social Science*, 173 (May 1934), 150.

which had nothing to do with telephones. Early in 1935 the Congress, led by Chairman Sam Rayburn of the Committee on Interstate and Foreign Commerce, authorized the new Federal Communications Commission to launch a full-scale investigation of the telephone company, including its overall policies toward research and development.[5]

According to the Commission's report, AT&T had patronized research in part to protect its telephone operations against intrusion from any technologically innovative competitor; for this reason it had aimed especially to gain control of key patents in radio. The company had also charged telephone subscribers for the cost of developing nontelephonic equipment, and it did not pay them any return on their forced investment, in the form of reduced rates, out of the profits on the new items. The commissioners applauded the Bell System for developing so many marvels of modern communications. But whatever AT&T's commitment to better service, Noobar R. Danielian, an economist and FCC staff member, flatly concluded in a widely praised book: "The most significant consideration in research, acquisition of patents, and pooling arrangements, has been the attainment and preservation of monopoly control in the principal fields of operation."[6]

Even before Congress voted the investigation, John B. Matthews, a vice-president of Consumers' Research, director of the League for Industrial Democracy, and a devout Marxist, declared AT&T guilty of having bought the loyalty of Robert A. Millikan with a $3,000,000 gift to Caltech. In point of fact, AT&T had given no such money to Caltech, and though Millikan himself had been on a $500 annual consultant's retainer since 1913, he had voluntarily terminated the arrangement in 1931.[7] The liberal press generally dismissed Matthews' indictment for the shrill piece of demagogy it was. All the same, the Federal Trade Commission itself had independently concluded that public utility corporations, by subsidizing university departments of engineering with money for fellowships and research, aimed for more than a technological corporate benefit. If the public utilities were to win college graduates over to their economic point of view, Mervin H. Aylesworth, the director of

[5] U.S. Congress, House, *Congressional Record*, 74th Cong., 1st Sess., LXXIX, pt. 3, March 4, 1935, p. 2912; *ibid.*, Senate, Feb. 12, 1935, pp. 1841–42. See also *New York Times*, Oct. 18, 1935, p. 4, and the previous testimony in defense of the company by its president, Walter S. Gifford, U.S. Congress, House, Committee on Interstate and Foreign Commerce, *Hearings, Federal Communications Commission*, 73d Cong., 2d Sess., April 10, 1934, pp. 173, 185.

[6] Federal Communications Commission, *Investigation of the Telephone Industry in the United States, Report*, House Doc. No. 340, 76th Cong., 1st Sess., June 14, 1939, pp. 191, 209, 582, 583. N. R. Danielian, *A.T.&T.: The Story of Industrial Conquest* (New York, 1939), p. 119.

[7] John B. Matthews and R. E. Shalcross, *Partners in Plunder* (New York: Covici Friede, 1935), pp. 347–48; Millikan to Frank Jewett, Aug. 2, 1929; Jewett to Millikan, March 26, 1935, and attached documents, especially the introduction by Senator Hinman, RAM, Box 38.

the National Electric Light Association, had told a utilities group, they had first to win over the college professor. "Once in a while it will pay you to take such men, getting five or six hundred or a thousand dollars a year, and give him a retainer of one or two hundred dollars per year for the privilege of letting you study and consult with him."[8]

Whatever industry's seemingly venal designs upon individual academics, reform-minded critics worried still more about what the Marxist Benjamin Stolberg, a frequent contributor to *The Nation*, called capitalism's general "degradation of learning." Liberals might not agree with Stolberg that every arm of American culture was controlled by an "impersonal, institutionalized, monopolistic plutocracy, cheap, vulgar, and empty"; like dissidents at the turn of the century, they did fear for the intellectual independence of the universities, especially private universities dependent upon wealthy donors.[9] More important, some recognized that the issue had taken on a new and more subtle dimension since the days when Mrs. Leland Stanford could dictatorially and successfully demand the ouster of the reformer Edward ·A. Ross. Now that the philanthropic foundations had assumed a major financial role in American higher education, just a few fabulously wealthy institutions could shape the objectives of a sizable fraction of American scholarship.

A statistical assessment of that point came from Eduard C. Lindemann, a contributing editor of the *New Republic*, staff member of the New York School of Social Work, and the director of the WPA Department of Community Organization for Leisure. Lindemann had studied a hundred foundations—another two hundred refused to cooperate—distilled his findings into a brief, densely written compendium, *Wealth and Culture*, and interpreted his results at the invitation of the *New Republic* editors. "While some critics believe . . . American foundations represent a gigantic conspiracy on the side of reaction," he had found "no evidence" to that effect. But Lindemann did stress that the legal control of the major foundations rested almost entirely with financially successful citizens of six eastern commercial states. "These accumulations of vested wealth," Lindemann went so far as to say, "cannot exert anything save a conservative influence in relation to American life."[1]

Charles Beard, in a review of Lindemann's book, extended the argument. The advance of science and technology, he noted, explaining an idea then widely current among reform-minded thinkers, produced a "cultural lag," a mismatch between the reality of material circumstances and the web of prevailing social practices and beliefs. The question was

[8] U.S. Congress, Senate, *Summary Report of the Federal Trade Commission . . . on Efforts by Associations and Agencies of Electric and Gas Utilities to Influence Public Opinion*, Sen. Doc. No. 92, part 71 A, 70th Cong., 1st Sess., 1934, p. 149.

[9] Stolberg, "Ballyhoo and the Higher Learning," *The Nation*, 132 (Feb. 18, 1931), 177.

[1] Lindemann, "Foundations," *The New Republic*, 89 (Dec. 16, 1936), 214. See also Lindemann, *Wealth and Culture* (New York, 1936), p. 58.

whether the philanthropic giants would—or could—help America catch up with herself. Beard, the veteran reformer, doubted it: "If a foundation throws its weight on the side of public or collective interests . . ., its trustees will be brought under the fire of [the] private interests adversely affected." Beard, the sensitive humanist, was no less dubious: The crisis of the day required "the assertion of moral values," and the foundations were inclined to support only those noncontroversial subjects that could be treated with "scientific assurance."[2]

Frederick P. Keppel, the head of the Carnegie Corporation of New York, attested that foundations had improperly forced the techniques of the natural sciences into the social sciences and the humanities; he also declared that the foundations had "over-stimulated certain fields" and "spoiled certain individuals."[3] Yet if the foundations had stimulated such natural sciences as physics, the physicists on their own part had been influential and prestigious supplicants for support; they had helped shape the programs of the foundations. And whatever Millikan's conservatism, his eagerness to fund the exploration of the atom had derived in no small measure from pure professional considerations: The field was intellectually exciting and the high-status object of interest among physicists around the world. The reformist critics failed to recognize that, at least in science, if the choice of academic research subjects hinged at all upon the propensity of philanthropic foundations to fund noncontroversial studies, it also depended upon criteria internal to the various professional disciplines. But while the reformers may have misunderstood the influence of such considerations in setting the balance of academic endeavor, the misunderstanding made their basic indictment no less telling. Even if the logic of the scientific discipline itself determined the professor's choice of research topic, the topic still did not necessarily serve any immediate social purpose.

In the early 1930s reformers of different stripes were uniting in a common demand—that the pursuit of science itself had to be reshaped to fit some moral definition of the nation's pressing social and economic needs. Marxists, sure that science had been tied to the apron strings of industry, insisted that it be used primarily in the interest of socialist reconstruction. Liberals, especially liberal humanists, emphasized simply that it should be governed, as Jacques Barzun said, "by choices arrived at on a humanistic basis."[4] Whatever the ideological persuasion, reformers

[2] Beard, "What Price Philanthropy?" *The Saturday Review*, 13 (March 14, 1936), 20–21.

[3] Keppel, "American Philanthropy and the Advancement of Learning," *School and Society*, 40 (Sept. 29, 1934), 408–9.

[4] Barzun, "Scientific Humanism," *The Nation*, 148 (April 29, 1939), 503. The general Marxist position was typically advanced in a collection of papers by Soviet scholars, *Science at the Crossroads* (London: Kniga, 1931), and formulated as a policy statement in the Soviet Union by Nikolai Bukharin in 1931. *New York Times*, April 9, 1931, p. 9. See also, "The Class War Enters Science," *The Nation*, 132 (April 22, 1931), 440; "Science—and Other Values," *New Republic*, 68 (Sept. 16, 1931), 114.

agreed that science had to be freed from the demands of profit-making and from the whims of private wealth. Some were prone to ask with Grosvenor Atterbury, the well-known architect, town planner, and pioneer in prefabricated housing: What in science should really interest us most today? Not the verification of Einstein's theory of relativity, Atterbury declaimed. In the present state of affairs, all that seemed "brilliantly useless, especially when you consider the millions who cannot afford decent homes because none of our great minds has ever been focused on the basic everyday problem of human shelter. . . . With a small amount of such brains as are now focused on the speed with which the neutron penetrates the nucleus of the atom . . . the cost of the poor man's housing today could be cut in half."[5]

The need for socially purposeful science often occupied the thoughts of Warren Weaver, who in the early 1930s became director of natural sciences at the Rockefeller Foundation. Weaver had grown up in Madison, Wisconsin, attending church on Sundays and sometimes playing with the La Follette boys during the week. His father, a prosperous but stormy and restless druggist, often traveled; his stepmother was a schoolteacher and a warmly sensitive comfort. At the University of Wisconsin, Weaver developed an engaging social presence, prompted in part by his courtship of Mary Hemenway of Junction City, Kansas. In 1920, married, a veteran of the wartime military and of a few years at the nascent California Institute of Technology, Weaver settled back in Madison. Finishing his Ph.D. in mathematics, he became a popular teacher, chairman of the department, and the coauthor of an authoritative textbook on electromagnetic theory with Max Mason. It was Mason, recently appointed president of the Rockefeller Foundation, who brought Weaver to his new post in New York, where the first order of business was to help plan the dispensation of the Rockefeller philanthropic millions over the next decade.[6]

Because of the collapse of the stock market, retrenchment was the order of the day at the Rockefeller headquarters, but in choosing the kind of retrenchment to make in the natural sciences program, Weaver was guided by a group of convictions. The sciences in general had come quite far in the United States, he was sure, far enough no longer to need major capital investment, either in facilities or fellowships, from the Rockefeller Foundation. The Foundation ought now to concentrate its resources on a few selected fields, as opposed to ordinary disciplines, of scientific interest. The choice of these fields, Weaver's argument con-

[5] Quoted in John Ely Burchard, "The Building Industry," in Karl T. Compton et al., *Physics in Industry* (New York: American Institute of Physics, 1937), pp. 145–46.

[6] Warren Weaver, "Reminiscences," COH, pp. 3–7, 12, 19–22, 36, 41, 48–49, 54, 57, 162, 242–43.

tinued, should be governed by two chief criteria. The first was ripeness for significant intellectual development. The second for Weaver, with his human sensitivity, moral convictions of a religious modernist, and loyalty to the progressivism of his friends the La Follettes, was the likelihood that the field would contribute to the "welfare of mankind."[7]

"The welfare of mankind," Weaver told his fellow Rockefeller officials, "depends in a vital way on man's understanding of himself and his physical environment. Science has made magnificent progress in the analysis and control of inanimate forces, but science has not made equal advances in the more delicate, more difficult, and more important problem of the analysis and control of animate forces." Weaver believed that certain fields of biology, especially those likely to exploit physics and chemistry, were ripe for major advance. Combining his sense of scientific opportunity with his reformist convictions, Weaver recommended that the Rockefeller philanthropies give major financial support to selected fields which dealt directly with man. "The past fifty or one hundred years," he summarized his case in 1933, "have seen a marvelous development of physics and chemistry, but hope for the future of mankind depends in a basic way on the development in the next fifty years of a new biology and a new psychology."[8]

Mason, who was notably interested in the physiological basis of mental illness, was all for Weaver's proposal. So were the Rockefeller trustees, especially those who had never been happy with Wickliffe Rose's emphasis on the sciences. Besides, during the depression social considerations that complemented Weaver's colored the outlook of many trustees, particularly Raymond B. Fosdick, Harry Emerson Fosdick's younger brother, a bookish lawyer, Princeton classmate of Norman Thomas, veteran of settlement house work with Lillian Wald, and devoted Wilsonian.

In 1919, while in London as Under Secretary General of the League of Nations, Fosdick got to know Frederick Soddy, a distinguished British physical chemist who would soon win the Nobel Prize and was already an outspoken social critic. Though his arguments were somewhat muddled, Soddy's essential points were worrisomely clear: Material want made war, and technology made war terrible. If man did not soon learn how to direct science to the creation of an economically more equitable world, science, especially science that commanded the power in the atom, would soon destroy mankind. Indelibly troubled by Soddy's forecast, Fosdick kept pondering the social ramifications of science while practic-

---

[7] *Ibid.*, pp. 280–82; Weaver to Lauder Jones, Jan. 19, 1933; Jan. 26, 1933; Weaver, "The Benefits from Science, Science and Foundation Program, The Proposed Program," Jan. 27, 1933, *passim*; RF, Series 915, Box 1. "The Rockefeller Foundation: Agenda for Special Meeting, April 11, 1933," pp. 76–77, RF, Series 900, Box 22.

[8] "Agenda for Special Meeting," pp. 76–77; *Annual Report of the Rockefeller Foundation, 1933*, pp. 199–200.

ing law in New York and listening with sympathy but some skepticism to the optimistic evangelism of Wickliffe Rose. In 1929 Fosdick commented in a book of gracefully written essays, *The Old Savage in the New Civilization*: "This divergence between the natural sciences and the social sciences, between machinery and control, between the kingdom of this world and the kingdom of the spirit—this is where the hazard lies. Science has exposed the paleolithic savage, masquerading in modern dress, to a sudden shift of environment which threatens to unbalance his brain."[9]

Now Fosdick and his fellow trustees endorsed the key elements of Weaver's program, and Mason officially announced that the principal support of the Foundation would go to special areas of the biological and, reflecting Fosdick's influence, the social sciences. Physics and chemistry would be eligible for support only insofar as they related to biological problems. In 1936 Fosdick, the new president of both the Foundation and the General Education Board, summarized the rationale of the departure. "Uneasiness and even alarm are growing as the belief gains ground that the contributions of the physical sciences have outstripped man's capacity to absorb them. . . . There can be but little question," he added, as though echoing Charles Beard, "that a serious lag has developed between our rapid scientific advance and our stationary ethical development."[1]

The shift in Rockefeller policy substantially enlarged the financial burdens of physical science in the private universities. The Foundation quickly reduced the money it had been making available for the National Research Council fellowships and stipulated that fully half the remaining awards had to go to the biological sciences. At the end of the 1920s the Foundation had pledged the NRC $100,000 a year for the support of academic scientific projects; in 1934 the commitment was terminated with a reduced final payment. Both the Foundation and the General Education Board, cutting their direct subventions to university science, cut those to the physical sciences the most. Not even Millikan could pry funds for cosmic ray research out of the Rockefeller philanthropies, while, at the same time, Thomas Hunt Morgan, the Caltech Nobel Prize biologist, had no trouble winning ample help for research in genetics.[2]

Philanthropic and private institutions aside, in 1933 William Wallace

[9] Interview with the late Raymond B. Fosdick; Fosdick, *Chronicle of a Generation* (New York, 1958), pp. 225–27; Fosdick, *The Old Savage in the New Civilization* (Garden City, N.Y., 1929), p. 37. See Frederick B. Soddy, *Science and Life* (London, 1920).

[1] *Annual Report of the Rockefeller Foundation, 1936*, p. 6.

[2] Max Mason to Isaiah Bowman, Dec. 31, 1934, RF, Series 200 E, Box 169; George Ellery Hale to Gano Dunn, Feb. 12, 1934, GEH, Box 14. The Rockefeller Foundation's support of genetic research proved to be crucially significant in the development of molecular biology. See Robert Olby, *The Path to the Double Helix* (London, 1974), pp. 440–43.

Campbell, now president of the National Academy of Sciences, lamented of the state universities: "The attitude of many, perhaps nearly all, of the legislatures toward research at public expense may fairly be described as unsympathetic and, in some cases . . ., as severely hostile." In Washington, D.C., earlier that year, Congressman William Stafford, a lame-duck Republican from Milwaukee, Wisconsin, who had learned a good deal about the Bureau of Standards during his two terms in the House, argued that it could stand a further reduction in its budget; Stafford saw no reason why the public ought to subsidize work for a few "favored industries." The Congress cut the Bureau's appropriation by about another 5 percent and, for good measure, established the requirement that industry would henceforth have to pay for industrial tests.[3]

After the inauguration of Franklin D. Roosevelt, it was rumored that the new President, who had acquired sweeping budgetary powers under the Economy Act of the previous year, intended to lop off even more funds slated for scientific research. A storm of protest from farm groups kept the reductions in the various agricultural agencies down to a minimum. But no dissent could save the Bureau of Standards from the fiscal ax of Budget Director Lewis Douglas. By the end of the summer the new administration had told the Bureau that it could spend no more than some 66 percent of its congressional appropriation. In August a Bureau physicist bitterly reported to an academic friend: "Research work here is badly crippled."[4]

So was the employment situation in science everywhere. By the summer of 1933 the Bureau of Standards had furloughed more than three hundred members—almost half—of its technical staff, and every federal scientist feared for his job. By the same year, General Electric had fired some 50 percent and AT&T almost 40 percent of their laboratory personnel.[5] In the universities salaries were cut, research and teaching assistants let go; if there were no wholesale firings of senior men, few faculty without tenure could look forward to a secure professional future. Professors had to scramble to get their students jobs. With the number of NRC fellowships reduced by 1933 to 165 and by 1936 to 49, it was next to impossible to gain one of the awards, especially for physicists. As early as 1931 Samuel Goudsmit reported that the spring meeting of the American Physical Society looked "much more like an employment

[3] Campbell, "The National Academy of Sciences," *Science*, 77 (June 9, 1933), 550; U.S. Congress, House, *Congressional Record*, 72d Cong., 2d Sess., LXXVI, pt. 3, Jan. 26, 1933, 2635–36; Lyman Briggs to Secretary of Commerce, Feb. 23, 1933, NBS, Entry 51, AG file.

[4] See the documents in FDR, OF 381, 1933/34; "News and Comment . . .," *Literary Digest*, 116 (Sept. 2, 1933), 10; *Science*, 77 (June 23, 1933), 600; William F. Meggers to George R. Harrison, Aug. 24, 1933, WFM.

[5] *Report of the Science Advisory Board, July 31, 1933 to Sept. 1, 1934* (Washington, D.C., 1934), p. 72; Danielian, *A. T. & T.*, p. 205; Kendall Birr, *Pioneering in Industrial Research: The Story of the General Electric Research Laboratory* (Washington, D.C., 1957), p. 120.

agency than a scientific gathering." By 1935, the job situation for fresh Ph.D.'s, Edwin C. Kemble, professor of theoretical physics at Harvard, told a colleague in Europe, resembled a "nightmare."[6]

The nightmare was not unlike that of the 1890s, when a coalition of budget-cutting conservatives and socially purposeful reformers had reduced appropriations for scientific luxuries. Now, a similar coalition had imposed a kind of moratorium on physical research in America—a moratorium all too manifest, Henry A. Barton, the director of the American Institute of Physics, remarked, for the simple reason that "the money ran out."[7]

[6] Goudsmit to John Wulff, May 4, 1931, SHQP, microfilm roll 65; Kemble to R. D. Present, Sept. 27, 1935, *ibid.*, microfilm roll 54.

[7] Barton, "Scientific Research in Need of Funds," *Literary Digest*, 119 (June 29, 1935), 18.

XVII

The New Deal and Research

The geographer Isaiah Bowman, a member of the National Academy and an entrepreneur in the world of learning, apprehensively watched the rising revolt against science, especially the questioning of the value of research in the offices of the major philanthropic foundations, and pondered how the scientific community might demonstrate its real "idealistic purpose."[1] In June 1933 Bowman found a hopeful answer in a proposal that his colleagues help the new Roosevelt administration through a Science Advisory Board.

Bowman's idea received enthusiastic backing from Secretary of Agriculture Henry Wallace, an accomplished plant geneticist who had long advocated increasing the productivity of the land through research and who had recently told an NBC radio network audience: "It is not the fault of science that we have unused piles of wheat on Nebraska farms and tragic breadlines in New York City." At Wallace's urging, on July 31, 1933, by executive order, President Roosevelt created the Science Advisory Board. It was to last for two years, was given no federal funds even for expenses, and was to act through the machinery and under the jurisdiction of the National Academy of Sciences and the National Research Council.[2] The President appointed nine members of the Academy to the Board, including Bowman, Frank B. Jewett, and Robert

---

[1] Bowman to Robert Millikan, May 9, 1932, RAM, Box 7.

[2] Wallace, "The Value of Scientific Research to Agriculture," *Science*, 77 (May 19, 1933), 479; copy of the Executive Order is in *Report of the Science Advisory Board, July 31, 1933 to September 1, 1934*, p. 7.

Millikan. As chairman of the Board he designated Millikan's counterpart on the east coast, the former head of the Princeton physics department and, since 1931, the president of MIT, forty-six-year-old Karl T. Compton.

Back in Wooster, Ohio, people knew Karl as the Compton boy who liked to dance. A good deal more amiable, easygoing, and likable than his younger brother, he displayed none of Arthur's religiosity or what some colleagues considered his obdurate self-righteousness. Still, Karl was no less the product of Elias Compton's stimulating parental regimen (for a while he was called "Psyche," because the father so often made him the guinea pig for experiments in child psychology). At the College of Wooster, Karl was a star scholar and captain of the football team; while taking his Ph.D. at Princeton, he was the first Compton to be singled out as the best student in the entire graduate school. A cut below his Nobel Prize sibling as a scientist, he nevertheless earned considerable professional standing for his research on the ionization of gases, including membership in the National Academy and the presidency of the American Physical Society. Karl was also a veteran of the World War I scientific mobilization. At Princeton he assiduously recruited faculty and funds, devoted time to the National Research Council, especially its fellowship board, and, through a consultantship at General Electric, linked the worlds of academic and industrial physics.

In 1931, anxious about the increasing financial difficulties of physical science, Compton helped found and became chairman of the governing board of the American Institute of Physics; though organized to reduce costs by centralizing the editorial operations of various physics journals, the Institute quickly took on the coloration of a trade association in defense of science and all its works. The author of a hard-hitting essay in a volume of rebuttals to Technocracy, Compton insisted that if civilization was destroyed, the blame would rest on "man's stupidity and 'cussedness' and not upon the machines which he has created." But Compton admitted that machines could cause severe economic dislocations, and, though a registered Republican, he thought that the nation was obligated to aid the unemployed through federal expenditures for public works and emergency relief. Compton was particularly enthusiastic about the Roosevelt Brain Trust, which he considered an attractive example of the reliance upon experts for the development of public policy.[3] Now the chairman of his own special kind of brain trust, Compton aimed to put the Science Advisory Board to work for both the New Deal and the beleaguered research community.

[3] Compton, "Technology's Answer to Technocracy," in J. George Frederick, ed., *For and Against Technocracy* (New York: Business Bourse, 1933), pp. 77–94; Compton, "Legislation for Reduction and Relief of Unemployment," in Charles F. Roos, ed., *Stabilization of Unemployment* (Bloomington, Ind.: Principia Press, 1933), pp. 242–43; Compton, "Science and Prosperity," *Science*, 80 (Nov. 2, 1934), 388.

The Board spawned a variety of committees, some to deal with the practice of research in such agencies as the Weather Bureau, the National Bureau of Standards, and the Bureau of Chemistry and Soils; others to consider such general issues as land use, patents, and mineral policy. Compton himself developed the case that the United States government ought to inaugurate a substantial program to put science, especially physical science, to work for the national welfare. Over half his Board's members were academics and, as such, eager to have the government come to the financial aid of science. In December 1934, with the Board's endorsement, Compton recommended to the President a tentative two-point federal program: first, $75,000,000 for research, mainly in the universities, to be spent over a trial period of five years; second, the presidential creation of a permanent advisory board of distinguished scientists to assist the Director of the Budget in determining the appropriations and activities of all governmental technical bureaus save those in the departments of War and Navy.[4]

The primary objective, Compton explained to FDR, was the "best possible advancement of science in America."[5]

Roosevelt sent Compton's program to Secretary of the Interior Harold Ickes for the scrutiny of his interdepartmental cabinet group, the National Resources Board, and it fell into the hands of the Board's working arm, a three-man advisory committee. The committee members included the economist Wesley C. Mitchell of Columbia University, a devoted student and ideological offspring of Thorstein Veblen and John Dewey, an authority on business cycles, and so commanding a figure in his profession that Herbert Hoover would later tell him in a get-well note: "I hear that you are laid up. This is not in the national interest"; and the equally prominent political scientist Charles E. Merriam of the University of Chicago, an expert in public administration, a candidate for mayor of Chicago in the progressive era—the young Ickes had managed his campaign—and a tireless reformer who had served on more boards and written more books than he cared to remember.[6] The chairman of the advisory committee was Frederick A. Delano, FDR's trusted old uncle—he had quietly helped arrange Roosevelt's transportation back to New York City when the polio struck at Campobello—a railroad executive and financier, a trustee of the Carnegie Institution of Washington, and one of the nation's prominent city and regional planners.

Mitchell and Merriam, who had been the chairman and vice-chairman

---

[4] Compton, "Science Still Holds Great Promise," *New York Times Magazine*, Dec. 16, 1934, pp. 6–7, 17; Science Advisory Board, "Minutes of Meeting, Dec. 9, 1934," RAM, Box 7; Science Advisory Board, "Transcript . . . of the Meeting . . .," Dec. 9, 1934," NRCM; Compton to the President, Dec. 15, 1934, and attachments, FDR, OF 330–A.

[5] Compton to the President, Dec. 15, 1934, FDR, OF 330–A.

[6] Hoover is quoted in Arthur F. Burns, ed., *Wesley Clair Mitchell: The Economic Scientist* (New York: National Bureau of Economic Research, 1952), pp. 3–4.

of Herbert Hoover's Committee on Recent Social Trends, believed that, to cope wisely with the complexities of a technological society, the government had to develop its policies on the basis of what they called "national planning." In contrast to Compton, who saw the advance of science as the drivewheel of the nation's economic development, they perceived the accumulation of knowledge as essential to develop plans for the nation's long-range future. So did Delano, who as the President's man on the committee, had to say that the request for $75,000,000 was somewhat "staggering in its size."[7] Whatever their special concerns, Delano, Merriam, and Mitchell all believed that the nation's difficulties were decidedly more social and economic than technological. Any program of research for planning, they advised Ickes, must include the social as well as the natural sciences and, for good measure, education.[8] When in January 1935 Ickes reported upon Compton's proposal to the President, he recommended adding to the National Resources Board a research and advisory group of scientists and engineers, social scientists, and educators to serve "a planning function,—planning for the full use of the research resources of the country."[9]

The next month the President had on his desk a letter to Ickes, approving the creation of the advisory group, applauding the development of research projects in aid of planning, and suggesting that these activities could be financed with some of the funds about to be appropriated for the new Works Progress Administration. The letter provided that the Resources Board would have to use at least 60 percent of the allotment from the work fund for "direct labor paid to persons taken from the relief rolls."[1]

Before sending the letter FDR evidently showed it to the quick, tough-minded man who would head the WPA, Harry Hopkins. Make the 60 percent into 90 percent, Hopkins was later said to have urged.

The President scratched out the initial figure, scrawled 90 percent in the margin, and mailed the letter to Ickes.[2]

The 90 percent requirement flatly blocked Compton's aim of funding the best possible advancement of science in America; most scientists, certainly most able scientists, were just not on the relief rolls. No less disappointing, the establishment of an advisory group for planning scarcely went far enough for Compton. In his judgment, the more the government intervened in the nation's urban, industrial, technological

[7] Delano to Compton, Jan. 17, 1935, NRPB, Entry 8, file 801.5.

[8] Ickes to Delano, Nov. 17, 1934, SIT, file 1–293; "Minutes of Advisory Committee Meeting, Jan. 21, 1935," NRPB, Entry 8, file 103.75; notes of the Century Club meeting, Jan. 21, 1935, by Delano and Charles Eliot, NRPB, Entry 8, file 103.1; Delano to Chairman [Ickes], Jan. 25, 1935, Entry 8, file 301.5.

[9] Ickes to the President, Jan. 31, 1935, SIT, file 1–293.

[1] Roosevelt to the Secretary of the Interior, Feb. 12, 1935, SIT, file 1–293.

[2] See the remarks of Charles Eliot, "Minutes, First Meeting of Science Committee, March 25, 1935," NRPB, Entry 8, file 103.1.

society, the greater its need for specifically programmatic, not just general planning, advice from outside scientific experts.

Compton blamed the failure of his program mainly on the President's failure to recognize its importance, on the unwillingness of the government's scientific bureaus to have their work "reviewed by competent people," and on a certain eagerness on the part of Delano, Mitchell, Merriam, and the White House staff "to put scientists in their place." But Compton knew that his program, especially its advisory aspects, had also come a cropper because of "timidity and jealousy" within the National Academy, especially on the part of President William Wallace Campbell.[3] Campbell, as the former head of the University of California, knew all about the political side of the administration of learning. To his mind, the Board, of which he was a member, should have been established by the apolitical Academy, not by the nation's political chief executive. Adding to Campbell's discontent, in 1934 the President enlarged the original Board to include Milton J. Rosenau, professor of epidemiology at Harvard, and Thomas Parran, whom then-Governor Franklin Roosevelt had appointed commissioner of health of New York State. While both men were highly distinguished physicians, neither was a member of the Academy, and Campbell counted the two appointments a clear-cut case of political interference in scientific affairs.[4]

Campbell complained repeatedly about such irregularities to administration officials, and Compton worried that this "nagging" of the President was undermining the Board. (Campbell's antics in fact cost the Academy standing in the Cabinet, especially with Henry Wallace and Harold Ickes.[5]) At one point, Compton was upset enough to pay Campbell a visit in Berkeley. Campbell adamantly defended the Academy's prerogatives to a nicety. Compton, as he himself recalled the gist of the encounter, "blew up." The Academy "could be damned and [Campbell] could be damned if they were going to insist on such trivialities and legalities in preference to grasping the opportunity to do a great public service and getting back of it wholeheartedly instead of throwing monkey-wrenches."[6]

[3] Compton to Isaiah Bowman, April 4, 1936, RAM, Box 7.

[4] Carroll W Pursell, Jr., "The Anatomy of a Failure: The Science Advisory Board, 1933–1935," *Proceedings of the American Philosophical Society*, 109 (Dec. 1965), 344–49; E. B. Wilson to Harvey Cushing, July 12, 1934, attached to M. McIntyre to Cushing, July 26, 1934, FDR, OF 330–A.

[5] Compton, memorandum of conversation with Campbell, June 21, 1934, attached to Compton to Millikan, June 26, 1934, RAM, Box 7; Rudolph Forster to the President, May 26, 1934, FDR OF 330–A; Rexford Tugwell to Marvin McIntyre, May 24, 1934, SAG, Science Advisory Board file; Harvey Cushing to Missy LeHand, March 14, 1935, FDR, PPF 2421.

[6] Compton, memorandum of conversation with Campbell, June 21, 1934, RAM, Box 7. Delano's science advisory group, while not foredoomed to failure, was also generally ineffective. The members decided at the outset that it was not their duty

But Compton's anger at Campbell masked his basic pro-Academy predilections. In his opinion, one of the major purposes to be served by an independent science advisory board was the "guardianship" of the government's scientific bureaus against exploitation by the party in power for "political ends." And any worthwhile scientific advisory group, Compton was certain, had to be staffed by "disinterested, non-partisan minds."[7] Whatever Compton's eagerness for a presidentially appointed advisory group, he had remained an Academy man, both in his loyalty to its posture of disinterestedness and in his penchant for its political elitism.

The political elitism of Compton and his colleagues was no less manifest in their approach to the federal funding of academic research. Compton's Board worried with Frank Jewett that federal subventions would inevitably mean "a large measure of bureaucratic control." The Board was willing to endorse such schemes—and Compton himself to advance them—only so long as the expenditure of the money was protected from "political" interference. But like his predecessors in their approach to the Smith-Howard bill in World War I, Compton insisted that the scheme had to be insulated not only from political interference with its administration but even from political determination of its programmatic purpose. If any considerations other than the scientific "value or merit" of the projects were to color the distribution of the funds, Compton frankly explained to FDR, then governmental support of research would be "definitely harmful."[8]

In his final report to Roosevelt, Compton recommended that the government establish a permanent science advisory agency as a successor to the Board and called for the allocation of $3,500,000 by those agencies of disinterested expertise, the National Academy and Research Council, in support of nationally important scientific research projects in nonprofit institutions.[9] The psychologist Edward L. Thorndike, who was a member of the Academy, explained in a meeting with Delano, Merriam, and Mitchell: You could trust the Academy to do research that would

---

"to determine policies," since that would risk embroilment in political controversy. Confining their activities to the collection and collation of information, they accomplished little more than the publication of a few thick reports, which few people read. "Minutes, Science Committee, April 9, 1938," NRPB, Entry 8, file 106.24; interview with Charles Wiltse.

[7] Compton to Bowman, April 4, 1936, RAM, Box 7; *Second Report of the Science Advisory Board*, pp. 18, 28.

[8] Jewett to Compton, Dec. 6, 1934; Science Advisory Board, "Minutes of Meeting, Dec. 9, 1934," RAM, Box 7; "Report to the President . . . upon a National Program . . .," *Second Report of the Science Advisory Board*, p. 39.

[9] Compton to the President, March 14, 1935, and attached "Report to the President . . . upon a National Program for Putting Science to Work for the National Welfare," printed in *Second Report of the Science Advisory Board, September 1, 1934 to August 31, 1935*, pp. 77, 79.

fit in with social planning. "Most of these people are sixty years or older. They have done science all their lives" and they would scarcely "waste a dollar." Instead of requiring that they decide beforehand what they would do with the money, Thorndike urged, simply ask them afterwards what they had done.[1]

Neither of Compton's recommendations gained the support of the White House. Harold Ickes opposed the appropriation of "a large 'free fund'" to any scientific group for "unspecified projects," and the Resources Board stipulated that all the Academy's research projects were to be "cleared" with Delano, Mitchell, and Merriam.[2] Yet Compton repeatedly refused to specify what he intended to spend the money on. You could not predict the future in science, he insisted, and it would be unwise to bind the funds to any specific projects far in advance. In his final report, Compton offered only a "suggestive" list of socially useful research items, and added that considerably "more important" than these utilitarian projects was the development of "fundamental knowledge."[3] In 1935 the Science Advisory Board went out of business on schedule, its accomplishments, beyond some changes in the programs of a few federal scientific agencies, merely a pile of reports.

Compton's program remained attractive to Lyman Briggs. Briggs was a Johns Hopkins Ph.D. in physics, a thirty-nine-year veteran of governmental science, and a practiced bureaucrat who had been nominated to the directorship of the Bureau of Standards by Herbert Hoover and appointed by Franklin Roosevelt. Briggs found ways to make the Bureau responsive to the demands of consumer advocates and pushed investigations into better housing. Eager to increase his budget and sensitive to the plight of the academics—he was an officer in the American Physical Society—Briggs suggested at the end of 1935 that the Congress should establish a well-funded federal program of research in the physical sciences. Franklin Roosevelt, declaring himself "wholly in sympathy" with the objective of such legislation, allowed that an appropriation was "justified on the general score of keeping the United States abreast of research conducted in other parts of the world."[4] In 1937 Briggs's proposal went to Congress as an administration measure.

Authorizing appropriations for research in the physical sciences,

[1] "Minutes of Meeting, National Resources Board Science Committee, March 25, 1935," NRPB, Entry 8, file 103.1.

[2] Ickes to the President, Jan. 31, 1935, SIT, file 1–293; "Work Relief Program of Planning Projects," April 1, 1935, NRPB, Entry 8, file 103.1. Compare this final draft, which was approved by the Resources Board, to the prior draft, attached to Delano to the President, April 1, 1935, copy in SIT, file 1–293.

[3] "Report . . . upon a National Program . . .," *Second Report of the Science Advisory Board*, p. 100.

[4] President Roosevelt to the Secretary of Commerce, Jan. 24, 1936, copy attached to Briggs to Secretary of Commerce, March 8, 1939, NBS, Entry 52, Blue Folder 788.

Briggs's bill provided for the Bureau of Standards to spend some of the money for work in its own laboratories and to allocate the rest, upon the advice of the National Academy, for the support of meritorious research projects in nonprofit institutions. This approach made Briggs's bill a best-science measure of a kind that raised a more difficult policy issue than its predecessor of World War I. The NRC measure of 1918 had recommended funding the best projects within each state, which made for an institutional but not a geographical conflict. Briggs's bill proposed to fund meritorious projects on a national rather than a state basis. Its grants would thus not only go to the relatively small number of well-equipped institutions with the best scientists; since these institutions were located mainly in the Northeast, the Midwest, and California, the allocations would also be concentrated geographically. Its regional implications manifest, Briggs's bill faced important competition from another measure for the support of research introduced by Representative Fritz G. Lanham of Texas. Lanham's bill proposed to encourage research nationwide by distributing the funds in equal amounts to each of the states for a publicly operated engineering station.[5]

Most scientists from the better universities of the Northeast and Midwest preferred Briggs's project approach to Lanham's geographical distributional scheme. Karl Compton, who had helped Briggs develop his bill, considered it a worthy legislative heir to the proposals of the Science Advisory Board. When hearings were held on the two bills in July 1937, Compton told the congressional subcommittee that supporting research on a "routine geographical basis" resembled "the firing of broadsides in a general direction," whereas funding meritorious projects was like "firing a sharpshooter directly at a target." But the sharpshooting approach was unacceptable to the southern and southwestern constituency of Richard C. Foster, the president of the University of Alabama and the secretary of the Association of State Universities. Summarizing the problem inherent in a best-science policy, Foster warned that measures like Briggs's would tend to favor the regions "where most of the country's research now is done . . . leaving the rest of the country to get along as best it can."[6]

United, the public college and university men might have pushed Lanham's measure through, but they were divided now as in World

<hr />

[5] Carroll W. Pursell, Jr., "A Preface to Government Support of Research and Development: Research Legislation and the National Bureau of Standards, 1935–1941," *Technology and Culture*, 9 (April 1968), 147, 152. Congressman Jennings Randolph also introduced a bill similar to the Briggs measure in its distributional scheme, which Briggs and other scientists opposed on administrative grounds. *Ibid.*, pp. 152–53; Briggs to F. K. Richtmyer, Nov. 2, 1937, NBS, Entry 52, Blue Folder 788.

[6] Compton to Jennings Randolph, July 16, 1937; Foster testimony, U.S. Congress, House, *Engineering, Scientific, and Business Research, Hearings before a Subcommittee of the Committee on Interstate and Foreign Commerce*, 75th Cong., 1st Sess., July 22–23, 1937, pp. 15, 28.

War I between the state-university and the land-grant forces.[7] At the close of the hearings, Congressman Edward C. Eicher of Iowa, chairman of the subcommittee and a proponent of the Briggs bill, had to say that, because of the clear-cut disagreements over the distribution of the funds, no legislation for research seemed possible in the current session. None proved possible in the next session either. If the land-grant colleges and the state universities could not agree on a measure that would distribute the funds on a geographical basis, they were politically powerful enough even in division to block any bill that would make the federal government the patron of a geographically concentrated best-science elite. In 1938 Director of the Budget David W. Bell, citing the opposition to the meritorious-project scheme, found the proposed legislation in disaccord with the program of the President, and Eicher, a Roosevelt loyalist, let Briggs's bill die in his subcommittee.[8] Denied FDR's support in the legislative crunch, the best-science elite simply did not command the political power to achieve its purpose.

In 1935, with the blessing of Secretary of the Treasury Henry Morgenthau, and in response to the recommendation of the medical committee of Compton's Science Advisory Board—its members were Parran, Rosenau, and Simon Flexner—Congress wrote into the Social Security Act an authorization for the National Institute of Health to spend $2,000,000 a year for research in chronic diseases and sanitation. Soon the administrators of the National Institute of Health were distributing some of their research funds in precisely the best-science fashion that Compton—and Briggs—had tried to achieve; when in 1937 Congress created the National Cancer Institute, it provided for the allocation of money to meritorious research projects no matter where they were located geographically.[9] Clearly Compton's difficulties ran deeper than his best-science elitism as such. At base the trouble, as Henry A. Wallace unintentionally suggested before the 1933 meeting of the American Association for the Advancement of Science, was the kind of best-science elitists the Comptons happened to be.

Wallace attacked the nation's scientists for their laissez-faire economic beliefs and estimated that at least half the research community still felt

[7] The Lanham bill, which had been drafted by Willis R. Woolrich, the dean of engineering at the University of Texas and the chairman of the deans of engineering of the Association of State Universities, promised to favor the state universities over their land-grant sisters. Albert R. Mann, speaking at the hearings for the Association of Land-Grant Colleges and Universities, emphatically declared that he would rather have no appropriations for research than see Congress authorize a "new policy in education"— meaning a departure from the Morrill Act—"full of danger and difficulty." Mann to Clarence F. Lea, *ibid.*, p. 86.

[8] *Ibid.*, p. 96; Eicher to Briggs, March 25, 1938, NBS, Entry 52, Blue Folder 788.

[9] Donald C. Swain, "The Rise of a Research Empire: NIH, 1930 to 1950," *Science*, 138 (Dec. 14, 1962), 1234.

that "the good old days will soon be back when an engineer or scientist can be an orthodox stand-patter." Of course, Compton's Board had scarcely manifested strict adherence to orthodox principles when it endorsed a program for federal aid to research, and with men like Thomas Parran, it was by no means uniformly anti-New Deal. But other members of the Board especially the group most interested in Compton's program, agreed with Robert A. Millikan, who on a national radio broadcast condemned federal meddling with private enterprise. Though Millikan's remarks neither diminished nor enlarged his standing as a physical scientist, Father John A. Ryan commented tartly, they showed that as a political scientist or economist he belonged "to the middle of the nineteenth century."[1] In the general sense that the New Deal meant liberalism, a sizable fraction of the Board was emphatically conservative.

As a humane pragmatist, Compton himself remained sympathetic to such measures as the legal protection of labor unions, social security, and even relief. But as a Republican and the president of MIT, which had close ties to big business, including major chemical, electrical, and utility companies, he grew increasingly discontented with the Roosevelt administration's imposition of what he considered excessive restraints on private enterprise. This particular reservation about the New Deal soon merged with his annoyance at FDR's failure to pay more attention to physical science. To Compton's mind, the New Deal could do more for the nation by investing in academic research rather than in so many public power projects—the one at Passamaquoddy was "uneconomically and politically" conceived—or by employing science to find new industrial uses for farm products instead of restricting production. With Millikan, Compton joined the governing board of the Farm Chemurgic Council. Sponsored by the Chemical Foundation, which was in turn controlled by the major chemical companies, the Council's program, in the enthusiastic assessment of the anti-Roosevelt San Francisco *Chronicle*, was "diametrically opposed to the program of the Tugwellian theorists."[2]

Whatever Compton's interest in an alternative to New Deal farm policy, his chief concern was a different approach to the nation's urban, industrial problems. Sure of the growth potential in private enterprise, he repeatedly asserted the argument of the 1920s: Support science because new knowledge would lead to new technology, to new industries, and ultimately, to new jobs. To Compton's mind, the administration ought to be aiming more at the enlargement of the economic pie and less at its

---

[1] Wallace, "Engineering-Scientific Approach to Civilization," *Science*, 79 (Jan. 5, 1934), 3; Millikan's address was reported in *The New York Times*, Aug. 7, 1934, pp. 1, 7; Ryan to the editor, Aug. 9, 1934, p. 16.
[2] Compton, "Investment for Public Welfare," *Science*, 83 (May 29, 1936), 507; the *Chronicle* is quoted in Carroll W. Pursell, Jr., "The Farm Chemurgic Council and the United States Department of Agriculture, 1935–1939," *Isis*, 60 (Fall 1969), 311.

redistribution. The New Deal, he once put the essential point of his antipathy, "has been played with the same old greasy cards. It has consisted of taking the aces and jokers from one group of players and handing them to another group. . . . But . . . through research, a new deck can be created with a lot more aces. . . . Research creates wealth, does not merely reshuffle it."[3]

The anti-New Dealism of Compton and his conservative colleagues was implied in the attitude of the Science Advisory Board toward Delano's insistence upon a joint program with the social sciences. The Board included no social scientists and had taken almost no serious steps to incorporate them into its general activities. Like Millikan, who at the 1932 meeting of the National Academy had flayed economics and sociology for having no commonly accepted first principles, the members of the Board, especially the physical scientists, tended to distrust the social sciences intellectually. Late in December 1934 Compton vigorously complained to Delano: Why hold back the natural sciences, when research in them was sure to yield dividends so much faster than research in economics or sociology?[4]

True enough, the social sciences were by no means as advanced as the physical sciences in either degree of rigor, freedom from normative judgments, or ability to predict behavior. But to Delano, Merriam, and Mitchell, indeed to many New Dealers, the social sciences were instruments to achieve what Henry Wallace considered the clear national need: "more science in distribution, and that means distribution of wealth as well as of physical products." The interest in the social sciences reflected the liberal concern for dealing with the impact of technology upon society. Such concern was generally absent among the more conservative members of Compton's Board, not least because the issue was identified with greater governmental intervention in society. If much of the Board disliked the social sciences, it was also because, with Millikan, they opposed "sociologists" and "political philosophers," some of them "in high places," who wanted to push man into the "soft bosom of the State."[5]

Going far beyond mere implied dissent from the New Deal, Compton repeatedly took his proscience case to the public, especially when, on Washington's Birthday 1934, he joined a group of scientists and business-

[3] Compton, "Science. Employment and Profits," *Vital Speeches*, 5 (Jan. 1, 1939), 183.

[4] For Millikan's Academy talk, see the Diary of Henry Crew, April 27, 1932, HC; *New York Times*, Feb. 23, 1934, p. 10; Delano to Compton, Jan. 17, 1935; Delano to E. B. Wilson, Aug. 26, 1937, NRPB, Entry 8, file 301.5, file 106.24.

[5] Wallace is quoted in Ray Tucker, "Noble Experimenting," *Collier's*, 94 (Nov. 10, 1934), 28. See also Wallace, "The Scientist in an Unscientific Society," *Scientific American*, 150 (June 1934), 285–87. Millikan is quoted in *The New York Times*, Aug. 7, 1934, p. 7.

men in an opulent symposium—it included an afternoon reception at the Museum of Science and Industry and an evening of dinner and speeches—sponsored by the American Institute of Physics in New York City. "Science struck back at its critics yesterday," *The New York Times* reported on the front page. Compton assailed the notion that over the long run science took away jobs, warned the critics against imposing a moratorium on research, and cautioned against governmental stifling of technical improvement, notably through certain NRA codes. Better that some of the federal funds being used for "artificial economic control," Compton insisted, be employed to "further the advance of science."[6]

Compton's proresearch arguments resounded in conservative circles. After a tour of industrial laboratories sponsored by the Research Council, Herbert Kohler, vice-president of the anti-union Kohler Company, told an audience of enthusiastic executives: "Science and research are . . . the first-line defense of a capitalistic dynamic economy as opposed to a static planned economy." High officers in the United States Chamber of Commerce, the American Cyanamid Company, and Du Pont, not to mention the 1936 Republican vice-presidential nominee, Frank Knox, were all members of the governing board of the Farm Chemurgic Council. The anti-New Deal *Saturday Evening Post* echoed Compton's reasoning, and his logic found its most prestigious spokesman in Mr. Conservative himself: The victims of technological unemployment, Herbert Hoover asserted, could be reabsorbed into the labor force only "by new industry producing new commodities and services. We can be sure of these new industries, if our pure science research is feeding its raw materials into the hoppers of applied science laboratories."[7]

But if Millikan's celebration of science and private enterprise had once commanded respectful attention, a contributor to the *New Republic* now wrote off the president of Caltech as merely subject to "pressure from big-business interests." Now *The New York Times*, normally friendly to science and by no means uncritically enthusiastic about the New Deal, commented on the Washington's Birthday symposium: "Neither the statistics nor the arguments are new. Nor did any of the protagonists of the laboratory explain why there is poverty amid plenty, and idleness where we expect to hear the hum of the machine." Now, when Governor George H. Earle of Pennsylvania cried: "We've got to face it; we've got to have a showdown with the machine," much of the liberal community preferred the approach summarized in the National

[6] *New York Times*, Feb. 23, 1934, p. 1. See also Compton, "Moratorium on Science Would Be Fatal to Recovery," *Electrical World*, 103 (March 3, 1934), 324.

[7] Kohler is quoted in *Time*, Nov. 4, 1935, p. 42; Pursell, "The Farm Chemurgic Council," pp. 309, 312; "True Business Picture," *Saturday Evening Post*, 209 (July 4, 1936), 22; Hoover, "Improving the Life of the Common Man," *Vital Speeches*, 5 (May 15, 1939), 454. Tom Girdler called technological unemployment an "utterly false diagnosis." *New York Times*, Oct. 12, 1938, p. 55.

Resources Committee's widely hailed report, *Technological Trends and National Policy*.[8]

Produced under the editorial leadership of the respected liberal sociologist William F. Ogburn, the document emphasized that, whatever its long-range benefits, in the short run the applications of science often provoked severe economic dislocations. To ease the maladjustments, Ogburn's group called for the government to anticipate and to head off the deleterious effects of developing technology. Without the intervention of the federal government, Franklin Roosevelt said in his second inaugural, America could not hope to "create those moral controls over the services of science which are necessary to make science a useful servant instead of a ruthless master of mankind."[9]

Adding the force of law to morality, the Bankhead-Jones Act of 1935 provided for basic research in agriculture *and* for investigations in conservation as well as in the distribution and marketing of farm products. Congressman Jennings Randolph of West Virginia, a liberal with a special devotion to labor, read the conclusions of *Technological Trends and National Policy* and wrote into a research bill the provision for an advisory commission to consider the social and economic effects of technological development. Instead of "complacent after-dinner speeches" before a group of "well-fed scientists and donors of funds for research," Henry Wallace said of the Washington's Birthday gala, we ought to see "our more articulate scientists insisting that the benefactions of science be used only in ways that are plainly in the general welfare."[1]

Compton agreed that Americans had reached the point in their national development where science had to be "consciously related to our social life and well-being." But while he had argued that research could deal with such technological challenges as sewage disposal or rural electrification, the way he elucidated his $3,500,000 request revealed his primary concern. Only a small fraction of the money, about 14 percent, was to be devoted to planning for socially purposeful projects. The rest, he told the President, was to assist "the most important scientific programs . . . now languishing" for lack of money. Compton was, after all, the product of a physics profession shaped increasingly along the lines of Rowland's program; its institutions, values, and status rankings lay with

[8] Benjamin Ginzburg, "Degradation of the Scientists," *New Republic*, 80 (Oct. 24, 1934), 311–12; *New York Times*, Feb. 24, 1934, p. 12; Jan. 1, 1938, p. 1. The National Resources Committee was, of course, a renamed variation of the National Resources Board.

[9] National Resources Committee, *Technological Trends and National Policy* (Washington, D.C., 1937). The report was summarized in *The New York Times*, July 18, 1937, pp. 1, 20; Roosevelt, press release of inaugural text, Jan. 20, 1937, NRPB, Entry 8, file 104.4.

[1] Randolph testimony, *Engineering, Scientific and Business Research Hearings*, pp. 9–10; Wallace, "The Scientist in an Unscientific Society," *Scientific American*, 150 (June 1934), 286.

research in pure, not socially purposeful, science. Watson Davis, the perceptive editor of Science Service, later commented that Compton's program was put forward by a group which had "a 'vested interest' in research done outside of the government"—a vested interest, Davis might have added, mainly in pure physical research done in the universities.[2]

Compton's emphasis on the physical sciences identified his program with the products—and problems—of industrial technology. The agricultural Bankhead-Jones Act was made possible to a significant extent by the pledge of the aging James P. Buchanan of Texas, chairman of the powerful House Appropriations Committee: "Tell [Henry] Wallace that I will give him all the money he wants for fundamental research. . . . I would like to leave some little monument to my years in Congress, and this shall be it." The political climate favored the construction of memorials through research programs in agriculture or medicine, which, it was believed, would help "the people" directly. It discouraged the building of monuments out of programs in the sciences assumed mainly to help industry, especially big industry. Budget Director Bell had opposed the Briggs bill particularly on grounds that industry could well afford to finance research for industrial development without the help of the government. In Bell's opinion, the federal money requested in the bill was "more urgently" needed for more tangible social purposes.[3]

Sensitive to the misery on the farm and the plight of the unemployed in the cities, liberals had good reason to consider Compton's principal argument for science—its usefulness, even indispensability, for economic recovery—just beside the pressing point. For some years the United States Department of Agriculture had been trying to find new industrial uses for farm products; the work had been largely unsuccessful, and further efforts were unlikely to yield dividends quickly enough in the future to save farmers from foreclosure in the present. Anticipating the views of latter-day economists, even Frank Jewett knew it was just "erroneous" to assume that industrial development neatly followed the advance of science. "New industries are simply not created that way," certainly not, Jewett added, in sufficient time to relieve current unemployment. Science might ultimately create new jobs. Ultimately, a *New Republic* writer

[2] *Report of the Science Advisory Board . . . 1933 to . . . 1934*, p. 11; *Second Report of the Science Advisory Board*, p. 39; Davis to Stuart A. Rice, Aug. 3, 1937, NRPB, Entry 8, file 106.24.

[3] Buchanan is quoted in Carroll W. Pursell, Jr., "The Administration of Science in the Department of Agriculture, 1933–1940," *Agricultural History*, 42 (July 1968), 235–36; Bell to Lyle T. Alverson, BOB, series 21.1, Commerce Department, National Bureau of Standards file. Wallace complained to Compton that the Science Advisory Board was "very heavily loaded on the side of physics, chemistry and the inorganic sciences." To Wallace, it seemed "more important at this stage of our civilization to emphasize the life side of the sciences." Wallace to Karl T. Compton, June 11, 1934, SAG, Science Advisory Board file.

remarked of Compton's program, would "not be too soon for the 20,-000,000-odd inhabitants of the country now on relief."[4]

Compton may have aspired to produce a disinterested, apolitical policy, but his program was irreducibly political—and politically conservative. If he failed to win a New Deal for the physical sciences, it was because key New Dealers recognized his arguments for what they essentially were. Did Compton attack relief as a mere palliative measure compared to research? Did he claim that the New Deal should invest in the physical sciences so as to enlarge the economic pie instead of paying so much attention to its redistribution? He was, Morris L. Cooke, the director of the Rural Electrification Administration, understood, calling for a trickle-down approach to prosperity reminiscent of the 1920s. By urging the administration to "stress scientific research in the hope of turning up new industries," Cooke told FDR's close aide Steve Early, Compton was advocating an approach to economic recovery that simply boiled down to "Mr. Hoover's point of view."[5]

[4] Pursell, "The Farm Chemurgic Council," pp. 314–17; Jewett to Compton, Dec. 6, 1934, RAM, Box 7; Harold Ward, "Science and the Government," *New Republic*, 84 (Sept. 11, 1935), 127
[5] Cooke to Early, Oct. 29, 1936, FDR, OF 2450.

XVIII

Recovery in Physics

K arl Compton kept complaining that Roosevelt's tax and spending policies would kill wealth and, with it, the philanthropic
support of science. Actually for every five philanthropic
foundations endowed in the 1920s, eight more were set up
in the 1930s. And in the Revenue Act of 1935 Congress authorized industrial corporations to deduct from taxable income up to 5 percent of net
profits for contributions both to social service agencies and in aid of
education and research.[1] The trouble for the Comptons was that none
of the new philanthropies chose to commit its resources to the physical
sciences, at least not in the major fashion of the Rockefeller agencies in
the preceding decade. On the industrial side, the National Association of
Manufacturers did join with the American Institute of Physics in an
effort to promote research. Despite the 5 percent clause, industrial managers still had to confront the possibility that giving away money to fund
the development of knowledge might help their competitors. Corporations and trade associations contributed equipment to university laboratories, fellowships to their graduate programs, and funds for well-defined
projects in applied science—but little for such esoterically impractical
areas as nuclear physics.[2]

[1] Between 1930 and 1939, 288 foundations were established; between 1920 and
1929, 173. *The Foundation Directory* (New York: 1964), p. 13; Thomas F. Devine,
Corporate Support for Education: Its Bases and Principles (Washington, D.C.:
Catholic University of America Press, 1956), pp. 55–76.

[2] Between 1936 and 1949 philanthropic expenditures for all purposes by industrial
corporations exceeded 1 percent of net income in only one year, 1945. Theodore

All the while many professors of science kept asking: Why not patent the technologically useful discoveries made in university laboratories and plow the royalties back into pure research? Early in the century Frederick G. Cottrell, a Berkeley chemist, had invented a process to recover valuable chemicals from the outpourings of industrial smokestacks and turned over his patent to the Research Corporation, a nonprofit institution formed in 1912 to handle the rights and dispense the profits as grants in aid of research. In the mid-1920s, Professor Harry Steenbock of the University of Wisconsin found that exposure to ultraviolet radiation would increase the quantity of vitamin D in certain foods. Steenbock gave his process to the Wisconsin Alumni Research Foundation, or WARF, which was created to hold and distribute the proceeds at the university for the benefit of scientific research.[3] In the 1930s the Research Corporation was able to make a number of grants to academic science, and WARF's income, almost $1,000 a day, an alumni reporter remarked, "saved Wisconsin as a great university."[4]

But however lucrative the success of the Research Corporation and WARF, many members of the academic world opposed the patent funding scheme. The chief objections included the argument that knowledge wrested from nature in universities ought to be bestowed upon the world free of charge, not exploited for profits by the universities. More important, professors cited cases in which staff in patent-minded university laboratories had become so proprietary about their research as to speak with colleagues elsewhere only with guarded caution. If that attitude spread through the entire academic system, the opponents warned, it would mean the destruction of the free exchange of information vital to the progress of science.[5] By the end of the 1930s, some two dozen leading colleges and universities had established procedures or agencies for patenting faculty inventions, mainly, it seems, to assure that the patents were properly used, not to earn profits for the support of research. Whatever the motive, only a few universities enjoyed any patent income.[6]

Lacking either governmental, industrial, foundation, or any other

Geiger, "Public Policy and the Five Percent," in Beardsley Ruml, ed., *The Manual of Corporate Giving* (Washington, D.C.: National Planning Association, 1952), p. 5.

[3] Frank Cameron, *Cottrell: Samaritan of Science* (Garden City, N.Y., 1952), pp. 149, 162–63, 175; Mark H. Ingraham, *Charles Sumner Schlichter* (Madison, Wis., 1972), pp. 176–93; interview with E. B. Fred.

[4] Quoted in George W. Gray, "Science and Profits," *Harper's*, 172 (April 1936), 546.

[5] *Ibid.*, 539–49; Alan Gregg, "University Patents," *Science*, 77 (March 10, 1933), 257–59. The pro and con arguments are summarized in The Committee on Patents, Copyrights and Trademarks [of the American Association for the Advancement of Science], *The Protection by Patents of Scientific Discoveries* (New York: The Science Press, 1934), pp. 20–28.

[6] The colleges and universities included Illinois, Minnesota, Caltech, MIT, Columbia, Princeton, Stanford, Yale, and Cornell. A. A. Potter, "Research and Invention in Engineering Colleges," *Science*, 91 (Jan. 5, 1940), 4–7.

kind of reliable fiscal relief, American physicists could only plead in the way of Arthur Holly Compton: "Isn't there someone in the city of Chicago who will match my [next] twenty years with the two million dollars . . . necessary to carry on the work [of nuclear physics] effectively and without delay?"[7]

No matter the revolt against science, people in Chicago and elsewhere remained interested in the doings in the laboratories, including the laboratories of nuclear physics. Science Service had a subscription list of some one hundred newspapers across the country and was estimated to be reaching a fifth of the reading public. In the description of a knowledgeable observer, the press included "a gallery of science writers, who follow the circuit of science meetings as faithfully and as expertly as the sports writers follow baseball and football teams."[8] When Arthur Compton spoke on recent physics at City College in New York, two thousand people jammed the lecture hall, and when Einstein lectured in Pittsburgh in 1934, scalpers sold tickets at $50 a seat.

Crowds visited the science and technology exhibits at the two world's fairs that bracketed Franklin Roosevelt's peacetime presidency. In the years between them, electrical engineers demonstrated a working television, chemical firms predicted evening gowns made of acetate, foundation garments bolstered with Lastex—the fiber was celebrated as "a chemical triumph of the . . . depression era"—and Du Pont announced that nylon stockings were just around the mass-market corner. In a properly regulated economy, it was predicted, technology would give labor increased leisure time and new jobs in the service industries. Even the liberal *Nation* was struck by the marvelous possibilities. "More power to the magicians. The politicians evidently cannot save us. Perhaps the scientists can."[9]

Engineers and conservationists were in fact beginning to turn dust bowls into granaries, control destructive waters with dams, and string power lines into the far corners of the rural United States. And Americans were starting to know the comfort and convenience of air conditioning in stores and theaters, of sleek new trains made of stainless steel or duraluminum, of the Douglas DC-3 which went coast to coast overnight, and of steering wheels, tableware, and even dice made of plastic. Watson Davis understandably applauded: "The bonanza days of science are with us still."[1] In 1936 doctors administered the new Prontylin, the first sulfanilamide drug, to Franklin D. Roosevelt, Jr., saved him from a lethal infection of streptococcus, and opened the age of the wonder drug in

[7] Compton Notebooks, VII, 59–63, AHC.
[8] Watson Davis, "Science and the Press," *Vital Speeches*, 2 (March 9, 1936), 361.
[9] The Lastex remark was quoted in *Journal of Home Economics*, 29 (Feb. 1937), 106; "Wonders Multiply," *The Nation*, 142 (Jan. 15, 1936), 62.
[1] Watson Davis, ed., *The Advance of Science* (Garden City, N.Y., 1934), p. 374.

the United States. And science was helping to wage war against an equally lethal social disease—the gang murders and kidnappings that swept across the country early in the decade.

Before, an authority on police methods noted, many detectives still preferred to rely on "a stool-pigeon rather than a microscope."[2] But the staple tools of the New York City Medical Examiner already included chemistry and toxicology, and, not long after the St. Valentine's Day Massacre, Northwestern University had opened one of the nation's first crime detection laboratories. In 1932 the FBI inaugurated a similar facility, both for its own use and to provide expert consultation for police departments throughout the country. When Bruno Hauptmann, the accused kidnapper of the Lindbergh baby, was apprehended in 1934, newspapers reported how New York City's chief toxicologist, Dr. Alexander O. Gettler, had examined recovered ransom gold certificates, found traces of glycerine, which was used to coat machine tools, and concluded that the culprit must be a mechanic. By studying the crudely written ransom notes, a psychiatrist determined that the kidnapper was "a very methodical person, very probably a German or Teutonic type . . . not given to personal display . . . reticent, very careful, very cautious."[3]

"The 'cracking' of the Lindbergh case," The New York Times applauded, "is a striking illustration of the new methods used in the relentless war on crime. In recent years science has furnished many useful instruments to the detective; often a crime can be solved in the laboratory." J. Edgar Hoover sent FBI agents to a meeting of the American Association for the Advancement of Science to discuss how physics might contribute to the war against crime. Soon the FBI was using X rays to check sealed packages for bombs, ultraviolet light to detect erasures in letters and ledgers, and the spectroscope to analyze whatever the criminal left behind. "The detection of crime can no longer be carried forward by so-called strong-arm methods," Hoover reported to the public. "There must be a scientific investigation at the scene of the crime, a search for fingerprints, for footprints, for dirt, for any article of evidence."[4]

Newspapers and popular magazines reported how physicists had added the neutron and the positron to the catalogue of fundamental particles, how they were building mighty atom-smashing machines and probing the enormous forces that bound the nucleus together. Ernest Lawrence kept announcing plans for ever-larger accelerators, with magnets weighing up to eighty-five tons. A colossal machine builder in a nation that

---

[2] Calvin Goddard, quoted in The New York Times, Nov. 20, 1932, sec. II, p. 6.

[3] Quoted ibid., Sept. 22, 1934, p. 3. Possibly the authorities apprehended and convicted the wrong man, but science may still have pointed to the right type of person. See Anthony Scaduto, Scapegoat: The Lonesome Death of Bruno Richard Hauptmann (New York, 1976), and The New York Times, Mar. 28, 1977, p. 1.

[4] Ibid., Sept. 30, 1934, sec. VI, p. 9; Hoover, ibid., July 8, 1934, sec. VIII, p. 18. See also J. Edgar Hoover, "Physical Science in the Crime-Detection Laboratory," Annual Report of the Smithsonian Institution, 1939, pp. 215–22.

revered the builders of colossal machines, Lawrence was the public per-
sonification of nuclear physics, the recipient of awards and honorary
degrees, a cover figure of *Time*, a prime—and willing—subject for
reporters (this "natural native of Canton, South Dakota," Bruce Bliven,
the editor of the *New Republic*, noted, remained "easy to talk to and
as completely American as you could imagine"). It was page-one news
when in 1936 Lawrence's staff realized the alchemists' dream of trans-
muting platinum into gold. Considering the high cost of platinum and the
inefficiency of the process, this nuclear reaction was of course no threat
to the U.S. Treasury. "Anyway," Lawrence commented cheerfully, "the
information we are getting is worth more than gold."[5]

Newsmen marveled more at the way the transmutation of one element into another actually released nuclear energy. Of course, the amount
of nuclear energy released by bombardment was minuscule compared to
the amount of electrical energy required to run the cyclotron itself, and
Ernest Rutherford cautioned that anyone who predicted an imminent
age of atomic power was "talking moonshine." All the same, Americans
had it on the authority of Karl Compton that atomic energy would soon
be harnessed; of Professor Bergen Davis of Columbia University that the
advent of atom-smashing machines amounted not only to "a landmark in
the conquests of physical science" but to "the beginning of an economic
revolution"; and of Ernest O. Lawrence that quite probably "in your
lifetime and mine" the enormous energy latent in the nuclear mass "will
play a vital role in technical development." The editors of *The New
York Times*, contemplating the prospects of nuclear physics, concluded:
"the imagination soars to Utopia."[6]

In the here and now, the cyclotron was already paying medical dividends, especially as a weapon in the war against cancer. To get at deeply
buried tumors with X rays or with gamma rays from radium, it was necessary to use a beam powerful enough to risk dangerously burning the
skin. Neutrons seemed likely to penetrate far into the body without such
a hazardous surface effect. Ernest's brother, Dr. John Lawrence, who
had come to Berkeley from the department of internal medicine at Yale,
found that neutron radiation destroyed a deadly kind of sarcoma in mice
four times as efficiently as X rays. The Lawrence brothers emphasized
that it was too early to tell whether the treatment would be safe and
effective for human beings. But in 1937 the Mayo Clinic pronounced
their mother, Gunda Lawrence, the victim of an inoperable cancer.
Desperate, the Lawrences subjected her to neutron therapy and she was
cured. By 1938 John was hailing neutron beams for doing "remarkable

[5] Bliven, *Men Who Make the Future* (New York, 1942), p. 174; Lawrence is
quoted in *Time*, Nov. 1, 1937, p. 40.

[6] "Talking Moonshine," *Scientific American*, 149 (Nov. 1933), 201; *New York
Times*, Jan. 30, 1933, p. 15; Jan. 31, 1933, p. 16; June 13, 1937, sec. II, p. 6; May 8,
1932, sec. III, p. 1.

things for animals suffering from malignant tumors or cancers," and radiation therapists at Berkeley were soon treating human beings on a regular basis.[7]

Ernest was also quick to understand the medical advantages in the remarkable discovery by the Joliot-Curies in 1934 of "artificial radio-activity." In the conclusion of the two French scientists, bombard-ment with alpha particles transmuted the nuclei of the lighter, normally nonradioactive elements into radioactive isotopes. Soon, at the Univer-sity of Rome, young Enrico Fermi, brilliant at both theory and experi-ment, a physicist's physicist who had been appointed to his professorial chair at age twenty-six, found that artificially radioactive isotopes could also be produced in prodigious variety by bombardment with neutrons. In Berkeley, Lawrence created numerous radioactive isotopes by hurling the nuclei of deuterium, which contained one neutron, at various targets.

Some artificial radioactive isotope, scientists supposed, might well do the therapeutic work of radium, and, with the cyclotron, it might be possible to produce the substitute far more cheaply than the naturally radioactive metal and in much greater quantity. By September 1934 Law-rence had found just the isotope, a radioactive variant of sodium. Pro-duced by the impact of deuterons upon normal sodium, an element in common salt, it could be obtained in abundant supply and introduced into the body without any poisonous chemical effects. It emitted gamma rays of an energy comparable to those given off by radium. It also had a half-life—the time required for its rate of radioactive emissions to decline to half their original intensity—of only fifteen hours, long enough for therapy, short enough not to do the body any harm. "In the biological field," Lawrence noted in his first published report of the new isotope, "radio-sodium has interesting possibilities that hardly need be emphasized here."[8]

With radio-sodium, it was reported, medical scientists could treat specific tumors by injection, explore how radiation affected living tissue, and trace the diffusion of chemicals and fluids in the body. In 1935 Lawrence announced that the production of the isotope had reached the commercially viable stage. While on the lecture circuit Lawrence would serve someone from the audience a radio-sodium cocktail, then use a Geiger counter to demonstrate the progress of the isotope through the bloodstream. Some physicists criticized Lawrence for the vaudeville and suggested he was taking up medical work merely to raise money. Whatever the case, by making clear that nuclear physics served medicine, the inventor of the cyclotron did a good deal to counter the social critics who considered research in nuclear abstrusities a wasteful irrelevance.

[7] Quoted in New York Times, Nov. 15, 1938, p. 25.
[8] Lawrence, "Radioactive Sodium . . .," Physical Review, 46 (Oct. 15, 1934), 746.

"The silly study of neutrons," Waldemar Kaempffert, the science editor of *The New York Times*, remarked, "has already given us artificially radioactive compounds and with them the hope of treating cancer inexpensively."[9]

With the partial recovery of the economy, academic income from both endowments and state appropriations for research rose back to the levels of 1929. Reflecting the general fiscal trend in federal research, the budget of the National Bureau of Standards bottomed out in 1934, then climbed to a level 50 percent higher in 1939, when the authorized staff reached the peak of the previous decade. Industrial executives were impressed by the advice of Charles F. Kettering of General Motors, the inventor of the automotive self-starter, a pioneer in the introduction of ethyl fuel, and something of a gas-pump sage on economic and technological matters. Research, Kettering emphasized to businessmen, was the way to win new markets, add to profits, and even "create a labor shortage." Despite the depression—and in part because the Revenue Act of 1936 explicitly made corporate expenditures for research tax-deductible—after 1933 the number of industrial research laboratories in the United States increased by some 12 percent.[1]

Business actually spent only about 75 percent as much on research as in the 1920s, but along with a continuing corporate demand for physicists in the electrical and communications fields, openings were appearing in metallurgy, textiles, and mining. Jobs for physicists were also burgeoning dramatically in aeronautics, which had no adequate supply of trained engineers, and in petroleum, which had discovered that geophysical techniques could be highly effective in prospecting for oil. In 1936 the MIT physics department reported that it had "many more [industrial] calls" for its graduates than it could hope to fill. Industrial salaries remained considerably higher than those in the universities.[2] The award of the Nobel Prize to Clinton J. Davisson in 1937 gave Bell Laboratories a specially appealing gloss, and Westinghouse announced that it was con-

[9] *New York Times*, Nov. 2, 1936, p. 20.

[1] Malcolm M. Willey, ed., *Depression, Recovery and Higher Education: A Report by Committee Y of the American Association of University Professors* (New York, 1937), pp. 183, 192; Rexmond G. Cochrane, *Measures for Progress: A History of the National Bureau of Standards* (Washington, D.C.: National Bureau of Standards, U.S. Department of Commerce, 1966), Appendices F and H; Kettering, "How to Create a Labor Shortage," *Saturday Evening Post*, 208 (May 30, 1936), 70. In 1938 there were 1,769 industrial laboratories; in 1933, 1,575. *Industrial Research Laboratories of the United States* (National Research Council, Bulletin No. 102, 1938); "Industrial Research in the United States," *Scientific Monthly*, 49 (Dec. 1939), 586.

[2] In 1938 business was reported to be spending $100,000,000 a year on research, as compared to $133,000,000 in the 1920s. "Industrial Research in the United States," 586; *Physics in Industry* (New York: American Institute of Physics, 1937), *passim*; Irving J. Saxl, "Don't Overlook the Physicist," *Nation's Business*, 25 (Jan. 1937), 26; *Report of the President, MIT, 1936*, p. 132; Edward J. v. K. Menge, *Jobs for the College Graduate in Science* (Milwaukee: Bruce, 1932), pp. 17–21.

structing a nuclear accelerator. But like their colleagues at the Bureau of Standards, most industrial physicists were still expected to concentrate on practical research. While the universities were turning out men in the new "applied physics," corporate laboratory managers continued to complain that graduates knew too much about atomic and nuclear physics and not enough about classical subjects relevant to the prevailing technology.[3]

The academic world still offered considerable professional freedom at a reasonable income. In 1929 the median academic salary had purchased some 7 percent more than in the comfortable paydays of 1914; in 1935, amid the deflation, it purchased almost 14 percent more. If academics complained of an economic pinch, it was because, a report of the American Association of University Professors noted, they were simply having greater difficulty living "in conformity with the assumed status and tastes"—which meant the high status and expensive tastes of a house in a better neighborhood, domestic help, good clothes, books, and vacations—"of a man on a college faculty."[4] Salaries aside, physics faculty who reached the higher rungs of the academic ladder could expect considerably more job security than in industry or government, and at any level they enjoyed decidedly more opportunities to do nuclear physics with cyclotrons.

At the Rockefeller Foundation, Warren Weaver recognized the biomedical advantages of neutron radiation and radioactive isotopes; he advanced his program for biology in part by granting subventions to nuclear research and contributing to the construction of cyclotrons. So did the managers of the medically oriented Macy and Markle foundations. Equally important, the biomedical advantages of nuclear accelerators often helped physicists raise money from local and national sources that had no special common purpose except the advancement of science. Where cash for the construction of a cyclotron was insufficient, nuclear physicists scrounged up outmoded industrial equipment or, in one case, vacuum tubes discarded by the U.S. Navy.

Exemplifying the crazy-quilt but strongly biomedical nature of cyclotron funding in the 1930s, Ernest Lawrence received money from his university, the Chemical Foundation, and the Research Corporation, from medically minded individuals and foundations, even from the newest medically minded agency of the federal government. When in 1937 the National Advisory Cancer Council awarded its first three grants, one went to Lawrence's laboratory for research in the biomedical applications of nuclear physics. By undertaking "this biological work," Lawrence frankly told Niels Bohr, he had been able to obtain "financial support for

---

[3] Kettering, "Research and Social Progress," *Vital Speeches*, 2 (March 9, 1936), 359; Henry A. Barton, "The Meeting of the Founder Societies of the American Institute of Physics," *Scientific Monthly*, 43 (Dec. 1936), 578.

[4] Willey, ed., *Depression, Recovery, and Higher Education*, pp. 60, 133.

all of the work in the laboratory." In the United States it was just "much easier to get funds for medical research" than for straight-out nuclear physics.[5]

The estimated operating costs of a cyclotron were $14,000 to $18,000 a year, more than the sum available for research in most departments in the 1920s. Still, by 1940 American physicists were running or constructing some twenty cyclotrons, and at least a dozen other nuclear accelerators of different types. Many of the accelerators were employed for medical purposes, especially the production of radioactive isotopes, but almost all were used at least part of the time for nuclear research as such.[6] The staff requirements of the machines helped to open new opportunities for jobs and doctoral training. Harry Hopkins even permitted the use of a small fraction of relief funds for assistants in the physical sciences, including cyclotron projects.

At least some research and teaching fellowships remained available. In part because the difficulty of finding jobs in the depression kept many students in school, between 1933 and 1939 American universities awarded physics Ph.D.'s at an average rate fully twice that of the halcyon days of normalcy. Of course, the academic world was such a buyer's market that the Princeton mathematician Oswald Veblen had "never known a [better] time for a college or university . . . to strengthen itself by the appointment of brilliant young men." Some of the new doctors of physics dropped out of the profession, others went into high-school teaching or government research, and an unprecedented fraction of them, about one in four, joined industry. But the majority—and quite probably the vast majority of the most talented elite eager to pursue nuclear studies— found academic jobs. In all, only a small fraction of the decade's crop of Ph.D.'s actually seems to have left physics.[7] In 1939 membership in the American Physical Society totaled 3,600 and the profession was larger by some 50 percent than in 1929.

The young sociologist Logan Wilson took a hard-nosed scholarly look at the professoriat and pronounced it a "floodgate for social advancement."[8] Physics continued to be just that, especially for the white males, many of them from middle- to lower-middle-class homes, who accounted for the vast majority of new Ph.D.'s in the field.

The leading women's colleges spent more money on scientific facil-

[5] Quoted in Mark L. Oliphant, "The Two Ernests—I," *Physics Today*, 19 (Sept. 1966), 44.

[6] "Cyclotron Installations," Jan. 1940, list in EOL, Carton 1, folder 8b.

[7] For 1933–40 the average per year was 144; for 1919–32, 67. Veblen to T. H. Taliaferro, April 16, 1935, OV, Box 15; Ernest V. Hollis, *Toward Improving Ph.D. Programs* (Washington, D.C.: American Council on Education, 1945), pp. 65, 74–75; Spencer R. Weart, "The Rise of 'Prostituted' Physics," *Nature*, 262 (July 1, 1976), 15.

[8] Logan Wilson, *The Academic Man: A Study in the Sociology of a Profession* (New York, 1942), p. 149.

ities than on any other capital improvement and, in a symbolic gesture, Bryn Mawr students went without dessert for a week to help finance a new laboratory. Just as large a fraction of undergraduates at Smith, Wellesley, and Vassar majored in science as at Harvard, Yale, and Princeton. But most female physicists remained unmarried, and many modern young women still agreed with the judgment in a forum for businesswomen that in any professional field the price of success—the loneliness and "disillusionment" of spinsterhood—was simply "too high." If in the depression women turned from literature and art back to political questions, their political attention did not center on the liberation of women from traditional occupational roles. Besides, the well-known author Helen Woodward reported, "The word 'career' has such an empty pompous sound that [young women] laugh at it. 'All I want to know,' says the girl of today, 'is how I'm going to make a living . . .?' "[9]

Many college women studied science because courses in such disciplines as psychology and zoology were considered useful for social work. If some aimed for jobs in science itself, many set their sights on employment as laboratory technicians, which like the work of stenographer, nurse, or teacher, the journalist Eunice Fuller Barnard enthusiastically noted, might well come to be considered a "woman's job." On the average after 1933, three women took scientific Ph.D.'s yearly for every two who had earned them in the previous decade, but the comparable ratio for men was four for two, or 33 percent higher. And whether by tradition or temperament, as Miss Barnard noted, women still preferred the disciplines with "immediate human applications."[1]

Four times as many doctorates were awarded to women in the biological and social as in the physical sciences. Women earned only about thirty-five, or little more than 3 percent, of the doctorates in physics and, at the end of the decade, represented a comparably insignificant percentage of the profession. Amid the sharp competition for university posts, female physicists faced particularly acute discrimination in academic hiring (in the 1930s the proportion of women on college faculties may well have declined).[2] A small fraction of the new female physicists turned

[9] Eunice Fuller Barnard, "Women's Rise in Science," *New York Times Magazine*, Oct. 27, 1935, p. 9; Ann Marie Lindsey, "Does Education for Independence Pay?" *Independent Woman*, 13 (Oct. 1934), 312–13, 332; Woodward, "Careers for Daughters," *The Delineator*, 124 (June 1934), 12.

[1] Barnard, "Women's Rise in Science," pp. 9, 10. Between 1933 and 1940, women took 342 Ph.D.'s in the physical, 642 in the biological, and 715 in the social sciences. Jessie Bernard, *Academic Women* (University Park, Pa.: Pennsylvania State University Press, 1964), p. 71.

[2] Bernard, *Academic Women*, p. 71. The statistics on physics are taken from M. Lois Marckworth, comp., *Dissertations in Physics: An Indexed Bibliography of All Doctoral Theses Accepted by American Universities, 1861–1959* (Stanford, 1961). According to Bernard, p. 40, the proportion of women on college faculties remained constant at about 27 percent, but census data suggest a decline from 33 percent in 1930 to 27 percent in 1940. See *Statistical Abstracts of the United States, 1932*, p. 60; *1942*, p. 75.

to industry, where the opportunities for pure research were of course limited. Most others joined the women's colleges, or lower-ranking coeducational institutions, where they endured inadequate research funds, no cyclotrons, and, often, professional isolation. Few, if any, had the opportunity to play a prominent role in the advance of nuclear physics or in the leadership of the profession.

Unlike women, Catholics still did not suffer overt discrimination in the academic world, and the Catholic Round Table of Science, a group of laymen and clerics, was trying to encourage more members of the faith into careers in research. So was the archbishop of Cincinnati, who in 1935 created the Institutum Divi Thomae, a well-appointed graduate scientific enclave, and the pope himself, who in 1936 reformed the ancient Pontifical Academy of Sciences and clearly reiterated that Catholics should assume no conflict between science and religion. But though the number of Catholics going to college rose steadily, the percentage of all undergraduates who were Catholic remained far smaller than the Catholic portion of the population.[3] Though the Catholic colleges continued to modernize, almost none save Catholic University of America or the Institutum Divi Thomae could claim the faculty and facilities for first-rate scientific training.

Aware of the inadequacies of scientific training in Catholic universities, Professor Hugh Scott Taylor of Princeton, born and educated in Britain, now a distinguished chemist and activist of the Catholic Round Table, argued in *Commonweal* that students ought to attend the schools, meaning the secular schools, best able to teach them what they wanted to learn. Taylor's brief elicited a number of vigorous clerical rebuttals contending, with Father Thomas Larkin, that Catholic undergraduates were unprepared to cope with the "insidious and sickly pagan atmosphere of morals and philosophy breathed at . . . secular institutions." Cardinal Patrick Joseph Hayes, archbishop of New York, warned that even graduate students who attended non-Catholic schools would not only risk "losing their faith"; they would "almost defy God." Buoyed by the humanist revolt against science, Father George Bull, the chairman of the philosophy department at Fordham University, not atypically defined the principal purpose of Catholic graduate education to be "the enrichment of human personality, by deeper and deeper penetration into the velvety manifold of reality, as *Catholics possess it.*"[4]

[3] James Arthur Reyniers, "Ways and Means of Developing Catholic Scientists" and George Speri Sperti, "The Institution Divi Thomae," in John A. O'Brien, ed., *Catholics and Scholarship: A Symposium on the Development of Scholars* (Huntington, Ind.: Our Sunday Visitor, 1938), pp. 126–28, 132–34. I am indebted to Father John Whitney Evans for the enrollment figures.

[4] *Commonweal*, 23 (Feb. 14; March 20; 1936), 427–29, 581; Cardinal Hayes, quoted in *Newsweek*, 7 (June 13, 1936), 42. Father Bull, quoted in Philip Gleason, "American Catholic Higher Education: A Historical Perspective," in Robert Hassenger, ed., *The Shape of Catholic Higher Education* (Chicago, 1967), p. 48.

More young Catholics enrolled at secular institutions anyway, but, like their brethren in the Catholic colleges, not to pursue careers in science. Just as eager for the more prestigious and financially rewarding careers as their predecessors, Catholic college graduates went into law or medicine or business. If they had a talent for science, they preferred engineering to academic research. In the 1930s, Catholics took at most only 3.5 percent of all the doctorates awarded in science, and doubtless no more than that fraction of the Ph.D.'s in physics.[5]

During the depression young Jews accounted for an increasingly disproportionate fraction of American undergraduates. As much children of the book as ever, they had special reasons to aspire to the professions. The New York *Daily News* approvingly remarked, "Plenty of people just now are exercising their right to dislike Jews."[6] The paper might have added that many were being refused employment. With quotas on the admission of Jews now a commonplace in law and, especially, medical schools, the percentage of Jews who studied to become doctors or lawyers declined substantially. So did the percentage who aimed to be engineers, not least because Jews found considerable anti-Semitism in industry. What kept rising considerably, as if by way of compensation, was the percentage of professionally ambitious Jews who aspired to careers in education, including, to an extent, higher education.[7]

The increase occurred when anti-Semitism in the universities was at an all-time high. The growing number of Jews on college faculties stirred the resentment of gentile professors and, in the severity of the academic market, administrators were frequently reluctant to fill positions with more Jews. Often identified with the stereotype of the crude and pushy Jew of the urban ghetto, new Jewish Ph.D.'s were believed neither to dress, talk, nor act in conformity with the prevailing standards of academic gentility. Besides, the tendency of Jewish professors to be politically liberal or radical could make them persona non grata to conservative administrators.[8]

In physics, academic anti-Semitism operated in a typical fashion against Eugene Feenberg, a talented theorist who had been raised in a nominally orthodox family in Texas, worked his way through the state university, compiled an outstanding academic record, and studied with Enrico Fermi in Rome. After Feenberg finished the Ph.D. at Harvard and

[5] See the figures for Catholic scientific doctorates in W. M. Cashin, "Catholics and Science Doctorates," *America*, 82 (Dec. 31, 1949), 388.

[6] Quoted in Carey McWilliams, *A Mask for Privilege: Anti-Semitism in America* (Boston, 1948), p. 45.

[7] *Ibid.*, pp. 40–41, 133; Nathan Goldberg, "Occupational Patterns of American Jews," *Jewish Review*, 3 (Oct.–Dec. 1945; Jan.–Mar. 1946), 170, 268, 270–72; Charles B. Sherman, *The Jew Within American Society* (Detroit: Wayne State, 1961), pp. 111, 175–80.

[8] See Samuel A. Goudsmit, "It Might as Well Be Spin," *Physics Today*, XXIX (June 1976), 42, and the letters of Goudsmit and R. D. Richtmyer, *ibid.*, XXIX (Dec. 1976), 42; B. A. Wooten to George B. Pegram, Sept. 21, 1935, GBP.

needed a job, Edwin C. Kemble praised his student to professional acquaintances in other universities and, perhaps trying to be especially helpful, added: Feenberg is "a tall rangy Texan and neither looks nor acts like a New York Hebrew." Responses came from the University of North Carolina: "It is practically impossible for us to appoint a man of Hebrew birth . . . in a southern institution"; from Purdue: No offer possible because of "anti-Semitism among the higher administrative officials."[9] Feenberg had to occupy himself for a year with a lectureship at the University of Wisconsin and spend two more as a "temporary" fellow at the Institute of Advanced Study before he was appointed to an assistant professorship at New York University.

But the extent of anti-Semitism in other fields helped make academic careers, including physics, attractive for technically talented Jews. It was a time when Thomas Wolfe, his perception made more acute by his own anti-Semitism, wrote of "the Jew boy" sitting and poring upon a tome in an East Side tenement. "For what this agony of concentration? For what this hell of effort? For what this intense withdrawal from the poverty and squalor of dirty brick and rusty fire escapes, from the raucous cries and violence and never-ending noise? For what? Because, brother, he is burning in the night. He sees the class, the lecture room, the shining apparatus of gigantic laboratories, the open field of scholarship and pure research, certain knowledge and the world distinction of an Einstein name."[1]

In the 1930s Jews took Ph.D.'s in physics certainly almost in proportion to—and probably in greater frequency than—their weight in the population. And though anti-Semitism operated in the academic world, in physics, where achievement was usually clear-cut, talent and brilliance could make their way. "To every man his chance," Wolfe went on to celebrate the promise of America.[2] Young Jewish physicists who were neither tall nor rangy nor Texan found academic jobs and seized their chance. By the end of the decade, though they accounted for only a small fraction of the profession, they occupied a disproportionately large place in its intellectual leadership—along with the Jewish refugees from Nazi Germany.

In Germany in 1933, Max Planck, now the president of the Prussian Academy of Sciences, had paid a visit to Adolf Hitler. "I want to

[9] Kemble to A. L. Hughes, July 16, 1937; Kemble to Louis C. More, Feb. 23, 1935; Arthur Ruark to Kemble, March 19, 1935; Hubert James to Kemble, Aug. 4, 1937, SHQP, microfilm rolls 53 and 54.

[1] Wolfe, *You Can't Go Home Again* (New York: Dell, 1960), p. 462.

[2] *Ibid.* Harry Cohen and Itzhak J. Carmin, eds., *Jews in the World of Science: A Biographical Dictionary of Jews Eminent in the Natural and Social Sciences* (New York, 1956), lists twenty-five Ph.D.'s in physics, or 2.6 percent of the total, for 1933–39, when Jews were estimated to represent some 3.5 percent of the general population, but this compilation is no doubt incomplete.

inform you that without the Jews there is no mathematics and physics possible in Germany," Planck reportedly warned.

"But we don't have anything against the Jews. . . . We fight only against the Communists," Hitler was said to have replied.

Harangued further, Planck left, sure that, barring Hitler's downfall, there was no way to stop the dictator's senseless destruction of German science.[3]

Hitler's strongmen were forcing the dismissal of hundreds of Jewish scholars from the German universities. Making a shambles of logic and decency, high Nazi officials, not to mention the Nobel laureates Johannes Stark and Philipp Lenard, justified the purification of "Aryan science" with infuriating arrogance. "Science, like every other human product," Lenard asserted in *German Physics*, a book he dedicated to the minister of the interior, "is racial and conditioned by blood." If scientists elsewhere believed Jews had contributed significantly to the advancement of physics, Lenard claimed that Einstein's was a "peculiar physics," too theoretical, unrelated to the facts of nature, the typical intellectual construction of a people who had little if any interest in "truth." To Lenard's twisted mind, "Jewish physics is merely an illusion—a perversion of basic Aryan physics."[4]

British professors mobilized to help their beleaguered German colleagues find refuge outside of Hitler's Reich, and in the United States in 1933, some two dozen academics, including seventeen university presidents, coalesced in the Emergency Committee in Aid of Displaced German—it was changed to "Foreign" after the invasion of Czechoslovakia—Scholars under the directorship of Stephen P. Duggan, the founder and head of the Institute of International Education. Duggan's young assistant director was an energetic and talented graduate in speech of Washington State College and the recent president of the National Student Federation, Edward R. Murrow. Frequently operating independently of the committee, American physicists circulated lists of displaced professors compiled by the Academic Assistance Council in London, asked for financial donations to the cause, and created an effective rescue network of their own. By 1936 more than 1,600 scholars, about one-third of them scientists, had been forced out of their posts. In the two years before he left Duggan's committee for CBS, Murrow received some fifty letters a day from displaced European scholars. The mathematician John Von Neumann reported from Europe: Pauli "would like to make another trip to America—

---

[3] Quoted in Charlotte Schoenberg to R. G. D. Richardson, July 27, 1933, copy in OV, Box 22, Emergency Committee file. See Planck's postwar account of the interview in Joseph Haberer, *Politics and the Community of Science* (New York, 1969), p. 132.

[4] Quoted in *Science*, 83 (March 20, 1936), 285; E. J. Gumbel, "Aryan Science," *Living Age*, 355 (Nov. 1938), 253.

but this is not a rare desire in mathematical and physical circles around here."[5]

The German immigration quota was never filled in any year before 1938, and professors who had jobs awaiting them in the United States were exempt from the quota anyhow. But many American academics, including physicists, considered hiring German refugees unfair when the country's own new doctorates were having such difficulties obtaining university posts. Others supposed that filling up professorships with foreigners would sharply limit the opportunities for advancement not only of America's younger but of her older scholars as well. Academic anti-Semitism also created problems, even for the Nobel laureate physicist James Franck, who accepted a professorship at Johns Hopkins in 1933, then had to leave for the University of Chicago a few years later because President Isaiah Bowman made life very difficult for Jewish faculty.[6]

Duggan's committee took cautious account of the lean job market for young Ph.D.'s and as a matter of policy helped only established scholars. The physicists tended to do the same as a matter of practice, since they were better acquainted with the prominent men in their field, and it was easier to persuade the administration to add an Enrico Fermi rather than an unknown to the faculty rolls. The Emergency Committee helped pay the university salaries of refugees with funds raised from benefit concerts, Jewish charities, and generous philanthropists; between 1933 and 1939 the Rockefeller Foundation, which matched most of the committee's grants, distributed about $500,000 in stipends for the émigrés. If anti-Semitism was an obstacle, the rescue advocates had increasingly powerful moral arguments on their side as the disease of Hitlerism grew more virulent and spread to Austria, Italy, and Czechoslovakia. By 1941 more than one hundred foreign physicists, including eight past and future Nobel Prize winners, had found academic posts in the United States.[7] Most were German, a considerable number of them at least partly Jewish or married to Jewish women.

Accustomed to societies where registration at a hotel as "professor" automatically brought deferential treatment, some of the refugees disliked the egalitarianism of the United States and talked contemptuously of the philistinism of its people. Some were also bothered by the extent to which American universities were so unlike their more autocratically

[5] *School and Society*, 46 (Aug. 7, 1937), 170–71; Alexander Kendrick, *Prime Time: The Life of Edward R. Murrow* (New York, 1969), p. 123; Von Neumann to Oswald Veblen, July 4, 1934, OV, Box 9.

[6] Charles Weiner, "A New Site for the Seminar: The Refugees and American Physics in the Thirties," in Donald Fleming and Bernard Bailyn, eds., *The Intellectual Migration* (Cambridge, Mass., 1969), pp. 214–15; Stephen Duggan and Betty Drury, *The Rescue of Science and Learning* (New York, 1948), pp. 99–100; interview with the late James Franck.

[7] Weiner, "A New Site for the Seminar," p. 217.

organized counterparts in Europe. But gradually, as the barriers of language and strangeness fell, most, like James Franck, decided that while democracy might not be the perfect form of government and society, it was the best. And particularly for the younger refugees the American academic system had its distinct advantages. In Europe one often had to wait years before obtaining a professorship and usually only the professors had graduate students. In the United States, everyone was a professor, seemed to have as many students as he could handle, and enjoyed easy access to cyclotrons, other physicists, and money for research. The refugees rapidly adapted to the American physics profession and ranked among its most productive members.

"The United States leads the world in physics," *Newsweek* remarked in 1936, and so it did.[8] The prominent graduate schools offered first-rate training in the fundamentals and on the frontier of the discipline; students no longer needed to go abroad to study. The country was dotted with well-appointed laboratories and had more cyclotrons than the rest of the world combined. The American Physical Society was a vital national forum, with some five hundred physicists usually attending its major annual meetings and one to two hundred the gatherings of its many regional sections. Prolific in research, American physicists submitted more articles annually to *The Physical Review* than the journal, which in 1939 bulked 2,500 pages large, could hope to publish in full. Louis de Broglie summarized the quality of the work: "Today scientific publications from the United States are awaited with an impatience and curiosity inspired by those of no other country."[9]

In later years it would become a commonplace that physics in the United States owed its ascendancy to the refugees, and, since Europe's loss was generally America's gain, the commonplace contained a germ of truth. Dispersed throughout the university system, the refugees enriched the major physics departments with the power of their mathematical techniques, experimental imagination, and frequently philosophical approach to the analysis of natural phenomena. Many of them theorists, they were particularly advantageous to cyclotronists, whose machines were spewing forth a confusion of data that demanded clarification and interpretation. Stanford experimentalists, who had been depending upon Oppenheimer at Berkeley, relished the arrival of the Swiss-born refugee Felix Bloch. A brilliant young theorist—he would later share the Nobel Prize—Bloch fell naturally into close collaboration with his colleagues in the laboratory, provided on-the-spot theoretical consultation, and like

[8] *Newsweek*, Nov. 7, 1936, p. 29.
[9] Quoted in *The New York Times*, May 11, 1935, p. 18. On *The Physical Review*, see P. D. Foote to W. F. G. Swann, Sept. 7, 1933, WFGS, American Physical Society file.

the refugees on other campuses, considerably stimulated the enterprise of physical research in Palo Alto.

What Bloch did for Stanford, Hans Bethe did for Cornell and, in a special way, for the profession as a whole. A Ph.D. at age twenty, Sommerfeld's best student after Heisenberg and Pauli,. a virtuoso in mathematical physics, Bethe left Germany in 1933; at Cornell, his exceptionally fluent English and sparkling humor and vitality made him a tonic to the department. Soon after his arrival Bethe decided that his colleagues in Ithaca and elsewhere needed an up-to-date summary of nuclear phenomena, especially on the theoretical side. Week after week, Bethe sat beneath a dim light, a pile of blank paper on one side and a pile of completed text on the other, and plowed steadily through the entire corpus of nuclear studies, clarifying what was known, spotlighting what was not. In the course of two years, with the help of Robert F. Bacher, he produced three massive review articles. A formidably comprehensive survey of the field, they were collectively called "Bethe's bible," and were soon to be found on the desk of virtually every nuclear physicist in the United States.

Yet for all the importance of the refugees, they would scarcely have had so much impact, Robert Oppenheimer accurately recalled, had there not already been "a rather sturdy indigenous effort in physics," including, Oppenheimer might have added, the sturdy institutional pyramid, local at its base and national at its top, that met the requirements of Rowland's best-science program.[1] The national peak of the pyramid operated more through the American Physical Society and *The Physical Review* than through the severely diminished NRC fellowship program. At the level just below the peak, the major departments continued to exercise considerable influence on the course of the profession, drawing the bulk of graduate students in the field and producing a significant fraction of the research. Here the leading physicists, native and refugee, stimulated students and colleagues alike into significant lines of research. Some did it by the traditional modes of example, advice, or teaching, but as the slightly though discernibly growing volume of papers published by two or more authors suggested, an increasing number did it as leaders or participants in organized research groups.

The transition to group research was made possible—and to a certain extent forced—by the growing ratio of students to faculty. Sometimes groups formed around physicists like I. I. Rabi at Columbia, a fount of ingenious ideas who liked to work with other people and avoided nuts-and-bolts details. While Rabi whittled and hummed, his young collaborators adjusted the apparatus, took measurements, and, under his deft direction, measured to remarkable accuracy the fundamentally important

[1] Quoted in Weiner, "A New Site for the Seminar," p. 191.

magnetic moments of the proton, the neutron, and whole nuclei. But group research developed primarily because of the increasing complexity of experimental equipment—even Rabi's required more than two hands—especially the cyclotron. Physicists kept building bigger cyclotrons to accelerate particles to the higher energies needed generally to probe the nucleus more deeply. A more powerful accelerator was also needed to explore experimentally Hideki Yukawa's arresting theory of the forces that bound nuclear particles together.

In 1935 Yukawa, a Japanese physicist, had advanced the idea that the particles in the nucleus were kept together by a force resembling the so-called exchange force which held two atoms in a chemical bond. In the chemical case, the two atoms exchanged one or more of their outer electrons. In the nuclear case, Yukawa argued, protons and neutrons exchanged another particle, singly charged and about two hundred times as massive as the electron. At Caltech, Carl Anderson and his coworker Seth Neddermeyer did not know about Yukawa's theory, but early in 1937 they reported the existence of a new singly charged particle in cosmic rays. It seemed to be heavier than the electron and lighter than the proton; by 1938 the most reliable estimates put its mass at between 130 and 240 times that of the electron—in short, within the range of the particle predicted by Yukawa. But there were serious discrepancies between the characteristics of the Anderson-Neddermeyer particle—it was eventually called the meson—and Yukawa's agent of nuclear exchange. To resolve the discrepancies, some physicists proposed to study mesons in cosmic rays; others argued that they might be investigated in a cyclotron powerful enough to produce mesons in quantity, a cyclotron of 100,000,000 electron volts.[2]

Ernest Lawrence intended to build such an unprecedentedly energetic cyclotron. Still boyishly enthusiastic, operating independently of the physics department and controlling a five-figure annual budget, Lawrence had successfully constructed increasingly energetic machines, the latest of them five feet across the magnetic pole face and producing 16,000,000 electron volt particles. But Hans Bethe had argued that, because of the relativistic increase of mass with speed, cyclotrons could not accelerate particles to 100,000,000 volts; long before reaching that energy the beam would become unfocused and fly off in all directions. Lawrence put a few of his theoretically able young colleagues to work on the problem raised by Bethe; they concluded that the relativistic focusing difficulty could be overcome. Bethe should realize, Lawrence confided to a fellow cyclotronist, that there are "many ways of skinning a cat."[3] In the late 1930s Lawrence and his staff designed a colossus of a cyclotron whose magnets

[2] See the report of the Washington Conference on Theoretical Physics, May 22–24, 1941, *Carnegie Institution of Washington, Yearbook*, 40 (1940–41), 92.

[3] Lawrence to Lee A. DuBridge, Dec. 4, 1937, EOL, Carton 6, folder 17. See also Lawrence to Arthur Compton, Dec. 7, 1937, EOL, Carton 12.

would measure more than fifteen feet in diameter and weigh more than 4,000 tons. Obtaining partial pledges of financial support from his patrons and the University of California, Lawrence requested the bulk of money for the machine, $1,500,000, from the Rockefeller Foundation.

Warren Weaver devoted considerable care to the evaluation of Lawrence's request, worrying about the wisdom of spending so much money for one project—the total capital cost of all the cyclotrons then operating in the world was estimated at $2,000,000—the likelihood that despite Bethe's prediction the machine would work, and the utility of the machine to science and the world. The questions of utility were easy enough to answer; the cyclotron was indispensable to explore the meson problem, and on the medical side, it could well produce in minutes artificially radioactive substances equivalent in the power of their radiations to thousands of grams, which was to say hundreds of thousands of dollars' worth, of radium. As to whether it would work, nine of the world's leading physicists, including Frédéric Joliot, Niels Bohr, and Marcus Oliphant, one of Rutherford's leading accelerator specialists, enthusiastically endorsed its construction. The project was a gamble, the consultants generally observed, but no one in the world was better qualified to pull off the gamble than Ernest Lawrence.[4]

At the top of the Rockefeller Foundation, President Raymond B. Fosdick had some special thoughts on the wisdom of the project. Back in the more confident days of the 1920s the Foundation had pledged $6,000,000 to Caltech for the construction of a two-hundred-inch telescope atop Mt. Palomar in California. Now, the more meanly the world seemed to behave, the more Fosdick stuck proudly, almost mystically, behind the Foundation's pledge, feeling that this "titanic tool of science . . .," he recalled, "might bring into fresh focus the enigma of the universe, its apparent order, its beauty, its power." The more the world was beset by rabid nationalisms, the more Fosdick was sure that the internationalism of science had to be maintained. The more the Nazis forced their irrational brutalities on Europe, the more Fosdick celebrated science for its method, its "confidence that truth is discoverable," its "faith that truth is worth discovering," in all for its historic aid to the democratic mind and spirit. Amid the seeming collapse of scientific civilization, Fosdick asked rhetorically: "Are bigger telescopes and cyclotrons needed in a world like this?" His answer: Yes, emphatically yes.[5]

In the spring of 1940, not many months after Lawrence was awarded the Nobel Prize, the Rockefeller Foundation granted him $1,150,000 for his giant cyclotron, which was to be housed, along with its sizable staff,

---

[4] Weaver to Lawrence, Feb. 13, 1940; Lawrence to Weaver, Feb. 20, 1940, Feb. 21, 1940, EOL, Carton 15; [Natural Sciences Division, Rockefeller Foundation], "Proposed Action," 1940, University of California Cyclotron, RF, file 205D.
[5] Fosdick, *Chronicle of a Generation* (New York, 1958), p. 281; *Annual Report of the Rockefeller Foundation, 1940*, pp. 45, 43.

on a hill high above the Berkeley campus. The venture signaled the coming of a new kind of physics in the United States. It was a physics of many practitioners and considerable reliance on theorists as well as experimentalists, a physics whose salient intellectual challenges led inexorably to expensive machines and massive organizations. In the language of an earlier era, this was physics in the large-project style of the Carnegie Institution of Washington. In the phrase of a later day, it was "big physics."

In the late 1930s big physics was already a fact of life in Berkeley. The Radiation Laboratory's polyglot staff—it included natives and refugees, gentiles and Jews—was large and seemed to grow constantly. There were physicists and engineers, theorists and experimentalists, research fellows, graduate students, and technicians. Much of the staff worked on projects and subprojects in physics; some specialized in the design and operation of nuclear accelerators. Fourteen of the cyclotrons then in the United States had been built or were run by experts trained in Lawrence's laboratory. Big physics promised to become the order of the day elsewhere as physicists more and more aimed to emulate Lawrence—and as events abroad increasingly compelled them to conclude, with Watson Davis: "The most important problem before the scientific world today is not the cure of cancer, the discovery of a new source of energy, or any other specific achievement. It is: How can science maintain its freedom and . . . help preserve a peaceful and effective civilization?"[6]

[6] Davis in *Science News Letter*, Aug. 20, 1938, p. 117. The fourteen-cyclotron statistic is in [Natural Sciences Division, Rockefeller Foundation], "Proposed Action," 1940, University of California Cyclotron, RF, file 205D.

XIX

Organizing for Defense

In 1938 almost 1,300 American scholars and scientists, representing 167 academic institutions and ranging ideologically from Robert A. Millikan to Dirk J. Struik, the mathematician who also edited the Marxist *Science and Society*, issued a manifesto. The document condemned the fascist suppression of science, castigated Nazi racial theories, asserted the legitimacy of modern theoretical physics, and ringingly concluded: "Any attack upon freedom of thought in one sphere, even as nonpolitical a sphere as theoretical physics, is an attack on democracy itself."[1]

Like other Americans, the nation's men of research differed over how democracy was to be defended. Various scientists, largely on the political left, some of them pro-Communists, others pacifists, disapproved of the preservation of intellectual freedom by military force. They proposed instead to ease the material wants of mankind, keep the international scientific community an apolitical beacon of peaceful cooperation, and, in the last resort, "'go on strike,'" as *Scientific American* urged, if the government demanded the use of their technical talents for military purposes. When the Nazis marched into Poland, a group in the American Association of Scientific Workers drew up a neutralist petition which denounced the military use of science as a "perversion." Circulated in

[1] The manifesto is quoted in *New Republic*, 97 (Dec. 21, 1938), 188–89.

the scientific community through the months of the phony war, the petition claimed a large number of signatories.[2]

But like Robert Millikan, the nation's generally conservative senior scientists had never flagged in their support of military preparedness. Neither had most of their junior colleagues in the scientific mobilization of World War I, including the chemist James B. Conant. The product of a comfortable Dorchester, Massachusetts home, where books and political talk were family staples, Conant had helped produce chemical weapons for the U.S. Army; he saw no difference between poisoning a soldier and blowing him to bits. After the war Conant had returned to Harvard with three ambitions: to become the leading organic chemist in the United States, the president of the university, and a member of the cabinet. In the 1920s, along with marrying Grace Thayer Richards, the daughter of the Nobel chemist Theodore Richards, who was his former professor, Conant had made considerable progress toward his first goal. In 1933 he had achieved the second. Now, "a square-shooting, level-headed liberal," in the description of a faculty physicist, and a well-read student of history and foreign policy, Conant joined William Allen White's Committee to Defend America by Aiding the Allies, and campaigned against the neutrality laws, convinced, as he told a CBS radio audience, that we could not "live as a free, peaceful, relatively un-armed people in a world dominated by totalitarian states."[3]

Some Jewish scientists may have sympathized with the antiwar attitudes of the political left, but native or refugee, scarred by anti-Semitism at home or abroad, most lost their tolerance for such sympathies after the rise of Hitler. Before 1933 Albert Einstein, a socialist and pacifist, repeatedly called for resistance to armed service and counseled his fellow scientists to refuse their expertise to the military. After 1933, still a socialist, he publicly declared that he saw no alternative to the rearmament of the western democracies.[4] Many American scientists

[2] *Scientific American*, 151 (Nov. 1934), 233; Robert S. Mulliken, "The Peace Resolution of the American Association of Scientific Workers," *Science*, 91 (May 3, 1940), 432. At the opening of the blitzkrieg the head of the Association, which had been founded as something of a professional scientific union, sent the petition to the President and the newspapers. Though it was reportedly signed by five hundred scientists, in the Boston-Cambridge area the document had been endorsed by only a small meeting of the local AASW branch, which cavalierly decided that it spoke for the entire membership. Many of the alleged signers quickly repudiated the petition. George Kistiakowsky, a veteran of the White Russian Army and future Science Advisor to the President, charged the branch with "Communist domination" and promptly resigned, as did some of his colleagues. *New York Times*, May 20, 1940, p. 6; June 30, 1940, p. 11; *Science*, 91 (June 21, 1940), 597. On the original aims and leadership of the Association, see "Provisional Program . . . ," [Spring 1939]; Donald Horton to Oswald Veblen, April 27, 1939, OV, Box 17.

[3] James B. Conant, *My Several Lives: Memoirs of a Social Inventor* (New York, 1970), pp. 49, 52, 212; Edwin Kemble to Conant, May 9, 1933, SHQP, microfilm roll 53.

[4] Otto Nathan and Heinz Norden, *Einstein on Peace* (New York, 1960), pp. 90, 95, 112, 117–18, 123, 135, 142, 158, 226, 229–30, 244, 252.

knew from first-hand visits, correspondents abroad, or the refugees the virulence of Hitler's Germany. They also knew that a number of able scientists remained in the Third Reich. The most dramatic scientific news of 1939 was the splitting of the uranium nucleus; the event occurred in Germany. Some German physicists, like the formidably able Werner Heisenberg, were believed to be fellow-traveling with the regime; others, like Philipp Lenard, had publicly given it their outright support. Thoughtful American scientists respected the potential of German military technology and worried about what horrendous uses the Nazis might make of it.

On the whole, the American scientific community moved away from isolationism rapidly, and at a faster pace than the nation at large. And when the phony war ended with the blitzkrieg, even many of the signers of the neutralist petition recanted, declaring with the University of Chicago chemist and later Nobel laureate Robert S. Mulliken that reasonable measures for the protection of intellectual freedom "now include immediate steps for our own defense."[5]

All through the 1930s many army and navy officers had acknowledged what Robert Millikan told a military gathering in 1934: Defense research ought to be "a peace-time . . . and not a war-time thing. . . . [It] moves too slowly to be done after you get into trouble."[6] In both armed services the technical bureaus now had well-equipped experimental facilities, and the technical chiefs considered ongoing programs of research and development a fundamentally necessary part of their bureaus' mission. Though virtually none of the technical agencies had developed radically new forms of military technology, most had at least brought the devices of World War I to a state of greater sophistication and effectiveness.

But isolationist sentiment, including the antimilitary attitudes reflected in Senator Gerald P. Nye's investigation of the munitions industry, kept appropriations at a low level, and between 1935 and 1939, research expenditures in each of the armed services never exceeded $7,000,000 a year, or merely 10 percent of the cost of a single battleship; in fiscal 1935 armed service expenditures for research and development totaled only 0.6 percent of the combined army and navy budgets.[7] Then, too, thick walls of secrecy separated the technical bureaus of one service from their counterparts in the other. And while the army and navy technical bureaus employed civilian scientists and engineers, the armed services

[5] Mulliken to the editor, *Science*, 91 (May 31, 1940), 525–26.

[6] "Report of Conference on Research . . . War Department, Science Advisory Board, National Research Council, Jan. 2, 1934," pp. 14–15, RAM, Box 7.

[7] U.S. Congress, House, Select Committee on Post-War Military Policy, *Hearings, Surplus Material–Research and Development*, 78th Cong., 2d Sess., Nov. 1944–Jan. 1945, pp. 228–29; *Report to the President of the United States upon a National Program for Putting Science to Work for the National Welfare* (mimeographed version; Washington, D.C., 1935, copy in Library of Congress), pp. 47–48.

did not have the money to pay for much outside civilian technical help, either industrial or academic, and as prerogative-conscious as ever, they did not like to rely upon it.

The small budgets, lack of interservice cooperation, and limited contacts with civilian science all helped retard the attack on what ranking officers in both services recognized as one of the most critical problems of defense, the detection of enemy aircraft. In 1930 Lawrence H. Hyland, a young staff member at the Naval Research Laboratory, or NRL, noticed that passing planes revealed their presence by reflecting back radio signals transmitted from the ground. Hyland's serendipitous observation opened possibilities for the accurate detection of aircraft by radio waves, or better yet—as Leo Young, a self-taught radio enthusiast who was a member of the NRL, realized—by radio pulses. In Young's scheme, a high-frequency beam of short pulses, each separated by relatively long intervals, would sweep the sky. By noting the bearing of the reflected pulses, one could determine the bearing of the aircraft. By measuring the time it took a single pulse to make the round trip to the plane and back, one could also deduce its distance from the transmitter. In the United States, Young's idea was the beginning of what the navy would eventually dub "radar," for *r*adio *d*etection *a*nd *r*anging.[8]

But while radio wave apparatus was well developed, Young had to break new electronic ground to design reliable high-frequency pulse equipment. Since the NRL funds were limited and its staff burdened with routine testing for the Bureau of Engineering, his pulse research program received considerably lower priority than detection by the more familiar radio waves. Collaborating with another staff member, Young managed to work on the pulse approach by scrounging time between other projects and staying after hours. His task was eased somewhat when in 1935 the House Naval Affairs Committee, its members impressed by the radar work, added $100,000 to the Laboratory's budget. In 1936 Young finally had equipment workable enough to detect aircraft two and a half miles distant, and, by 1938, with pulse research now at the top of the NRL priority list, a set reliable enough for use at sea. Tested on maneuvers in 1939, it performed so well that a high-ranking officer quite accurately called it "one of the most important military developments since the advent of radio itself."[9]

For all its importance, the army did not learn about the possibilities of pulse radar until 1936, and then only because a Signal Corps scientist happened to visit the naval laboratory. The Corps promptly inaugurated

[8] Henry E. Guerlac, "Radar," unpublished manuscript, copy in LAD, Section A–III, pp. 42–43, A–IV, pp. 9–14; Harold G. Bowen, *Ships, Machinery, and Mossbacks: The Autobiography of a Naval Engineer* (Princeton, N.J., 1954), pp. 138–43.

[9] Guerlac, Section A–IV, pp. 19–32; Bowen, pp. 141–43, and for the quotation, p. 145.

its own skimpily funded radar research program at Fort Monmouth, New Jersey, expecting to receive progress reports from NRL. The reports suddenly stopped coming when the navy reclassified pulse radar research from confidential to secret. The exchange of information was ultimately reestablished, but the two services never created a joint radar research program. Neither for some time did they make any serious attempt to exploit the growing corps of civilian experts in such radar-relevant fields of electronics as early television and nuclear accelerators.[1]

By themselves the army and navy did manage to develop radar equipment comparable to what the British deployed so effectively in the Battle of Britain. Eventually put into production, the Signal Corps' equipment would reveal the attack squadrons approaching Pearl Harbor almost an hour before their arrival. The Naval Research Laboratory version, its shipboard presence made manifest by a seventeen-foot-square rotating antenna called the "flying mattress," would help turn back the Japanese navy at the Battle of the Coral Sea. But had the armed services enlisted the aid of civilian experts earlier, and had they cooperated with each other, they might have achieved an operational pulse radar sooner than the eight years since the advent of Leo Young's promising idea, perhaps even soon enough for the defense command at Pearl Harbor to learn how to use it.

There was simply no overarching organization within the military establishment designed to link the technical bureaus of the two services with either each other or the civilian world. Outside of it there was only the National Research Council, whose relations with the army and navy had never recovered from the desuetude into which they had fallen at the end of World War I. The military's failure to maintain working contact with the Council troubled Captain Clyde S. McDowell, the veteran of the submarine detection project at New London. Eager to rectify the situation, McDowell got nowhere until 1933, when Admiral William H. Standley, an enthusiast of research, became Chief of Naval Operations.[2]

With Standley's backing, McDowell enlisted the NRC for the advancement of naval technology—on the understanding that the arrangement would cost the navy no money in peacetime. Isaiah Bowman testily insisted that helping the service required "something more than the voluntary donation of time by scientists already fully occupied in their respective laboratories." Following a remedial suggestion of Bowman's, the navy agreed to induct a small group of young scientists into the reserve. When called for active duty part of each year, they were to familiarize themselves with naval requirements, do naval research, and, all the while, assist navy liaison with the NRC. Thirty scientists were

[1] Guerlac, Section A–IV, pp. 45–48, Section B–I, p. 19, n. 23.
[2] McDowell to Robert Millikan, Feb. 2, 1933, May 11, 1933, July 27, 1933, RAM, Box 7.

ultimately commissioned. Short of funds, the navy called only one to active duty.[3]

Millikan offered the services of the NRC to the army in 1934 at a conference of generals. Chief Signal Officer Irving J. Carr allowed that he would rather take his complex technical problems to the "wonderful staffs" of the nation's scientific laboratories than "carry high priced physicists on our rolls." Central responsibility for the procurement and development of new material rested with Major General Robert E. Callan, the assistant chief of staff for supply, and he was by no means enthusiastic about joining hands with the NRC. Echoing the views of his predecessors in World War I, Callan made clear that the principal determining factor in the development of new weapons had to be tactical needs, not the imagination of civilian scientists. Though the army might turn to the NRC if it needed help on a special project, it did not, the general implied, require the Council's ongoing advice, and, in any case, there was no money to pay for it.[4]

Some of the leaders of the NRC, aware that it did not possess the organizational weight to be militarily effective, looked respectfully at the contrary example of the National Advisory Committee for Aeronautics. This Committee had enjoyed the enthusiastic support of the navy's air officers, though not of all the army's. Billy Mitchell, who thought virtually all of military aviation ought to belong to the War Department, told a congressional probe that the Committee had "distinctly retarded the better development of aviation" by delving into the "political side of the . . . question." "Its offices," he added darkly, "are in the Navy Building."[5] But, knowing that today's plane might be obsolescent tomorrow, most ranking army air officers, including Mitchell, wanted a vigorous program of research and development and were friendly to the Committee. Faced with congressional parsimony, they needed the aeronautical agency to do the research they could not afford to do themselves.

Led by the Johns Hopkins physicist Joseph S. Ames, who remembered the production fiasco of wartime aviation, the Committee on its part often put in good words for keeping the aviation industry in business with ongoing military procurement programs. It also developed excellent facilities for aerodynamical research at Langley Field and

[3] Bowman to Millikan, Nov. 30, 1934; Capt. S. C. Hooper to Millikan, Oct. 5, 1937, RAM, Box 7. It is interesting to note the cautious observation of George Ellery Hale, who, aware of the "greatly enhanced public attitude against war" and certain that the campuses harbored "an extraordinary number of extreme pacifists," suspected that if anything threatened to embroil the National Research Council in controversy it was "participation in military or naval affairs." Hale to Isaiah Bowman, Aug. 3, 1933, RAM, Box 7.

[4] "Report of Conference on Research . . . War Department, Science Advisory Board, National Research Council, Jan. 2, 1934," pp. 20, 16–17.

[5] U.S. Congress, House, Select Committee of Inquiry into the Operation of the United States Air Service, *Hearings into the Operations of the U.S. Air Services*, 68th Cong., 1925, p. 1890.

sponsored studies of the fundamental principles of flight. Aeronautical engineers around the world acclaimed the Committee's technical reports. So did army and navy air officers after the Committee developed a streamlined engine cowling that made possible greater speed without greater engine power or fuel consumption. Facilitating the exchange of technical information between the two air services, the Committee got them to cooperate, even frequently persuaded one to avoid unnecessarily duplicating the research of the other. It also brought about civilian contributions to military aeronautical development programs by letting research contracts to academic as well as industrial laboratories.

The Committee, which owed so much of its origins to Charles D. Walcott, was rooted in the operational tradition of the Powell Survey and the Carnegie Institution of Washington. By stimulating expansively cooperative efforts, it took a large-project approach to the advancement of aeronautics. But by letting research contracts to academic and industrial laboratories, it also embodied the tradition of grants to individuals. The National Research Council had aimed to combine both traditions. The National Advisory Committee for Aeronautics succeeded where the Council had failed because, unlike the NRC, the Committee was a governmental agency with independent administrative status, its own congressional appropriation, and authority to conduct militarily important research. As such, it was a compelling model to Vannevar Bush, who in the spring of 1940 was decidedly interested in how best to mobilize civilian science generally for national defense.

"I'm no scientist, I'm an engineer," Bush would assert.[6] Reedy, flint-faced, his gaze shrewd and impatient behind the thin-rimmed spectacles, Bush looked like a type-case no-nonsense Yankee. A grandfather had been a whaling captain. The father, Perry Bush, had left his strict Cape Cod Methodist home to crew on a mackerel boat. He eventually studied at the liberal Tufts School of Theology and pursued a career in the Universalist Church. But Bush grew up across the river from Boston, in Chelsea, a hilly town of simple frame houses which at the turn of the century was filling up with immigrants from eastern Europe. The YMCA, the recreational center for Chelsea boys, refused to admit Jews and Catholics. More ecumenical, the Reverend Perry Bush joined with the local Catholic priest in efforts at civic betterment, especially for the young. He befriended saloonkeepers, mastered the game of pool—he could sink more than one rack of balls without a miss—and opened his church to acolytes of other faiths. Young Vannevar Bush, who as a liberal Protestant was also barred from the YMCA, found all of his boyhood friends—and some of his long-lasting attitudes—among Chelsea Jews and Catholics.[7] Whatever his Yankee heritage, Bush held upper-class,

[6] Interview with the late Vannevar Bush, July 24, 1965.

[7] Transcript of Eric Hodgins interview with Vannevar Bush, 1966, p. 85A, copy in VBM; Vannevar Bush, *Pieces of the Action* (New York, 1970), pp. 238–39.

especially Brahmin, Boston in considerable contempt and Vannevar Bush himself in high, often assertively high, esteem.

By the time Bush followed his father and two older sisters to Tufts, there was little left to pay his educational expenses. The Reverend Bush had worked his way through the college by hauling coal into the dormitories. Vannevar, a sickly boy who even now was sometimes lame with a rheumatic leg, earned money mainly by tutoring. He charged each student fifty cents, except shills like the college football star, who got his fifty cents back for loudly extolling the indispensability of Bush's tutoring in order to pass the course.[8] Bush excelled at mathematics. But engineering, a more lucrative discipline, seemed preferable, and in 1916 Bush received a doctorate in that field jointly from Harvard and MIT.

In 1919 Bush was appointed an associate professor of power transmission in the MIT department of electrical engineering. Department chairman Dugald C. Jackson would tell new faculty that they were expected to double their income by industrial consulting. During World War I Bush had worked at New London developing a submarine detector under the sponsorship of a J. P. Morgan firm. Now a young man with a family to support—his wife, a onetime member of the choir in his father's church, was the daughter of a prosperous Chelsea grocer who lived above the Reverend Bush in a house at the top of the hill—he consulted to the Morgan and other companies. He also developed patentable devices of his own. He never exploited the patents himself—Bush preferred the creative to the commercial side of invention—but made money by assigning them to firms that would. More important, he helped Laurence K. Marshall, his college roommate and best man at the wedding, turn inventions into profitable ventures, including a successful thermostat company and a new vacuum tube concern, the Raytheon Corporation.

At MIT, Bush earned popularity for his stimulating teaching and a special reputation for insisting that even engineering students had to master the art of writing crisp, clear, precise sentences. He pushed electrical engineering graduate work, which was a new development in the profession of his day. In research, Bush pioneered the application of advanced mathematical techniques to the analysis of power circuits. The resulting equations defied solution by ordinary methods. To deal with them, Bush invented a crude but successful mechanical computer. Aided by a $25,000 grant from the MIT administration and the talents of many younger faculty and assistants, he drove forward the development of his rudimentary device into the sophisticated differential analyzer, a remarkable mechanical antecedent of the electronic analogue computer. The differential analyzer found a wide range of industrial and scientific uses and brought special professional renown to Bush and his department.

[8] Transcript of Hodgins interview, p. 7.

At the end of the 1920s, MIT was said to have at least one third of all the electrical engineering graduate students in the country; many of them worked for Bush.[9]

Since chairman Jackson was often absent on his own extensive consulting business, Bush assumed increasing responsibility for the general management of the department, including its relations with electrical companies, trade associations, and wealthy donors. As an administrator, he could be quick-tempered, brusque, and intolerant. To some of his colleagues, he also seemed arrogant, autocratic, and ferociously ambitious, with a tendency to confuse the interests of his own computer development program with those of the department as a whole. But Bush also inspired admiration and loyalty in other, especially younger, colleagues; among some members of his research group the loyalty bordered on adulation. Even Bush's critics appreciated his technical genius, wide-ranging intellectual curiosity, willingness to boost other people's promising research, and considerable executive talents. Bush caught the eye of the new MIT president Karl Compton, who appointed him dean of engineering and vice-president of the university.

The two men advantageously complemented each other, Compton with his genial manner and tact, Bush with his slashing decisiveness and drive. They bought neighboring farms in New Hampshire, primarily for recreation, though Bush also intended to raise turkeys and make his acres pay. Bush worked under Compton on the Science Advisory Board. Appointed to the National Advisory Committee for Aeronautics, he, too, became increasingly involved in the affairs of governmental science. He shared Compton's faith in the economic benefits of technological innovation. No friend of the New Deal—and a great admirer of Herbert Hoover—he attacked federal restraints on business for rendering the creation of new products and industries "nearly impossible." But along with the entrepreneurial propensities and political conservatism, Bush maintained a legacy from his father. He spoke seriously of the engineer's obligation to be disinterested and hailed professionalism at its best as "an anchor . . . [of] democracy." Like Perry Bush, he was usually prepared to question orthodoxies, defy tradition, master offbeat skills, and learn how to deal with people not normally found in his own professional circles. Bush richly deserved the compliment he once offered the educator, businessman, inventor, organizer, and engineer Elihu Thomson: "You have showed us that a man may be truly a professor and at the same time very practical."[1]

[9] Bush to Samuel Wesley Stratton, Oct. 24, 1927; Jan. 4, 1928; June 4, 1930, SWS, file 535; Bush to Warren Weaver, Jan. 6, Jan. 19, April 15, July 7, 1933, VBM. The graduate student estimate is in *MIT, Report of the President, 1929*, p. 40.

[1] "Patent System and Stimulation of Industries," *Report of the Science Advisory Board . . . 1934 to . . . 1935*, pp. 339–40; Bush, "The Engineer and His Relation to Government," *Science*, 86 (July 30, 1937), 91; Bush, "In Honor of Elihu Thomson," *Science*, 77 (May 5, 1933), 418–20.

Bush assumed the chairmanship of the National Advisory Committee for Aeronautics when in 1939 he moved to the capital as president of the Carnegie Institution of Washington. His post in the Committee drew him increasingly into the defense program and he frequently discussed the general state of American military preparedness with his friends Frank Jewett, Karl Compton, and James B. Conant. All of them emphatically pro-Ally, they were drawn together, Bush recalled, by "one thing we deeply shared—worry." They knew that the war would be a highly technological contest. The discovery of nuclear fission in 1939 had provoked speculations in the press about the possibility of an atomic bomb, and the Bush group found the prospect of such a weapon in Nazi hands terrifying.[2] As veterans of the technical mobilization during World War I, they knew enough about the military's system of research and development to distrust its capability for the innovation of new weapons. In their opinion, something had to be done to reduce the insularity of the military's technical bureaus both from each other and, more important, from the world of civilian science.

Bush, though a member of the National Academy, saw no point in relying upon the National Research Council either to mobilize civilian science or to coordinate the research programs of the armed services. He remembered how in World War I the Council's lack of money and authority had hamstrung its scientists at New London. Knowing all about the sorry record of the Science Advisory Board, he was persuaded that in dealing with the government any agency of the Academy was doomed to futility. Like George Ellery Hale before him, Bush was prepared to be an entrepreneur of organization. Unlike Hale, he was convinced of the need to mobilize science under a new *federal* agency, one resembling his National Advisory Committee for Aeronautics—an official body, funded by the government and reporting directly to the President.[3] Jewett, Compton, and Conant endorsed such a scheme. Though it represented a radical departure from the apolitical Academy tradition, Bush intended the new agency to last for the duration only and he aimed to man it with enough members of the Academy to keep science safe from politics.[4]

[2] Bush, *Pieces of the Action*, p. 33. Hitler was said to have ordered Germany's best scientists to work on nuclear fission. William L. Laurence, "The Atom Bomb Gives Up," *Saturday Evening Post*, 213 (Sept. 7, 1940), 61. Bush to Frank Jewett, May 2, 1940, VB, Box 55, file 1375.

[3] Interview with the late Vannevar Bush; Frank Jewett to Frank R. Lillie, April 2, 1936, RAM, Box 7; Frederick Delano to Steve Early, May 25, 1940, FDR OF 4010. For Bush's admiration of the National Advisory Committee for Aeronautics, see Bush to Harold G. Moulton, May 29, 1940, NACA, Bush file.

[4] Compton was no less discontented than Bush with the Academy after the fiasco of the Science Advisory Board. Jewett worried that the Academy lacked the working capital to prosecute large-scale military research projects. Compton to Bowman, April 4, 1936, RAM, Box 7; Jewett, "A Review of the Years 1939–1947," *Annual Report of the National Academy of Sciences, 1946/47*, pp. 2–3.

In May 1940 the influential governmental lawyer Oscar Cox arranged for Bush to discuss his proposal with Harry Hopkins. Now the secretary of commerce, Hopkins was considering the mobilization of the nation's technical genius through a new inventors' council. ("The people of the United States," Admiral Harold G. Bowen, the director of the Naval Research Laboratory, disapprovingly remarked, "still believe that . . . some great revolutionary invention in the art of war may be the product of a solitary unsuccessful inventor working in an attic." Still, Bowen added, what "the public believes . . . must always be a concern of Government.") But it was of course 1940, not 1915, and Harry Hopkins was no Josephus Daniels.[5] Aware that the trained scientist and engineer had long since won the day against the Edisonian inventor, Hopkins agreed to keep the administration of inventions separate from the organization of science and gave Bush the green light for what by now had been named the National Defense Research Committee, or NDRC.

Early in June, Hopkins took Bush into the Oval Office. Bush had a single sheet of paper containing a four-paragraph sketch of his proposed agency. Some ten minutes later Roosevelt said, "That's okay. Put 'OK, FDR' on it."[6] On June 27, 1940, the President approved the official order establishing NDRC. Hopkins assured Bush that the agency would have FDR's full support and as much money as it needed from the President's emergency funds.

The chairman of NDRC was Bush himself. The membership included Conant and Compton, and Richard C. Tolman, the respected theorist of physical chemistry and dean at the California Institute of Technology, a committed advocate of preparedness who was already on his way to Washington. In addition, there was Frank Jewett, the president of the Academy and, as a vice-president of AT&T, a key figure in the circles of industrial research; Conway P. Coe, the commissioner of patents, whom Hopkins had designated to advance the scheme for an inventors' council; and a delegate each from the army and navy.

Bush assigned himself the responsibility of dealing with the President, the Congress, and the armed services. He designated each of the other civilian members the head of one of the five general-purpose research divisions (they covered armor and ordnance, chemistry and explosives, communications and transportation, instruments and controls, patents and inventions). Each man was in turn to establish as many sections as necessary to handle his division's specific military problems. It was a wise organizational design, since it put the day-to-day technical work in the hands of the sections and left the NDRC leadership free to deal with broad issues of policy. Taking care to carve out a politic position among

[5] Bowman to Max Mason, April 20, 1940, RAM, Box 7. See the staff memo, C. S. Guthrie to E. J. Noble, May 20, 1940, HLH, Book #2: "Mobilization of Scientists."

[6] Interview with the late Vannevar Bush.

the existing research bureaus in the government, Bush had already decided that NDRC would supplement rather than replace the work of the military's own technical bureaus. He also proposed to leave aeronautical activities to the National Advisory Committee for Aeronautics and, in order to keep the Academy happy, to make use of its advisory services. Bush was no less certain about another major policy point—just how the National Defense Research Committee was actually to conduct defense research.

Conant expected the NDRC to follow the example of the NRC in World War I—to oversee the creation of new central laboratories, like the poison gas establishment in which he had worked, and to staff them with scientists from around the country. But Jewett, who evidently remembered the struggle to create the experimental station at New London, urged that NDRC use the nation's existing research establishment "to the limit." That way the special competence of existing research groups could be exploited, men could stay on the campuses to teach, and there would be no time and expense wasted in constructing new facilities from scratch. Bush himself had written into the founding documents for NDRC an authorization to use the research contract in the manner of the National Advisory Committee for Aeronautics.[7] At the outset he and his colleagues agreed to conduct defense research primarily by letting contracts for military purposes to industrial corporations and universities. Their decision committed NDRC, like the NACA, to both the large-project and individual grant approaches and extended the combination beyond aeronautics to federal sponsorship of all the militarily relevant sciences. It was a decision with major implications for the future.

In the summer of 1940, NDRC joined hands with Frederick Delano's resources board, which was constructing a National Roster of Scientific and Specialized Personnel, to inventory the nation's research facilities and technical manpower. Thousands of scientists and engineers responded, including Albert Einstein ("I was always interested in practical technical problems," the onetime patent office employee reminded the Roster officials). Karl Compton made the rounds of the military agencies and compiled a list of critical projects. Somewhere in the middle of it all the NDRC started letting research contracts. By December 1940 it had authorized 126 of them, to 32 academic institutions and 19 industrial corporations.[8]

By the following spring NDRC, which now had a London office

[7] Conant, *My Several Lives*, p. 236; Jewett to Bush, June 21, 1940, OSR, Entry 39, NDRC, Jewett OF 49.00 II. See the draft documents Bush turned over to Hopkins' office, attached to E. J. Noble to Hopkins, June 13, 1940, HLH, Book #2: "Mobilization of Scientists"; A. Hunter Dupree, "The Great Instauration of 1940," in Gerald Holton, ed., *The Twentieth-Century Sciences* (New York, 1972), pp. 457–59.

[8] Einstein is quoted in "Report to the Nation—'National Roster of Scientists,'" NRPB, Entry 8, file 106.24; *Review of Scientific Instruments*, 12 (Feb. 1941), 114.

for liaison with the Allies, was emerging as a pivotal organization in the military research and development complex. Its staff jammed the buildings of the Carnegie Institution and the National Academy, and was spilling over into Dumbarton Oaks, the Georgetown estate owned by Harvard University. Bush ran the operation from his wood-paneled Carnegie office, pleased at the progress—yet, despite it all, increasingly aware that NDRC was less than completely adequate to cope with the military in the arena of defense research.

Some army and navy officers disliked the independence of the NDRC approach to forging new weapons. Fearing that secrecy could not be maintained on the campuses, they preferred to have the research done in their own facilities. The army and navy technical services were more ably staffed than a quarter of a century before; with their appropriations for research and development now suddenly rising to unprecedentedly high levels, they were also perhaps even more jealous of maintaining their prerogatives in military matters. Major General Gladeon M. Barnes, an accomplished University of Michigan graduate in engineering, the former chief of the Ordnance Department Technical Staff, and now the head of the whole of Army Ordnance, had no doubts about the proper role of Bush's organization. It was to do the bidding of the army and navy by working on such weapons and devices as the generals and admirals deemed militarily necessary. It was not to plunge ahead on its own, especially in the development of prototypes.[9]

While NDRC could conduct *research* on new military devices, it did not possess the authority to approve the *development* of its preliminary models into prototypes ready for production. Neither was NDRC as closely involved as Bush would have liked in the military's overall planning of weapons research. Besides, NDRC was outgrowing the President's emergency funds and needed direct access to the Treasury. And though NDRC was making considerable progress in weapons research, there was no comparable organizational focus for research in military medicine. To Bush's restless administrative mind, NDRC had to be reconstructed, and the chance to accomplish just that came his way when in May 1941 the Bureau of the Budget requested a general report on the organization of defense science.[1]

The report led to an executive order—Bush probably wrote it himself—which, in the exaggerated phrase of the *New York Times* science

---

[9] Constance McLaughlin Green et al., *The Ordnance Department: Planning Munitions for War* (Office of the Chief of Military History; United States Army in World War II, The Technical Services; Washington, D.C., 1955), pp. 220–21, 228. Karl Compton, "Scientists Face the World of 1942," in Ruth Nanda Anshen, ed., *Science and Man* (New York, 1942), pp. 23, 25.

[1] Interview with the late Vannevar Bush; Irvin Stewart, *Organizing Scientific Research for War: The Administrative History of the Office of Scientific Research and Development* (Boston, 1948), p. 35; Bush, "Organization of Defense Research," attached to Harold D. Smith to the President, March 17, 1941, FDR, OF 2240.

editor, seemed to crown him a "czar of research." The document made Bush's organization eligible for a direct congressional appropriation. It also awarded him all the additional bureaucratic powers he wanted, including the authority to advance ideas for weapons from the germinal to the production stage. And organizationally the order relegated NDRC to an advisory role, created an NDRC-like medical committee, and located both in a new operating agency under Bush's directorship, the Office of Scientific Research and Development, or OSRD.[2]

As OSRD brought more scientists into the defense research effort and awarded more contracts, a journalist remarked of Bush: "He has done a tremendous job, and made less fuss about it than Harold Ickes would make about brushing his teeth."[3] On Capitol Hill he was usually an impeccable witness, even-tempered, straightforward, and as informative as possible. Within the maze of the executive branch, he tried to avoid strong-arm tactics, relying instead on reason, persuasion, and as much patience as Vannevar Bush could muster.

The armed services might have an exaggerated attachment to secrecy (nuclear physicists chuckled over the story that the army had tried to suppress a Superman cartoon showing the hero under bombardment by a cyclotron), but Bush respected the military's proper concern with security. We will listen but we won't talk, the OSRD maxim went. To assure that if anyone did talk it would not hurt the war effort seriously, Bush imposed a policy of compartmentalization on all of OSRD; no scientist, engineer, or technician was allowed any more information than was necessary to the performance of his task.[4] While OSRD was exhilaratingly free of military direction, Bush held the reins tightly enough to keep the energies of his scientists funneled into weapons likely to be of practical value in the current war, not in the next. While the civilian-dominated OSRD did spectacularly well at the military's business, Bush often complimented the armed services on public occasions, and he kept OSRD, which operated only a nominal public relations office and issued virtually no press releases, out of the limelight.

From time to time Bush did tend to treat high-ranking armed service officers as though they were technical dunderheads, to operate in a steamroller manner, to allow that his civilian scientists were singlehandedly remaking the defense posture of the army and navy. But if Bush provoked antagonisms among the military, no matter: He had his budget, his legal authority, his access to the President—and a readiness to use them all as he saw fit. Meeting with some admirals to hash out a navy refusal to

[2] Waldemar Kaempffert, "American Science Enlists," *New York Times Magazine*, Nov. 2, 1941, p. 3. The pertinent sections of the order are in *Review of Scientific Instruments*, 12 (Aug. 1941), 420.

[3] J. D. Ratcliff, "War Brains," *Collier's*, 109 (Jan. 17, 1942), 28.

[4] Louis Ridenour, "Science and Secrecy," *The American Scholar*, 15 (Spring 1946), 140; Stewart, *Organizing Scientific Research for War*, p. 28; Karl Compton, "Scientists Face the World of 1942," p. 22.

provide OSRD with certain crucial information, Bush typically did not argue. He simply reminded everyone: You have your orders from the President and I assume the navy will cooperate.[5] Bush generally refrained from going over the heads of other bureaucrats; he just as generally let them know that he could do so if they failed to yield. In 1917, by the time Congress declared war, the jealously private National Research Council had accomplished virtually nothing for national defense; by that fateful Sunday in December 1941 NDRC, then the more potent OSRD, had provided the country with almost eighteen precious months of military research and development.

More than most scientists, Bush understood the main source of power in Washington. "I knew that you couldn't get anything done in that damn town," he recalled, "unless you organized under the wing of the President."[6]

[5] Bush, *Pieces of the Action*, p. 281.
[6] Interview with the late Vannevar Bush.

XX

A Physicists' War

In the summer of 1940, when NDRC was getting started, Prime Minister Winston Churchill appointed a small technical mission to visit the United States, under the leadership of Sir Henry Tizard, an influential defense scientist and a vigorous proponent of the full exchange of technical information with America. Late in August, Tizard flew across the Atlantic. The rest of his mission followed on the *Duchess of Richmond*, shepherding a black box containing some of Britain's most vital secrets in military technology.

The Tizard mission met with Bush's experts, including the head of his radar section, Alfred L. Loomis, a first cousin of Secretary of War Henry L. Stimson and a retired investment banker who practiced physics in a research laboratory at his home in Tuxedo Park. Loomis knew that considerable advantages were to be gained from the development of the then-primitive radars in the centimeter wavelength range. Like the British radar currently maintaining its crucial watch on the English Channel, the U.S. Army and Navy radar sets all operated at the relatively long wavelengths of one to two meters. In the centimeter, or "microwave," region, radar would require much smaller antennas, locate aircraft, especially low-flying aircraft, more accurately, and better discriminate between adjacent targets. Loomis' radar section, naming itself the Microwave Committee, had decided to leave the further development of long-wave systems to the armed services and to concentrate on centimeter devices. In the summer of 1940 the Microwave Committee saw no quick way to overcome the chief obstacle to centimeter radar: the lack of a vacuum tube capable of generating sufficiently energetic radiation at such short wavelengths. But then the Tizard Mission produced from its black box the astounding resonant cavity magnetron.

Developed by a British team early in 1940, the magnetron emitted ten-centimeter radiation at an intensity some thousand times as great as the most advanced American tube. Loomis excitedly told Stimson that the British device had advanced the U.S. radar development program by a full two years.[1] The British urgently needed special radar equipment to help RAF fighters cope with the massive German campaign of night bombing. Loomis' Microwave Committee agreed to develop the magnetron at high priority into an airborne intercept system, code-named the AI-10. Drawing upon the successful organizational example of the British radar enterprise, the Committee also decided to establish a major radar research laboratory run and largely staffed by physicists.

The radar project was promptly established under an NDRC contract at MIT, whose staff already included a small group of microwave experts and whose proximity to the ocean and the Boston airport would ease the conduct of airborne tests. Like Ernest Lawrence's cyclotron operation at Berkeley, the enterprise was named the Radiation Laboratory —and soon nicknamed the "Rad Lab"—not least to mislead outsiders into thinking that its resident physicists were engaged in something then considered so remote from military needs as nuclear physics. And in short order, Ernest Lawrence himself, acting for the Microwave Committee, offered the directorship of the laboratory to his friend and protégé, Lee A. DuBridge.

Like Lawrence, whom he admired enormously, DuBridge had simple tastes, boundless energy, and an uncomplicated eagerness to do something important. Growing up a good student and good Methodist in various cities from the Midwest to the west coast, he found his métier in physics while on a scholarship at Cornell College, Iowa, where his father, a YMCA instructor, had once been athletic coach. After his Ph.D. and an NRC fellowship at Caltech, DuBridge gained a national professional reputation for his research in photoelectricity. In 1934, age thirty-three, he was appointed a full professor and chairman of the department at Rochester University. At the time the department was a backwater. DuBridge brought in able young men, including—it was a precedent for Rochester physics—Jewish faculty.[2] He assured Rochester a place on the map when, after an inspirationally suggestive visit from Lawrence, he constructed the university's first cyclotron, then put it to productive use in nuclear research.

Students and faculty in and out of physics liked DuBridge. His boyish friendliness was warmed by the wonder that the world should be treating

---

[1] "The Diary of Henry L. Stimson," Oct. 2, 1940, HLS.
[2] DuBridge to Rush Rhees, Jan. 14, 1935, RR.

him so well. Mrs. Doris May Koht DuBridge—she was the daughter of a farmer-merchant from upstate Iowa who had married her husband despite her parents' suspicion that all physicists were doomed to penury—went out of her way to help new faculty get settled and generally made the DuBridge home an unassumingly hospitable place. Increasingly popular on the campus, DuBridge was appointed dean of the Faculty of Arts and Sciences. Word got around the country that he was an excellent administrator and Ernest Lawrence kept recommending him as the best man he knew for any number of jobs. Now, here was Lawrence on the telephone with the offer of a major post in defense research. DuBridge was a Republican who after World War I had seen no sense in getting mixed up with Europe again, but by the fall of 1940 he was no longer an isolationist. Besides, he recalled his decision to accept, if Ernest Lawrence was interested, "I wanted to be in."[3]

With Lawrence's help, DuBridge recruited a scientific staff, and soon some three dozen physicists were tuning up experimental equipment in MIT's massive buildings and fitting out a new rooftop microwave laboratory constructed of wood and covered with gray-green tarpaper. Their average age in the mid-thirties—forty-three-year-old I. I. Rabi was one of the old men of the enterprise—the group included some of the best nuclear physicists in the country. They knew something about high-frequency radiation from their work on cyclotrons, but the magnetron confounded even them at first.

"It's simple," Rabi told the theorists who were seated around a table staring at the disassembled parts of the tube. "It's just a kind of whistle."

"Okay, Rabi," Edward U. Condon asked, "how does a whistle work?"[4]

Rabi was at a loss for a satisfactory explanation. The mysteries of the magnetron had to be solved promptly along with getting a practical microwave system quickly into the field. By December, DuBridge's physicists were shivering in their unheated rooftop penthouse—one of them wore a coonskin coat against the cold—fiddling with a mock-up of the AI–10.

The pressure to succeed was notably high. Along with considering the "doubledomes"—so this generation of army and navy officers called the scientists—incapable of martial practicality, some military men, especially Admiral Bowen of the Naval Research Laboratory, resented the Rad Lab decision to concentrate on microwave radar when the full potential of the military's own long-wave systems had not yet been tapped. In fact, the Rad Lab physicists could not coax the penthouse mock-up of the microwave AI–10 to detect aircraft in the neighborhood

[3] Lawrence to DuBridge, March 14, 1940, EOL, DuBridge file; DuBridge is quoted in Guerlac, "Radar," Section B–II, pp. 13–14.

[4] Quoted in interview with the late Edward U. Condon, Oct. 1967, AIP, p. 22.

until February 1941. Some of Loomis' committee, including even Ernest Lawrence, wondered whether the magnetron could ever be made into an operational airborne unit.[5]

Frantically improvising, DuBridge's physicists finally achieved a workable prototype of an AI–10 in the spring of 1941, when the British had already successfully fought back the German night-bombing offensive. Now the demand for the AI–10 was less pressing in Britain or the United States, where after a series of demonstrations of a prototype, an army review board had concluded that the device did not meet specifications. The future of the Rad Lab looked uncertain; Karl Compton wondered whether NDRC would even fund the enterprise at MIT for another fiscal year. But John D. Rockefeller, Jr. anonymously underwrote the salaries, and the AI–10 seemed highly adaptable to the crucial task of locating enemy submarines.[6]

Faster and capable of descending to considerably greater depths than their predecessors of World War I, German U-boats were difficult to track down with sonar. They were also equipped with chemicals that, when emitted beneath the sea, created great clouds of sonically confusing bubbles. Still, submarines had to surface to take in fresh air and recharge their batteries. When they did, even at night, they could be detected by aircraft equipped with radar. In the spring of 1941, using long-wave-equipped planes, the Allies were already driving German U-boats away from the near approaches to the British Isles. Now, in response to a request from England, the Rad Lab made the transformation of the AI–10 into an airborne submarine detection unit its new high-priority project.[7]

By August 1941 the Rad Lab ASV, for *a*ir to *s*urface *v*essel radar, was detecting capital ships twenty to thirty miles away and surfaced submarines two to five miles distant. The equipment also displayed the location of the vessels on a circular oscilloscope screen called the Plan Position Indicator, which allowed the pilot to determine at a glance their range and bearing relative to his plane. Intrigued, the U.S. Navy ordered ten ASV sets for its own experimental use and a hundred shipboard versions of microwave radar for its prime submarine hunter, the destroyer. Eager to deploy the Rad Lab ASV as quickly as possible, the British diverted two B-24 Liberator bombers which were going abroad under Lend-Lease for installation of the units. In December, flight trials of the first—it was dubbed Dumbo I because the bulbous radar dome in its nose gave it an elephantine look—proved successful enough to impress the U.S. Army. In January 1942, the United States now at war, the Air

[5] "The Diary of Admiral Julius A. Furer," JAF, Jan. 1, 1942; Guerlac, Section C–1–I, pp. 2–3; Section B–II, p. 19.
[6] Compton to Rockefeller, May 8, 1941, VB, Box 26; Guerlac, Section B–II, pp. 20–21.
[7] Admiral Bowen delayed giving Bush a set of cables dating back to March 8, one of which contained the British request, until March 22. Caryl P. Haskins to Karl Compton, March 22, 1941, OSR, Entry 39, Compton file, Liaison.

Corps urgently requested ten ASV's from the Rad Lab to equip an equal number of B-18s for patrol along the country's critically important coastal sea lanes.[8]

While the army was fitting out the planes with the microwave search units, losses on the eastern sea frontier reached critical proportions. Scarcely a day went by without the sinkings of two or three tankers or cargo vessels bringing essential raw materials from South and Central America (by June the American merchant marine's accumulated losses would exceed the total of its tonnage sunk in World War I). In January and February army air crews, unequipped with radar, managed to attack only four submarines in eight thousand hours of patrol. At the end of March the ten ASV-equipped B-18s started operating out of Langley Field as the 1st Sea Air Attack Group. On the first night of patrol, one of them detected three U-boats and sank the last. By the end of the year the B-18s, together with navy destroyers, a strengthened convoy system, and other aircraft equipped with long-wave ASV search radar, had driven enemy submarines three hundred miles away from the coast.[9]

After Pearl Harbor and the success of the ASV, all the uncertainty about the Rad Lab's future had disappeared. Scores of army and navy officers established residence at MIT to keep their respective service branches up to date on the military possibilities of microwave systems and to inform the physicists of current military needs. Increasingly the armed services called upon DuBridge's men for emergency service. The Rad Lab spawned the nearby Research Construction Corporation, a shop that could quickly assemble prototype models of radar sets for experimental or even operational use. Over the course of 1942 the laboratory diversified its microwave research efforts and pushed ahead with its one long-wave project, the Loran, for *long-range* navigation. An ingenious elaboration of a British idea, Loran broadcast a vast net of crisscrossed radiations in the sky so that a plane or ship a thousand miles from home could determine its position to an accuracy of 1 percent; by the summer it was winning praise from captains and pilots ferrying convoys across the North Atlantic. By the end of 1942 the Rad Lab was an equal partner in an Anglo-American radar effort marked on both sides of the Atlantic by generous cooperation and dazzling technical ingenuity. Its budget had

[8] Karl Compton to Harvey Bundy, Aug. 29, 1941, OSR, Entry 1, Radar; Compton to Irvin Stewart, Aug. 29, 1941, OSR, entry 39, Compton file, General; Guerlac, Section B–II, p. 40.

[9] Samuel Eliot Morison, *History of United States Naval Operations in World War II* (15 vols.; Boston, 1947–62), I, 198–200; Wesley Frank Craven and James Lea Cate, eds., *The Army Air Forces in World War II* (7 vols.; Chicago, 1948–58), I, 527; Guerlac, Section C–1–II, p. 6; Albert Vorseller, "Science and the Battle of the Atlantic," *Yale Review*, 35 (June 1946), 671–74; Guerlac, Section E–II, p. 12; Stimson Diary, March 26, 1943; Elting E. Morison, *Turmoil and Tradition: A Study of the Life and Times of Henry L. Stimson* (Boston, 1960), p. 574, n. 14.

reached $1,150,000 monthly, and its staff had multiplied to almost two thousand people. By 1945 it would contain almost four thousand, about one quarter of them academics, almost five hundred of them physicists.[1]

The Rad Lab eventually worked out of a score of Cambridge buildings and grew into a dozen divisions, eight of them devoted to such subjects as fire control, beacons, and long-range research. Back in 1940 old Robert Millikan had pronounced it "a mistake . . . to concentrate fifty prima donnas in physics at any one spot." Some of the prima donnas, accustomed to following independent lines of research, did provoke administrative headaches. Like every successful organization, a veteran of the laboratory observed, the Rad Lab "needed—and had—a son of a bitch."[2] He was F. Wheeler Loomis, a spectroscopist and the long-time physics chairman at the University of Illinois, where he had built a first-rate department with an amiable manner and autocratic hand. Now in his mid-fifties, which made him senior by a decade even to Rabi, he handled the nasty personnel problems, shifted skilled manpower from one project to another, and generally kept the prima donnas in harness.

DuBridge on his part maintained peace, at times with a laissez-faire tolerance of his physicists' preference for independence. When necessary he exercised his director's authority, particularly in key policy matters. Ordinarily he wisely and modestly ran the laboratory more like a dean than like a director—some of the staff were more distinguished professionally than DuBridge himself—aided by a steering committee of the operating division heads. Every week the steering committee revised and set the general technological tasks to be accomplished. The execution of the jobs was left up to the day-to-day initiative of the various system research groups. Throughout the Rad Lab, which resembled a university in a pressure cooker, new ideas got a full hearing and eccentric brilliance was welcomed. The staff put in a six-day workweek and volunteered extra time on Sundays. Many could have left for lucrative jobs in industry. Few did, even though, in accord with a no-profit-no-loss rule standard in OSRD, they earned only their academic salaries. The work was highly stimulating, on the leading edge of military technology, and done with memorable esprit de corps.

To most American physicists before 1943, and to the many who even after that time never joined the Manhattan Project, physics in World War II meant the Rad Lab—and two other essential projects. One, at the California Institute of Technology, was for solid fuel rockets; the other, at Johns Hopkins, was for the proximity fuze, a small, rugged radar set installed on and designed to explode an artillery shell at a set

---

[1] Guerlac, Section D–II, pp. 13–14; "Longhairs and Shortwaves," *Fortune*, 32 (Nov. 1945), 169.

[2] Millikan to H. Victor Neher, Dec. 2, 1940, RAM, Box 21; interview with Edward L. Bowles.

distance from its target. The production costs for the Rad Lab radar systems alone amounted to some $1.5 billion; the combined bill for radar, proximity fuzes, and rockets far exceeded the $2 billion spent on the atom bomb. More important, soldiers and sailors appreciated the rockets that blasted enemy bunkers or the proximity fuzes that shot down dive bombers and kamikazes. One of the Rad Lab ten-centimeter beacons earned a simple accolade at a coastal airport. "Congratulations from the Colonel and everyone else," the telegram read. "Your squawker saved plane lost at sea in ice storm. No visibility. No radio. No navigation. Only you."[3]

Physicists in the war could respect the conviction of the Rad Lab staff, some of whom eventually worked on the Manhattan Project: The atom bomb only ended the war. Radar won it.[4]

In March 1941 Vannevar Bush, a keen student of the impact of new weapons like radar upon the fortunes of war, called Harry Hopkins' attention to a point of "greatest importance"—that the planning of defense strategy was best conducted with the participation of scientists. Bush did not mean grand strategy but the design of such military operations as bombing campaigns or antisubmarine warfare. Only by drawing upon scientific experts familiar with the latest laboratory products, he was sure, could military planners know the best way to exploit new weapons. Only by having access to the military's strategic requirements could defense scientists best understand the kinds of weapons that needed developing. Bush never argued that civilian scientists as such should make professional military decisions. He did hold that the military ought to make a place for civilian scientists as strategic advisers from the theater commands all the way up to the highest planning councils in Washington.[5]

Bush wrote into the OSRD executive order clauses designed to open the door to his own participation in strategic planning, particularly as the President's adviser on long-range defense matters. But the military was not prone to open up its strategic councils to a civilian scientist, even if his entrance was sanctified by an executive order. Besides, Bush's cussedness—and the NDRC emphasis on microwave radar—had earned

[3] Quoted in Guerlac, Section C–1–IV, p. 15. By the end of the war about 25 percent of the entire American electronics industry was devoted to the production of proximity fuzes. Joseph C. Boyce, ed., *New Weapons for Air Warfare* (Boston, 1947), p. 103. The Caltech rocket project received more than $80,000,000 in contract funds from OSRD, about the same as the Rad Lab, and by V–J Day the navy was ordering rockets at the rate of $1.2 billion a year. John E. Burchard, ed., *Rockets, Guns and Targets* (Boston, 1948), pp. 36, 5.

[4] Interview with Lee A. DuBridge.

[5] Bush, "The Organization of Defense Research," n.d., with annotation "sent to Harry Hopkins, 3/3/41," VB, Box 51, file 1269; Bush to the President, March 16, 1942, OSR, Entry 2; Furer Diary, March 20, 1942.

Bush the enmity of Admiral Bowen, who was an influential and able antagonist. While Bush's relations with the army brass were much better, he sometimes provoked even Army Chief of Staff General George C. Marshall and Air Force Chief of Staff General Henry H. Arnold to suspect him of maneuvering to advance himself more than the war effort. Early in 1942, almost a year after the executive order, Bush complained that his scientists remained outside the military's strategic councils. "In a scientific war," he told Harvey H. Bundy, "the scientists should aid in making the plans."[6] There was no dissent from Bundy, a boardroom Boston lawyer of supple and discreet intelligence who was the secretary of war's closest personal assistant, general troubleshooter, and representative on OSRD. And, whatever the attitude of his generals, neither was there dissent from Secretary Stimson, the patrician public servant.

In his seventies now, remarkably open and agile-minded, and awed by the spectacular evolution of military technology since his stint as secretary of war under William Howard Taft, Stimson was ready to challenge the dogma that warfare was exclusively the business of the professional military. He found regular army officers too conservative in their attitude toward new weapons. He was at home with Bush, whom he considered a man of "very ingenious as well as practical mind" and often treated like a son. Like his cousin Alfred Loomis, who stopped in to chat from time to time, Stimson marveled especially over the innovations coming out of the Radiation Laboratory at MIT. When military officials displayed little interest in the newly developed airborne radar, ASV, Stimson took a ride on the microwave-equipped experimental bomber Dumbo II and watched it quickly zero in on a distant ship. The next day Generals Marshall and Arnold found identical notes on their desks from the secretary of war: I've seen the new radar equipment. Why haven't you?[7]

In April 1942, taking his cues from Bush, Stimson pushed through the creation, under the new Joint Chiefs of Staff, of a Joint Committee on New Weapons and Equipment. JNW was chartered principally to keep civilian research agencies informed of the requirements of the armed services, the armed services in turn aware of promising developments in the civilian laboratories. Its members were a rear admiral, a brigadier general, and, as chairman, Bush.[8] In April 1942 Stimson also decided to acquire his own adviser on radar matters. Everyone seemed to recommend Edward L. Bowles, a forty-five-year-old professor of electrical engineering

[6] Interview with Edward L. Bowles, June 15, 1975. Bowles recalled Arnold's handing him a letter from Bush and saying: Bowles, what does this mean? I know what he says; I can read. But what is he up to? Bush, "The Situation on New Weapons," n.d., with annotation "original to Bundy, 1/26/42," VB, Box 17, file 389.

[7] Stimson Diary, July 23, 1942; Guerlac, Section C-1-II, p. 9.

[8] K. Maertens, "The Joint Committee on New Weapons and Equipment," June 30, 1942, OSR, Entry 1, Cooperation: JNW.

at MIT, a pioneer in microwave research, and now the secretary of the NDRC Microwave Committee. One day in April 1942 Bowles, nervous and uncertain why the secretary should want to see him, was ushered into Stimson's office and asked to criticize aloud, then and there, a highly adverse British report on American radar installations in the Canal Zone. Bowles saw some reasons to say that the report should not have been so adverse. The next Monday he was established in Pentagon Room 4–E–936, just above Stimson's office, as Expert Consultant to the Secretary of War.

As a young man in the Missouri Ozarks, where his father was a country doctor, Bowles had earned extra money by trapping. He mastered the cunning ways of his quarry; in his last season he beat the local professional three minks to none. Making it through college on one odd job or another, Bowles took his Ph.D. in electrical engineering and moved up from instructor to assistant professor on the MIT faculty in 1925, when Vannevar Bush was already a full professor. The more power Bush acquired in the department and the more young faculty he brought into his computer project, the more Bowles saw the Bush operation as an imperious, consuming, threatening juggernaut. No match for Bush technically, Bowles moved into microwave research because there he could get money to maintain an independent base. His patrons included the army and navy, with whom he worked on the use of radar to land aircraft in bad weather. At the Rad Lab, despite his microwave expertise, Bowles felt himself something of an outsider in a club of physicists, and he did not get along in particular with Alfred L. Loomis. When Bowles took the job with Stimson, it was with the understanding that Loomis would not interfere.

In the Pentagon, Bowles characteristically cultivated the barber who, having cut the army hair for years, knew all about the habits, manners, idiosyncrasies, and conceits of the regular officer corps from Generals Marshall and Arnold on down. Bowles may have been an independent civilian consultant, but he never sent an important report to Stimson without first asking Marshall or Arnold for his comments. With Stimson himself, who was sometimes crotchety by midday, Bowles made a point of bringing up key policy issues after the secretary's late-afternoon game of deck tennis, when his mood was usually much improved.[9] In all, if Bush tended to seek a strategic role for himself mainly by frontal maneuvering, Bowles pursued the same end with his trapper's craft and cunning—and with an astute perception of the military's outlook and requirements.

Bowles was especially aware of the wide gap between the advent of a technological innovation and its effective use in warfare. Bush wrote

[9] The preceding discussion is based on an interview with Edward L. Bowles and on Bowles to the author, April 24, 1973.

first-rate JNW reports on the implications of OSRD devices, notably radar, for antisubmarine strategy. Despite Bowles's dislike of Bush, he respected Bush's abilities, including his perceptions—they were acute—of the strategic implications of new weapons and devices. But in Bowles's opinion, Bush's JNW reports, which Marshall usually sent over for evaluation, tended to pay too little attention to the details of how the new devices were to be militarily deployed. The Rad Lab ASV, Bowles noted, could not be treated "simply as a magic gadget" whose installation in an aircraft armed with depth charges "ipso facto creates an effective anti-submarine weapon." "Scientific aids," he explained to Stimson more generally, "are helpless or at most only partly effective without a command and operational framework expressly and objectively set up to make them 'click.'" It was Bowles who, eager to make things click in antisubmarine warfare, recommended the establishment of the full-scale ASV-equipped bomber unit at Langley Field that did so much to help drive the German submarines away from the coastal shipping lanes.[1]

Asked by Stimson to pay close attention to the U-boat war, Bowles was specially interested in the reports on operations research by the handful of scientists who had recently begun working for the navy's Anti-Submarine Warfare Unit in Boston. Operations research had originated in Britain early in the war when teams of physicists and mathematicians started applying statistical analysis and the laws of geometry to figure out the optimal bomber formation for minimizing losses from enemy flak and fighters, or the optimal search and attack pattern for going after a submarine hidden in the vast expanse of the sea.[2] In Boston, the operations research scientists showed mathematically that, whatever the tactics employed, the chances of finding a submarine were considerably greater with radar-equipped aircraft than with destroyers. In August 1942 Bowles carried that point to its ultimate organizational conclusion. If U-boats were to be driven from the ocean, he argued to Stimson, there ought to be a regular land-based antisubmarine air force operating in coastal waters; the control of that force ought to be centered

[1] Interview with Edward L. Bowles. The quotations are from Bowles, "Memorandum for the Secretary of War: Vitalization of ASV Submarine Destruction Program," May 20, 1942, copy supplied by Mr. Bowles; Bowles, "Memorandum for the Secretary of War: A. Submarine Destruction Program," Aug. 7, 1942, OSW, Entry 77, Box 82, submarine file. Striking examples of the differences in the approaches of Bush and Bowles are: JNW, "Report on Anti-Submarine Devices," July 4, 1942, OSW, Entry 77, Box 75, report on ASW, 1942 file, and Bowles, "Memorandum for the Secretary of War: B. JNW Report . . .," Aug. 7, 1942, OSW, Entry 77, Box 82, submarine file.

[2] The chief force behind the early development of operations research was the British physicist Patrick M. S. Blackett, who argued that too much scientific effort had gone into the *production* of new devices" and "too little into the *proper use* of what we have got." Quoted in Morison, *History of United States Naval Operations in World War II*, I, 221.

in one service, the army.[3] In essence, Bowles was calling for a radical shift from the prevailing antisubmarine strategy—defensive reliance on the convoy—to a new offensive strategy: air search.

Stimson, Marshall, and Arnold readily backed the strategic departure of going after submarines with army aircraft. So did Bush, who, though scarcely an uncritical enthusiast about Bowles, agreed entirely with his analysis of the requirements and opportunities in antisubmarine warfare.[4] But in the navy there was no Bowles and of course no great liking for Bush, at least not as a strategic adviser. While Stimson often consulted Bush on strategic matters, no similar requests for his expert advice came from Secretary of the Navy Frank Knox; Colonel Knox, as he had been known since his Rough Rider days, was considerably more inclined than Stimson to take the military point of view. Nor did any come from Admiral Ernest J. King, the commander-in-chief of the U.S. fleet and Chief of Naval Operations.

King "shaves every morning with a blowtorch," a young woman remarked to FDR. Capable, tough, demanding, cold, austere, humorless, the admiral frequently seemed beyond approach and uninterested in contrary opinions. Bush, who detected a stubborn conservatism in King's attitudes toward modern naval operations, listened with disbelief when, together with a group of young officers, King went over the design of a new cruiser that bristled with radar. "There's too much radar on this ship," King exploded. "We've got to be able to fight a ship with or without radar." To King, a forty-year veteran who knew his navy from the torpedo room to the flattop deck, Bush was trying to interfere in matters that were not his business and about which, as a civilian, he could have no sound opinions.[5]

The conflict between Bush and King came to a head when in the

---

[3] A/S/W Operations Research Group, "Preliminary Report on the Submarine Search Problem," May 1, 1942; Bowles, "Memorandum for the Secretary of War: Submarine Destruction Program," Aug. 7, 1942, OSW, Entry 77, Box 82, submarine file. The key point of the operations analysis was this: Call the time between contact with the U-boat and the beginning of the search T_s. There would always be a T_s after which the search vessel could not cover the entire area in which the submarine might lie. The reason: Simply that the size of the area would be increasing faster than the search vessel could cover it. For an ASV-equipped plane, T_s was reasonably long. For a destroyer equipped with sonar, it was only three minutes, a slim margin.

[4] V. Bush, W. R. Purnell, and R. G. Moses [JNW] to the Joint Chiefs of Staff, Jan. 6, 1943, ibid.

[5] The young woman is quoted in Ernest J. King and Walter Muir Whitehill, Fleet Admiral King (New York, 1952), p. 413; King is quoted in Vannevar Bush, Pieces of the Action (New York, 1970), p. 91. For Bush's problems with the navy and King, see also pp. 50, 75–76, 87, 89–90, and Furer Diary, May 26, 1942. Bowles reflected on the difficulties a civilian scientist faced when dealing with the army and the navy: Compared to the navy, the army was technologically unsophisticated and "just conservative in its habits. The Navy was not so much technologically conservative. It was more like an insular aristocracy and resented intruders, scientists or otherwise, who tried to mix in Navy business." Interview with Edward L. Bowles.

spring of 1943 losses to enemy submarines, this time on the North Atlantic shipping routes, suddenly reached critical proportions. The RAF, using hand-built microwave ASV sets from the Rad Lab, was already taking a significant toll of the German marauders in the Bay of Biscay, their home waters, and various new OSRD devices, including airborne rockets and magnetic detectors, promised to improve still more the effectiveness of antisubmarine aircraft. The time had definitely come, Bush argued, when antisubmarine strategy had to shift to an offensive centered on assaulting the German wolfpacks in mid-ocean with hunter-killer groups of long-range and carrier-based bombers fitted out with the products of the OSRD arsenal. Now, too, Bush also insisted, scientists generally, and the operations research scientists in particular, had to be integrated into the navy's planning of overall submarine strategy.[6]

Stimson, especially impressed by a brilliant report on the subject from Bowles, argued vigorously for the strategic departure; Admiral King would have none of it, neither of the integration of the scientists nor of the airborne offensive. To his mind, the presumption of operations research scientists to determine how ships and planes should operate strategically added up to another case of civilians mixing improperly in the navy's business. Further, there was the question of control, navy or air force, of the land-based long-range bombers, and King would not have the army air force directing what he deemed a navy mission. Above all, however devastating science promised to make an air offensive, in King's iron-clad opinion the best way to protect shipping against the submarine was the convoy. In March, King told a naval group what in April he repeated to the secretary of the navy himself: "I see no profit in searching the ocean [with hunter-killer groups] or even any but a limited area such as a focal area—all else puts to shame the proverbial 'search for a needle in a haystack.' Let me say again, by way of emphasis, that anti-submarine warfare—for the remainder of 1943 at least—must concern itself primarily with escort of convoys."[7]

Exasperated by King, Bush sometimes vented his spleen to Admiral Julius A. Furer, a technologically imaginative career officer who had graduated first in his class at Annapolis, won the Navy Cross in World War I, and now, as the coordinator of Naval Research and Development, was the naval member of the OSRD Advisory Council. Furer shared the view that strategic decisions had to be made by regular navy men; he also typically suspected Bush of a primary interest in the advancement of Vannevar Bush. All the same, Furer judged it a mistake to exclude scientists altogether, simply because they were civilians, from the planning

[6] Bush, *Pieces of the Action*, pp. 89–90; Furer Diary, April 9, 1943; Bush to Admiral King, April 12, 1943, OSR, Entry 1, Cooperation: Navy, Scientific Council.
[7] King, remarks, "ASW Conference, March 1, 1943 . . ."; King to Secretary of Navy, April 4, 1943, EJK, chronological file. Furer Diary, May 8, 1942, indicates King's attitude toward civilian scientists as operational advisers.

of the strategic response to the submarine. By temperament and position —the coordinator could only persuade, not coerce—Furer was a mediator. Knowing King, his Annapolis classmate, for the closed-minded martinet he could be, Furer urged him to consult Bush at least occasionally, if only to protect the navy's flank against the army. If the army kept picking up the high-powered strategic talents like Bowles, Furer confided to his diary, the navy could "lose a lot of friends" and harm its long-run "prestige and interests."[8]

Bush, on his part, wrote King a long letter, shrewdly careful in word and tone, to illuminate how the U-boat crisis had raised the general question of the proper relationship in a wartime democracy between the armed services and the nation's scientists. "Antisubmarine warfare," Bush emphasized, "is notably a struggle between rapidly advancing techniques." Neither scientists nor admirals alone could develop the most advantageous strategy. The challenge required the professional judgment of both groups working together. The principal responsibility for the prosecution of military operations, Bush acknowledged, of course rested with the professional military. But to exclude scientists from the high strategic councils in this or any other subject was simply a "mistake" and one over which, as Bush made respectfully yet firmly clear, he was exceedingly troubled.[9]

The extent of the crisis in the North Atlantic, the urgings of Furer, and now Bush's letter—doubtless for all these reasons King called in the director of OSRD for a wide-ranging talk, which Bush found one of the most stimulating of his experience, that ended in an amicable agreement. King had recently appointed Admiral Francis S. Lowe as his assistant chief of staff for antisubmarine warfare. Three scientists would join Lowe's group as advisers. They would have access to all naval information pertinent to antisubmarine matters and they would enjoy the right to make tactical and strategic recommendations to King himself.[1] In mid-May, King established Lowe as commander of a new Tenth Fleet. It consolidated all the antisubmarine units in the Atlantic and included a Civilian Scientific Council under the chairmanship of Dr. John Tate, professor of physics at the University of Minnesota, the managing editor of The Physical Review, and the head of the OSRD division on submarine warfare. In July 1943, Tate persuaded the navy to incorporate into the Tenth Fleet the group of operations research scientists that had originated in Boston.

Months before the creation of the Tenth Fleet, in October 1942, the

[8] Furer Diary, March 20, May 26, 1942; April 9, 1943.

[9] Bush to King, April 12, 1943, OSR, Entry 1, Cooperation: Navy, Scientific Council.

[1] Bush, "Memorandum of Conference with Admiral King," April 19, 1943, *ibid.*; Bush to King, April 20, 1943, EJK, chronological file.

Germans had begun deploying U-boats equipped with receivers capable of detecting the radiations from long-wave ASV radar. Warned of an approaching plane, the submarine could dive before the Allied aircraft flew close enough to attack. But the German receivers could not detect the pulses from search radars emitting microwaves. By the time of the creation of the Tenth Fleet, long-range army bombers armed with microwave ASV were patrolling in mid-ocean and surprising U-boat crews on the surface at night. In April 1943 convoy losses dropped sharply. Through the late spring, hunter-killer groups in both the army and the Tenth Fleet helped destroyers drive the U-boats away from the North Atlantic. By early summer Allied convoys were passing safely through lanes that had been perilous only a short time before. In Berlin, Admiral Karl Doenitz wrote in his diary that U-boat losses had reached "impossible heights"—because of the "increased use of land-based aircraft and aircraft carriers, combined with the possibility of surprise through the enemy radar location by day and night."[2]

In the end, the work of the operations research scientists impressed the commander of the Tenth Fleet. So did their leader, the mathematical physicist Philip Morse, who, perceptive about navy egos, instructed his staff never to claim any credit for the fleet's successes.[3] Admiral King emerged an all-out enthusiast of hunter-killer groups (so long as the navy controlled every wing of antisubmarine aircraft). For a while there was some doubt about King's enthusiasm for the Scientific Council; Bowles remembered the admiral's confiding that he had set it up as a ruse, to get Bush off his back. But Bush recalled that John Tate handled King and his officers with "tact, diplomacy, and success"; through Tate's efforts "a tense disordered situation evolved into one . . . of cordial cooperation." Whatever the case, by late 1943, the operational analysts were assuming general, not simply antisubmarine, duties in the navy's strategic planning. In the fall of 1944, King established the Scientific Council in the navy staff at the right hand of the commander of the fleet himself.[4]

In April 1943, when Bush wrote his letter to King, the Joint Chiefs of Staff sent a special radar mission to England. Headed by Karl Compton, the group included Lee DuBridge and military officers from Admiral King's staff, the army air force, and the Signal Corps. The Compton

[2] Quoted in Morison, *History of United States Naval Operations in World War II*, X, 83.

[3] *Ibid.*, I, 223.

[4] Interview with Edward L. Bowles; Bush to Fred Fassett, Oct. 4, 1950, VB, Box 110, file 2598; Commander-in-Chief to the U.S. Fleet, June 22, 1944, CNO–CIC. Evidence that the navy's initial antagonism to civilian strategic advisers continued is in Bush to Furer, Jan. 17, 1944, OSR, Entry 1, Cooperation: Navy, Scientific Council, and Furer Diary, Jan. 15, Jan. 19, Oct. 9, 1944.

mission, as it became known, was to iron out problems of coordination in the Allied radar program—and in its supersecret complement, radar countermeasures. Not long after Pearl Harbor, Bush had established a countermeasures laboratory at Harvard for the development of techniques and devices to fool, jam, and confuse enemy radar sets. Since the Germans inaugurated a jamming program of their own, the countermeasures war, as a member of OSRD predicted, had soon turned into "a real battle of wits." The Allied air forces now had high stakes in the battle because they were revving up for a full-scale strategic bombing offensive against continental Europe.[5]

Countermeasures aside, the bombing offensive posed a special challenge for radar itself. Because of the prevailing weather conditions in Europe, Allied bombers could expect to zero in on their targets visually no more than about one day out of every five during the year, even less frequently during the fall and winter months. The rest of the time the targets were obscured by clouds. The British had developed various microwave radar aids for bombing through overcast, including a ten-centimeter set that could peer through the clouds and draw a map on a radar scope of the terrain below. But these devices were far from accurate. Besides, the British did not have enough sets to supply both their own needs and those of the U.S. strategic bombing group in England, the 8th Air Force. The situation was worrisome enough to bring to England in the latter days of the Compton mission Assistant Secretary of the Army for Air Robert A. Lovett and David T. Griggs, a physicist, member of Bowles's staff, and a knowledgeable enthusiast of radar aids to bombing.

Griggs and Lovett worked separately on the strategic bombing problem; the Compton mission, abroad for a full month, ranged through the entire web of Allied needs and practices in radar, including the relationship between civilian radar experts and RAF planners. Every Sunday scientists at the British equivalent of the Rad Lab, the Telecommunications Research Establishment, met informally—the gatherings were called "Sunday Soviets"—with air marshals and members of the Air Ministry to discuss strategic and tactical issues and map out the main lines of technical development in accord with operational needs. Both the air marshals and the scientists were enthusiastic about bombing through clouds by radar; the accuracy of hitting military and industrial targets was said to be remarkable.[6] Lovett, Griggs, and the Compton mission came home with a set of proposals that amounted to a major radar policy package.

On the technological side of the package, they called for the devel-

[5] Frank D. Lewis to Frederick L. Hovde, Aug. 4, 1942, OSR, Entry 1, London Mission.

[6] Lee A. DuBridge, "Classified Diary," April 29, 1943, May 6, 1943, in possession of DuBridge.

opment of advanced radar aids for strategic bombing. On the advisory side—even the military officers in the Compton mission were impressed by the Sunday Soviets—they urged the adoption of various measures to involve civilian scientists more deeply in the operational planning of the radar war at home and abroad. Eager to facilitate bombing through overcast, the military endorsed the prompt development by the Rad Lab of a three-centimeter, and hence considerably more accurate, version of the British device for targeting through clouds. While rejecting a greater advisory role for civilian radar experts stateside, the Joint Chiefs did authorize OSRD scientists to work directly with the command of the 8th Air Force in England. Members of the Rad Lab would go abroad to help install, maintain, and modify the radar bombing equipment, and the Harvard countermeasures laboratory would also establish a branch in Britain.[7]

The Rad Lab went to work on radar aids to bombing, sidelining virtually all other projects in a feverish drive to construct twenty of the three-centimeter sets—they were code-named H2X—for shipment to England before the cloudy weather of autumn. In the meantime, the Harvard countermeasures group inaugurated ABL-15, for American British Laboratory of OSRD Division 15, at Great Malvern, a spa in the hills of Worcestershire where the Telecommunications Research Establishment was located. Through the summer, H2X specialists from the Rad Lab joined the experts of ABL-15 in the laboratories nestled among Great Malvern's well-clipped hedges and high walls. By October, when a squadron of B-17s equipped with H2X arrived in England, the contingent had grown large enough to warrant formal organization as BBRL, the British Branch of the Radiation Laboratory.[8]

As the strategic bombing campaign intensified in the fall of 1943, both BBRL and ABL-15 mushroomed, their staffs crowding into billets at the County Hotel, their ebullient presence discomfiting the elderly ladies in residence. The laboratories were energetically, though informally, linked to the air force command through Griggs, a private pilot as well as a physicist, who was rapidly turning into an emphatic proponent of strategic air power. When Edward Bowles visited the European theater at the end of 1943, he thought that Griggs's services ought to be established on an expanded and more regular basis. He took up the matter with General Carl A. Spaatz, who in January 1944 returned to England from the African theater to take command of the newly organized U.S.

[7] "Report of the United States Special Mission on Radar to the Joint Chiefs of Staff, June 12, 1943," JCS, file 334, Sections 1–4, Box 250; Guerlac, Section E–III, pp. 17–23, Section E–IV, pp. 19–20; DuBridge, "Report on Visit to England . . .," OSR, Entry 1, Loomis Division 14: Radar.

[8] DuBridge to Bush, "Proposal for a British Branch of the Radiation Laboratory," Sept. 9, 1943, Oct. 18, 1943, OSR, Entry 39, K. T. Compton Office File: BBRL and ABRL, and Entry 1, London Office: General.

Strategic Air Forces. In March 1944 Spaatz officially endorsed Bowles's proposal for the creation of a small Advisory Specialist Group of civilian experts in radar and countermeasures.[9]

Like Griggs, whose energies seemed demonic, the specialists were almost all physicists. The group included Louis Ridenour and H. P. Robertson, both tactical experts as well as able military analysts; Victor Fraenkel, the imaginative head of ABL-15; and John G. Trump, the charming, patient director of BBRL, who kept his outfit at work even on British holidays—"partly because of my feeling that holidays are immoral, at least for Americans"—and who, along with Fraenkel, helped integrate the work of the specialists with the laboratories at Great Malvern. All the specialists had offices in Spaatz's headquarters, but in general they worked separately, fanning out over a wide variety of sensitive operational problems, including the highly sensitive ones associated with the coming invasion of Europe.[1]

Despite some initial resistance on the part of the military, the influence of the specialists was evident on D-Day—in the jamming and deception of German coastal radars; in the radar beacons used to mark the paratroop drop zones behind the beaches; in the H2X bombing of the beaches themselves ahead of the assault forces. After D-Day, as the Allied armies pushed eastward toward the Rhine, the specialists helped devise effective methods for knocking down the buzz bombs (the machines whined right over their living quarters on the way to London).[2] They developed methods of radar control of fighter-bombers from the ground, which permitted close tactical air support of infantry and armored columns even in bad weather. The military had its doubts about the innovation, but the doubts substantially ended after the day in October 1944 when clouds and rain grounded most Allied fighters and General George S. Patton found the Germans relentlessly driving back a key part of his tank force. Ground control radar operators guided two volunteer fighter squadrons to the battle. The planes broke down through the clouds and, catching the German tanks out in the open—the tank commanders never expected to see aircraft on a day like that—destroyed at least six of them and forced the rest to flee.[3]

All the while the specialists wrestled with what in 1944 proved to be considerably more difficult to accomplish than first expected: bombing

[9] Interview with the late David T. Griggs; Spaatz to Commanding General, 8th Air Force, March 4, 1944, OSW, Entry 81, Photostats file; Guerlac, Section E-V.

[1] John G. Trump, "A War Diary, 1944–45," Aug. 7, 1944, in the possession of Trump; Guerlac, Section E-V, p. 12.

[2] Typescript of D-Day cables, OSR, Entry 1, Loomis Division 14: Radar; Cable, CY to Signal, June 13, 1944, AAF, Entry 60A, Radar-Airborne file; Trump to DuBridge, June 20, 1944, OSR, Entry 1, London Mission: 9B. LAB 15; interview with the late David T. Griggs; Guerlac, Section E-V.

[3] Griggs to Bowles, Oct. 17, 1944, OSR, Entry 39, K. T. Compton Office File, Rad Lab; Trump Diary, Nov. 27, 1944.

accurately through overcast by H2X radar. When relying on H2X, B-17 crews sometimes navigated wide of their objectives—at least one mission raided Switzerland—and on the whole dropped almost 90 percent of their bombs more than half a mile from target; in general, for pinpoint bombing attempts, the accuracy of H2X was lower by a factor of 150 than that of visual aiming under the best atmospheric circumstances.[4] To deal with the problem, the specialists proposed technological modifications to improve the "seeing" power of H2X equipment. But Louis Ridenour also noted that under comparable conditions, and using the same H2X equipment, the 15th Air Force did a much better job of bombing than did the 8th, in part because of superior training. "All of us technical people," Ridenour declared, have tended "to overemphasize the importance of gadgetry" and ignore the importance of "its operational use."[5] The Advisory Specialist Group devised more effective techniques for bombing with radar. They also stressed to the air force that B-17 bomber crews had to be better trained in H2X and conditioned in a variety of ways to regard the equipment not as an exotic device to be used only in the last resort of overcast but as a normal requirement of their business.

Spaatz and his subordinates paid attention to the advice, but before the defeat of Germany the accuracy of H2X bombing improved only slightly. Various air force commanders resented having to risk—and too frequently lose—crews on missions likely to bomb areas two miles or more away from the target. All the same, the air force decided, better to drop bombs wide of the target than to drop no bombs at all. That was an important way to maintain pressure on the enemy. And whatever the inaccuracy of the bombing, the raids, Spaatz told Trump, forced the Luftwaffe into the air and accelerated the destruction of the German fighter force. By the end of the war the strategic air forces would drop half again as much tonnage over Europe via H2X as via visual sighting.[6]

Whatever the merits of the blind bombing program, the commitment to it brought the air force and the Advisory Specialist Group into an increasingly close relationship. So did the contributions of the specialists to D-Day and to ground control radar for tactical air support. When in March 1944 the Group was created, Secretary Stimson marveled that civilian scientists were "now thoroughly in vogue with our Army." None, whether specialists or not, managed to achieve a close relationship with the army ground forces, and in the Pacific the navy tended to regard civilian scientists as more of an irritation than they were worth.

[4] T. R. Murrell to G. E. Valley, May 16, 1944, AAF, Entry 60A, Radar-Airborne file; Guerlac, Section E–IV, pp. 60, 64.

[5] Ridenour to General McClelland and Lee DuBridge, Oct. 25, 1944, AAF, Entry 60A, Radar-Airborne file.

[6] T. R. Murrell to G. E. Valley, May 16, 1944, *ibid.*; Trump Diary, Dec. 7, 1944, April 15, 1944; Craven and Cate, eds., *Army Air Forces in World War II*, III, 20, 25; Guerlac, Section E–IV, p. 58.

But in January 1943 there had been only one operations analysis section in the entire U.S. Army Air Force; by January 1945, in USAAF commands around the world, there were seventeen such groups, employing 32 mathematicians, 21 radio and radar engineers, 14 terminal ballisticians, 11 physicists, and some 100 other analytic experts.[7] OSRD sent civilian scientific advisers to every service and every theater of war. By V–J Day, from Africa to Southeast Asia and on to the Aleutians, civilian scientists were in vogue as strategic and operational advisers to a degree without precedent in the annals of American military history.

All along, outsiders and insiders, whether privy to newspaper snippets or to classified information, knew that, in the way World War I had been a chemists' war, this was a war of physicists. By early 1942 OSRD had spent four times as many contract dollars in physics as in chemistry.[8] Chemists and engineers were of course indispensable to the development, and, especially, the production of weapons, but physicists comprised the elite corps of the advanced research and advisory army.

Even before the bombs fell on Pearl Harbor some 1,700 physicists were in defense work. The group represented a quarter of the profession but included more than three quarters of its eminent leadership, plus a sizable percentage of its lower-ranking members at all accomplished in modern research. By fall 1942 the demand for more physicists had risen to three to four times the annual output of the universities. By spring 1943, J. Hammon McMillen of Kansas State College remarked: "Almost overnight, physicists have been promoted from semi-obscurity to membership in that select group of rarities which include rubber, sugar and coffee."[9] No one knew better than Bush and Conant just how select a rarity they were. In the summer of 1943 the army, navy, and OSRD combined counted themselves 315 physicists, as opposed to 17 chemists, short for high-priority projects, and Bush needed the bulk of them.[1]

To ease the shortage, the War and Navy departments, the U.S. Office of Education, state superintendents, and even local American Legion posts propagandized for better high-school courses and more students in physics. The Westinghouse Company inaugurated its famous science talent search, fueled by the award of twenty scholarships a year.

[7] Stimson Diary, March 22, 1944; Furer Diary, June 1, 1944, May 11, 1945; Col. W. B. Leach, "Two Year Report on Operations Analysis," Jan. 1, 1945, OSW, Entry 82, Box 68.

[8] Statistics of total contracts awarded, attached to George W. Bailey to Frank Jewett, OSR, Entry 39, Jewett Office File 49.01.

[9] *Life*, XI (Oct. 20, 1941), 75; Chairman NDRC to Director OSRD, Oct. 24, 1941, OSR, Entry 39, Jewett Office File 49.01; McMillen, "Relation of Physics to the War Effort," *Science and the War* (Lawrence, Kan.: Kansas Academy of Science, 1943), p. 15.

[1] "Summary of Shortages of Scientific Personnel as of August 28, 1943," attached to Ward F. Davidson to Karl T. Compton, Sept. 3, 1943, OSR, Entry 39, K. T. Compton Office File: General Correspondence.

The army, navy, and Congress established various specialized training programs, mainly to produce technicians in such fields as radar. Eager to go further, scientists and educators, James B. Conant prominent among them, urged the administration to send students to college at federal expense. They also stressed that the aim of training and recruiting young scientists must not be defeated by the operations of the military draft.[2]

Bush grumbled about the constant threat of the draft to OSRD scientists when one day in 1943 he stopped to chat in the Pentagon with the secretary of war. Stimson promised to help in any way he could. Bush declined the offer, then added with a parting flourish, "I'm so cocky I could beat your army single-handed."[3] But Bush needed all the bravado he could muster when in late 1943 the army, which was beginning the buildup for D-Day, started to demand many more younger men.

The army was 200,000 G.I.'s short of its scheduled strength, and the average age of its force was climbing above twenty-five. Men from eighteen to twenty-five made the most capable combat soldiers, yet virtually all the remaining single, physically able male civilians in that age group were occupationally deferred. Such deferments had been allowed scientists in defense research. They had also been granted to graduate and undergraduate students in fields like physics who would obtain their degrees by July 1945.[4] Army manpower officers complained about the young men in college, and they enjoyed the emphatic support of Secretary Stimson's trusted Under Secretary of War, Robert P. Patterson, who had charge of manpower and procurement. Patterson consistently opposed student deferments on grounds that they conferred an economically undemocratic privilege upon an entire occupational class. Once the war entered its hard, bitter phase, he told Stimson, college students should be drafted like everyone else.[5]

In November 1943, now that a decidedly hard, bitter phase was coming to the beaches of Normandy, Congress shifted the manpower policy balance against occupational deferments. Through the winter and spring of 1944, the congressional action was followed by a series of announcements and directives, including a public presidential statement chiding local boards for their "overly lenient" treatment of younger men; the clamping of tight restrictions on the occupational deferment of

[2] American Institute of Physics, War Policy Committee, "Third Report: Wartime Training in Physics," Dec. 7, 1942, attached to Henry Barton to President Roosevelt, Dec. 10, 1942, FDR, P.P.F. 1280; Conant to the President, March 28, 1942, FDR, O.F. 107.

[3] Interview with the late Vannevar Bush.

[4] Selective Service System, *Occupational Bulletin No. 10*, March 1, 1943; *Occupational Bulletin No. 11*, March 1, 1943.

[5] Albert A. Blum, *Drafted or Deferred: Practices Past and Present* (Ann Arbor: Bureau of Industrial Relations, Graduate School of Business Administration, University of Michigan, 1967), pp. 27, 64–65; William B. Tuttle, Jr., "Higher Education and the Federal Government: The Triumph, 1942-1945," *The Record*, 71 (Feb. 1970), 495–96.

all men under twenty-six; the rescission of all inessential occupational bulletins and the elimination of all undergraduate deferments. General Lewis B. Hershey, the director of Selective Service, announced that one of the few ways anyone under twenty-six could qualify for occupational deferment was by the endorsement of a federal office represented on a new Inter-Agency Committee. The membership of the Committee, first announced in mid-April, did not include a representative of OSRD.[6]

The crackdown on the deferment of young men threw the OSRD leadership into turmoil, in part because the elimination of student deferments cut off the supply of new technically trained manpower. Considerably more important, the restrictions on deferments for men under twenty-six threatened to wreck every major scientific project, research and advisory, from high-octane aviation gasoline to radar. At the Rad Lab alone there were almost five hundred scientists under twenty-six, virtually every one of them irreplaceable. Paralyzing uncertainty swept through OSRD laboratories, demoralizing the staffs, diverting energies away from the workbench and into anxious discussions about the draft. The editor of a leading chemistry journal smashed at the new draft regulations: "Sabotage!"[7]

Bush, who had two sons headed for combat service, emphasized to the OSRD Advisory Council: "There should be no shielding, behind a scientific façade, of young men who are not specially qualified, by reason of experience or outstanding technical competence, to carry on vital war research and development." Still, it was essential to assure the continued vitality of OSRD laboratories. Taking his case to the Pentagon, Bush insisted vociferously—some staff officers pronounced him berserk—upon the inclusion of OSRD on the Inter-Agency Committee for the certification of deferments. He also demanded the continued protection from the draft of the scientists on his Reserved List. Established in 1943, the Reserved List was authorized to include up to 7,500 key scientists whose requests for deferment were endorsed by a high-ranking group of military and civilian officials.[8]

In the end, the secretaries of both war and navy assured Bush that Selective Service would fully protect the Reserved List ("the number . . . is not large," Under Secretary Patterson himself acknowledged, "and their value is so great in their present activities that they should not be

---

[6] FDR is quoted in Blum, *Drafted or Deferred*, p. 127. Hershey's announcement was Selective Service System, *State Director Advice No. 225-H*, April 11, 1944.

[7] "Rape of the Laboratories," *Time*, April 24, 1944, p. 76; "Unselective Service," *Science Newsletter*, 45 (April 22, 1944), 259–60; Irvin Stewart, *Organizing Scientific Research for War: The Administrative History of the Office of Scientific Research and Development* (Boston, 1948), pp. 272–75. The editor is quoted in the *Time* article.

[8] Bush to Members of the Advisory Council, April 13, 1944, OSR, Entry 39, Jewett Office File 49.41; Bush, *Pieces of the Action*, p. 290; Stewart, *Organizing Scientific Research for War*, pp. 264–66.

interrupted or disturbed on any account"). In addition, OSRD was appointed to the Inter-Agency Committee. From the spring of 1944, virtually no Reserve List scientist under twenty-six was drafted. And virtually none under thirty was called up after February 1945, when the restrictions on occupational deferments were extended up to twenty-nine-year-olds.[9]

In the middle of the draft crisis a journalist had claimed of OSRD's threatened researchers, "Many of these men . . . are worth their weight in generals."[1] The generals and the admirals did agree that Bush's scientists, especially his physicists, seemed indispensable to the national defense—even before their indispensability as both progenitors of military technology and advisers upon its use was thrown into bold relief by the atomic fires of Hiroshima and Nagasaki.

[9] Patterson is quoted in Bush, circular letter to all OSRD officers, May 23, 1944, OSR, Entry 39, Jewett Office File 49.41; Leonard D. Carmichael, "The Nation's Professional Manpower Resources," in Leonard D. White, ed., *Civil Service in Wartime* (Chicago, 1945), p. 111.

[1] "Unselective Service," *Science Newsletter*, 45 (April 22, 1944), 259–60.

XXI

The Bomb
and Postwar Research
Policy

I n early 1939 Enrico Fermi, now a Nobel laureate and refugee from
Italy, gazed thoughtfully out his office window high in the Columbia physics building. In the laboratory Fermi had been investigating
the recent report of the German and Austrian scientists Otto Hahn
and Lise Meitner that uranium atoms fissioned upon impact by neutrons.
Now, looking toward the expanse of Manhattan beyond the pane of glass,
Fermi shaped his hands into a large-sized ball. A little bomb like that, he
remarked, and it would all disappear.[1]

Hahn and Meitner's startling discovery stimulated a number of
physicists, native and refugee alike, to think of a nuclear explosive. In
the process of fission, a large quantity of energy was released, and so
were more neutrons, possibly enough to make a chain reaction. Certain
that the Germans would recognize the same possibility, in the summer of
1939 the refugee physicists Eugene Wigner and Leo Szilard prevailed
upon Einstein to call FDR's attention to the military implications of
nuclear fission. A committee was formed, and $6,000 appropriated. But
the committee members, thinking of fission as a potential source of peaceful industrial power, gave the government's uranium project little push
or direction. Anxious to explore the possibilities of a nuclear weapon,
Wigner and his colleagues felt as if they were "swimming in syrup."[2]

---

[1] Interview with George Uhlenbeck, who at the time shared the office with
Fermi.
[2] Wigner, "Are We Making the Transition Wisely?" *Saturday Review of
Literature*, 28 (Nov. 17, 1945), 28.

In June 1940, authority over the program was transferred to NDRC and Vannevar Bush. Mindful of maintaining balance in the defense research program—and half hoping that an atom bomb might be impossible —Bush refused to embark on a major nuclear weapons program, especially since it could prove to be a wild goose chase.[3] Chief among the numerous obstacles to the bomb was the natural occurrence of uranium as a mixture of two isotopes, U-235 and the slightly heavier U-238. The fissionable isotope was U-235, which amounted to only 1 part in 140 of ordinary uranium. An explosive chain reaction would require a still undetermined number of kilograms of nearly pure U-235, yet no one knew an obviously easy way to separate even a few grams of U-235 from the more abundant U-238.

But in mid-1941 optimistic word about the prospects of a uranium bomb came from British scientists. They estimated that the amount of uranium for a "critical mass," in the phrase used to denote the minimum material necessary for a self-sustaining chain reaction, totaled only about 10 kilograms. The British also believed that sufficient U-235 could be separated from U-238 by the method of gaseous diffusion. Already under investigation at Columbia University by Harold Urey and the physicist John Dunning, this technique took advantage of the physical fact that two gaseous substances of different weights diffuse through a porous barrier at different rates. Forced through a sufficiently long cascade of porous barriers, the lighter U-235 would eventually be separated from the heavier U-238. Adding to the encouraging news from the British, in Berkeley Ernest Lawrence had begun work on separation by electromagnetic means. Moving in a magnetic field, identically charged ions of different weights follow different trajectories. If uranium ions were sent through the magnetic field of a cyclotron, the ions of U-235 would be eased away from those of U-238 and made available for collection.

Lawrence's imagination was also fired by the results of experiments with U-238 in his cyclotron. Early in 1941, Glenn T. Seaborg, a young physical chemist at Berkeley, had demonstrated that, after bombardment with neutrons, U-238, which was element 92 in the periodic table, eventually transformed itself into a new element 94, which Seaborg christened plutonium. Furthermore, as expected on theoretical grounds, plutonium fissioned just like U-235. The striking implication: Fashioned into a chain-reacting pile, all the abundant, nonfissioning U-238 could be transformed into fissionable plutonium suitable for an atomic explosive.

During the spring and summer of 1941 Lawrence, soon joined vigorously by Arthur Holly Compton, insisted to Bush and Conant that, if only because of what the Germans might be doing, prudence required stepping up atom bomb research. In October a persuaded Bush obtained President Roosevelt's go-ahead for a major attempt to test whether a bomb could

[3] Bush to Jewett, May 2, 1940, VB, Box 55, file 1375.

actually be developed in time for use in the current war. Shortly after Pearl Harbor, his sense of urgency now acute, Compton decided that the scientific side of the question, then under investigation by geographically isolated groups, would be better explored in a concerted fashion. In January 1942 he brought various teams of physicists, including Fermi's from Columbia, to the University of Chicago and organized them into the Metallurgical Laboratory. The misleading name veiled its main immediate purpose: the achievement of a chain reaction to test the feasibility of producing plutonium in an atomic pile.

In June 1942, with the favorable scientific evidence steadily mounting, FDR gave Bush the green light for a full-scale effort to build the bomb. Aware that the operation would require an industrial effort far beyond the managerial capacities of OSRD, Bush and Conant arranged for the assignment of process development, engineering design, procurement of materials, and the selection of plant sites to a new Manhattan District of the Army Engineers. In September, command of the Manhattan Project was given to Brigadier General Leslie R. Groves, the builder of the Pentagon, brisk, efficient, supremely self-confident, and decidedly able. Under Groves, the direction of physical and chemical research for the bomb remained in the hands of three civilian program chiefs: for gaseous diffusion, Harold Urey at Columbia University; for electromagnetic separation and plutonium studies, Ernest Lawrence at Berkeley; and for weapon theory and chain reactions, Arthur Holly Compton at Chicago.

While Compton handled administration and policy at the Chicago Metallurgical Laboratory, Enrico Fermi masterminded the construction of a chain-reacting pile in a squash court under the west stands of Stagg Field, the university's unused football stadium. Physicists hand-hauled into place heavy graphite blocks—they were needed to slow the neutrons —seeded in uranium slugs, and gradually built up an arrangement resembling a flattened sphere about twenty-six feet in diameter. Finally, on December 2, 1942, almost four hundred tons of graphite were in place, together with six tons of uranium metal and fifty tons of uranium oxide. Under Fermi's direction, the reaction control rods were slowly withdrawn. The clicks of the neutron counters rose steadily, like a mounting frenzy of crickets. Fermi, raising his hand, announced that the pile had gone critical. A while later, Arthur Compton telephoned Conant at Harvard: "Jim, you'll be interested to know that the Italian navigator has just landed in the new world."[4]

Later that month, President Roosevelt approved the expenditure of $400,000,000 for uranium separation plants and a plutonium-producing pile. In 1943, moving decisively, Groves let contracts for the construction

[4] Arthur Holly Compton, *Atomic Quest: A Personal Narrative* (New York, 1956), p. 144.

of gaseous diffusion and electromagnetic facilities at Oak Ridge, Tennessee, a 59,000-acre governmental reservation that the army had acquired for the purpose. He also ordered the construction of the plutonium works on 400,000 federal acres near the Columbia River at Hanford, Washington.

When representatives of Stone and Webster, the prime contractor for the electromagnetic plant, first visited Berkeley, they came away shaking their heads in awe at the way Lawrence's skyscraper plans rested so far on the successful separation of only a few millionths of a gram of U-235.[5] Lawrence, his gusto for the impossible intensified by the war emergency, drove his staff day and night. In a cut-and-try fashion, Berkeley physicists and engineers designed vacuum tanks, ionization apparatus, U-235 collectors, and magnets. Forced by the shortness of time to skip the pilot-plant stage, Stone and Webster went from the Berkeley designs directly to the building of the vacuum separation chambers and the huge steel magnet ovals. The chambers contained an unprecedentedly enormous volume of evacuated space; the magnet coil windings required more than 28,000,000 pounds of silver (the metal, worth $400,000,000, was borrowed from the Treasury Department). When the racetracks, as the electromagnetic system was dubbed, were first operated in late 1943, electrical shorts and vacuum leaks erupted everywhere. In 1944, after major repairs, breakdowns still plagued the system, so Lawrence brought squads of scientists and technicians from Berkeley to trouble-shoot the equipment. The racetracks would work, Lawrence assured a discouraged General Groves.[6]

Even more prodigious efforts were required for the gaseous diffusion process, the key to which was the development of a suitable diffusion barrier. The barrier had to have millions of tiny holes, be impervious to the exceptionally corrosive effects of uranium hexafluoride gas, and possess sufficient structural strength to withstand high pressures. After months of research at Columbia, Bell Telephone Laboratories, and elsewhere, Percival C. Keith, in charge of the diffusion process for the prime contractor, the Kellex Corporation, despaired that the barrier problem would ever be solved; so even did Ernest Lawrence, who urged deemphasis of the diffusion method. But not Bush, Conant, or Groves, who boldly pushed ahead the construction of literally acres of diffusion cascades at Oak Ridge. Industrial physicists, chemists, and engineers performed prodigious feats of design and development to fabricate corrosion-resistant valves, pipes, and pumps. In the fall of 1943 Clarence A. Johnson, a young engineer working for Keith's corporation, hit upon a promising idea for the barrier, one that fused elements of the existing

[5] Stephane Groueff, *Manhattan Project: The Untold Story of the Making of the Atomic Bomb* (Boston, 1967), pp. 36–37.

[6] Herbert Childs, *An American Genius: The Life of Ernest Orlando Lawrence* (New York, 1968), p. 345.

approaches with an originality of its own. At Columbia, Urey, grown increasingly pessimistic and temperamental under the stress of the barrier project, refused to divert his laboratory to the pursuit of Johnson's idea. Keith maneuvered into control of the project the Princeton chemist Hugh Scott Taylor, who, aided by academic and industrial scientists as well as his own ingenuity and sense, inched barrier design and production successfully forward.[7]

Still, in mid-1944 the barrier problem was not yet completely solved. In desperation, Groves ordered the installation at Oak Ridge of yet another type of separator, a thermal diffusion plant. The idea of the physical chemist Philip Abelson, who had worked on it independently for three years with naval support, thermal diffusion exploited the propensity of atoms of different weights to separate when caught between hot and cold surfaces. Thermal diffusion could increase the percentage of U-235 in natural uranium by only a small amount. But the degree of enrichment might be sufficient to replace the gaseous diffusion method as the expected source of material for the electromagnetic separators.

The plutonium project required a grit and ingenuity all its own. The cyclotron at Washington University in St. Louis was enlisted to make experimental samples of the new element by bombarding U-238 with neutrons night and day. Working with mere micrograms of the material, chemists under Glenn Seaborg at the Metallurgical Laboratory devised production techniques for the chemical separation of plutonium from uranium. In 1943 physicists and chemists left Chicago for Oak Ridge, where they designed and built an experimental pile and separation facilities; at the end of the year the equipment yielded the first milligrams of plutonium. Among the Oak Ridge staff was Eugene Wigner, who came up with a basic design for the pile at Hanford which was adopted by the scientists and engineers of the prime contractor, the Du Pont Company. In 1944, the completed Hanford pile was turned on. It went critical for some minutes, then the chain reaction stopped.

Tests showed that the pile was producing an isotope of the element xenon, which poisoned the chain reaction by absorbing large quantities of neutrons. The effect had never showed up in the Oak Ridge pile because, contrary to Groves's instructions, the staff had never run it at full power. Groves was angry and, along with Compton, decidedly worried. But the young physicist John Wheeler of Princeton University, who with Bohr had written the fundamental paper on the theory of fission, had warned the Du Pont staff about the possibilities of poisoning. Over the opposition of a number of scientists at Oak Ridge and Chicago, and at the insistence

[7] Richard G. Hewlett and Oscar E. Anderson, Jr., *The New World, 1939/1946* (*A History of the United States Atomic Energy Commission*, Vol. I; University Park, Pa.: Pennsylvania State University Press, 1962), pp. 133–36; Richard G. Hewlett, "Nuclear Physics in the United States during World War II," unpublished manuscript, pp. 20–21.

of George Graves, an assistant technical adviser in the company, Du Pont engineers had designed room in the pile for excess uranium slugs. Stuffing the pile to the maximum, the Hanford staff overwhelmed the poisoning effect. Du Pont people could be forgiven their celebratory doggerel about Graves, who "with baleful glare . . . /Cried, 'Dammit, give it stuff to spare— / The longhairs may have missed.' "[8]

By early 1945 the Hanford facilities were beginning to turn out pure plutonium. By the same time, the barrier problem had been overcome at Oak Ridge. There, operating in tandem, the thermal diffusion assembly, the racetracks, and the gaseous diffusion cascades were yielding U-235 of high purity in steadily increasing quantities. But long before either plant reached full production, both had started sending their small first yields of plutonium and enriched U-235 to the apex of the Manhattan Project, the special weapons laboratory that Groves had established under J. Robert Oppenheimer at Los Alamos, New Mexico, a lonely mesa nestled 7,500 feet up in the majestic Sangre de Cristo Mountains some seventy-five miles north of Albuquerque.

From its opening in March 1943, Los Alamos absorbed physicists like a sponge. The Rad Lab was losing its best theorists to the bomb effort, Rabi complained to Bush, and if the losses continued, advanced radar research would be seriously impaired. Nevertheless, the high priority of the Manhattan Project permitted the Los Alamos scientific and technical staff to grow, by spring 1945 to well more than two thousand, including some six hundred army enlisted men assigned to the laboratory as part of a Special Engineering Detachment. The scientists ranged from first-degree apprentices—the median age of the technical staff was twenty-seven—to native, refugee, and British physicists whose names were already textbook fixtures. To Frederic de Hoffman, a new Ph.D. from Harvard, his arrival at Los Alamos seemed like "one grand final exam day, with all the senior faculty members of all the U.S and European physics faculties assembled to give . . . that final exam."[9] At the Fuller Lodge on the mesa, one could sometimes see as many as eight Nobel laureates dining at once.

The physicists at Los Alamos lived with numerous irritations, including the isolation—it was a hazardous eighty-mile round trip to Santa Fe—the jerry-built housing, and above all the army, with its pervasive security restrictions. Famous physicists played charades with pseudonyms—Niels Bohr was "Nicholas Baker"—and the birth certificates of newborn babies were inscribed "P.O. Box 1663, Sandoval County Rural." Nothing was

[8] Quoted in Compton, *Atomic Quest*, p. 193.

[9] Rabi to Bush, June 1, 1943, OSR, Entry 1, Manhattan District: Personnel Matters; Smith, "Los Alamos: Focus of an Age," in Richard S. Lewis and Jane Wilson, eds., *Alamogordo Plus Twenty-Five Years* (New York, 1971), p. 34; John H. Manley, "Organizing a Wartime Laboratory," in Jane Wilson, ed., *All in Our Time: The Reminiscences of Twelve Nuclear Pioneers* (Chicago: Bulletin of Atomic Scientists, 1975), p. 135; de Hoffman, "A Novel Apprenticeship," *ibid.*, p. 163.

permitted that might indicate to the outside world the identity of the inhabitants. In the laboratories, army engineers sometimes interfered with technical matters in which they had no competence.[1] To the physicists and their families, the army personified was General Groves, who could be infuriatingly pompous, behave as though his judgment were infallible, and take unassailable positions on matters as small as whether the Los Alamos compound should have a west gate. Still, Groves's dedicated attention to the project produced welcome results that ranged from the miracle of getting a scarce B-29 for tests to the thoughtful favor of supplying every household with an electric hotplate when he learned that the wood-burning stoves were difficult to operate. Sometimes reluctantly, Groves also eased military rigidity to accommodate the freewheeling ways of science. And on their part, the scientific staff learned to cope with the isolation, the army, the security restrictions, and each other.

Los Alamos scientists skied and hiked in the remote beauty of the surrounding mountains, gathered at intellectually stimulating social evenings, and kept the maternity ward busy. Way down in the hierarchy, a homesick army recruit from New York hung a bagel from the ceiling so that he could lie back on his bunk and admire it. The future Nobel laureate Richard P. Feynman, then in his early twenties, dealt with the strain by playing the bongo drums, challenging the censors with coded letters, or picking the combination locks of safes containing classified documents. Higher up the ladder, the moody Edward Teller, who was apt to spend his leisure in long walks, insisted at work upon pursuing his own scientific demon, a thermonuclear weapon, and drove his neighbors to distraction by rhapsodizing on the piano at odd hours of the night. Oppenheimer relaxed over writings from the Sanskrit or the poetry of John Donne.

Oppenheimer, the director, his hair cropped efficiently short, mediated between Groves and the scientific staff, easing army regulations, housing problems, and personal crises. At times overwhelmed to the point of despair by the administrative burden, he depended heavily on the hardheaded executive talents of his division leaders and his early assistant John H. Manley, who had helped drive him to an acrimonious outburst one night in Berkeley by nagging about the need for an organizational plan at Los Alamos. Later on, Samuel K. Allison, the Chicago physicist and former head of the Metallurgical Laboratory, became Oppenheimer's associate director and kept track of the numerous details involved in the fabrication of a weapon. But the scientific tone at Los Alamos was shaped principally by Oppenheimer. Despite Groves's vigorous objections at the hazard to security, Oppenheimer refused to compartmentalize the laboratory; he insisted upon holding a weekly progress review open to everyone who wore the white badge that signified an academic degree. "Here

---

[1] Lansing Lamont, *Day of Trinity* (New York, 1965), pp. 54–55; Robert R. Wilson, "A Recruit for Los Alamos," in Wilson, ed., *All in Our Time*, p. 157.

we have assembled the greatest bunch of prima donnas ever seen in one place," Groves remarked.[2] Through some combination of tough judgment, mystique, and a melancholic sense of the laboratory's terrible goal, Oppenheimer dampened the scientific factionalism, persuaded both apprentices and prima donnas to feel privileged to be cooped up on the mesa, and inspired the entire staff to remarkable feats of research and technical ingenuity.

Much of the technical challenge centered on finding out how a critical mass of U-235 or plutonium would behave in the microsecond between the start of the chain reaction and the explosion. An experimental physics division under Robert F. Bacher laboriously used accelerators as neutron sources to measure the fission properties of the two elements. Exploiting their expertise in the electronic techniques of microwave radar, they devised sensitive equipment to determine the reaction speeds at different neutron energies. Shaping the experimental data into a prediction of critical mass behavior was the task of the theoretical division under Hans Bethe. With his formidable theoretical powers, Bethe, in the observation of Feynman, resembled a battleship surrounded by an escort of smaller vessels, the younger theorists, moving majestically forward through the ocean of the unknown.[3]

Both experimentalists and theorists pondered the key problem of bomb design. To obtain a powerful explosion, sufficient fissionable material to form a critical mass had to be brought together quickly, then kept together long enough to release a lot of energy. The most direct assembly method consisted of using a gun to fire one subcritical mass of fissionable material into another with the speed and force of an artillery shell. But while experiments suggested strongly that the gun method would work for U-235, they also indicated by spring 1944 that it would not work for plutonium.

The trouble was that plutonium emitted many more neutrons spontaneously than did U-235, enough possibly to predetonate the bomb assembly before it could release more than a minute fraction of its potentially explosive energy. But an alternative to the gun method had been proposed in April 1943 by the young Caltech physicist Seth Neddermeyer. Called implosion, this approach relied upon surrounding a mass of subcritical plutonium with high explosives which, when fired, would produce a spherically symmetrical shock wave traveling inward toward the center of the bomb. The shock wave would compress the plutonium into a critical mass and, keeping it compressed while a rapid chain reaction developed, would maximize the energy released. The great speed of compression would also eliminate the danger of predetonation. At first re-

[2] Robert F. Bacher, "Robert Oppenheimer," *Proceedings of the American Philosophical Society*, 116 (Aug. 1972), 281–82; Groves is quoted in Groueff, *Manhattan Project*, p. 204.

[3] Stanislaw M. Ulam, *Adventures of a Mathematician* (New York, 1976), p. 150.

jected as too difficult by most, and scoffed at by some, Neddermeyer's idea acquired increasing luster at Los Alamos as the dangers of a plutonium fizzle became more apparent.

But the array of imploding explosives around the plutonium core would have to yield a spherical shock wave of absolute uniformity. The slightest nonuniformity, and the plutonium would tend to squirt away, halting the chain reaction. In February 1944, Oppenheimer brought in the physical chemist George Kistiakowsky to head a new implosion effort at the laboratory. In early March, Kistiakowsky drew up a day-by-day chart of desirable progress. His final entry, projected for the end of the year: "The test of the gadget failed. Project staff resumes frantic work. Kistiakowsky goes nuts and is locked up!"[4] The pressure on Kistiakowsky's people rose even more after it became clear that, because of the likelihood of spontaneous fission, a plutonium bomb could not be set off by the gun method; it would have to be detonated by implosion.

Mainly to meet the challenge of implosion, Oppenheimer reorganized the laboratory in mid-1944, shifting its emphasis from research to the design and fabrication of both a U-235 and, especially, a plutonium bomb. When Hanford and Oak Ridge began sending production quantities of plutonium and U-235, the tension on the mesa rose precipitously. In January 1945, aware that the cost of the project was close to $2 billion, Groves warned his aides that if the Manhattan Project failed, "each of you can look forward to a lifetime of testifying before Congressional committees."[5]

By early March, the design of the implosion bomb was set in principle, and Oppenheimer appointed a Cowpuncher Committee to ride herd on the final development of the system. By mid-April, Kistiakowsky had overcome the critical problem of the asymmetrical shock wave. In early July, the laboratory moved ahead toward the assembly of the U-235 gun bomb for shipment to the 509th Composite Air Group, a specially trained B-29 unit in the South Pacific. No precombat explosive test of the uranium bomb, code-named Little Boy, appeared necessary; the Los Alamos scientists had high confidence in the gun method. But a trial of the implosion weapon had long seemed imperative. The Cowpunchers worked frantically to ready a plutonium bomb for detonation in the desert near Alamogordo, New Mexico.

On July 16, in the dark before morning, the plutonium gadget rested atop a lone tower in a sheet-metal shack three hundred feet above the desert floor. Outside a ten-thousand-yard radius, scores of men waited, tense, nervous, apprehensive. Oppenheimer, standing in the doorway of the South-10,000 bunker, seemed to totter on the verge of nervous collapse. Kistiakowsky had bet him a month's salary against ten dollars that

[4] Quoted in Hewlett and Anderson, *The New World*, p. 248.
[5] Quoted in Lamont, *Day of Trinity*, p. 76.

the implosion device would work. Samuel Allison brought the countdown to zero with a scream. A spot of light burst through the darkness, then boiled upward, exploding into a rainbow of fire that colored the desert wastes and dazzled the mountain ranges in the distance. Moments later the shock wave blasted through with the roar of twenty thousand tons of TNT. After the first exhilarating cheers of success, there was an awesome silence. Through Oppenheimer's mind flashed a line from the Bhagavad-Gita: "I am become Death, The shatterer of worlds."[6]

Twenty-one days later, at 2:45 A.M. August 6, 1945, three B-29s belonging to the 509th rose from the island of Tinian in the Marianas and headed for Japan fifteen hundred miles to the north. In the belly of the lead plane, the *Enola Gay*, was the uranium bomb, Little Boy. By 7:30 A.M. Navy Captain William S. Parsons, the head of the ordnance division at Los Alamos, had completed arming the bomb. Fifteen minutes later the plane was over the Japanese mainland. "It is 8:50," Captain Robert A. Lewis, who had been jotting a log of the flight, soon noted. "Not long now, folks. . . . There will be a short intermission while we bomb our target." The plane lurched upward when the five-ton bomb fell away toward its detonation point two thousand feet above the city. A few moments later Captain Lewis exclaimed: "My God!"[7] Hiroshima was covered with a great swirling column of smoke that reached thirty thousand feet in less than three minutes and could still be seen when the *Enola Gay* was four hundred miles away.

In the United States later that day, the White House released the news: "Sixteen hours ago an American airplane dropped one bomb on Hiroshima. . . . That bomb had more power than 20,000 tons of TNT. . . . It is a harnessing of the basic power of the universe."[8] Three days later, August 9, 1945, another B-29 from the 509th dropped an implosive plutonium bomb, code-named Fat Man, on Nagasaki.

Within weeks Americans knew fully the grim results of the atomic bombings—from the pictures of Hiroshima and Nagasaki devastated beneath the towering mushroom clouds, from the cold statistics of fire, blast, destruction, and death, from the prompt surrender of Japan. Here and there, notably in liberal religious circles, the atomic bombings of Japan were publicly denounced on moral grounds as cause, in the phrase of *Commonweal*, for "American guilt and shame." Among most people, the nuclear end of the war provoked a jumble of ambivalent feelings, not only joy and relief but doubt, and fear that perhaps science had finally gone too far. There was no comfort to be found in Oppenheimer's flat response to a reporter's query about nuclear weapons: "If you ask: 'Can

[6] Quoted *ibid.*, p. 235.

[7] Quoted in William L. Laurence, *Dawn over Zero: The Story of the Atomic Bomb* (New York, 1946), p. 221.

[8] *Public Papers of the Presidents . . . Harry S Truman, 1945* (Washington, D.C., 1961), p. 197.

we make them more terrible?' the answer is yes. If you ask: 'Can we make a lot of them?' the answer is yes. If you ask: 'Can we make them terribly more terrible?' the answer is probably."[9] Yet pollsters reported that the public backed the use of the atomic bombs against Japan overwhelmingly because it brought the war to a speedy end. And whatever ambivalence the bomb provoked, it sealed a consensus, rapidly emerging even before Hiroshima, about the future requirements of national security.

Vannevar Bush himself publicly attested that, if the enemy's early technological superiority had been slightly greater or more diverse, "Nazi Germany would not now be prostrate." The United States, defense analysts said, managed to surpass German military technology only because the grace period before Pearl Harbor had allowed the nation sufficient time to mobilize her civilian scientific resources. After Hiroshima—as well as the German V-1s and V-2s, those harbingers of intercontinental missiles —it was widely said that there would be no grace period again. The White House had called the atomic bomb the "greatest achievement of organized science in history."[1] Now military officials, civilian analysts, congressmen, and thoughtful observers across the partisan and ideological spectrum agreed: In the atomic age, the United States could not do without a national policy of scientific, especially nuclear, research.

Americans debated the shape of that policy with special attention to the opinions of a newly influential group: the generation of atomic scientists —the Los Alamos generation—that Hiroshima and Nagasaki had brought to the center of the public stage. From the days of John Wesley Powell to those of Robert Millikan, American scientists had scarcely been concealed from public notice. But unlike any of its predecessors, this generation was dominated by physicists who seemed to wear the "tunic of Superman," in the phrase of a *Life* reporter, and stood in the spotlight of a thousand suns.[2]

Long before the glare of publicity, the role of martial superman had provoked a certain queasiness among the atomic scientists. I. I. Rabi had refused to take a full-time post at Los Alamos—was this to be the "culmination of three centuries of physics"?—and in the spring of 1945 he warned: "If we take the stand that our object is merely to see that the next war is bigger and better, we will ultimately lose the respect of

[9] *Commonweal*, 42 (Aug. 24, 1945), 443; Oppenheimer is quoted in *Time*, 46 (Oct. 29, 1945), 30. It is interesting to note that, speaking for the Federal Council of Churches, John Foster Dulles declared of Hiroshima and Nagasaki: "If we, a professedly Christian nation, feel morally free to use atomic energy in that way . . . the stage will be set for the sudden and final destruction of mankind." Quoted in *Time*, 46 (Aug. 20, 1945), 36.
[1] Bush testimony, U.S. Congress, House Committee on Military Affairs, *Hearings, Research and Development*, 79th Cong., 1st Sess., May 1945, p. 4; *Public Papers of . . . Harry S Truman, 1945*, p. 198.
[2] Francis Sill Wickware, "Manhattan Project," *Life*, 19 (Aug. 20, 1945), 100.

the public. In popular demagogy we [will] become the unpaid servants of the 'munitions makers' and mere technicians rather than the self-sacrificing public-spirited citizens which we feel ourselves to be." After the test explosion at Alamogordo, Kenneth T. Bainbridge, the chief of the operation and a shrewd, sensitive human being, turned to Oppenheimer: "Now we're all sons of bitches."[3] Hoping to maintain their scientific, political, and moral integrity, the Los Alamos generation on the whole declared with the Manhattan Project group under the leadership of the Nobel laureate James Franck that scientists could "no longer disclaim direct responsibility for the use to which mankind . . . put their disinterested discoveries."[4]

The Franck group was located at the University of Chicago Metallurgical Laboratory, where in the spring of 1945 the staff, its work for the bomb finished earlier than at other sites, had debated the use of the weapon in the war. Some of the Chicago scientists sided with Leo Szilard, the mercurial, effusive, perceptive Hungarian refugee, who was already appalled at the mass bombings of civilian populations. Sure that atomic bombs were "primarily a means for the ruthless annihilation of cities," Szilard opposed the atomic bombing of Japan on the moral ground that it would open the door "to an era of devastation on an unimaginable scale."[5] Many Metallurgical Laboratory scientists endorsed the earlier, more pragmatic report drawn up by the small group—Szilard among them—under Franck. If the United States were to release this new means of "indiscriminate destruction" without warning, the Franck Report argued eloquently, she would sacrifice her moral position in the world, precipitate a nuclear arms race, and "prejudice the possibility of reaching an international agreement on the future control of such weapons." Better, the report concluded, to warn the Japanese of what they faced, and to reveal this awful weapon to the world, "by a demonstration in an . . . uninhabited area."[6]

[3] Rabi is quoted in Oppenheimer to Rabi, Feb. 26, 1943, JRO, Box 59, Rabi file; Rabi to the Research Board for National Security, April 3, 1945, OV, Box 33; Bainbridge is quoted in Lamont, *Day of Trinity*, p. 242.

[4] James Franck et al., "A Report to the Secretary of War, June 1945," in Morton Grodzins and Eugene Rabinowitch, eds., *The Atomic Age: Scientists in National and World Affairs* (New York, 1963), pp. 19–20.

[5] Leo Szilard, "A Petition to the President of the United States, July 3, 1945," appended to Szilard, "Reminiscences," in Donald Fleming and Bernard Bailyn, eds., *The Intellectual Migration* (Cambridge, Mass., 1969), pp. 150–51; Szilard, "A Petition to the President of the United States, July 17, 1945," Grodzins and Rabinowitch, eds., *The Atomic Age*, pp. 28–29. Szilard told Oppenheimer that from the "point of view of the standing of the scientists in the eyes of the general public one or two years from now it is a good thing that a minority of scientists should have gone on record in favor of giving greater weight to moral arguments" than to pragmatic ones. Quoted in Martin J. Sherwin, *A World Destroyed: The Atomic Bomb and the Grand Alliance* (New York, 1975), p. 217.

[6] Franck et al., "A Report to the Secretary of War, June 1945," Grodzins and Rabinowitch, eds., *The Atomic Age*, p. 27.

But not all Manhattan Project scientists, at Chicago or elsewhere, agreed with the Franck Report. Particularly important were the high-level scientists giving advice on nuclear policy to Secretary of War Stimson, who had become accustomed to taking seriously the opinions of scientists on strategic matters. A scientific panel consisting of Oppen-heimer, Arthur Compton, Enrico Fermi, and Ernest Lawrence concluded in mid-June 1945: "We can propose no technical demonstration likely to bring an end to the war; we can see no acceptable alternative t direct military use."[7] Neither could Bush, Conant, or Karl Compton, who, at an advisory level closer to Stimson, had already joined in a recommendation that the bomb should be dropped without warning upon a military instal-lation surrounded by workers' housing. Stimson and his scientific advisers expected that using the bomb in this manner would most likely achieve their primary aim of bringing the war to a speedy end. Some also be-lieved that such a course might further, not impair, the prospects of post-war amity, since the indisputable proof of the bomb's destructiveness would force the world to recognize the imperativeness of peace.

Whatever their differences, even before Hiroshima the scientists who supported the Franck Report and those who sided with Stimson's advisers did agree on two points: The postwar peace would hinge on the establish-ment of some system of international control of nuclear energy, and reaching that goal would require the cooperation of the Allied powers, especially the Soviet Union. The two points commanded all the more unity among the Los Alamos generation after Hiroshima, when many of the nation's atomic scientists seemed to regard themselves, in the phrase of *Time*, as the "world's guilty men."[8] Guilty or not, World War II physicists knew, contrary to what various outspoken laymen claimed, that there was no "secret" of the bomb; the laws of nature were available to anyone. Since the technology to exploit those laws could be developed sooner or later by a scientifically capable nation, the current Anglo-American monopoly of the atomic bomb was no guarantee of long-term American security, certainly not against the Soviet Union.

For years scientific visitors to the USSR had all returned with the same report: Russian science was on the march. The second five-year plan had called for considerable investments in research—$200,000,000 a year according to congressional testimony in 1936—and the Russian gov-ernment paid its physicists well, provided them with well-stocked librar-ies, ample research facilities, and even rest houses by the sea or in the country. During the war, the Soviet government had maintained a high level of pure research; it had also husbanded its scientific manpower, keeping young men in school and producing, an observer remarked, "a

[7] Quoted in Alice Kimball Smith, *A Peril and a Hope: The Scientists' Movement in America, 1945–1947* (Chicago, 1965), p. 50.

[8] *Time*, 46 (Nov. 5, 1945), 27.

veritable army of young physicists."[9] In mid-June 1945 Soviet planes brought hundreds of foreign scientists, including a dozen Americans, to Russia to celebrate the two-hundred-twentieth anniversary of the Soviet Academy of Sciences. The itinerary included technical conferences and laboratory inspections and climaxed in an enormous banquet in the Great Hall of the Kremlin, where Kalinin, Molotov, and Stalin himself toasted Russian scientists for their contributions to the national defense and designated thirteen of them Heroes of the Socialist Order. Irving Langmuir, a member of the American delegation and not inclined to admire anything Communist, returned home convinced: The Russians would be "capable within a very few years of constructing atomic bombs."[1]

The mushroom clouds had hardly cleared over Hiroshima and Nagasaki before atomic scientists started publicly, passionately, and persistently to plead for the international control of atomic energy. Many of them expected that their Soviet brethren, aware that there could be no adequate defense against nuclear weapons, might help bring about an international accord, and make it work. On the world level, Austin M. Brues, a Manhattan Project veteran, maintained, scientists "see that nationalism means war and nothing else; that wishful thinking along the old pattern can destroy us." Yet now that science was so obviously an instrument of national power, it was even more inseparable than after World War I from the politics of national interest.[2] Most Los Alamos generation scientists tacitly treated it as such. So, quite explicitly, did the leadership of OSRD, in their policy toward former Axis scientists, international scientific affairs generally, and particularly the international control of atomic energy.

Unlike Hale's contemporaries, the leaders of World War II research neither condemned nor sought to ostracize Axis scientists on grounds of war guilt. They may have empathized with the plea of Hugo R. Kruyt, the Dutch president of the successor agency to Hale's ill-fated Inter-

---

[9] Testimony of Lyman J. Briggs, Subcommittee of the House Committee on Appropriations, *Hearings, Department of Commerce Appropriations Bill for 1937*, 74th Cong., 2d Sess., p. 141; Caryl P. Haskins, "Cooperative Research," *The American Scholar*, 13 (April 1944), 218.

[1] Langmuir to Robert M. Hutchins, Sept. 10, 1945, IL, Box 5.

[2] Brues is quoted in Smith, *A Peril and a Hope*, pp. 180–81. The theologian Reinhold Niebuhr declared: "There is . . . no 'scientific method' which could guarantee that statesmen who must deal with the social and political consequences of atomic energy could arrive at the kind of 'universal mind' which operated in the discovery of atomic energy. Statesmen who deal with this problem will betray 'British,' 'American,' or 'Russian' bias, not because they are less intelligent than the scientists but because they are forced to approach the issue in terms of their responsibility to their respective nations." Niebuhr, "A Faith to Live By . . .," *The Nation*, 164 (Feb. 22, 1947), 204. What Niebuhr said of statesmen tended, of course, to be true of scientists, too.

national Research Council: "keep politics as far from science as possible."[3] But they found their seemingly apolitical posture toward Axis scientists convenient to take, in part because they deemed some of them not guilty. A number of German academics had deeply resented the Nazi perversion of the universities. They and most leading Japanese physicists had contributed nothing to the Axis war machine. Matters of justice aside, the leadership of OSRD was sure, any attempt to suppress basic research in Germany or Japan would provoke a spirit of retaliation and revenge. The national interest of the United States, they insisted, called for encouraging democratically inclined Axis scientists to do basic research, publish their results, and contribute to the economic and political rehabilitation of their countries.[4]

On a more global level, during the war Joseph Needham, the British scientific liaison in Chungking, had proposed the establishment of an international scientific corps to help disseminate, independent of national and commercial interests, advanced scientific knowledge to the less industrialized nations of the East. Anton J. Carlson, the politically liberal president of the American Association for the Advancement of Science, pronounced the scheme "purely wishful thinking," because the great powers, the military services, and even the scientific members of the proposed corps would behave in just as nationalist a fashion as everyone else. Bush, who liked the thrust of the idea, transformed it into a recommendation to FDR for the inclusion of an international scientific section in the charter of the United Nations. The section would, among other duties, police the scientific activities of aggressor nations to the end that no nation should have cause to fear the scientific activities of another.[5] That aim also colored strongly the American plan for the international control of nuclear energy which took shape during the spring of 1946 with contributions of major significance from Bush, Conant, and especially Oppenheimer.

In this program, a world authority established under the United

---

[3] *The Fourth General Assembly of the International Council of Scientific Unions, Held at London, July 22nd to 24th, 1946, Reports of Proceedings* (Cambridge, England, 1946), p. 3.

[4] *Study of the National Academy of Sciences . . . on the Treatment of German Scientific Research and Engineering from the Standpoint of International Security,* July 2, 1945, OSR, Entry 1, Control of Germany file. Arguing to Bush that German atomic scientists should not be treated as dangerous persons, Oppenheimer declared: "You and I both know that it is not primarily men of science who are dangerous, but the policies of Governments which lead to aggression and to war. You and I both know that if the German scientists are treated as enemies of society, the scientists of this country will soon come to be so regarded." Oppenheimer to Bush, Dec. 11, 1945, *ibid.* The outraged response of Lee DuBridge to the destruction of the Japanese cyclotrons in the fall of 1945 was not atypical. DuBridge, "Recommendation on the Japanese Cyclotrons," [fall 1945], LAD.

[5] Needham, "Memorandum on an International Science Cooperation Service," attached to Needham to Bush, July 25, 1944; A. J. Carlson to Ross T. Harrison, Sept. 25, 1944; Bush to the President, Feb. 22, 1945, OSR, Entry 1, Postwar Planning, Needham Plan.

Nations would eventually control the production and use of all fissionable materials and oversee research and development in atomic energy for peaceful purposes only. But there would have to be a period of transition, Bush and Conant believed, between the signing of the agreement and the full operation of the authority, between 1946, when the United States commanded a nuclear monopoly, and the point in the distant future when no nation would possess a nuclear arsenal. In this transition period, they argued, the agreement would have to go into effect by stages, each accompanied by a quid pro quo from the Soviet Union. On the side of the United States, the stages might proceed from sharing of information about basic nuclear science, then to the release of material concerning the practical, industrial uses of nuclear energy. On the side of the Soviet Union the quid pro quo for each such stage would necessarily be the opening up of Russian basic research laboratories to free, unlimited access by foreign scientists, then acceptance of international inspection of all her nuclear facilities. Only after the inspection system was successfully in force would the United States turn its bombs into power plants.

Whatever the particular stages, Bush, Conant, and Oppenheimer recognized, any international control agreement had from start to finish to protect American security. To that end, Bernard Baruch, whom President Truman appointed to carry the American plan to the United Nations, insisted that the arrangement must include sanctions against violators, sanctions which must not be thwarted in the Security Council. Truman agreed. On June 14, 1946, in a portentous speech, Baruch presented the American plan—the Baruch Plan, it was quickly dubbed—to the newly created United Nations Atomic Energy Commission. "There must be no veto," he stressed, "to protect those who violate their solemn agreements not to develop or use atomic energy for destructive purposes."[6]

Three days later, June 19, Andrei Gromyko declared that the Soviet Union flatly rejected any tampering with the veto power. Otherwise ignoring the Baruch Plan, Gromyko proposed instead an international convention that would ban the bomb, of course meaning the American bomb, almost immediately. Over the next six months, in meetings of scientists and diplomats, the United States and Russia negotiated; the negotiations deadlocked over the issue of the veto. By January 1947 the Baruch Plan was a dead issue.

The veto quarrel doubtless clouded the real reason for the failure of the Baruch Plan. The Soviet Union, keenly aware it was a nonnuclear power, sought to neutralize the American advantage. The Russians preferred either to see the American nuclear arsenal destroyed forthwith, or to develop their own atomic bombs as quickly as possible, or both. Whatever the Soviet reasoning, the United States, its armies swiftly demobilizing, knew that massive Russian forces were poised on the borders of

[6] Quoted in Hewlett and Anderson, *The New World*, p. 578.

eastern Europe. Under such conditions, America would scarcely agree to immediate outlawing of the atomic bomb, the principal deterrent to Russian expansion.[7]

Like Bush and Conant, the Los Alamos generation at large generally endorsed a staged international atomic energy agreement because in the postwar era they emphatically agreed with its premise—that American security had to be protected. As citizens, the majority of atomic scientists read the lessons of the 1930s in the same way as the majority of their fellow Americans: Isolationism was perilous. As scientists, they were particularly sensitive to an argument that had first emerged in response to the repressions of the Nazis, then gathered force during the war. Science, the argument ran, needed freedom to survive, and so did freedom need science. Now, in the immediate postwar years, the Soviet scientific capability not only contributed to the plea for the international control of nuclear energy; it also helped etch deeper the belief that there was another side to the social responsibility of science—that scientists, in the words of Ernest Lawrence's old friend Merle Tuve, had a "definite, irreducible and nontransferable responsibility" to help maintain the national defense.[8]

Like Tuve, the head of the OSRD section that developed the proximity fuze, most of the nation's atomic scientists did not want to fulfill that responsibility as subordinates to the military. They insisted upon a degree of autonomy in peacetime defense research similar to what they had enjoyed under OSRD. But restless under the centralized direction of research, few wanted to remain in anything resembling OSRD itself.[9] While some intended to stay in such civilian-oriented laboratories as Los Alamos, the vast majority preferred to contribute to the national defense from the independent havens of campuses. There they would be free of excessive red tape and unreasonable or, most hoped, any security restrictions. There they could contribute to defense research on a part-time

[7] *Ibid.*, pp. 598, 619; Barton J. Bernstein, "The Quest for Security: American Foreign Policy and International Control of Atomic Energy, 1942–1946," *Journal of American History*, LX (March 1974), 1023–24, 1030, 1032, 1039.

[8] Tuve, "Notes for First Meeting of Committee on Planning for Army and Navy Research," June 22, 1944, and "Addendum to Minutes of Meeting, June 22, 1944," attached to Tuve to Bush, June 27, 1944, OSR, Entry 1, Cooperation: Postwar Military Research Committee.

[9] Bush told a congressional committee that in postwar defense research military officers and civilian scientists "must be equals and independent in authority, prestige, and in funds." U.S. Congress, House, Select Committee on Postwar Military Policy, *Hearings, Surplus Material—Research and Development*, 78th Cong., 2d Sess., Nov. 1944–Jan. 1945, pp. 248–49. Jewett to Bush, Sept. 2, 1943, OSR, Entry 1, Cooperation: Kilgore Committee; Conant, "Science and Society in the Post-War World," *Vital Speeches*, 9 (April 15, 1943), 396–97; Frank Jewett testified in May 1945: "The whole thing is repugnant to the ordinary civilian-life ways of scientists—the restrictions under which they have to operate and the cellular structure. The uniform experience in talking to all of the men who have given a lot of time and effort, to OSRD, is that they want to get out of this thing and get back to their work as soon as possible." *Hearings, Research and Development*, May 1945, p. 58.

basis, through advisory committees or actual research contracts. And there they could foster what virtually all of them considered essential to the maintenance of national security—the training of new scientists and the advancement of science as such.

In James B. Conant's estimate, the war, including the draft, had cost science some 17,000 advanced and 150,000 first degree graduates. The production of scientific doctorates was not expected to reach normal levels again until the mid-1950s. Manpower analysts studied the wartime halt in the prewar growth rates and predicted that by 1953 there would be a deficit of two thousand Ph.D.'s in physics, a sizable figure measured against demands for physicists in industry alone.[1] Besides, the nation was commonly said to have drawn heavily during the war upon its balance in the bank of pure science, and it no longer could expect substantial replenishment from Europe.

In the past, scientific training and research had been paid for by private wealth, but now private wealth, Karl Compton typically reemphasized, could no longer manage the task. The war effort had given professors the heady taste of doing research with few financial restraints. Typically, the young physicists at the Rad Lab had grown accustomed merely to signing an order for a new instrument whose purchase might have required a major faculty debate before the war. University administrators doubted that able scientists would be satisfied to return to the routine of heavy teaching loads, limited equipment budgets, and low salaries. In the wake of this physicists' war, the Los Alamos generation insisted that if it was the responsibility of the civilian scientist to contribute his expertise to defense research, it was also the responsibility of the federal government to finance the basic research and training on which the nation's security ultimately depended.

During the war the average annual federal investment in scientific research and development had shot up from $48,000,000 to $500,000,000, from 18 percent to 83 percent of the total public and private expenditure for the purpose.[2] The bulk of the increase had naturally gone to the military. Like the leaders of the Los Alamos generation, thoughtful Americans, especially those of a liberal political persuasion, were bothered about the degree to which defense research had come to overshadow basic research and training, and they were bothered by a good deal more in the legacy of wartime science, especially by the specter of monopoly.

Even the probusiness *Fortune* Magazine had concluded that dollar-a-

[1] Vannevar Bush, *Science—The Endless Frontier . . . July 1945* (Washington, D.C.: National Science Foundation, 1960), p. 158.

[2] The President's Scientific Research Board, *Science and Public Policy: A Report to the President, 1947* (5 vols.; Washington, D.C., 1947), I, 10. Part of the material in the rest of this chapter previously appeared in *Isis,* copyright © 1977 by *Isis,* and is reprinted here in different form with the permission of the editor.

year men in the war agencies sometimes rejected the suggestions of small businessmen and inventors for the self-interested purpose of protecting the market positions of their own corporations. Speeches, articles, and advertisements had frequently celebrated, in the language of *Collier's*, "the amazing new world . . . in the making in the plants and laboratories . . . devoted to war."[3] But liberals wondered who would control this amazing new world and its technological marvels. They found disturbing that some 66 percent of wartime contract dollars for research and development went to only sixty-eight corporations, some 40 percent to only ten. OSRD spent about 90 percent of its funds for principal academic contractors at only eight institutions, about 35 percent at the MIT Radiation Laboratory alone. And more than nine out of ten of the contracts from OSRD and the military granted the ownership of the patents deriving from this publicly funded research to the contractors, not to the public.[4]

Critics called the patent practice an unwarranted and dangerous giveaway. Before the war giant American firms had entered into patent agreements with foreign, including German, cartels. These contracts, it was alleged, had kept the United States from developing sufficient supplies of such strategic raw materials as beryllium, magnesium, optical glass, chemicals, and above all, synthetic rubber. Besides, the Temporary National Economic Committee had concluded that big business tended to use the patent system "to control whole industries, to suppress competition, to restrict output, to enhance prices, to suppress inventions, and to discourage inventiveness."[5] Assistant Attorney General Thurman Arnold, the administration's chief trustbuster and a central figure in the wartime patent controversies, explained to a congressional committee: If science and technology were to serve the public interest, it was necessary to do more than just break up patent pools. Something had to be done about the concentrated private control of the underlying knowledge and know-how from which patent pools germinated.[6]

[3] "The Bottleneck in Ideas," *Fortune*, 27 (May 1943), 82–85, 177–86; *Collier's*, 111 (May 8, 1943), 40.

[4] Carroll W. Pursell, Jr., ed., *The Military-Industrial Complex* (New York, 1972), pp. 165, 168–69; James Phinney Baxter, *Scientists Against Time* (Cambridge, Mass., 1968), p. 456.

[5] For salient aspects of the controversy surrounding the relationship between the control of patents and shortages of critical materials, especially synthetic rubber, see *The New York Times*, February 23, 1942, p. 30; March 26, 1942, pp. 1, 16; March 29, 1942, p. 9; April 12, 1942, p. 48; April 15, 1942, pp. 1, 7; April 18, 1942, p. 32. U.S. Senate, *Investigation of Concentration of Economic Power: Final Report and Recommendations of the Temporary National Economic Committee*, 77th Cong., 1st Sess., Sen. Doc. No. 35, 1941, p. 36. Assistant Attorney General Wendell Berge cast doubt on the academic use of patents when he attacked the Wisconsin Alumni Research Foundation for a wide range of abuses in the administration of the Steenbock vitamin D patent. U.S. Senate, Subcommittee of the Committee on Military Affairs, *Hearings, Scientific and Technological Mobilization*, 78th Cong., 1st Sess., 1943, pp. 741–42.

[6] U.S. Senate, Subcommittee of the Committee on Military Affairs, *Hearings, Scientific and Technological Mobilization*, 1943, pp. 11, 25.

In the increasingly powerful Bureau of the Budget, where Director Harold Smith had set a small staff group to work analyzing plans for postwar federal research, young James L. Sundquist pondered the effect of private control upon the public interest in science. In the private sector, if profit was not immediately apparent, Sundquist asserted, new fields of research languished, including smoke control, stream pollution control, low-cost housing, low-cost rural electrification, and the uses of substitute materials. In the governmental sector, established research programs tended to fare better than such new socially purposeful projects. The trouble was, Sundquist argued, that the federal research budget reflected not the *"national need"* but the "national *pressures*" for research. He found the principal pressures arising from economic interest groups and the government's own research agencies.[7] He might have added the increasingly insistent pressures from scientists who were interested in areas of research determined not by social standards of national need but by professional criteria of scientific significance. To balance the emphasis on economically and, now, militarily useful research, liberals like Harold Smith and his staff considered it in the national interest for the government to respond to the mounting demand for federal investment in basic research and training. They also regarded it as equally essential for the government to identify and invest in such areas of research as pollution control for which no pressure groups existed but which would likely pay a high social return.[8]

In the Congress the liberal concern with the control and purpose of national research found a spokesman in Senator Harley M. Kilgore of West Virginia. A small-town lawyer, National Guardsman, Legionnaire, Mason, and past Exalted Ruler of the Elks Lodge, Kilgore was quick to admit "utter, absolute ignorance" of science and technology. The son of a wildcat oil prospector, he was equally quick to discourse on the power of big business, including its power to deprive ordinary people of a fair chance. While a judge of the county criminal court, Kilgore won a reputation for his willingness to help juvenile offenders get a new start. In 1940 the pro-Roosevelt faction in the bitterly divided West Virginia Democratic Party found him a natural, if last-minute, choice for the senatorial nomination. Kilgore squeaked through a three-way primary with less than 40 percent of the vote, winning mainly because of the support of the CIO.[9] Handily elected on FDR's coattails, he went to Washington a down-the-line New Dealer eager to do all he could for the plain folks back home, especially the prolabor folks who had put him into office.

[7] James L. Sundquist to Arnold Miles, Feb. 16, 1945, BOB, series 39.33, file 91.

[8] Smith testimony, U.S. Congress, Senate, Subcommittee of the Committee on Military Affairs, *Hearings on Science Legislation*, 79th Cong., 1st Sess., Oct. 1945, p. 97.

[9] Kilgore, speech, May 10, 1943, copy in OSR, Entry 1, Cooperation: Kilgore Committee; *New York Times*, May 16, 1940, p. 24.

His suits wrinkled, his pockets crammed with papers, pencils, and a good luck horse chestnut, Kilgore soon established himself as an outspoken member of the Truman Committee, decrying the rubber shortage, patent abuses, and the power of dollar-a-year men. In 1942, at the urging of Herbert Schimmel, a congressional staffer who had a Ph.D. in physics, Kilgore introduced a bill for the better mobilization of the nation's technological resources.[1] In the course of extended hearings, Kilgore gradually learned about the critical importance of civilian science to national defense, the concentration of contracts for research and development, the pro-contractor patent policy of the armed services and OSRD. Moving steadily away from the problems of war mobilization, by the end of 1945 Kilgore had put together a full legislative program for peacetime science.

In Kilgore's program, the federal government was to finance basic research and training in the major fields of science, including a civilian-controlled effort in fundamental defense research. Kilgore called for the evaluation and, in the form of recommendations to the President, planning of all federal research activities to good social purpose. He also urged federal support of the social sciences, insisted that at least part of the money in all fields be distributed on a geographical basis, and proposed that the ownership of patents arising from federally sponsored research normally go to the government. In exceptional cases the contractor might be awarded ownership of the patent, but the government would always receive a royalty-free license for its use. To carry out his program, Kilgore proposed the establishment of a new federal agency, a National Science Foundation, which would of course be a regular part of the executive branch under the direct control of the President.[2]

In a way, Kilgore aimed de facto to extend the social reformism pioneered by John Wesley Powell to virtually the whole of federally sponsored science. In part his program responded to the long-standing demands of public colleges and to the professional self-interest of social scientists; in part it drew upon the preference for national planning of current Budget Bureau staffers, not to mention the pleas in the 1930s of Delano, Merriam, and Mitchell; in part it reflected the issues raised by the wartime mobilization of research. Whatever the roots of its specific elements, Kilgore's package added up to a major liberal initiative for peacetime science.

But while Kilgore's program was evolving, some armed service officers and the National Association of Manufacturers denounced it as a threat to socialize all of science in the United States. Both the army and navy objected to the patent provisions on grounds that they would raise the

---

[1] Henry H. Collins, Jr. and Herbert Schimmel to Senator Kilgore, Aug. 14, 1942, HMK, file A & M 967, series 1; Schimmel to the author, Oct. 3, 1973.

[2] U.S. Senate, Subcommittee of the Committee on Military Affairs, Report No. 7, *Preliminary Report [on Science Legislation]*, [Dec. 1945], 79th Cong., 1st Sess., copy in OWMR, Box 275, Legislation . . . Scientific Research file.

cost of, if not make impossible, industrial contracts for military research and development.[3] Like their employers, industrial scientists vociferously attacked Kilgore's patent policy, too, but the chief scientific opposition to his overall program came from the leadership of OSRD, especially Vannevar Bush.

Unlike many industrial scientists, Bush endorsed governmental curbs on the domination of markets by large industrial combinations through patents or any other device.[4] He had a variety of proposals in mind to prevent large firms from using the patent system to bar the entry of small firms into lucrative new technological markets; none of his ideas included the measures in Kilgore's program, especially not the stipulation that patents arising from federally sponsored research should belong to the government.[5] To Kilgore's call for the geographical distribution of some fraction of federal research funds, Bush advanced the traditional best-science objection: All the money should be awarded to the best investigators, wherever they were located. Kilgore's proposal to foster the social sciences antagonized the conservative Bush; he did not respect them intellectually and regarded them for the most part as just so much political propaganda masquerading as science. Besides, the OSRD staff worried that their inclusion might place the natural science program in jeopardy. Irvin Stewart, one of Bush's closest aides, advised a dean in Texas: Any provision for the support of the social sciences would be "dynamite."

[3] *Hearings, Scientific and Technological Mobilization*, 1943, pp. 309, 240–41, 245, 271; Walter S. Landis [Vice-President of American Cyanamid Co.], *What's Wrong with the Kilgore Bill* (privately printed, October 1943); *Legislation . . . That Could Radically Change Our Whole Economic Structure* (Commerce and Industry Association of New York, November 1943), copy in HMK, file A & M 967; Julius A. Furer, "Memo for Files," March 26, 1943, and Ralph Bard to Chairman, Senate Committee on Military Affairs, June 2, 1943, NCONR, Kilgore bill file; Harvey H. Bundy to Colonel Ege, April 3, 1943, and attached documents, OSW, Entry 82, Box 73, Technological Mobilization bill.

[4] At the end of World War I Bush's fledgling Raytheon Corporation had almost been forced out of business when General Electric, RCA, Westinghouse, and AT&T formed an exclusive patent pool and marketing agreement for vacuum tubes. In the coming postwar period, he believed, the nation's prosperity would depend to a considerable extent upon the appearance of similar small businesses applying new knowledge in a useful manner. Bush to Oliver E. Buckley, July 25, 1945, OSR, Entry 2, Report to the President, #3; Bush, *Pieces of the Action* (New York, 1970), pp. 198–99.

[5] Bush thought that the government might make it less expensive for small firms to defend a patent by supporting the firm's case in court. The law might also be modified to help small firms to defend themselves against infringement suits by big firms. It would be a valid defense to show that, by exchange of patent rights in combination with others, the big firm had acquired more than a reasonable fraction of the market; if the plaintiff won, he would be awarded only royalties, not an order to cease and desist. Bush was even willing to go so far as to compel the licensing of patents held by large producers who were found to be participants in an exclusive cross-licensing agreement. Bush to Conway P. Coe, November 4, 1943, VB, Box 24, file 567; Bush to John T. Tate, February 8, 1945; Bush to Delos G. Haynes, February 21, 1945; Bush to Robert E. Wilson, March 10, 1945, OSR, Entry 2, Report to the President.

Social scientists could not easily convince Congress to spend federal funds "for studies designed to alleviate [the] conditions of the Negroes in the South or to ascertain the influence of the Catholic Church on the political situation in Massachusetts."[6]

Nothing in Kilgore's program disturbed Bush more than its direct control by the President, especially since the control was to be exercised in conjunction with planning for socially purposeful research. Here was not so much a threat of political interference with intellectual freedom as a prospect of permitting the political system to determine the kind of research to be encouraged. Homer W. Smith, the director of the physiology laboratory at New York University School of Medicine and an adviser to Bush on postwar policy for science, warned: We stand in danger of losing our intellectual heritage "not to personal dictators, antiscientific prejudices or other antiquated tyrants, but to an as yet unnamed horror of [a] national policy wherein men will be 'requested' . . . to devote their lives to artificial and imposed intellectual interests." Of course, Smith implied, the scientist need not heed the requests, but he would be "forced" to do so through the "pressures of personal and professional financial needs, advancement, security and scientific opportunity." What the scientist required was the opportunity for "free" research, which meant "the intellectual and physical freedom to work on whatever he damned well pleases."[7]

Smith's analysis recalled Alexander Agassiz's assault on John Wesley Powell—and it merited something akin to Powell's response. Without Kilgore's program, were not scientists subject to personal pressures, financial and otherwise? Did they not worry about advancement, security, and scientific opportunity? Did not professional pressures help to determine their research topics? And was not the trend, in other disciplines as well as physics, toward group efforts, in which junior scientists scarcely enjoyed the same latitude as their seniors? Under these circumstances scientists were hardly as "free" as Smith claimed. They were free only from any government encouragement to invest their talents in one field or another. They were free only to operate within a framework defined by private decisions.

Kilgore's program rested on James Sundquist's fundamental assumption: Science had become too important to leave the formation of national scientific policy entirely to the haphazard play of interest groups, scientific or otherwise. Like Kilgore, Bush, too, saw a need for the planning and

---

[6] Bush to Isaiah Bowman, Jan. 7, 1946, OSR, Entry 1, NSF: Public Interest, Committee for the Bush Report; Bush to D. C. Josephs, Sept. 12, 1946, VB, Box 60, file 1416; U.S. Congress, House, Subcommittee of the Committee on Interstate and Foreign Commerce, *Hearings, National Science Foundation Act*, 79th Cong., 2d Sess., 1946, p. 53; Stewart to A. P. Brogan, Sept. 24, 1945, OSR, Entry 1, Postwar Planning.
[7] Smith to Bush, Aug. 29, 1945, OSR, Entry 1, NSF: Public Interest, Educational Institutions.

evaluation of the national research program. But Bush objected to vesting control over much of the nation's science in a centralized federal bureaucracy, and not without justification. Policy decisions could all too easily favor the bureaucratic over the scientific or even the national interest. Of course, more than in the private arena, public decisions would be open to scrutiny and debate. Nevertheless, while Kilgore aimed to subject the shaping of federal scientific policy to the political process, Bush, a political elitist, wanted the immediate determination of that policy left essentially in private hands, even if its execution involved the expenditure of public funds. "Powerful writing," he exclaimed of Smith's fusillade.[8]

During the fall of 1944, ready to seize the initiative from Kilgore, Bush successfully maneuvered to obtain a letter from FDR requesting a report to the President on postwar policy for science, and in July 1945 he sent the completed document to President Truman under the title *Science —The Endless Frontier*.[9] Like Kilgore, Bush called for a program of federal aid to basic scientific research and training, along with a civilian-controlled effort in fundamental defense research. Unlike the senator, he proposed no remedy for what he acknowledged were "uncertainties" and "abuses" in the patent system, omitted the social sciences from consideration, and made no mention of distributing the funds in his program in accord with either a geographical scheme or the planning of socially purposeful research. Bush conceded that his program had to be responsible— but, in a politically elitist fashion, not responsive—to the President and Congress. Stressing the importance of insulating it from political control, he proposed to operate his program through a new federal agency, which, unlike Kilgore's National Science Foundation, would, in the apt summary of an OSRD staff aide, be "a new social invention—of government sanction and support but professional guidance and administration."[1]

Bush's report, in the remark of an OSRD staff member, was "an instant smash hit."[2] But while applauded in scores of editorials across the partisan and geographical spectrum, it was no smash hit with the *New Republic*, which commented: "Research needs to be coordinated carefully and the projects should be selected in terms of our national necessities, and not the accidental interests of various scientific groups." The politically insular posture of Bush's program was certainly not a smash hit in the

[8] Bush to Carroll L. Wilson, n.d., attached to Smith to Bush, Aug. 29, 1945, *ibid.* To Bush, the primary question was "whether science in this country is going to be supported or whether it is also going to be controlled." Bush to Bernard Baruch, Oct. 24, 1945, OSR, Entry 1, NSF Legislation.

[9] Daniel J. Kevles, "FDR's Science Policy," *Science*, 183 (March 1, 1974), 798–800; Diary of Oscar S. Cox, Oct. 24, 1944, OSC; "News Stories and Editorial Comment on the President's Letter . . .," FDR, OF 4482.

[1] Bush, *Science—The Endless Frontier*, pp. 32–40, 21–22, 38; Laurence K. Frank, "Comments on Bush Report . . .," OSR, Entry 2, Misc.

[2] Lyman Chalkley to Carroll L. Wilson, July 20, 1945, OSR, Entry 2, Distribution; "News Stories and Editorial Comment on the Postwar Program for Scientific Research," OSR, Entry 1, Postwar Planning Reports.

Budget Bureau or with Maury Maverick, the former mayor of San Antonio, Texas, and the head of the Smaller War Plants Administration, who recognized the degree of concentration of defense research contracts and who in October 1945 exploded to a congressional committee: "I get a little tired of these hired hands of the monopolies and some of the professors . . . piously abrogating [sic] to themselves all the patriotism. . . . I'm not sure but that the office holder has been, and is, more conscious of the public welfare than many scientists. . . . Let us all bear in mind that we have a political Government and that our Constitution is a political instrument. The political character of our Government guarantees democracy and freedom, in which the people, through their Government, decide what they want."[3]

At the opening of the atomic age, the revolutionary needs of national security had joined the older requirements of economic development to force an end to what had long been, de facto, a federal policy of laissez-faire in physics. Bush was willing to endorse that end insofar as he was willing to put the government into the business of funding academic scientific research outside of agriculture. But if Kilgore's program was directed at mobilizing scientific research in a fashion politically responsive to the nation's best needs, Bush's aimed to enlist the nation's resources, through a politically elitist mechanism, to satisfy the scientific community's traditional goal of advancing the best science. Like George Ellery Hale at the end of World War I, Bush may have thought that his program disinterestedly sought the apolitical national interest. Set against Kilgore, Bush insisted upon a politically conservative interpretation of what the end of laissez-faire was to mean in postwar America for the Los Alamos generation and its science.

[3] "A National Science Program," *New Republic*, 113 (July 30, 1945), 116; *Hearings on Science Legislation, 1945*, pp. 368–69. In the administratively sensitive Budget Bureau, Harold Smith, who wondered, only half puckishly, whether the "Endless Frontier" implied the "Endless Expenditure," declared most scientists had "learned to accept governmental funds with ease, and . . . they can adapt themselves to governmental organization with equal ease." Smith to Carroll L. Wilson, July 18, 1945, BOB, series 39.27, Box 82, OSRD file; Smith to Julius A. Furer, June 14, 1945, JAF, Research and Development file. Budget staff members found Bush "pompous" and the research policy area filled with "prima donnas." Donald Stone, "Notes Relating to Meeting of Vannevar Bush . . .," Jan. 1945"; Donald Miles to Stone, Feb. 12, 1945, BOB, series 39.32, files E8—21/44.1 and E8–20/45.1.

XXII

Victory for Elitism

or weeks after the end of the war, scientists waited impatiently
for a presidential message to Congress on the domestic control
of atomic energy. The message finally was delivered on October
3, 1945, accompanying an administration bill. The measure had
been drafted in the War Department and it was introduced that day by
the chairmen of the two military affairs committees, Congressman
Andrew J. May of Kentucky and Senator Edwin C. Johnson of Colorado.

The May-Johnson bill proposed to vest virtually complete authority
over nuclear research and development in an Atomic Energy Commission, which, in a politically elitist fashion, was to consist of part-time
members appointed by the President but insulated to an extraordinary
degree from his removal power. The bill included some weakly worded
clauses designed to discourage monopolistic practices in the exploitation
of atomic energy by private industry. Beyond that, the measure displayed
little concern with the social and economic impact of nuclear fission and
it ignored the task of balancing the Commission's program between civilian
and military needs. The May-Johnson bill went at the control of nuclear
energy more in the manner of Bush, who had suggested a number of its
features, than of Kilgore. In tone, and to a degree in substance, it also
made the chief object of the nation's nuclear energy program seem to
be not the peaceful but the military atom.[1]

[1] Richard G. Hewlett and Oscar E. Anderson, Jr., *The New World, 1939/1946*
(*A History of the United States Atomic Energy Commission*, Vol. I; University Park,
Pa.: Pennsylvania State University Press, 1962), pp. 412–15, 427–28. Text of the bill,
H.R. 4280, 79th Cong., 1st Sess., kindly supplied by Richard G. Hewlett.

In the Truman administration, the implications of the bill were recognized by Don K. Price at the Bureau of the Budget and James R. Newman at the Office of War Mobilization and Reconversion. Price, a Rhodes Scholar and an authority on public administration, was Harold Smith's chief staff man on policy for science. Newman, a product of City College of New York and Columbia Law School, a former counsel for the American Jewish Committee and the Anti-Defamation League, was an irrepressible, outspoken liberal, with a sparkling mind and considerable literacy in science, philosophy, and mathematics. To both Price and Newman, the May-Johnson bill's egregiously independent, part-time Commission flagrantly violated the principle that the President should exercise direct control over all executive agencies. Newman, an enthusiast of Kilgore's program, especially the senator's commitment to planning socially purposeful research, thought that the bill also gave too little emphasis to civilian uses of atomic energy, had too few teeth in its antimonopoly clauses, and left too much of the nation's nuclear future, including its advancement of basic nuclear research, up to the vague good faith of private institutions.[2] Inside the executive branch, Newman and Price launched an offensive for an atomic energy program based on sound principles of public administration and the imaginative adaptation of liberal doctrine to a revolutionary field.

The May-Johnson bill also angered many of the nation's atomic scientists. It seemed to permit domination of the proposed Commission by the military. It also subjected almost all nuclear research in the country to rigid security restrictions, with penalties for violators ranging upward from a fine of $100,000 and a ten-year prison term. To many atomic scientists, the promilitary and security features of the bill threatened to interfere with the conduct of nuclear studies; like every other branch of science, nuclear exploration required a reasonably free flow of information. More important, both features seemed sure to signal to the rest of the world that the United States' primary nuclear interest was in weapons, a posture that would diminish the chances for preventing a nuclear arms race.[3]

These objections were felt most strongly among the rank and file of atomic scientists. The May-Johnson bill was endorsed by Bush, Conant, Lawrence, Fermi, Arthur Compton, and even Oppenheimer. To all save Oppenheimer, who vacillated, the faults of the bill were exaggerated. Bush privately labeled as "absurd" the charge that the measure would lead to military domination of atomic energy. In certain quarters of the Manhattan Project, notably the Metallurgical Laboratory, a conviction that the Franck Report had not been given a full hearing before

[2] Hewlett and Anderson, *The New World*, pp. 437–38.
[3] Alice Kimball Smith, *A Peril and a Hope: The Scientists' Movement in America, 1945–1947* (Chicago, 1965), p. 130.

Hiroshima had already made the wartime administrators of science suspect. Now a number of Project scientists concluded with Herbert Anderson, a highly respected physicist who had been at the Metallurgical Laboratory: "I must confess my confidence in our leaders Oppenheimer, Lawrence, Compton, and Fermi . . . who enjoined us to have faith in them and not influence this legislation, is shaken. . . . Let us beware of any breach of our rights as men and citizens."[4] As men, citizens, and socially responsible scientists, Project veterans like Anderson decided to take legislative matters into their own hands.

In the fall of 1945, scores of atomic scientists descended upon Washington to buttonhole congressmen, lobby within the administration, and educate the public to the necessity for a civilian-controlled program of atomic energy free of unreasonable security restrictions. They started in a single fourth-floor walk-up with a desk, telephone, ancient typewriter, and $20 worth of stationery. Backed by a growing number of study and discussion groups at Project sites and campuses around the country, by early 1946 they formed the active center of the new Federation of American Scientists, with their own magazine, the *Bulletin of the Atomic Scientists*. The leader of the Washington group was thirty-five-year-old William Higinbotham, a physicist who had left his graduate studies at Cornell to join the Rad Lab, then finished the war as an electronics specialist at Los Alamos. Like most of the Washington activists, he was not one of the nation's prominent atomic scientists, but Higinbotham and his young allies, "quiet, modest, lucid and impellingly convincing," in the judgment of the liberal newscaster Raymond Gram Swing, did a good deal to shift influential public opinion against the May-Johnson bill.[5]

In the administration, the scientists' lobby linked up with Price and Newman, who soon added to his program the complete exclusion of the military from the Commission; in the Congress they found a champion in Senator Brien R. McMahon. A former assistant attorney general of the United States, a Democrat and New Dealer, McMahon was a determined freshman from Connecticut. In December 1945 he introduced a bill—it had been drafted by Newman and an associate—for an atomic energy program civilian in control, liberal in purpose, and responsive to the political system. Soon President Truman, who increasingly relied on Newman as his special adviser on atomic energy, shifted to the McMahon bill. In the Congress, many moderates and even conservatives, following the lead of Senator Arthur H. Vandenberg, gathered behind the measure because the field of atomic energy was so new, revolutionary, and fraught with urgent considerations of national security. In July 1946,

[4] Anderson is quoted *ibid.*, p. 140; Bush to Conant, Nov. 4, 1946, VB, Box 27, file 614.

[5] Quoted *ibid.*, p. 171.

after the bill had been modified to give the military a certain—some said significant—voice in the atomic energy program, the McMahon bill passed the Congress with bipartisan support.[6]

The act created a full-time, civilian Atomic Energy Commission, whose members were to be appointed by and responsible to the President. It awarded the Commission complete control over the production, ownership, and use of fissionable materials; it also prohibited the issue of patents for inventions applicable solely in the atomic energy field. Thus rendering atomic energy virtually a state monopoly, the act not only instructed but vigorously equipped the Commission to sponsor pure and applied research in its field, to foster socially as well as militarily purposeful investigations, to encourage competition in the private sector through licensing, and to prevent any industrial corporation from cornering and possibly suppressing a new development threatening to its market position. No less important to most of the nation's scientists, the act enabled the Commission to assure the widespread dissemination of nuclear information, pure and practical, subject to appropriate international agreements and security restrictions which they found acceptable.

In October 1946 President Truman nominated as the first chairman of the Commission David E. Lilienthal, the former head of TVA, who in the course of his confirmation hearings delivered one of the memorable lectures of the day on the relationship between civil liberties and security in a democratic state. The other commissioners included Robert F. Bacher, who had left the Rad Lab to accept an appointment at Los Alamos on the understanding that he would depart the mesa the moment he had to put on a military uniform. If the Atomic Energy Act of 1946 was a victory for civilian control of a singularly important branch of physical research, later critics of the Commission's practices would declare it a triumph for private over public interests. But at the time, in the context of the broader issues in postwar policy for science, Newman could justly call it a victory for a basic principle—"the proposition that the self-regulating mechanism of the market place cannot always be depended upon to produce adequate results in scientific research."[7]

The Atomic Energy Act left most of the vast field of nonnuclear defense research to the armed services, which were eager to follow the dictum of the journal *Army Ordnance*: The maintenance of military superiority in peacetime would require "the marshaling of the best scientific brainpower of our country . . . from our great private research laboratories . . . educational institutions and . . . technical and scientific societies."[8] Armed service planners aimed to enlist civilian scientists as

---

[6] Hewlett and Anderson, *The New World*, pp. 441–43, 453–55, 489, 502–4, 506, 510–13, 530.

[7] James R. Newman and Byron S. Miller, *The Control of Atomic Energy: A Study of Its Social, Economic, and Political Implications* (New York, 1948), p. 17.

[8] *Army Ordnance*, Nov.–Dec. 1944, p. 485, copy in OSR, Entry 1, Postwar Military Policy Committee.

advisers upon the substance, the conduct, and the strategic implications of military research and development programs. They also intended to continue awarding contracts to industrial and academic laboratories for applied military research. And the more thoughtful planners wondered whether the War and Navy departments ought not to go beyond the nuts-and-bolts development of hardware and help overcome the depletion in the nation's bank account of fundamental scientific knowledge and manpower.

Climaxed by Hiroshima, the miracles of OSRD had driven home to the armed services the military importance of two types of fundamental research. The first was the study of subjects clearly related to military technology, such as the behavior of electromagnetic radiation at the frequencies of microwave radar. The second was the familiar pure science, most commonly exemplified in the discussions of 1945 by the nuclear explorations that had made possible the atomic bomb. The first would contribute directly to the improvement of military technology; either could yield radically new weapons in the future. Conducted on the campuses with the aid of graduate students, both would enlarge the pool of trained scientific manpower. Now, when even Secretary Robert Patterson admitted that the army might have treated draft-eligible scientists too strictly, the armed services strongly endorsed peacetime federal programs outside of atomic energy in both types of fundamental research, but especially in the militarily pertinent kind so obviously in the long-range interest of national defense.[9]

Admiral Furer's wartime staff had included especially keen advocates of such research—a group of bright, imaginative, resourceful young naval officers, most of them Ph.D.'s in science, who acted, in the argot of their outfit, as the admiral's "Bird Dogs," his cocky troubleshooting ambassadors to the naval operating arms. In the middle of the war the Bird Dogs began advancing a plan for a peacetime central office of naval research. Its chief duty in their prospectus was to sponsor, not only in the navy

---

[9] Early in 1945, the secretaries of war and navy had established a joint military-civilian group in the National Academy of Sciences to carry on a militarily related program of fundamental research when OSRD went out of business. The resort to the Academy was for the interim, until a permanent, peacetime agency could be established. But the interim was no excuse for such an arrangement in the opinion of the Budget Bureau, including Harold Smith, who recognized that under the control of the politically elitist Academy the group would "not be responsible to any part of the Government." At Smith's urging, FDR directed his secretaries of war and navy to withhold all funds and projects from the Academy group. Not long after FDR's death, following a Smith visit to the Oval Office, President Truman reiterated to Stimson and Forrestal that full control of military research and development "must at all times be lodged solely within the framework of the government." Smith to Franklin D. Roosevelt, March 31, 1945; Roosevelt to Bush and Roosevelt to Secretaries of War and Navy, March 31, 1945, FDR, OF 330–8; Truman to Secretary of War, June 8, 1945, attached to Truman to Bush, June 8, 1945, OSR, Entry 1, Cooperation: RBNS; Smith, "Conference with President Truman, June 8, 1945," HDS.

but in the best qualified civilian laboratories, the fundamental research essential for the creation of radically new weapons.[1] The Bird Dogs and their allies got a chance to turn their idea into policy when they became part of a new research office that Secretary Forrestal, acting under his temporary war powers, created in May 1945. In 1946 Congress made the agency permanent as the Office of Naval Research, or ONR.

The new Office was headed by Admiral Harold G. Bowen, the forceful head of the Naval Research Laboratory. Bowen was no friend of the regular bureaus; they had banished him to the Research Laboratory, a kind of purgatory in the prewar years, after he fought to introduce the use of superheated steam. Now Bowen wanted to push the development of a nuclear-powered navy. He was also still smarting at the high-handed and supercilious way that the OSRD hierarchy had treated the navy's scientific arms during the war, and smarting all the more now that the Rad Lab was getting all the credit for wartime radar. Bowen found congenial to his own predilections the Bird Dogs' advocacy of a program of fundamental naval research conducted independently of the bureaus. And he was ready to carry out the program with zeal. The zeal possibly reflected an eagerness to enlist academic physicists in his drive for nuclear-powered ships.[2] It almost certainly revealed a determination to show that his part of the navy could manage civilian scientific research at least as well as Bush and OSRD.

In the fall of 1945 Bowen and his staff traveled around the country with promises of research funds and promptly raised suspicions among academic scientists. Would not military support involve irritating red tape, crippling security restrictions, and projects of primarily military, not scientific, interest? But Bowen's Office minimized the red tape and allowed university scientists virtually complete freedom in the conduct and publication of their research. Equally important, Bowen's Office funded not only militarily relevant but even pure research projects. It also left the initiative for proposing the projects up to the academics. The navy chose which projects to support with the help of a cadre of civilian scientific advisers, whose members included Warren Weaver as chairman, Lee A. DuBridge, and the Compton brothers.[3]

[1] The Bird Dogs [Bruce S. Old et al.], "The Evolution of the Office of Naval Research," Physics Today, 14 (August 1961), 32; R. D. Conrad to Coordinator of Research and Development, Sept. 3, 1943, OSR, Entry 1, Postwar Planning; Bruce S. Old, Ralph A. Krause, and John T. Burwell to Secretary of the Navy, Sept. 23, 1944, NCONR, Postwar Research file, Vol. IV.

[2] Bowen to Burt Nanus, April 8, 1959, copy kindly supplied by Edward L. Bowles; Lewis Strauss to Secretary of the Navy, May 25, 1945, JFN, file 39–1–8; Furer Diary, May 29, 1945, June 4, 1945, JAF; Harvey M. Sapolsky, "The Origins of the Office of Naval Research," chapter two, ONR: Science and the Navy (in preparation).

[3] John E. Pfeiffer, "The Office of Naval Research," Scientific American, 180 (Feb. 1949), 12–13; The Bird Dogs, "The Evolution of the Office of Naval Research," p. 35.

In February 1946 Bowen's staff announced that they had negotiated contracts with forty-five schools and industrial firms; by August 1946, when the bill establishing ONR was signed into law, Bowen already had in force 177 contracts, totaling $24,000,000, with eighty-one universities or private and industrial laboratories. ONR was supporting more than 602 academic research projects, which together involved some two thousand scientists and an equal number of graduate students. It was building cyclotrons and betatrons, signing up astronomers, chemists, physiologists, botanists, branching out into such unmilitary studies as meteors, the rare earths, and plant cells.[4]

Yet at least some scientists worried with Harlow Shapley, the prominent Harvard astronomer and political activist, who declared: The government's "intercession in American science . . . has altered, and perhaps become ominous . . . because of the Navy's great move in support-ing science on a wide basis. . . . Those who were worried about domina-tion of freedom in American science by the great industries, can now worry about domination by the military."[5] ONR may have permitted scientists to pursue and publish what they wanted, but it did decide what to support in part on criteria of utility to the navy. In any case, the degree of scientific freedom allowed by Bowen's operation was scarcely the whole issue. Earlier, Admiral Furer had argued that naval support of civilian research would win the service a cadre of distinguished scientists who would act as "ambassadors of national preparedness" in the civilian scientific community. Now, the army chief of staff for re-search and development observed, the military was said to be making "prisoners of war" of the nation's scientists.[6] Whatever Furer's en-thusiasm, such overwhelmingly military patronage of academic science —the Office of Naval Research accounted for three out of four of all federal dollars spent for such fundamental research—scarcely made for sound public policy. It might produce fine science, but it threatened to bind too many of the nation's scientists to the military's purposes and bureaucratic self-interest.

The military domination of academic science was a matter of some

[4] *Newsweek*, 27 (Feb. 11, 1946), 89; Bowen to the Secretary of the Navy, Aug. 30, 1946, HGB, Box 2; Karl T. Compton to John R. Steelman, Oct. 23, 1947, attached to Karl T. Compton to James Webb, Oct. 27, 1947, VB, Box 26, Karl T. Compton file.

[5] Harlow Shapley to Isaiah Bowman, Nov. 6, 1946, VB, Box 13, Bowman file. Louis N. Ridenour, Albert Einstein, Philip Morrison, et al., "Military Support of American Science, A Danger?" *Bulletin of the Atomic Scientists*, 3 (Aug. 1947), 221–30. Henry Wallace declared the military were "going far beyond their legitimate concern—the development of weapons in accord with our defense needs—and thus unconsciously and without intent crippling science and corrupting scientists." Wallace, "Science and the Military," *New Republic*, 116 (Feb. 3, 1947), 26.

[6] Naval Coordinator [Furer] to All Bureaus and Offices, Feb. 23, 1945, JFN, file 39–1–8; Major General H. S. Aurand to the Secretary of War, WDGS, Legislative and Liaison Division, Bill File S. 1850.

concern in the Bureau of the Budget, which had approved the entrance of Bowen's office into the large-scale support of fundamental research only as an interim arrangement. The navy had agreed that, once the National Science Foundation was created, the bulk of the pure and even some of the militarily related research projects of ONR would be turned over to the new agency, where they belonged.[7] In the Budget Bureau and among thoughtful members of the scientific community, the way to halt the increasing military role in academic science was for Congress to create the foundation as soon as possible.

Early in the summer of 1945, to advance his version of such a foundation, Bush had a bill drafted in OSRD and introduced by Senator Warren G. Magnuson of Washington, a freshman New Deal Democrat. At hearings on the Kilgore and Magnuson bills in the fall, all ninety-nine witnesses save one—Frank B. Jewett—endorsed the creation of a single federal agency that would award grants, contracts, and fellowships to sponsor training and fundamental research, including militarily relevant research, in all fields of the natural sciences. But there was no such consensus about the key issues dividing Kilgore and Bush: planning, social sciences, geographical distribution, patents, and, above all, the degree of programmatic control by the President. The issue of presidential control, which drew the most attention at the hearings, found specific expression in a dispute over the proposed foundation's administrative structure. In Kilgore's politically responsive scheme, the foundation was to be managed by a director appointed by and responsible to the President. In Bush's politically elitist approach, the foundation was to be governed by a presidentially appointed part-time board of private citizens, mainly scientists, who would choose a director responsible not to the President but to themselves.

To Don K. Price and James R. Newman, the dispute over the foundation raised much the same issues as the battle over atomic energy. Together with Harold Smith they formed a coalition against Bush with considerable influence in the White House, and in the President's first postwar legislative message, paragraphs written by Newman made Kilgore's general program the program of the Truman administration. Kilgore, who faced an election in the fall of 1946, seemed to want a passable bill. Early in 1946 he introduced a revised measure—it was

---

[7] Blessed with an abundance of military funds, colleges were rapidly building research groups and laboratories, a Budget staff member noted. "Several years from now the military budgets may be cut sufficiently so that the military cannot afford to support so much basic research. . . . When this time arises the colleges will be faced with the unpleasant dilemma of finding other sources of funds, firing some of their staff, and closing down laboratories, or accepting applied research on weapons with the necessary security restrictions." Unsigned memorandum, n.d. [April 1947], BOB, series 39.33, file 93; W. L. Clark III to Director, Planning Division, Office of Naval Research, May 29, 1946, ONR, Planning Division file.

designated S. 1850—which compromised on details but maintained his program's social purposefulness and degree of presidential control.[8] In July, S. 1850, intact save for its social science clause—the social sciences, it was contended on the floor, were merely elaborate ways by which some people told many others how to behave, especially in politics—passed the Senate.[9] But in the House, the young Arkansas Congressman Wilbur D. Mills had introduced a bill along the lines of the original Bush program. Anxious to get home to campaign for reelection, Mills and his colleagues declared themselves unable to resolve the complex differences between the Mills bill and S. 1850. The foundation failed in the 79th Congress.

When the Republican-dominated 80th Congress convened in January 1947, leadership for a foundation bill was assigned to Senator H. Alexander Smith of New Jersey. A first-term senator in his mid-sixties, Smith was a Princeton graduate and an acolyte of Woodrow Wilson in international affairs, a successful corporation lawyer and devotee of his alma mater—in the 1920s he interrupted his legal career to be secretary of the university—a conservative Republican disposed to sympathize with the views of professors from private colleges in the East, including the politically elitist, best-science views of Vannevar Bush.[1] Smith drew up a new bill that closely followed Bush's approach on key points, including geographical distribution and insulation from presidential control. In the summer of 1947, despite objections of senators from scientifically have-not states, Congress passed the Smith measure. But President Truman vetoed the bill, explaining that he could not approve the establishment of an executive agency so far beyond the control of the nation's chief executive.[2]

The veto message emphasized that the President would gladly sign into law an act establishing a science foundation subject to the President's

[8] J. R. Newman and M. E. Doyle to Fred Vinson, June 26, 1945, OWMR, Box 277, Research Board for National Security file; "Special Message to the Congress . . . Sept. 6, 1945," *Public Papers of . . . Harry S Truman, 1945*, pp. 292–94; copy of S. 1850 in OWMR, Box 275, Legislation . . . Scientific Research file.

[9] U.S. Congress, Senate, *Congressional Record*, 79th Cong., 2d Sess., July 1, 1946, p. 8048; Kenneth M. Jones, "Science, Scientists, and Americans: Images of Science and the Formation of Federal Science Policy, 1945–1950" (unpublished Ph.D. dissertation, History, Cornell University, 1975), p. 325.

[1] Arguing in 1946 for a foundation insulated from the President's control, Senator Smith had claimed that scientists were "not at all controlled by the consideration of who pays them the money, whether it comes from the Government, or whether it comes from private sources. They are inspired by their interest in the subject, by their dedication to the pursuit of knowledge." U.S. Congress, Senate, *Congressional Record*, July 2, 1946, p. 8113.

[2] It is interesting to note that Senator Wayne Morse of Oregon, who had urged the Senate, unsuccessfully, to incorporate a clause mandating at least some geographical distribution of the funds, urged Truman to veto the bill, which, he asserted, was "fostered by monopolistic interests" and was opposed by "a great many educators and scientists associated with state-supported educational institutions." Morse to Truman, HST, OF 192–3. It is also interesting to note that the chief of the Senate Majority Policy Committee staff had predicted that the President would never veto the bill. George H. E. Smith to H. Alexander Smith, April 24, 1947, HAS, Box 132, NSF file.

direct control. But though a compromise to meet that stipulation passed the Senate, Robert Taft had said that no bill creating such a foundation would pass the 80th Congress, not if he had anything to say about it.[3] None did.

When the Democratically controlled 81st Congress convened in January 1949, the Senate quickly passed the same compromise bill, only to have it bottled up in the House Rules Committee by a coalition of Republicans and Southern Democrats. Relentlessly economy-minded, they argued: why spend additional money for scientific research when the armed services were already funding all the research they needed?[4] The bill remained in the Rules Committee until February 1950, when the House leadership extracted it by a parliamentary maneuver. In March finally, the Congress passed and the President signed the act establishing the National Science Foundation.

The act of 1950 was considerably watered down from the S. 1850 of 1946. Like S. 1850, it made the director of the Foundation a presidential appointee and included a mandate for planning the overall federal research program. Unlike S. 1850, it omitted any specific scheme for geographical distribution and made no change in the wartime policy for patents arising out of federally financed research and development. While S. 1850 had authorized the Foundation to conduct militarily related basic research, the 1950 act left the control of all such research to the armed services. As created, the Foundation was to pursue primarily pure research and training. At last, Lee DuBridge, now the president of the California Institute of Technology, optimistically cheered, there was an agency to "free basic science from the danger of becoming a stepchild of military technology."[5] But if the Foundation had been originally conceived as the government's chief sponsor of fundamental research, in 1950 it was only a puny partner in an institutionally pluralist federal research establishment.

In 1949 the Public Health Service, the Defense Department, and the Atomic Energy Commission together spent more than $63,000,000 on the campuses for applied and fundamental research.[6] The National Science

---

[3] Paul Scherer to Bush, attached to John Teeter to Bush, Aug. 29, 1947, VB, Box 110, Teeter file; U.S. Congress, Senate, *Congressional Record*, 80th Cong., 2d Sess., May 3, 1948, pp. 5178, 5182–83.

[4] Robert P. McCune, "Origins and Development of the National Science Foundation and Its Division of Social Sciences, 1945–1961" (unpublished Ph.D. dissertation, Education, Ball State University, 1971), pp. 146–48; William D. Carey to Elmer Staats, Aug. 18, 1949, BOB, series 39.32, file E8–21/49.1.

[5] Quoted in McCune, "Origins of the National Science Foundation," p. 156.

[6] Stephen P. Strickland, *Politics, Science, and Dread Disease: A Short History of United States Medical Research Policies* (Cambridge, Mass., 1972), pp. 32–54; Richard G. Hewlett and Francis Duncan, *Atomic Shield, 1947/1952 (A History of the United States Atomic Energy Commission*, Vol. II; University Park, Pa.: Pennsylvania State University Press, 1969), pp. 80, 82–83, 108–9, 111–12, 255–59; "Federal Research and Development Activities at Colleges and Universities," July 17, 1950, BOB, series 47.8a, Research and Development, College and University file.

Foundation, limited by its organic act to a ceiling of $15,000,000 in any one year, was struggling along on a piddling appropriation of $350,000. Budget Bureau planners may have expected the Foundation to be the chief federal sponsor of basic research, but the Public Health Service successfully fought to keep its rapidly growing programs of such research in medicine. Although at least one-third of Office of Naval Research dollars were spent on pure research projects only remotely related to naval purposes, ONR had reneged on its agreement to cede any of its fundamental research, militarily relevant or even pure, to the new Foundation.[7] Unless the navy was in direct contact with the men actually conducting basic research, Secretary Forrestal argued to a congressman, the fleet and air arms could not remain abreast of current scientific discoveries. In 1949 the Defense Department together with the Atomic Energy Commission accounted for 96 percent of all federal dollars spent on the campuses for research in the physical sciences. For every two of those dollars spent by the AEC, the military spent at least three.[8]

Of course the military enjoyed considerable power because of the mounting Cold War concern with national security. Military requirements compelled even the civilian Atomic Energy Commission to stress research and development for weapons, notably the hydrogen bomb. But the AEC never permitted itself to become a mere nuclear weapons contractor for the Department of Defense. By virtue of its mandate and its civilian control, it could and did maintain an ongoing investment in the peaceful uses of atomic energy, including the development of nuclear power; late in 1951, at an AEC laboratory, a nuclear reactor first transformed atomic power into 45 kilowatts of electrical energy.[9] In the absence of a National Science Foundation between 1945 and 1950, no comparable institutional sponsor of fundamental physical research for civilian needs had existed, nor had any comparable institutional mechanism to plan, if not to enforce, a research program in the physical sciences better balanced between civilian and military purposes.

Once created, the Foundation was legally authorized to slip an evaluative check rein over agency research programs, but the defense agencies were determined to resist such judgment. Their opposition was not lost upon the first director of the Foundation, the physicist Alan Waterman, a student and protégé of Karl Compton's, a member of the Yale faculty, and a wartime veteran of the OSRD Office of Field Services. Shortly

[7] William D. Carey to Elmer Staats, May 10, 1950, July 17, 1950, BOB, series 39.33, unit 95; series 39.32, file E8–33/44.2; R. C. Atkinson, "Memorandum of Conference with Willis Shapley," July 12, 1950, BOB, series 39.32, file E8–20/48.1; Shapley to Ramsey, Aug. 8, 1950, BOB, series 39.33, unit 91.

[8] Forrestal to Charles A. Wolverton, March 7, 1947, HAS, Box 132, NSF file; "Federal Research and Development Activities at Colleges and Universities," July 1, 1950, BOB, series 47.8a, Research and Development, College and University file.

[9] Hewlett and Duncan, *Atomic Shield*, pp. 31–32, 182, 411, 498.

after the Japanese surrender, Waterman was appointed chief scientist at the Office of Naval Research. He was deliberate, firm, modest, patient, and above all prudent. He consistently opposed the transfer of most of the military's fundamental research programs to the National Science Foundation before its creation. Now, after its establishment, he proposed to ignore the Foundation's mandate for planning. Along with considering any centralized evaluation of federal research impossible and inappropriate, he was also worried about pitting his infant, penurious, and decidedly vulnerable agency against such giants as the Department of Defense.[1]

Perhaps under the circumstances Waterman was wise to avoid taking on the defense research agencies. Whatever the merits of his judgment, the task of planning and evaluation was certainly much more difficult in 1950 than it would have been in 1946, when even the armed services had been willing to concede the primary sponsorship of fundamental research to civilian hands. In retrospect, the delay in the establishment of the National Science Foundation was critically important in the evolution of postwar policy for research and development, not least because it cost the nation a program balanced between civilian and military patronage and purpose.

James B. Conant, who from the beginning worried about the increasingly military patronage of American science, later blamed the delay in the passage of the bill for a National Science Foundation upon President Truman, mainly for his 1947 veto.[2] Conant might have enlarged his reckoning of responsibility to include the congressional conservatives who blocked the passage of the bill before as well as after the veto and who in 1947, persuaded by Senator Taft's declaration that he would prefer no foundation at all to one under the President's control, sent the White House a politically elitist measure. True enough, in 1946 the

---

[1] Carey to Staats, Feb. 24, 1950; James L. Grahl to files, April 3, 1951, BOB, series 39.33, unit 94 and file 93; Milton Lomask, *A Minor Miracle: An Informal History of the National Science Foundation* (Washington, D.C., 1976), pp. 72–73, 83. After the Korean War broke out, the defense research agencies demanded still higher —some observers thought exorbitantly higher—funding. In the Bureau of the Budget William D. Carey watched the trend apprehensively. One day in the spring of 1952, Carey went over to the Foundation's headquarters on California Street in Washington—"a rambling old house with a ripe brick exterior and a kind of mouldy antiquity," he described it—to prod Waterman into making active use of his planning mandate. He found Waterman running the Foundation like "a dean of studies" in "an atmosphere of unhurried deliberateness, and good if not fashionable living." The Foundation was "settling down cozily with the Defense Department . . . in joint planning activities. . . . Waterman's idea is to operate on the basis of the scientist-to-scientist approach. . . . Out of this spiritual communion will emerge gentlemen's agreements leading to the formation of spheres of influence for research sponsorship." Carey to Staats, May 15, 1952, BOB, series 39.33, file 93.

[2] James B. Conant, "An Old Man Looks Back, Science and the Federal Government: 1945–1950," *Bulletin of the New York Academy of Medicine*, 47 (1971), 1248; *New York Times*, Jan. 9, 1971, p. 13.

President scarcely threw the full weight of his office behind the passage of S. 1850, and afterward he displayed much less concern with the substance of the program than with its control. But while in 1947 his advisers in the Budget Bureau understood that the veto might prolong indefinitely the military ascendancy over academic science, they also agreed with Don K. Price, by then at the Public Administration Clearing House in Chicago, who argued in a special memorandum that the President must veto the bill. Along with giving control over a federal granting agency to people who would be getting the grants, Price pointed out, the measure would improperly vest the responsibility for planning an overall federal science program in officials largely beyond the reach of presidential power.

Price's analysis called up much the same objections that Secretary David Houston had advanced a quarter of a century before against George Ellery Hale's proposed executive order for the National Research Council. Yet if in key respects the issues were similar, in the post-World War II context, the stakes were much higher—"the safeguarding," in Price's phrase, "of the public's interest in [science]." So long as the choice was posed between a politically elitist and a politically responsive system, the President, who based his veto message on Price's memorandum, opted for letting the military remain the principal patron of basic research in the physical sciences.[3]

No single individual did more to keep the choice framed between those two alternatives than Vannevar Bush. Few scientists were better placed to affect policy than Bush, who remained in Washington as head of the Carnegie Institution and who headed the successor to the Joint New Weapons Committee in the defense establishment. No administrator of science was said to know better than Bush how to call forth the best efforts of the nation's scientific talent or to organize scientific research. No one save Oppenheimer, who was preoccupied with atomic energy policy, commanded more respect in congressional circles and among the public at large. With all his authority and prestige Bush—in the angry view of James R. Newman—systematically "sabotaged" the President's program for science.[4]

In the fall of 1945, even after the President had endorsed the main lines of Kilgore's program and Harold Smith had spelled out the President's policy to Bush in a special letter, Bush still testified for the Magnuson bill. To help advance the bill, he even supplied Magnuson with the services of John H. Teeter, an engineer and technical aide on the

[3] "D. K. Price's Memo on the National Science Foundation," BOB, series 39.33, file 93; James E. Webb to Mr. Latta, Aug. 1, 1947, HST, bill file (S. 526); Truman, "Memorandum of Disapproval," Aug. 6, 1947, BOB, series 39.32, file E8–33/44.2.

[4] Newman to Hans Klagsbrunn, Dec. 14, 1945, OWMR, Entry 16, Legislation Scientific.

OSRD payroll. A former student and then colleague of Bush's at MIT before going off to industry, Teeter was the kind of conservative who often confused liberalism with socialism, a political novice tending to approach congressional politics with a conspiratorial air. He tirelessly advocated Bush's program and kept a watchful eye on the Kilgore forces, whose real aim, he was sure, was to subject science to basely political control. It was Teeter who induced Congressman Wilbur Mills to introduce his pro-Bush measure in the House.[5] In the spring of 1946, despite broad academic and scientific support for the compromise S. 1850, Bush urged Senate Republicans to amend the measure back to his original program and endorsed the Mills bill in public hearings. Howard A. Meyerhoff, professor of geology at Smith College, a liberal advocate of S. 1850 and the secretary of the American Association for the Advancement of Science, bluntly pronounced the death of the Senate compromise a "homicide" committed by Bush and Teeter.[6]

From the beginning of the battle through the veto, Bush behaved with an unusual degree of independence for a presidentially appointed federal official. Admonished by Don K. Price and Harold Smith to get in harness with the President, Bush queried Truman: Should I consider myself a member of the official family bound not to speak publicly against administration policy? Consider yourself a member of the official family, Truman shot back, so long as you are head of OSRD.[7]

If Bush persistently violated the President's wish, his friends claimed that he was only acting as a private citizen. The claim had its merit, since Bush remained head of OSRD only to wind up its affairs and was not drawing a government salary in his defense post. Besides, save on some atomic energy matters, the President and his advisers did not consult Bush about policy for science; under Truman, in contrast to Franklin Roosevelt, Bush was far from the center of White House power. His postwar exclusion from the orbit of the Oval Office rankled. Conant

[5] Smith to Bush, Oct. 1, 1945, Entry 1, NSF Hearings; Smith, "Daily Record," Oct. 15, 1945, HDS; Teeter to Ruebhausen et al., Oct. 8, 1945, OSR, Entry 1, NSF, Inter-Office Memoranda; Teeter to Ruebhausen, June 4, 1946; Teeter to Peter Edson, June 4, 1946, OSR, Entry 1, NSF Legislation.

[6] Bush to Conant, June 19, 1946, OSR, Entry 1, NSF Legislation; Meyerhoff, "Obituary: National Science Foundation 1946," Science, 104 (Aug. 2, 1946), 97. In June 1946, Teeter brought Father Hugh O'Donnell, the president of Notre Dame and a contributor to Science—The Endless Frontier, to the House of Representatives for lunch. "The entrance into the House dining room was like high mass," Teeter reported to Bush. "Every man we needed to see came to the table . . . and Father O'Donnell sold each one on the [Mills] bill and on a militant attitude towards communists." Teeter to Bush, June 10, 1946, OSR, Entry 1, NSF Legislation.

[7] Contrary to prevailing canons of practice, Bush had Magnuson introduce his bill without first asking the Budget Bureau whether it was in accord with the President's program. Bush to Truman, Oct. 13, 1945; Truman to Bush, Oct. 19, 1945, HST, OF53; Newman to Snyder, Dec. 4, 1945, OWMR, Entry 16, Legislation Scientific.

believed, according to Teeter's report, that the National Science Foundation was "V.B.'s baby," that Bush approached the issue as a "personal matter, with much personal feeling."[8]

Still, Bush was convinced that the National Science Foundation could—and should—be kept safely insulated from political accountability. No one need worry, he told President Truman, that the Foundation's distinguished governing board of private citizens, many of them accomplished scientists, would act improperly in awarding grant money; it would make no difference that they might come from the institutions receiving the grants.[9] With equal ingenuousness, Bush expected to prevent the military from dominating academic science. Like the Bureau of the Budget, he backed the navy's support of research in the universities only as a temporary measure. Bush acknowledged that the armed services, having entered the field of basic research, seemed to want to stay in it. But he was "not at all fearful on this matter." Bush assumed that Secretaries Patterson and Forrestal undoubtedly saw the inappropriateness of permanent military patronage in the same disinterested way as himself. And when the time came, Congress would surely recognize how "quite unreasonable" it would be to continue funneling federal money into academic science primarily through the armed services.[1]

Bush failed to win the battle over atomic energy legislation because most atomic scientists contested him with single-minded vigor. They engaged in no comparable campaign for Kilgore's program, or even, the more time passed, for the speedy creation of any kind of National Science Foundation. Physical scientists who had worried about military domination in the atomic energy program in 1945 worried about it much less in physics generally after the armed services, especially ONR, proved that they would fund science tolerantly and richly. In 1948 nearly 80 percent of the papers presented at the American Physical Society meetings were said to have been supported by Office of Naval Research

[8] Teeter to Ruebhausen et al., Oct. 8, 1945, OSR, Entry 1, NSF Inter-Office Memoranda. In 1947 Bush complained to the President that when Franklin Roosevelt had wanted a report on science policy, he had called in Vannevar Bush; when Harry Truman wanted one, he gave the responsibility to Presidential Assistant John R. Steelman. Remaining outside of presidential favor, Bush also lost the support of Secretary of Defense Forrestal while chairman of the Research and Development Board in the Defense Department. Evidently frustrated at his powerlessness, Bush developed severe psychosomatic headaches, which were first diagnosed as a brain tumor. His physician told him to get out of the government, which Bush did, never to return, and the headaches went away. Bush, *Pieces of the Action* (New York, 1970), p. 303; Bush, "Memorandum of Conference with the President on Sept. 24 [1947]," VB, Box 112, file 2675; transcript of Eric Hodgins interview with Bush, 1966, pp. 179A–183, VBM.

[9] Bush to Conant, Oct. 20, 1945, VB, Box 27, file 614; "Memorandum of Conference with the President on Sept. 24 [1947]," VB, Box 112, file 2675.

[1] Bush to Ruebhausen, Sept. 27, 1946, VB, Box 100, file 2309; Bush to Forrestal, Dec. 11, 1946, VB, Box 85, NSF file.

money. In 1949 ONR paid for more than eleven hundred projects at more than two hundred institutions, at a cost of $29,000,000. About three out of every four of these navy research dollars went to the physical sciences—and to the scientists who in the wake of Hiroshima enjoyed the most public standing and political leverage. Even Bush noted that American scientists were decidedly happy with ONR, happy enough, possibly, not to mind the absence of a National Science Foundation.[2]

In the contest over atomic energy legislation the scientific community was united in its opposition to Bush, and its members displayed relatively little interest in patents, social impact, or the politically elitist nature of the May-Johnson proposal. In the dispute over the National Science Foundation program, the nation's atomic scientists were either neutral or divided with respect to such issues as patents or the social sciences. More important, whatever views physical scientists held on those matters, like Harlow Shapley most tended to share in greater or lesser degree Bush's predilection for a National Science Foundation insulated from political control. And in the postwar years their traditional concern for freedom in science was strengthened by the memory of Hitler's oppression, not to mention the current specter of Stalinism.[3]

In the final legislative compromise, the nation's scientists won a victory for political elitism. As they conceded it had to after 1947, the law empowered the President to appoint the director of the National Science Foundation, but it also provided for the director to share control of policy with Bush's part-time private board. For the most part, the members of the first board were spokesmen for the nation's leading institutions of academic and industrial science. When the board first met early in 1951, it declared that the principal function of the Foundation was to advance basic scientific research and training, and that alone. It soon also endorsed Waterman's determination to avoid the tangled

[2] John E. Pfeiffer, "The Office of Naval Research," *Scientific American*, 180 (Feb. 1949), 11, 14; Bush to Ruebhausen, Sept. 27, 1946, VB, Box 100, file 2309. *Business Week* noted in 1946 apropos the effect on the prospects of a National Science Foundation of massive military support of basic science: "Even though many scientists deplore the warping of the direction of research inevitable in a military program, some of the university people—with their fields well established, their file of telephone numbers organized, and the money flowing freely—may hesitate to upset a going operation." *Business Week*, Sept. 14, 1946, p. 24.

[3] Shapley's Committee for a National Science Foundation went no further on the control issue than to call for a compromise on the question of a board vs. a single administrator. Committee press release, Dec. 28, 1945, OWMR, Box 275, Legislation . . . Scientific. Shapley himself was willing to go along with a division of responsibility between a part-time board and a presidentially appointed administrator. While an Inter-Society Committee for a Foundation opted for a presidentially controlled agency, the vast majority of its members were drawn from the affiliated scientific societies of the American Association for the Advancement of Science, most of which were in the life and social sciences. Fred D. Schuldt to Carey, Dec. 5, 1947, BOB, series 39.33, file 93; Lomask, *A Minor Miracle*, pp. 52–53.

thickets of evaluation and planning in federal research. On his part, Waterman, who had advocated Bush's politically elitist approach to the control of the Foundation, deferred to the board on all major policy issues.[4]

Waterman and his board agreed in particular on the key policy point of how to distribute the Foundation's funds. While the final law called upon the Foundation to help advance science in the scientifically under-developed regions of the country, the governors of the Foundation generally ignored the injunction, in part because their funds were limited, but, more important, because, like Bush, they were strongly disposed to a best-science approach.[5] So were most other managers of federal programs for research and development that relied on academic and industrial laboratories. Though paying some attention to spreading their largess around the country, the Office of Naval Research and the Atomic Energy Commission tended to award their contracts to the people most qualified to pursue the required research, which meant in a distributional pattern concentrated among the leading institutions of scientific research.[6]

In 1945 liberals like Kilgore may have wanted a federal program balanced between civilian and military needs, between big business and small business, between the leading and the less developed universities, between the welfare of science and the welfare of the nation. By the early 1950s the outcome was quite different: a program for the physical sciences dominated outside of atomic energy by the military, its dispensa-tions concentrated geographically in the major universities, its primary energies devoted to the chief challenges of national defense and funda-mental physics. While the Cold War had made such an outcome possible, the result was produced not by design but by a combination of demands and defaults, including the demands and defaults of a new political power group, the nation's physical scientists. Whatever the cause of the outcome, it constituted a revolution in the relationship of American physicists to

[4] The vice-chairman was the biochemist Detlev W. Bronk, the new president of the Johns Hopkins University, Jewett's successor as head of the National Academy of Sciences, and a forceful advocate of Bush's views. The chairman was Conant, who had defended Bush's program and attacked Kilgore's: "There is only one proven method of assisting the advancement of pure science,—that of picking men of genius, backing them heavily and leaving them *to direct themselves*." Conant to the Editor, *New York Times*, Aug. 3, 1945 [published Aug. 10], OSR, Entry 2, President's Letter, Nov. 17, 1944; Lomask, *A Minor Miracle*, pp. 72–73; Schuldt for files, Dec. 18, 1950; National Science Board, press release, Feb. 15, 1951, BOB, series 39.33, unit 94.

[5] The clause was the work of William D. Carey in the Budget Bureau. Eager to temper Bush's best-science propensity for "putting the money where it will produce the most efficient research," Carey had argued that the Foundation should not only be instructed to avoid the geographical concentration of its largess; it should also be told actively to broaden "the research base of the Nation instead of feeding the 'fat cats' who are most likely to come up with the best answers." Carey to Staats, Jan. 26, 1949, BOB, series 39.32, file E8–21/48.1.

[6] Don K. Price, *Government and Science* (New York, 1954), p. 89.

their society and government. Now, through the Office of Naval Research, the Atomic Energy Commission, and the National Science Foundation, they were supplied with what they had been seeking for the better part of a century—a system of federal support for basic research and training insulated from political control and focused on the advancement of the best possible physics.

XXIII

The Physicists Established

In 1950 the new journal *Physics Today* was warmed to note: "The springtime of Big Physics has arrived."[1]

The Los Alamos generation had come out of World War II disposed to think big—and expensively. Now federal money was flowing almost as freely as during the war, and the nation's leading scientists enjoyed considerable say about how the research dollars were to be spent. On the advice of physicists, the Atomic Energy Commission transformed some of its wartime research enterprises into permanent national laboratories, including Argonne, south of Chicago; Oak Ridge; and Berkeley. Argonne and Oak Ridge were to be centers of reactor research, and chain reacting piles were prolific sources of what physicists were calling "low energy" neutrons. The Berkeley laboratory, which Ernest Lawrence wanted to make a "paradise of physics," set its sights on experiments with high energy particles.[2]

During the war physicists had mastered a remarkable variety of sophisticated electronic techniques, including circuits fast enough to monitor nuclear phenomena occurring in millionths of a second. Some of the Berkeley staff were adapting outmoded radar equipment to nuclear experiments, even to the acceleration of particles to energies high enough for nuclear bombardment. But what most excited Lawrence was the

---

[1] *Physics Today*, III (July 1950), 35.
[2] Quoted in Herbert Childs, *An American Genius: The Life of Ernest Orlando Lawrence* (New York, 1968), p. 371.

prospect of completing the 100,000,000 electron volt accelerator, the construction of which had been launched with the Rockefeller grant of 1940.

Hans Bethe may have warned that a beam of particles accelerated to such relativistic energies would eventually smash into the walls of the machine. No matter: While at Los Alamos during the war, the young physicist Edwin M. McMillan, Lawrence's brother-in-law and a staff member at the Berkeley laboratory, had conceived a new type of cyclotron, a synchrotron, which eliminated the relativistic difficulty. In 1945 McMillan began building his synchrotron with $170,000 from the Manhattan District, a parting gift to Lawrence from General Groves. Using the 184-inch magnet bought with the Rockefeller money—it had been diverted during the war to the task of separating uranium isotopes—the huge machine went into actual operation shortly after midnight on November 1, 1946. Immediately it reached a beam energy of 200,000,000 electron volts, twice Lawrence's ambitious prewar goal.[3]

On the east coast, I. I. Rabi and Norman Ramsey, a former student, were growing restless. Columbia had contributed to the war drive yet, unlike Berkeley or the universities associated with Argonne and Oak Ridge, it had little to show for its efforts. Keen to remain competitive, Rabi and Ramsey successfully created in the northeastern region still another nuclear research laboratory. Located at Brookhaven, Long Island, on the six-thousand-acre site of a former army camp, the complex was financed by a contract from the Atomic Energy Commission to a consortium of nine universities.[4] The Brookhaven National Laboratory, which opened in 1947, enjoyed a universitylike atmosphere and included departments of physics, biology, medicine, and engineering. Its staff was expected to total some one thousand people, including two hundred visitors from surrounding academic and industrial laboratories. The equipment was to include at least one nuclear chain reactor and a 500,000,000 electron volt synchrotron.

At Berkeley and Brookhaven, the scientific rationale behind the quest for such powerful accelerators was the ongoing drive to explore nuclear forces. The chief explanation of these forces remained Hideki Yukawa's theory that the constituents of the nucleus were bound together by the exchange of a fundamental particle, a meson. By now the evidence was considerable that the meson detected during the 1930s in cosmic rays at sea level was not Yukawa's. But in 1947 a team in Britain found, in cosmic rays high in the atmosphere, a second meson, one that seemed to meet Yukawa's requirements handsomely. Then, early in 1948, came a report

[3] Ibid., pp. 370, 386–87; Richard G. Hewlett and Francis Duncan, *Atomic Shield, 1947/1952 (A History of the United States Atomic Energy Commission,* Vol. II; University Park, Pa.: Pennsylvania State University Press, 1969), p. 232.
[4] Norman F. Ramsey, *Early History of Associated Universities and Brookhaven National Laboratory* (Brookhaven Lecture Series, BNL 992 [T–421]; Brookhaven, N.Y., March 30, 1966).

from Lawrence's Radiation Laboratory: The Berkeley synchrotron, by accelerating alpha particles to 380,000,000 electron volts, had produced both types of mesons in prodigious quantities.

The achievement not only ratified the expectation that mesons could be produced in accelerators; it meant that the study of mesons no longer needed to depend on the low intensity source of cosmic rays, and that mesons aplenty would be available for experiments in the laboratory. "Just as we were apparently closing one door [in research] . . .," the young theorist Richard P. Feynman exulted, "another was being opened wide by the experimenters, that of high energy physics."[5] Eager to press on to higher energies, to the unknown particles and phenomena that higher energies might yield, influential physicists advised the Atomic Energy Commission to fund the construction of still more powerful accelerators. In March 1948 the AEC authorized the construction at Brookhaven of a 2- to 3-billion electron volt accelerator; at Berkeley, of another machine designed to operate between 6 and 7 billion electron volts.

In the immediate postwar years the leaders of the Los Alamos generation could all glow with Rabi: "Isn't physics wonderful?"[6]

In most respects, physics continued to be wonderful through the postwar decade. With the outbreak of the Korean War, overall federal expenditures for research and development passed the $1-billion mark, kept edging upward after the armistice, and in 1956 reached over $3 billion. In the physical sciences, the bulk of the funds was supplied almost entirely by the Department of Defense or the Atomic Energy Commission and went mainly to industrial firms or the national laboratories. But in 1952 the physical sciences at nonprofit institutions received almost $250,000,000, including more than $17,000,000 for unclassified basic research in physics.[7]

Samuel K. Allison, the respected University of Chicago physicist, a liberal isolationist in the 1930s who had wound up tolling the countdown at the nuclear test in the desert at Alamogordo, found the reason for the bonanza—"that physics and physicists are important for waging war"—dismaying. We should, Allison urged with a certain anguish, "justify physics as a worthy intellectual activity of a world at peace."[8] Edward U. Condon, the director of the National Bureau of Standards, added the trenchant warning: "If in the years to come science and the scientists are closely identified in the public mind as the wizardry and the wizards who have made all the fantastic new weapons of mass destruction . . ., [a] horror and revulsion of war may, in that illogical and irrational way that

[5] Richard P. Fenyman, "Pocono Conference," *Physics Today*, I (June 1948), 10.
[6] Rabi to Ernest Lawrence, March 1, 1948, EOL, Carton XII.
[7] *Physics Today*, VI (Jan. 1953), 22; J. Howard McMillen, "Government Support of Basic Research in University Physics Departments, 1952–53," *ibid.*, VII (May 1954), 7–9.
[8] Samuel K. Allison, "The State of Physics; Or the Perils of Being Important," *Bulletin of the Atomic Scientists*, VI (Jan. 1950), 2, 27.

so many things go in politics, be extended to science and the scientists."[9] But in the postwar decade such admonitions and warnings provoked little manifest concern among most physicists and university officials. There was no gainsaying the fact of the Cold War. Besides, for the university officials, federal research contracts included overhead payments, which helped defray the costs of less favored departments. And the outpouring of funds created what Gaylord P. Harnwell, the chairman of physics at the University of Pennsylvania, unabashedly called a "bull market" for the members of his discipline.[1]

The industrial demand for physicists rose sharply during the Korean War, then, after the armistice, climbed higher still. Government laboratories, from the established Bureau of Standards to the new Oak Ridge, Argonne, and Los Alamos, could not get enough physicists. The greater the nonacademic demand, the greater the demand for professors to teach the discipline. Physics job registers typically listed two to four times more openings than the number of physicists looking for jobs. Selective Service, which generally exempted young scientists from the draft, sometimes singled out physicists for special dispensation. Recruiters combed the campuses for physics majors and Ph.D.'s. In 1956, when the American Physical Society held its meetings in New York City, the recruiters mobbed the fifth and sixth floors of the hotel, enticing and pirating candidates for industrial, governmental, and academic positions.[2]

By the late 1940s the number of students majoring in physics had more than doubled over prewar levels. Numerous fellowships and assistantships made graduate work more accessible than ever; they also made it possible, Merle Tuve noted, for "graduate students . . . to live, and to have wives and children like normal people." In 1953, typically, sixty students were pursuing graduate degrees in physics courtesy of the National Science Foundation, another six hundred through contracts from the Atomic Energy Commission and the Office of Naval Research.[3] In 1949 American universities awarded 275 doctorates in physics; in 1955, more than 500. During the postwar decade American graduate schools produced on the average more than three times as many Ph.D.'s each year as in the prewar period.

Socially, the profession continued to move along. The war had brought married, including middle-class, women into the work force to stay. But while educated women ranked female scientists high in esteem, higher even than women writers and artists, they generally preferred to

[9] Edward U. Condon, "Some Thoughts on Science in the Federal Government," *Physics Today*, V (April 1952), 9.

[1] Gaylord P. Harnwell, "The Quality of Education in Physics," *Physics Today*, IV (May 1951), 4.

[2] Paul D. Foote, "Physics Then and Now," *ibid.*, IX (Sept. 1956), 21.

[3] Merle Tuve, "Technology and National Research Policy," *ibid.*, VII (Jan. 1954), 7; Thomas H. Johnson, "The AEC's Physics Research Program," *ibid.*, VI (Aug. 1953), 8–12.

find their own primary fulfillment as mothers of accomplished children and wives of prominent husbands. On the whole, women of the postwar era went to work to help raise the family standard of living; they had jobs, not careers. In any case, professionally oriented women still aspired to the more "womanly" professions. Classes in high-school chemistry, which could open the door to careers in such fields as home economics, nutrition, or nursing, enrolled almost as many girls as boys; in physics courses, boys outnumbered girls three to one. While more women, almost ninety, took doctorates in physics over the postwar decade, the fraction of female Ph.D.'s in the discipline declined to one in forty, the fraction of women in the profession generally to one in twenty-five. Proportionately fewer female than male physicists found employment in industry.[4] In the academic world, where some graduate departments still refused to admit female applicants, women were still mainly consigned either to the women's colleges or, at other institutions, to second-class posts on the research, as opposed to the professorial, staffs.

Catholic commentators increasingly asked why there were so few Catholics in the sciences, including physics. Catholic colleges still devoted only a minuscule portion of their budgets to the advancement of science and science remained suspect in the eyes of many Catholic clerics. While a steadily rising percentage of Catholic students attended secular institutions, the Catholic community continued to suffer, Father John Tracy Ellis observed, from its "frequently self-imposed ghetto mentality." On the whole, professionally ambitious Catholics still aspired to careers in law, business, engineering, or medicine rather than pure science. In the mid-1950s fewer than 5 percent of National Science Foundation fellowships in physics went to Catholics.[5] But manifestly no longer immigrants, Catholics were attending college in growing numbers. Catholic colleges were also enlarging their investments in research, and a few institutions, notably Catholic University and Notre Dame, were earning respected reputations in certain fields of science. By the end of the postwar decade, though the percentage of Catholics in the physics profession remained far

[4] Alice S. Rossi, "Barriers to the Career Choice of Engineering, Medicine, or Science Among American Women," in Jacquelyn A. Mattfeld and Carol G. Van Aken, eds., *Women and the Scientific Professions: The MIT Symposium on American Women in Science and Engineering* (Cambridge, Mass., 1965), pp. 126, 66; William H. Chafe, *The American Woman: Her Changing Social, Economic, and Political Roles, 1920–1970* (New York, 1972), pp. 178, 182–83, 192, 206; W. C. Kelly, "Physics in the Public High Schools," *Physics Today*, VIII (March 1955), 12–14; Bernard C. Murdoch and Marsh W. White, "Young Professional Physicists," *ibid.*, III (Sept. 1950), 20–22; Lindsey R. Harmon and Herbert Soldz, comps., *Doctorate Production in United States Universities, 1920–1962* (Pub. No. 1142; Washington, D.C.: National Academy of Sciences–National Research Council, 1963), p. 52; "Conference on the Role of Women's Colleges in the Physical Sciences [Bryn Mawr College, 1954]," pp. 6, 7, 9, 10–11, copy in Bryn Mawr College library.

[5] John Tracy Ellis, "American Catholics and the Intellectual Life," *Thought: Fordham University Quarterly*, 30 (Fall 1955), 386, 380, n. 49.

below the Catholic percentage of the general population, it was on the upswing.

Jews entered physics in greater numbers than before World War II. *Fortune* ventured to account for the "disproportionately high percentage of outstanding scientists with Jewish backgrounds" by the "scholarly tradition" characteristic of the group.[6] For contemporary young Jews, there was also the alluring prominence of such names as Oppenheimer, Rabi, Teller, Franck, and Szilard, not to mention Einstein. Then, too, the revelation of Hitler's barbarities had joined with the ongoing social revolution in the United States to knock down anti-Semitic barriers in universities and professional schools. More than ever before in America, physics offered Jews a clear welcome, one of the few high-paying, high-status careers in which they could get ahead rapidly, and the prospects of national prestige, possibly even of power in Washington.

To a lesser degree, the barriers were falling for other minority groups, too. Before World War II, about a dozen blacks had taken doctorates in physics. The historically obvious social and economic disadvantages of black Americans accounted for the situation, Herman R. Branson, the Howard University physicist, observed. So did the culture of the community, he added, which put no special value on scholarly or scientific careers. Still, Branson declared, there were more than 75,000 students in black colleges, and one might "confidently expect a fraction, at least 5 or maybe 10 percent, would be attracted into science." The trouble was, Branson could only conclude, that the colleges "attended by Negroes have been singularly ineffective as agencies for stimulating interest in science as a career." In the postwar decade, the number of black physics Ph.D.'s rose to perhaps twenty-five. This number was higher for the black portion of the population in comparison to the parallel figures for women, but not to those for Oriental-Americans. Although three out of four of these new Ph.D.'s were foreign-born, the proportional contribution to the profession of the native group alone represented more than twice their weight in the general population. And although natives and immigrants together accounted for only about 2.5 percent of the doctorates, Oriental-Americans were rapidly assuming the significant role in American physics symbolized by the 1957 Nobel Prize awarded Chen Nin Yang and Tsung-Dao Lee, who had both left China for doctoral training and careers in the United States.[7]

[6] Francis Bello, "The Young Scientists," *Fortune*, 49 (June 1954), 143.

[7] Herman R. Branson, "The Negro Scientist," in Julius H. Taylor, ed., *The Negro in Science* (Baltimore, Md.: Morgan State College Press, 1955), pp. 3, 8; James M. Jay, *Negroes in Science: Natural Science Doctorates, 1876–1969* (Detroit: Balamp Publishing, 1971), p. 50; Commission on Human Resources of the National Research Council, *Minority Groups Among United States Doctorate-Level Scientists, Engineers, and Scholars, 1973* (Washington, D.C.: National Academy of Sciences, 1974), pp. 19, 21–22. The figures given in this last report are for all fields of science; I have assumed that the percentages are the same for physics in particular.

Whatever their origins, many more physicists streamed into industry in the postwar years. By the mid-1950s, for the first time, industry employed as large a fraction of American physicists, two out of five, as did the academic world, though the universities still hired the large majority of Ph.D.'s. In the late 1940s it was a standard faculty club joke that there were two academic salary levels, one for physicists, another for everyone else. Taken together in 1951, academic and industrial physicists, including the holders of mere bachelor's degrees, earned at the median $7,500 a year, more than half again as much as the national median income. One in four physicists earned $9,000, one in six more than $10,000 a year. While the majority of physicists were occupied with research and development, one out of eight was now in management, compared to one out of twelve in 1951. Even the academic physicist, Gaylord Harnwell remarked, was "caught up in the big business of research and finds himself the administrator of a training program under all the pressures of a production line. He is more concerned with the negotiation of contracts than with the solution of experimental difficulties; the telephone is more often in his hand than the voltmeter."[8]

But the voltmeter was in his hand often enough. In 1949 *The Physical Review* bulged with 70 percent more pages than in 1939, the prewar record year; in 1956, it was 370 percent larger still, and was published biweekly. Before the war, Samuel Allison recalled, American Physical Society meetings were held in a small lecture room at the National Bureau of Standards, with no more participants than could fit into a single group photograph. "Nowadays a shuttle taxi service has to be provided to haul the panting members from auditorium to auditorium, and important papers have matinee and evening performances." By the mid-1950s there were about 21,000 practicing physicists in the United States. At a meeting of the Physical Society, Harrison M. Randall, the former chairman of the Michigan department who had taken his Ph.D. in 1901, looked out over the sea of faces. Isn't it terrible, he lamented to a friend, how we don't know any of these new men? (Yes, the friend replied. But what's worse, they don't know us.)[9]

Many senior members of the profession could not recognize the new recruits in other ways, too. The younger physicists seemed interested too little in physics and accomplishment, too much in pay scales and security. They also expected a remarkable degree of luxury in the conditions of research. At Brookhaven, everyone seemed to want a new $3,500 oscilloscope. "Someone walked into my office the other day," reported Samuel Goudsmit, who headed the laboratory physics department, "and com-

[8] Marsh W. White, "American Physicists in the Current Quarter Century," *Physics Today*, IX (Jan. 1956), 32–36; Harnwell, "The Quality of Education in Physics," *ibid.*, IV (May 1951), 4.

[9] Allison, "The State of Physics," p. 2; interview with Samuel Goudsmit, Randall's former colleague.

plained that he had to share the one we'd got for him with another researcher." The new recruits also seemed disturbingly content to submerge themselves in team research, publish papers jointly, often with as many as twenty other authors—to become, in short, organization men. To Merle Tuve, regardless of age, much of the profession seemed to be "losing sight of the higher goals and values of pure science." Too many of its members were finding their satisfactions "in the size of their operations, with large funds and large groups of technicians."[1] But whatever the nostalgia for the past or discontent with the present, team research, costly equipment, and large expensive operations were simply indispensable in the glamour fields of the discipline—low and, especially, high energy physics.

These were the fields that dominated the journals and drew a disproportionately large portion of the new recruits into the profession. While in 1953 two out of five American physicists were under thirty-five, the figure was three out of five in quantum theory and nuclear studies. In low and high energy physics, the money was located. In 1953 the Office of Naval Research and the Atomic Energy Commission together supplied $8,000,000 for meson physics and another $7,000,000 for general nuclear studies. That year projects supported by the two agencies involved eleven nuclear reactors and thirty-five accelerators, fifteen of them operating at high energies.[2] By 1954 the 3- and 6-billion electron volt machines at Brookhaven and Berkeley were in operation, and in 1955 the Atomic Energy Commission authorized the construction of a 25-billion electron volt accelerator at Brookhaven, estimated to require at least five years, not to mention $20,000,000, to build.

High energy physics was also a field of remarkable intellectual excitement. In the early 1950s, wholly unexpectedly, entire families of new particles—"strange particles," they were named—were detected in cosmic rays. By the mid-1950s physics seemed headed toward Greek alphabetical, if not phenomenological, bedlam. Thirty elementary particles were known, not only the familiar protons, electrons, and neutrons but the strange lambdas, sigmas, K's, and xis. In the United States in 1953, twenty-three-year-old Murray Gell-Mann, who had a puckish sense of language, imposed some semblance of understanding on the interactions of these particles with his theory of "strangeness." The powerful accelerators built to probe nuclear forces by producing mesons were appropriated to the production of elementary particles, and gradually, the study of elementary particles emerged out of nuclear physics as a field unto itself.

[1] Quoted in Daniel Lang, *From Hiroshima to the Moon* (New York, 1959), pp. 217–18; speech by Tuve, 1947, copy in OV, Box 14.

[2] Theresa R. Shapiro and Helen Wood, "The New Physicists," *Physics Today*, VI (Feb. 1953), 14–16; Thomas H. Johnson, "The AEC's Physics Research Program," 8–12.

Low or high energy physics, the vast size of the American physics community—its large number of practitioners, sizable complements of theorists, numerous accelerators, and ample funds—inevitably made for high productivity in research. It also made for high quality, not least because in the postwar era physics doubtless tended to draw not only more but also more talented people into the profession. Yet responsible too for the high quality was the spectacular fulfillment of Henry Rowland's best-science program.

Though ninety institutions now offered the Ph.D. in physics, the bulk of federal funds was funneled into a relatively small number of academic enclaves. In 1952/53, typically, 72 percent of all federal dollars for unclassified basic research in the discipline went to only seventeen institutions, which enrolled 65 percent of all the physics graduate students. Similarly, some 62 percent of the National Science Foundation doctoral and postdoctoral fellows for 1954/55 attended just eleven institutions.[3] Such elite institutions tended to contain the best physicists, many of them heading large, organized research groups. Locally they could set standards of significance for students and staff members alike. Nationally, the better physicists played their traditional key role in the activities of the American Physical Society and their professional journals. But now, as referees, consultants, and panelists, they also advised federal agencies upon which graduate students, postdoctoral fellows, physicists, and projects to support.

It was this system, commonly known as "peer review," which did the most to permit the best-science elite to set the course and standards of their profession. Much to the advantage of the system, the best-science elite was divided along lines of specialty, geography, and training background, and the different groups had access not to just one but to many journals, fellowship boards, and, especially, granting agencies. Applications for support might be disapproved by one agency; they could always be submitted to another. What one peer review group might find mediocre, another might judge highly meritorious.

In postwar America, physics drew added strength from the fulfillment of Rowland's program by a system that was institutionally plural from bottom to top and commanded by a best-science elite that was plural, too. Of the ten Nobel Prizes awarded in the discipline between 1946 and 1955, Americans won or shared in three. Nobel Prizes or not, American physics increasingly led the world.

"Physical scientists are the vogue these days," a *Harper's* contributor commented just after the war. "No dinner party is a success without at least one physicist to explain . . . the nature of the new age in which we

[3] J. Howard McMillen, "Government Support of Basic Research in University Physics Departments, 1952–53"; National Science Foundation, *Fourth Annual Report . . . June 30, 1954* (Washington, D.C., 1954), pp. 97, 112–17.

live." In the postwar decades, neither was any public forum on the issues of the nuclear age complete without a physicist. Physicists were asked to address women's clubs, lionized at Washington parties, and paid respectful attention by conventions of theologians or social scientists. The vogue provoked its warnings, including those from scientists who knew that success in one field scarcely guaranteed wisdom in another. In Cambridge, from the podium of Charles William Eliot himself, Harvard President James B. Conant disputed the claim that the study of science necessarily produced "impartial analysts of human affairs." Scientists, Conant cautioned, were "statistically distributed over the whole spectrum of human folly and wisdom." The more the opinions of physicists seemed to count, the more some Americans, especially those of a humanistic cast of mind, appreciated the admonition of Judge Jerome Frank of the U.S. Circuit Court of Appeals: "Let us beware of the new technocracy. We want no dictatorship of physicists."[4] But in the Cold War most Americans, including humanists, considered physicists essential to the shaping of public policy, especially in defense.

In the government, physicists served on ad hoc panels, summertime study projects—"Summer Studies and some are not," the MIT physicist Jerrold R. Zacharias quipped—and advisory boards. They took on consultantships to defense agencies and participated in the proliferating activities of the National Academy of Sciences, now rich with federal contracts for its expert advice. Plane tickets, not to mention the office safe with classified documents, became the staples of the physicist's life. "We brief the President . . . on the nation's nuclear stockpile," Samuel Goudsmit marveled. "We're at Eniwetok or Las Vegas, or we're talking with troop commanders in Europe or Japan. . . . Air Force generals used to be just newsreel figures to us, but now they're fellows [with whom] we have to talk over atomic-driven planes and plan offensive and defensive tactics." Goudsmit shook his head, too, at the way young physicists would head for atomic bomb tests in the Pacific apparently as "jaunty" as though off to a "holiday," then return "full of jolly little reminiscences," seemingly undisturbed by the implications of the test explosions. Rabi or Oppenheimer, Goudsmit added pointedly—"a detonation leaves them awed and anxious."[5]

Not atypically among the leadership of the Los Alamos generation, Rabi served on key scientific advisory committees of the Defense Depart-

[4] The *Harper's* contributor and Judge Frank are quoted in Kenneth M. Jones, "Science, Scientists, and Americans: Images of Science and the Formation of Federal Science Policy, 1945–1950" (unpublished Ph.D. dissertation, History, Cornell University, 1975), pp. 93, 168; James B. Conant, "The Scientific Education of Laymen," *Yale Review*, 36 (Sept. 1946), 2, 19.

[5] Zacharias is quoted in Daniel S. Greenberg, *The Politics of Pure Science* (New York, 1967), p. 128; Goudsmit, in Lang, *From Hiroshima to the Moon*, pp. 217, 219.

ment and the Atomic Energy Commission. A consultant at Brookhaven, in the early 1950s he also managed to father, in part as a delegate to UNESCO, the giant European Center for Nuclear Research—CERN, in the French acronym—the joint venture in high energy physics of eleven European nations. In 1953 Rabi spent an estimated 120 days advising the government in one capacity or another. An enthusiast of Dwight Eisenhower as President—they came to know each other when Eisenhower was president of Columbia—Rabi exercised considerable power as a quiet insider. So did Lee DuBridge, another Republican, who for a few years chaired the most significant advisory group in the government outside of atomic energy. A part of the Executive Office of the President, this Scientific Advisory Commitee met with Eisenhower about twice a year, and conducted analyses of the nation's defense posture on a continual basis. An early member was J. Robert Oppenheimer. In the postwar years, the sociologist Philip Rieff wrote, "Oppenheimer became a symbol of the new status of science in American society. His thin handsome face and figure replaced Einstein's as the public image of genius."[6]

For Rabi, Oppenheimer, and their colleagues, the dominant policy issue for advice was nuclear weapons. The official forum for consideration of the subject was the General Advisory Committee of the Atomic Energy Commission, a body established by the law of 1946, whose membership, reading like a Who's Who of wartime physics, included Rabi, DuBridge, Fermi, and as chairman through the early 1950s, Oppenheimer. In the immediate postwar years, the Committee devoted its energies to methods of improving the nation's atomic arsenal through the development of more efficient nuclear weapons. Among the innovations it endorsed and pushed was the development of a primarily fission bomb boosted in its explosive power by the fusion of small additional quantities of thermonuclear material.[7] But after the Soviet Union exploded its own atom bomb in 1949, the Committee found itself saddled with a dire question: whether to proceed with a crash program for a completely thermonuclear weapon, not primarily a fission but a fusion hydrogen device, a superbomb with a likely power at least a thousand times as great as the bombs of Hiroshima and Nagasaki.

The Committee soon delivered its advisory answer: There should be continued research in thermonuclear weapons, but the United States should

---

[6] Greenberg, *The Politics of Pure Science*, p. 143; Robert Jungk, *The Big Machine* (New York, 1968), pp. 38–40, 51–52; interviews with I. I. Rabi and Robert F. Bacher; John Walsh, "The Eisenhower Era: Transition Years for Science," *Science*, 164 (April 4, 1969), 50–51; Detlev W. Bronk, "Science Advisor in the White House," *Science*, 186 (Oct. 11, 1974), 119–20; Rieff, "The Case of Dr. Oppenheimer," in Philip Rieff, ed., *On Intellectuals* (Garden City, N.Y., 1969), p. 319.

[7] Herbert York, *The Advisors: Oppenheimer, Teller, and the Superbomb* (San Francisco: W. H. Freeman, 1976), pp. 9, 22–23, 26.

not inaugurate a crash program akin to the Manhattan Project for the superbomb. Unlike the situation with fission weapons in 1942, no one yet saw the way to build a militarily practical thermonuclear weapon. More important, by avoiding the development of such a bomb, the majority of the Committee saw "a unique opportunity of providing by example some limitations on the totality of war and thus of eliminating the fear and arousing the hope of mankind." Fermi and Rabi added a minority annex to that opinion: "The fact that no limits exist to the destructiveness of this weapon makes it[s] very existence and the knowledge of its construction a danger to humanity as a whole. It is necessarily an evil thing considered in any light. For these reasons . . . we think [it] wrong on fundamental ethical principles to initiate the development of such a weapon."[8]

But in and out of the government, powerful groups were demanding a decisive response to the Soviet atomic advance. The demand was made all the more clamorous by the revelation that Alan Nunn May and Klaus Fuchs, key physicists in the British nuclear weapons program, had committed espionage for the Soviet Union. And inside the American nuclear establishment, a small group of scientists led by Edward Teller, not only a brilliant physicist but a militant anti-Communist, were arguing passionately for a thermonuclear crash program. Early in 1950, overruling both the Advisory Committee and the Atomic Energy Commission itself, President Truman ordered the superbomb developed with all possible speed. By early 1951 Teller and the mathematician Stanislaw Ulam had conceived an entirely new, ingenious method for a militarily practical hydrogen bomb. The idea was "technically so sweet," Oppenheimer himself later said, that you could no longer argue about building it.[9] When the device was first tested in 1954, it exploded with a force not of thousands but of millions—15,000,000—tons of TNT, enough to obliterate New York City and then some.

To the chagrin of most physicists, and the apprehension of some, the Cold War not only produced an escalation of the arms race; it also put barbed wire and guarded gates around the Radiation Laboratory at Berkeley. It kept under security wraps research on fissionable elements, including thorium, uranium, and plutonium. In the year following November 1947, three out of every four research reports produced in Atomic Energy Commission laboratories were deemed to require security restrictions. Samuel K. Allison complained that such secrecy, by preventing open discussion of new types of nuclear reactors, was retarding the development of atomic power. And who could say, Allison added, whether "these [nuclear] facts, if known to another Yukawa or another Dirac,

---

[8] Quoted in Robert Gilpin, *American Scientists and Nuclear Weapons Policy* (Princeton, 1962), p. 94.
    [9] Quoted *ibid.*, p. 122.

might not prove the missing link in the formulation of a successful theory of nuclear forces?"[1]

After the perfidy of May and Fuchs was revealed, security measures in science grew more stringent. In 1950 some conservative congressmen tried, unsuccessfully, to require security clearances for all fellowship recipients of the National Science Foundation. That year the Joint Committee on Atomic Energy successfully imposed the same requirement on all applicants for AEC fellowships, whether they would be engaged in classified research or not. There is always the chance, Senator William Knowland, the conservative California Republican, declared, that some student, even if engaged in nonsecret studies, might "hit upon a 'superduper' atom bomb and be off to Russia." Albert Einstein and Oswald Veblen protested the fellowship restrictions: "The opportunity of the young scientist to develop his ideas should not be purchased at the expense of his human dignity." Others protested the State Department's refusal to grant foreign scientists, including Paul A. M. Dirac, visas to attend scientific congresses in the United States. The Department even ignored the visa request of a British physicist, wanting to attend a conference on unclassified nuclear physics, whom the British government designated a member of a diplomatic mission to help declassify highly sensitive information on nuclear energy.[2]

Worse than the visa and fellowship restrictions were the assaults against the loyalty of individual physicists. In 1948 the House Un-American Activities Committee pronounced Edward U. Condon of the National Bureau of Standards "one of the weakest links in our atomic security." Condon, who only two years before had been elected president of the American Physical Society, was said to have entertained or associated with Soviet agents. Passionate protests erupted from the scientific community, notably the American Physical Society Council, whose members included Karl Compton, Lee DuBridge, and Robert Millikan. "In the close fraternity of physicists," *Fortune* declared, "Dr. Condon is one of the most eminent and respected of men." After a thorough investigation, the Atomic Energy Commission reported that the FBI files contained some "unfavorable information" about a few of Condon's acquaintances, but the Commission found "no question whatever concerning Dr. Condon's loyalty" and cleared him fully.[3] During the 1948 campaign, in a speech

[1] Hewlett and Duncan, *Atomic Shield*, p. 247; Greenberg, *Politics of Pure Science*, p. 138; Allison, "The State of Physics," p. 4.

[2] Quoted in J. Stefan Dupré and Sanford A. Lakoff, *Science and the Nation: Policy and Politics* (Englewood Cliffs, N.J., 1962), p. 128; Einstein and Veblen to Alfred N. Richards, Oct. 19, 1949, OV, Box 9, National Academy file; *Physics Today*, VII (July 1954), 7; Edward A. Shils, "America's Paper Curtain," in Morton Grodzins and Eugene Rabinowitch, eds., *The Atomic Age: Scientists in National and World Affairs* (New York, 1963), p. 423.

[3] The House Committee is quoted in Jones, "Science, Scientists, and Americans," p. 392; "The Scientists," *Fortune*, 30 (Oct. 1948), 166; the AEC report is quoted in *Time*, 52 (July 26, 1948), 16.

before the American Association for the Advancement of Science, President Truman attacked the rumor-mongering and slander that jeopardized science, then shook hands with Condon in a public gesture that no one could misunderstand.

Suspicion and security regulations drove scores of scientists to quit government laboratories, and in the late 1940s recruiters had a hard time persuading physicists to come to work at Los Alamos, Brookhaven, Berkeley, or Argonne. In the mid-1950s a survey found that almost half the graduate students in a somewhat random sample said that they would accept a lower salary if the job required no security clearance; only one in twelve wanted a job in government, where scientists had to pay the double penalty of both lower salaries and security clearances.[4] Because of what it symbolized, the Condon case struck deeply at U.S. scientific morale, but not as deeply as the later blow of 1954, when the Eisenhower administration ordered the suspension of the security clearance of J. Robert Oppenheimer.

In December 1953 the Atomic Energy Commission filed official charges against Oppenheimer in a 3,400-word letter. While the document was chiefly concerned with allegations that Oppenheimer had indulged in Communist and left-wing associations, it also called him to account for having opposed the development of the hydrogen bomb.[5] In the spring of 1954, at Oppenheimer's request, a three-man personnel review board headed by Gordon Gray, the president of the University of North Carolina, inquired into every aspect of Oppenheimer's character, associations, and policy advice. Numerous witnesses testified in his favor; they ran the gamut from Rabi and DuBridge to George F. Kennan and John J. McCloy. Yet at the hearings Oppenheimer twice contradicted what he had told security officers during the war about his knowledge of suspected Communists, a type of inconsistency in which he had engaged on a few other occasions. Most troubling were his accounts of a wartime conversation with Haakon Chevalier, a Berkeley friend whom he had since seen from time to time, as recently as December 1953. In 1943 Chevalier had mentioned to Oppenheimer that he knew an engineer, George Eltenton, who could pass technical information to the Soviet Union. Oppenheimer had declared that he would have no part of any such activity. He had reported the Chevalier encounter to security agents, but in 1946 and at the hearings in 1954 he said that his report had been elaborately mendacious.

A handful of scientists, notably Edward Teller, bore witness against Oppenheimer. Teller considered Oppenheimer a Communist and an advocate of Soviet appeasement. (When later asked how he accounted for the complete lack of evidence that Oppenheimer was ever a card-

[4] M. Stanley Livingston, "Employment Problems of Young Physicists," *Physics Today*, VIII (March 1955), 20–21.

[5] Philip M. Stern, *The Oppenheimer Case: Security on Trial* (New York, 1969), pp. 233–35.

carrying Communist, Teller declared that the Communists did not make their most important people party members; it hampered their work.) At the hearing, unwilling to accuse Oppenheimer openly of disloyalty—the real issue, Teller was persuaded—he testified instead that "his [Oppenheimer's] actions frankly appeared to me confused and complicated. . . . I would like to see the vital interests of this country in hands which I understand better and therefore trust more. . . . I would feel personally more secure if public matters would rest in other hands."[6]

The Gray board found Oppenheimer clearly, unquestionably loyal but, by a two-to-one vote, recommended against the restoration of his security clearance. In June the Atomic Energy Commission upheld the board four to one, the lone dissenter being the Princeton physicist Henry DeWolf Smyth. By his associations, the AEC majority concluded, Oppenheimer had exhibited a "willful disregard of the normal and proper obligations of security." On occasion, he had also lied or stretched the truth, which evidenced "fundamental defects" in his character.[7]

The decision outraged the majority of nuclear scientists. Bitter at Teller and his anti-Oppenheimer colleagues, they found infuriating the star-chamber quality of the proceedings, which, though in principle an inquiry, had resembled a prosecution. They distrusted the Commission's motives, especially those of Commissioner Lewis Strauss, who was widely believed to dislike Oppenheimer and to have used the security issue as a vehicle for vendetta. While the Commission said that it attached "no importance" to Oppenheimer's position in the H-bomb debate—"Dr. Oppenheimer was, of course, entitled to his own opinion"—most nuclear scientists nevertheless took the outcome to mean that Oppenheimer had been persecuted for his onetime opposition to building a hydrogen bomb. Most important, they emphatically disputed the Commission's standards of judgment.

The Gray board concluded that Oppenheimer possessed an "unusual ability" to keep vital secrets. Yet, like the board, the Commission majority felt itself impelled to declare him a security risk under the requirements

---

[6] Interview with Edward Teller, May 19, 1963; U.S. Atomic Energy Commission, *In the Matter of J. Robert Oppenheimer: Transcript of Hearing Before Personnel Security Board, Washington, D.C., April 12, 1954 through May 6, 1954* (Washington, D.C., 1954), p. 710. Teller told his biographers that he distrusted Oppenheimer's motives. See Stanley A. Blumberg and Gwinn Owens, *Energy and Conflict: The Life and Times of Edward Teller* (New York, 1976), p. 340. Harold P. Green, the lawyer assigned to draw up the AEC charges against Oppenheimer, said in 1975 that "a very substantial portion of the charges, certainly most of them related to the H-bomb, were drawn from F.B.I. interviews with Teller." Quoted in Martin Sherwin's review of *Energy and Conflict, New York Times Book Review,* Aug. 8, 1976, p. 2.

[7] The majority opinion, signed by three of the five commissioners: Lewis L. Strauss, Eugene M. Zuckert, and Joseph Campbell, in the Commission Ruling, June 29, 1954, pp. 1, 4, copy in EOL, Carton XIV. Commissioner Thomas E. Murray concurred in a separate opinion.

of security—they had been made more stringent by an Eisenhower executive order in 1953—as officially defined. The Commission's ruling stressed that Oppenheimer consistently placed himself outside the *rules* governing others. In the judgment of the majority, he fell "far short" of the "exemplary standards of reliability, self-discipline and trustworthiness" to be expected of someone with his access to the nation's innermost defense secrets.[8]

On their part, the leaders of the Los Alamos generation conceded that Oppenheimer had violated the rules, that he had committed indiscretions, that he had even stretched the truth or lied. But like Henry DeWolf Smyth, they did not regard him as a security risk in the essential sense that his access to classified information risked the nation's security. On the contrary, the denial of such access to so talented a physicist, a committee of the Federation of American Scientists concluded, would "hurt national security and science in America."[9]

Indeed, the leadership of the Los Alamos generation was prepared to believe, with Philip Rieff, that the Oppenheimer case signaled "not merely the personal defeat of a leading scientist, but the removal of scientists as a group from their high place in the political order." Actually, as Gordon Gray argued, the Oppenheimer ruling was not an attack against scientists as such.[1] The Eisenhower administration kept appointing scientists to serve on advisory panels, boards, and committees. And physicists still enjoyed a voice at the highest levels of government, including the White House. For Oppenheimer's advocates, the trouble was that the voice most listened to in strategic matters seemed to be Edward Teller's, whose views occupied a minority position among leading nuclear scientists.[2]

For the liberal majority of physicists, it was scarcely a halcyon time. The Eisenhower administration pursued its policy of brinkmanship, and the arms race continued to escalate. In 1953 the Russians exploded their first thermonuclear device, and in 1954 deadly radioactive ash, fallout from the detonation of the first American thermonuclear weapon, poisoned the bodies of Japanese fishermen—one later died—on the tuna trawler the *Lucky Dragon No. 5*.[3] During the 1956 presidential campaign Adlai Stevenson, following the advice of a group of scientists, urged considera-

---

[8] Majority opinion, pp. 3, 4; the Gray board's finding is quoted in Smyth's opinion, Commission Ruling, p. 27. Reflecting general editorial opinion, the normally liberal New York *Post*, two of whose executives had studied the hearing transcript, concluded: "Dr. Oppenheimer is clearly guilty of arbitrariness and deceit." Quoted in Stern, *The Oppenheimer Case*, p. 425.

[9] Quoted in Princeton *Herald*, July 3, 1954, clipping in OV, Box 9, Oppenheimer file.

[1] Rieff, "The Case of Dr. Oppenheimer," pp. 316–17; Gray's admonition is quoted in *Physics Today*, VII (July 1954), 7.

[2] Gilpin, *American Scientists and Nuclear Weapons Policy*, p. 13.

[3] Twenty years later, inhabitants of Rongelap Atoll, 125 miles downwind from the explosion, were still suffering ill effects from the blast fallout. *New York Times*, June 3, 1974, p. 19.

tion of a ban on nuclear tests, and Eisenhower attacked him for advocating a "moratorium on ordinary common sense." Earlier, Secretary of Commerce Sinclair Weeks had tried to fire the head of the Bureau of Standards, the physicist Alan V. Astin, because the Bureau had determined that a battery additive produced by a California firm added nothing to the life of batteries. "I am not a man of science," Weeks snapped, ". . . but as a practical man I think that the National Bureau of Standards has not been sufficiently objective because they discount entirely the play of the marketplace."[4]

In the mid-1950s a Monsanto Chemical Company recruitment film displayed a group of scientists in white coats and assured the audience: "No geniuses here; just a bunch of average Americans working together." Secretary of Defense Charles E. Wilson delivered himself of the observation: "Basic research is when you don't know what you are doing." Soon Wilson ordered a 10 percent cut in Defense expenditures for research and development, which inflation had already reduced since 1952 by an estimated 25 percent.[5] The rate of Ph.D. production seemed to be leveling off at a steady five hundred or so per year. Not even Robert Millikan at the height of Babbittry would have been happy with the sentiments of the Monsanto film or Secretary Wilson, but for the leaders of the Los Alamos generation the ambience of the day was especially cloying.

If American physicists had long been accustomed to stress the utilitarian benefits of research, the refugees, in the recollection of Edward Teller, tended to find such appeals "nauseating."[6] As members of an elite professional class in their own countries, they had been in a position to think of science as its own reward, and, for all the utilitarian appeals, many native American physicists of course still liked to think of it as such. In the mid-1950s, manifesting a cultural elitism reminiscent of Henry Rowland's, a significant part of the physics community felt all the more out of place in an America which was in no mood for cultural elitism.

Liberal physicists, like liberals generally, talked worriedly about complacency, conformity, and fear. The Federation of American Scientists, demoralized after the failure of the Baruch Plan, fell into a state of desuetude. "Many scientists," the Harvard physicist Edwin Kemble lamented, "have felt it prudent to avoid all political entanglements. Join nothing, sign nothing, and you are pretty sure not to get into bad company!" Despite the strength of American physics, a touch of gloom was

[4] *Public Papers of . . . Dwight David Eisenhower, 1956*, p. 976; Weeks is quoted in James L. Penick, Jr., et al., eds., *The Politics of American Science, 1939 to the Present* (rev. ed.; Cambridge, Mass., 1972), p. 199.

[5] Quoted in William H. Whyte, Jr., *The Organization Man* (New York: Anchor, 1957), p. 235; Wilson is quoted in Greenberg, *The Politics of Pure Science*, p. 273; George A. W. Boehm, "The Pentagon and the Research Crisis," *Fortune*, 57 (Feb. 1958), 154.

[6] Interview with Edward Teller.

descending over the Los Alamos generation's advisory elite. Eugene Rabinowitch, the editor of the *Bulletin of the Atomic Scientists*, pronounced his colleagues "a harassed profession, occupying a defensive position in the political arena. Their early hopes of playing an important role insuring world peace and prosperity are still in abeyance."[7]

On October 5, 1957, Americans learned that the Soviet Union had launched the world's first artificial earth satellite, Sputnik. Twenty-nine days later, November 3, Sputnik II went up, weighing 1,120.9 pounds, packed with a maze of scientific instruments and signaling back the condition of a live dog. On December 6, the United States' attempt to launch its own satellite from Cape Canaveral fizzled in a cloud of brownish-black smoke. "Who wants to overtake whom in science?" sneered Soviet Premier Nikita Khrushchev.[8]

The Sputniks demonstrated incontestably that the Soviets possessed the rocket and guidance capability for intercontinental ballistic missiles; the live dog, that they were well on the way toward putting a man into space. Even after the United States managed successfully to launch a satellite early in 1958, commentators wondered whether the nation's scientists were not paid too little attention at the highest levels of the government, whether the disparately organized American rocket program could meet the Soviet challenge, whether American science could compete with its Soviet counterpart in any field. Liberals and conservatives alike worried whether Edward Teller's prediction might be right: "Ten years ago there was no question where the best scientists in the world could be found—here in the U.S. . . . Ten years from now the best scientists in the world will be found in Russia."[9]

Thousands of sermons, editorials, and articles pronounced America not only prosperous but chrome-plated, complacent, and purposeless. John Gunther's *Inside Russia Today* warned that Soviet children crammed into ten years or less what American children did in twelve, and for both boys and girls the "main emphasis is on science and technology." In addition to ten years of mathematics, every child took four years of chemistry, five of physics, six of biology. The Soviet Union, President Eisenhower himself added already had more scientists and engineers than the United States and was producing them at a faster rate. American scientists declared that the opportunities in basic research for Soviet physicists, chemists, and biologists far exceeded their own. In science especially, American schools, with their undue emphasis on adjustment over learning,

[7] Edwin C. Kemble, "Physicists and Politics," *Physics Today*, VI (June 1953), 18; Rabinowitch is quoted in Donald W. Cox, *America's New Policy Makers: The Scientists' Rise to Power* (Philadelphia: Chilton Books, 1964), p. 39.

[8] Quoted in Eric F. Goldman, *The Crucial Decade—and After* (New York, 1960), p. 311.

[9] Quoted in William Manchester, *The Glory and the Dream: A Narrative History of America, 1932–1972* (Boston, 1974), p. 792.

were found to be an amalgam of inadequate facilities, shoddy instruction, and teachers who were ill paid and poorly prepared. About 75 percent of American high-school pupils studied no physics at all; many who did graduated without even a slim understanding of Newton's laws. "The nation's stupid children get far better care than the bright," *Life* remarked. "The geniuses of the next decade are even now being allowed to slip back into mediocrity."[1]

In the two weeks following the launching of the Soviet satellite, Eisenhower conferred with numerous scientists. Among them was Rabi, who urged the President to bring the part-time scientific advisory group located in the Executive Office directly into the White House, with full-time responsibilities.[2] The group included James R. Killian, the president of MIT, who had headed up several key studies on defense technology. On November 7, Eisenhower went on television to announce Killian's appointment to the new White House post of Special Assistant to the President for Science and Technology. Killian's duties would include helping the President on the scientific improvement of defense with the aid of a President's Science Advisory Committee. An elevated version of the old Executive Office group, the first Committee was dominated by Los Alamos generation physicists. His "scientific advisers," the President shortly reported, were warning that the country's most "critical" problem was the need for scientists, "thousands more" in the next ten years than the nation was currently planning to produce.[3]

In 1958 the President signed into law the National Defense Education Act, which aimed to help needy and capable students in critically important fields of the humanities as well as the sciences, and which made available $250,000,000 for the improvement of public school facilities, including laboratories. Not long after Sputnik, numerous high schools around the country adopted the high-powered Physical Science Study Committee course in physics, a program designed by a group of leading physicists representing more than a dozen universities, government agencies, and commercial laboratories. And in 1958 the federal government established the National Aeronautics and Space Administration to oversee and coordinate all the nation's nonmilitary activities in space research and development. "How much money would you need to . . . make us even with Russia . . . and probably leap-frog them?" Congressman James G. Fulton of Pennsylvania asked the head of the space agency. "I want to be firstest with the mostest in space, and I just don't want to wait for years. How much money do we need to do it?"[4]

[1] Gunther is quoted *ibid.*, pp. 791–92; *Life*, in Goldman, *Crucial Decade*, p. 314.
[2] Jeremy Bernstein, "I. I. Rabi: Physicist," *The New Yorker*, 51 (Oct. 20, 1975), 82.
[3] *Public Papers of . . . Dwight David Eisenhower, 1957*, p. 814.
[4] Quoted in Vernon Van Dyke, *Pride and Power: The Rationale of the Space Program* (Urbana, Ill., 1964), p. 22.

In the wake of Sputnik, both the Congress and President Eisenhower were also willing to invest a good deal to keep the United States first in all of science. Just a few days after the news of the Russian launch, the new Secretary of Defense, Neil McElroy, rescinded his predecessor's order to reduce basic defense research funds by 10 percent and issued a directive endorsing the support of basic science as a key element in Defense Department policy.[5] Between 1957 and 1961 federal expenditures for research and development more than doubled, to $9 billion annually, including more than a twofold increase in outlays for basic research to nonprofit institutions; in the same period overall federal obligations for basic research more than tripled, to $827,000,000. Symbolizing the trend, the budget of the National Science Foundation, the haven of basic research, rose from $30,000,000 to $76,000,000. And in the same period Sputnik produced similar advantages for a field that had no direct relationship to space or national defense, high energy physics.

Early in 1957, at Stanford University, the physicist Wolfgang K. H. Panofsky had submitted a proposal to the Atomic Energy Commission for a giant linear accelerator—it would run for two miles straight through the hills near Palo Alto—at a cost of $100,000,000. Even before Sputnik, American physicists had been warning, in the words of the physicist Robert Marshak, that the Soviet objective seemed to be to "overtake American science."[6] But Panofsky's proposal had been languishing. In 1958, an advisory panel of physicists pointed out that the Soviet Union proposed to build a 50-billion electron volt synchrotron, a machine twice as energetic as the most powerful yet budgeted for in the United States at Brookhaven. The panel also endorsed the proposed Stanford linear accelerator. In May 1959, President Eisenhower announced at a symposium on basic research that he would ask Congress for the $100,000,000 necessary to build the Stanford machine. American progress in the field of high energy physics, he declared, was vitally important to the nation.[7]

With the election of John F. Kennedy, who had campaigned to get the country moving again and close the missile gap, the drive for first-rate science took on a supercharged tone. Early in 1961, the Russian astronaut Yuri Gagarin became the first man to orbit and return safely to earth. In May, President Kennedy called for the United States to achieve a manned lunar landing and return before the end of the decade. Inflated by Project Apollo, the budget of the National Aeronautics and Space Administration

[5] Boehm, "The Pentagon and the Research Crisis," p. 154.

[6] Quoted in Greenberg, Politics of Pure Science, p. 217.

[7] Ibid., pp. 226–27, 230–33; Public Papers of . . . Dwight David Eisenhower, 1959, p. 107. Eisenhower's second science adviser, George Kistiakowsky, helped persuade John A. McCone, the chairman of the AEC, to back the Stanford machine by emphasizing its importance for the competition in national prestige with Russia. George B. Kistiakowsky, A Scientist at the White House: The Private Diary of President Eisenhower's Special Assistant for Science and Technology (Cambridge, Mass., 1976), p. 265.

leaped almost fivefold by 1964, to more than $4 billion a year. In 1962 at Houston, Texas, where the Manned Spacecraft Center brought $150,000,000 a year into the city, Kennedy put the rhetorical question to an audience at Rice University: Why go to the moon? "Why does Rice play Texas?" Because the moon is there, because the task would be "hard," not "easy," because it would "organize and measure the best of our energies and skills."[8]

Many scientists, especially those whose specialties derived no particular benefit from Project Apollo, objected that the manned lunar landing was not the best way to organize and measure the nation's scientific energies and skills. Manned space exploration promised to yield little more scientifically than unmanned probes, which could radio back data to the earth from the moon or even the planets. And the money needed to send a man to the moon, the critics claimed, could be better spent on attractive scientific projects ranging from molecular biology to a variety of subjects in physics. Still, despite Project Apollo's drain on the budget, federal funds for research and development continued to climb, to almost $15 billion by 1965. So did funds for basic research, which that year accounted for 14 percent of the expenditures, or about one third more than the 1960 percentage. In 1965 basic research in physics commanded $320,000,000, a two-and-a-half-fold increase since 1959 both in dollars and portion of the total federal budget.[9]

The federal money created a sharp demand for more scientists of all types. Once again young European scientists—and even some of their distinguished seniors—migrated to jobs in the United States, enough to raise cries abroad of an alarming "brain drain." "The Americans . . . are proving that you can be rich and clever," John Davy, the science correspondent of the London *Observer*, remarked. "The enormous, wealthy campuses of Stanford, Berkeley, the Massachusetts Institute of Technology, Harvard and others are not only rolling in dollars—they are fizzing with intellectual excitement." At home, the National Academy predicted that industrial and international competition, especially with the USSR, would keep the demand for scientists and engineers at a high level indefinitely.[1]

Drawn by the appeal and the seemingly assured future of post-Sputnik science, still more Americans went into physics. For decades the number of Ph.D.'s awarded in the field had been increasing at an average annual rate of about 5 percent; after 1958 the rate jumped to more than 10 percent. Reversing the industrial trend of the post-Hiroshima decade, the new crop of physicists went disproportionately into the academic world,

[8] *Public Papers of . . . John F. Kennedy, 1962*, p. 669.

[9] Physics Survey Committee, National Research Council, *Physics in Perspective* (2 vols.; Washington, D.C.: National Academy of Sciences, 1972), I, 2.

[1] Quoted in Spencer Klaw, *The New Brahmins: Scientific Life in America* (New York, 1968), p. 35; *New York Times*, July 12, 1964, p. 1.

where federal money combined with local aspirations to make demand high in all fields but higher still in physics. Between 1959 and 1967 the number of institutions offering doctorates in physics almost doubled. In 1965 only 180 new physics Ph.D.'s entered industrial laboratories, while 500 went into universities. In the late 1960s, one out of two doctors of physics was employed in an academic institution, only one out of four in an industrial firm.[2]

The academic competition for Ph.D.'s drove salaries to unprecedented heights. In 1964 physicists earned at the median $1,000 a year more than scientists in all other fields. One in two professors of physics in research earned almost $12,000 a year, slightly less than twice the median family income nationally; one in four earned $15,000 a year and up.[3] And summer grants were readily available from federal agencies to supplement the standard nine-month academic salary. There was a certain campus envy over the physicists' high standard of living, personal and professional, their jetting off to conferences abroad, their insider's access to Washington councils, their ample largess for graduate students. At Columbia during the summer months, a journalist was surprised to find few physics graduate students. Some were at an institute in Norway, a professor explained, more at another in Italy. Such institutes always seemed to be held at pleasant places, the professor added. "It's almost unheard of now to spend the summer here in New York."[4]

Following in part the accustomed social pattern, the new physics Ph.D.'s included only a trickle of blacks, about 2 percent women, and a disproportionately large number of Jews; Jews occupied one out of seven physics faculty positions generally, one out of four in the ranking universities. But Oriental-Americans, five out of six of them now foreign-born, accounted for about 6 percent of the doctorates, far more than their proportion of the general population, and the new crop also included a growing number of Catholics. Since the late 1950s the siege mentality among Catholic Americans had been diminishing, the spirit of ecumenicism taking hold. Now, with a Catholic having been elected President, Catholic youngsters were attending the better secular colleges in proportion to their weight in the population. In the late 1960s, in all fields combined, one out of five graduate students was Catholic. So was one out of every seven academic physicists.[5]

[2] *Physics in Perspective*, I, 808, 607, 850, 426–27, 89–90.

[3] Sylvia Barisch, "Who Are the Physicists . . .?" *Physics Today*, XIX (Jan. 1966), 70–71.

[4] Quoted in Klaw, *The New Brahmins*, p. 87.

[5] Stephen Steinberg, *The Academic Melting Pot: Catholics and Jews in American Higher Education* (New York, 1974), pp. 120, 101, 103; Commission on Human Resources of the National Research Council, *Minority Groups Among United States Doctorate-Level Scientists, Engineers, and Scholars, 1973*, pp. 19, 21–22. Again, the figures for Oriental-Americans were obtained by assuming that the statistics for

Whatever the sex or social background, physicists of the mid-1960s did still more of their research in groups. On the average, every paper published in *The Physical Review* was signed by two authors. In virtually every major field of the discipline, Americans published more papers annually than the physicists of any other country. By now the field with the most physicists, about a quarter of the profession, and with the highest production of papers was "condensed matter," which meant especially matter in the solid state suitable for such innovations as the transistor and magnetic memories. But the glamour field remained high energy physics. While it employed about one out of ten physicists, it received one-third of federal physics funds.[6] No physics department could call itself self-respecting without high energy practitioners, not least because the field remained one of the most intellectually exciting in the discipline. By the mid-1960s more than ninety elementary particles were known. To impose order on them, much as Mendeleev had ordered the elements into the periodic table, the theorist Murray Gell-Mann had proposed a scheme known as the Eightfold Way. Physicists applauded when in 1964 a Brookhaven team announced the confirmation of the scheme—the paper had 31 authors—by detecting a key particle, the omega-minus, whose existence Gell-Mann's theory predicted. The 1946–1955 Nobel Prize statistics may have been impressive, but they now paled. Between 1956 and 1965, Americans won or shared in eight out of the ten Nobel Prizes awarded in physics, five of them for work in or related to high energy.

The Nobel Prizes only added to the prestige that American physicists enjoyed at home, a prestige enlarged by their seemingly considerable influence in the Kennedy White House. Kennedy's Presidential Science Advisor was Jerome B. Wiesner of MIT, a native of Dearborn, Michigan, where his father ran a dry goods shop, and graduate of the state university, an electrical engineer who had joined the leadership of the Los Alamos generation at the MIT Radiation Laboratory and had served a postwar stint at Los Alamos itself. Like most of the nation's leading physicists, he combined a vigorous commitment to a strong national defense with an equally vigorous advocacy of arms control. He advanced this view as a defense consultant through the postwar years and then was appointed a member of Eisenhower's first President's Science Advisory Committee. A Democrat and an outer-circle adviser to Kennedy during the 1960 campaign, early in 1961 he submitted a report to the President-elect on the national space program. Among its recommenda-

physics were the same as for science generally. These results are consistent with the statistics gathered specifically for physics in the 1970s. See *ibid.*, p. 27, and Susanne D. Ellis, *Report, Manpower Statistics Division: 1975–76 Graduate Student Survey* (New York: American Institute of Physics, 1976), p. 2.

[6] *Physics in Perspective*, I, 91, 118.

tions: The United States should not commit itself to a crash effort for man in space.

After the Kennedy administration's commitment to Project Apollo, the perceptive reporter Meg Greenfield wrote: "In Washington these days, the definition of a truly hip science adviser is one who knows that the moon money could be better spent on other scientific projects and who also knows that Congress won't appropriate it for any of them. The kind of passive in-betweenness this suggests is more or less the state of science advising now."[7] But whatever Wiesner's in-betweenness, much of the Los Alamos generation's leadership counted the Kennedy White House as its own.

Wiesner himself presided over a newly created Office of Science and Technology, which enlarged the number of scientists working on technical issues at the White House level. It was even said that he agreed to give special consideration to all the important public policy reports of the National Academy of Sciences. No matter Wiesner's setback on man in space: Kennedy's Special Counsel Theodore C. Sorensen later remarked, perhaps with unintended irony, that, "by learning . . . to accept philosophically [the President's] decisions contrary to their advice," Wiesner and his fellow academic Walter Heller "greatly raised the stature of their offices." And Wiesner, whose way had been paved by Killian, then Killian's successor under Eisenhower, George B. Kistiakowsky, and the mounting fear of fallout, certainly helped the Kennedy White House achieve the nuclear test ban treaty with the Soviet Union. "You mean that stuff is in the rain out there?" Kennedy queried Wiesner one day.[8]

Whatever the degree of Wiesner's power, there was no gainsaying Kennedy's appointment to the Atomic Energy Commission chairmanship of the Nobel laureate chemist Glenn T. Seaborg, the first scientist to hold the post. There was also the certain glow of Kennedy's White House dinner for Nobel Prize winners. American scientists liked his calling the group "the most extraordinary collection of talent, of human knowledge, that has ever been gathered together at the White House . . .," and they hardly minded his addition, "with the possible exception of when Thomas Jefferson dined alone." There were Kennedy's addresses to the National Academy of Sciences, including his celebration of pure science. "We realize now that progress in technology depends on progress in theory," he told the audience, embracing the scientific community's orthodox rhetoric, "that the most abstract investigations can lead to the most concrete results."[9] He was the first President to address the Academy since

[7] Meg Greenfield, "Science Goes to Washington," *The Reporter*, 29 (Sept. 26, 1963), 26.

[8] Sorensen, *Kennedy*, pp. 264–65; Kennedy is quoted in Schlesinger, *A Thousand Days*, p. 455.

[9] *Public Papers of . . . John F. Kennedy, 1962*, p. 347; *ibid., 1963*, p. 802.

Lincoln, many scientists claimed happily (he was actually the first since Coolidge). And then there was Kennedy's decision to present the coveted Fermi Award of the Atomic Energy Commission to J. Robert Oppenheimer.

John Kennedy was not long in his grave when, in a cabinet room gathering including his secretaries of defense and state, as well as Henry DeWolf Smyth and even Edward Teller, President Lyndon B. Johnson presented the Fermi citation to Oppenheimer. Oppenheimer responded quietly: "In his later years, Jefferson often wrote of the 'brotherly spirit of science which unites into a family all of its votaries. . . .' We have not, I know, always given evidence of that brotherly spirit of science. This . . . is in part because . . . we are engaged in this great enterprise of our time, testing whether men can . . . live without war as the great arbiter of history. . . . I think it is just possible, Mr. President, that it has taken some charity and some courage for you to make this award today."[1]

Charity and courage, no doubt, but Lyndon Johnson also possessed no small political sense. In 1964 the vast majority of the Los Alamos generation would join Scientists and Engineers for Johnson. The first such political association of its kind, the venture helped considerably in convincing voters that Senator Barry Goldwater was unfit to control the nation's nuclear forces. In the mid-1960s, the high priest of the relations between science and government was Sir Charles P. Snow, the British physicist, novelist, and man of affairs, who in various essays, especially his celebrated *Two Cultures and the Scientific Revolution*, was understood chiefly to argue how essential it was for citizens and statesmen to equip themselves better to understand the advice of scientists. Whatever advice Lyndon Johnson may have heeded in giving the Fermi Award to Oppenheimer, he understood that it was politically advantageous to have the leading scientists of the nation on your side, especially if they were physicists.

It was a time when Americans ranked nuclear physicists third in occupational status—they had been fifteenth in 1947—ahead of everyone except Supreme Court Justices and physicians; when physicists, among other scientists, were identified not only as the makers of bombs and rockets but as the progenitors of jet planes, computers, and direct dial telephoning, of transistor radios, stereophonic phonographs, and color television; when research and development in what President Clark Kerr of the University of California called this "age of the knowledge industry" were believed to generate endless economic expansion; when electronic and computer firms were assumed to follow close upon the heels of local Ph.D. programs; when Governor Edmund G. Brown of California reported that, on the basis of an experiment in his state, space and defense

[1] Quoted in Nuel Pharr Davis, *Lawrence and Oppenheimer* (New York, 1968), p. 354.

392 | THE PHYSICISTS

scientists could solve problems of smog, sewage or waste disposal, and transportation.[2]

The widely respected commentator Richard Rovere applauded physicists for bringing about a *pax atomica* ("the bomb and its diplomatic consequences have had, by and large, a stabilizing effect on our time").[3] Government planners talked of cutting back the production of nuclear weapons—the nation already had enough—and expert commentators declared that the age of nuclear power had arrived. Atomic explosions were discussed as ways to dig tunnels and canals. Nuclear reactors were already lighting more than a million homes in the United States. Placed by the shores of underdeveloped countries, they could generate electricity for development while desalinating sea water for irrigation; the breeder reactor promised to produce not only power but new fissionable plutonium even while it consumed uranium fuel.

Alvin Weinberg, the director of the Oak Ridge National Laboratory, recalling the mighty pyramids, the magnificent cathedrals, the glories of Versailles, had predicted that history would find in the "monuments of Big Science—the huge rockets, the high-energy accelerators, the high-flux research reactors—symbols of our time." History might also conclude, Weinberg had added, that "we build our monuments in the name of scientific truth . . .; we use our Big Science to add to our country's prestige . . .; we build to placate what . . . could become a dominant scientific caste." Whatever history would conclude, in the mid-1960s American physicists headed a community of scientists who, in the analysis of Don K. Price, now a dean at Harvard, had collectively become "something very close to an *establishment*, in the old and proper sense of that word: a set of institutions supported by tax funds, but largely on faith, and without direct responsibility to political control."[4]

[2] Kerr is quoted in *Time*, 82 (Oct. 4, 1963), 108; Governor Brown's testimony is in *The New York Times*, July 23, 1965, p. 13.
[3] Rovere, "The Bomb and International Politics," in *Hiroshima Plus 20*, prepared by *The New York Times* (New York, 1965), p. 110.
[4] Alvin M. Weinbℓg, "The Impact of Large-Scale Science in the United States," *Science*, 134 (July 1961), 161; Don K. Price, *The Scientific Estate* (Cambridge, Mass., 1965), p. 12.

XXIV

New Revolt Against Science

The year after Sputnik an Eisenhower aide, Malcolm Moos, began filing away ideas for what the President might say when he ultimately left office. A political scientist from the Johns Hopkins University, Moos was generally knowledgeable about the increasingly close relationship between the academic community and the federal government. From his vantage point in the White House, he was also learning what an enormous business American industry did with the Pentagon. Troubled by these trends, Moos eventually drafted passages about them for the President's farewell address, and in January 1961 Eisenhower told the nation: "In the councils of government, we must guard against the acquisition of unwarranted influence . . . by the military-industrial complex." We must "gravely" regard the "prospect of domination of the nation's scholars" by federal largess. "We must also be alert to the equal and opposite danger that public policy could itself become the captive of a scientific-technological elite."[1]

Amid the post-Sputnik boom, Eisenhower's warning both expressed and advanced an emerging uneasiness. Scholars and journalists were call-

---

[1] *Public Papers of . . . Dwight David Eisenhower, 1960–61*, pp. 1038–39. Moos's role is described in the Los Angeles *Times*, March 31, 1969, pp. 1, 6. Told that scientists were distressed by his speech, Eisenhower assured Kistiakowsky that he was not anti-science but concerned only with the rising power of the military in American life. George B. Kistiakowsky, *A Scientist at the White House: The Private Diary of President Eisenhower's Special Assistant for Science and Technology* (Cambridge, Mass., 1976), p. 425.

ing alarmed attention to the political elitism of the federal scientific establishment. Significant influence in the government's scientific advisory apparatus was said to be confined to about a thousand scientists, and James R. Killian estimated the consistently influential group at no larger than two hundred. The insiders tended to be the leaders of the Los Alamos generation, predominantly physicists who had taken their Ph.D.'s at about a dozen graduate schools and held faculty positions at a handful of universities. About one third of the members of the President's Science Advisory Committee were professors in Cambridge, Massachusetts. "We all know each other . . .," the decidedly influential MIT advisory physicist Jerrold R. Zacharias once said. "People always think that because the U.S. has a population of 170 million and there are a lot of people in the Pentagon, it all has to be very impersonal. Science isn't. It's just us boys."[2]

Since the battle over the control of atomic energy, advisers like Zacharias had tended to avoid the normal political process of open pressure, advocacy, and debate. "The overwhelming majority of scientists have never wanted any part of this," John Fischer, the editor of *Harper's*, wrote with dismay. "Typically, they regard the political process as something sinister if not dirty; often they treat politicians . . . and sometimes the ordinary voter as well—with scarcely veiled contempt." The President's Science Advisor kept the names of his consultants secret, in part because, as Meg Greenfield observed, to reveal them would put them "under public scrutiny, which was exactly where they did not want to be."[3] Of course, security requirements accounted for some of the secrecy. Secrecy was also defended as necessary to protect the advisers from the kind of outside pressure that might distort their disinterested formation of advisory opinions. But secretiveness, as distinct from secrecy, was also convenient.

Washington supplied a sizable percentage, in some cases more than half, of the research budgets of the leading universities. Echoing Hilary Herbert's warning of the 1880s, Philip Abelson, the editor of *Science*, found the contemporary scientist actually reluctant to question publicly "the wisdom of the establishment. . . . He stirs the enmity of powerful foes. He fears that reprisals may extend beyond him to his institution. Perhaps he fears shadows, but in a day when almost all research institu-

[2] Robert C. Wood, "Scientists and Politics: The Rise of an Apolitical Elite," in Robert Gilpin and Christopher Wright, eds., *Scientists and National Policy-Making* (New York, 1964), pp. 48–49; Carl William Fischer, "Scientists and Statesmen: A Profile of the President's Science Advisory Committee," in Sanford A. Lakoff, ed., *Knowledge and Power: Essays on Science and Government* (New York, 1966), p. 323; Daniel S. Greenberg, *The Politics of Pure Science* (New York, 1967), pp. 15–16; Zacharias is quoted in George A. W. Boehm, "The Pentagon and the Research Crisis," *Fortune*, 57 (Feb. 1958), 160.

[3] John Fischer, "Why Our Scientists Are About to Be Dragged, Moaning, into Politics," *Harper's*, 234 (Sept. 1966), 22; Greenfield, "Science Goes to Washington," *The Reporter*, 29 (Sept. 26, 1963), 22.

tions are highly dependent on federal funds, prudence seems to dictate silence." Whatever the reasons, the secrecy of the advisory system kept its members virtually immune, David Lilienthal complained, "from the tough, essential and distinguishing characteristic of the democratic process . . ., the essence of which . . . is direct *accountability*, in the open air, to public lay scrutiny."[4]

Most Americans—liberals, conservatives, and the millions who were not particularly either—had been willing to tolerate such lack of accountability since Hiroshima. They identified the scientific elite with the industrial machine that produced the goods and prosperity of the affluent society. During the Cold War, they counted the same elite as indispensable to the national defense. The leaders of the Los Alamos generation also commanded respect as paragons of social responsibility, seeking to maintain the nation's nuclear arsenal while attempting to control, if not end, the nuclear arms race. "Physicists have known sin; and this is a knowledge which they cannot lose," J. Robert Oppenheimer had said in a widely remembered remark about the atomic bomb.[5]

But by the mid-1960s, people were saying that the test ban treaty might herald a reduction in the strategic arms race, and the imperfections in the affluent technological society, including urban decay and racial discrimination, were rapidly forcing themselves upon the nation's attention. A growing number of Americans were asking with Meg Greenfield: "As presiders over the national science purse, are the scientists speaking in the interests of science . . . government or . . . their own institutions? Is their policy advice . . . offered in furtherance of national objectives—or agency objectives—or their own objectives?"[6] By the end of the decade, in the tumultuous years of the war in Vietnam, millions of Americans doubted the social responsibility of any group so closely identified with the military-industrial complex as the nation's physical scientists.

The sheer bigness of Big Science was enough to arouse questions. Since 1940 the federal budget had risen elevenfold; the budget for research and development, or R & D in the now common acronym, some two hundred times. As early as 1963 Congressman Albert P. Thomas of Texas, the chairman of the House appropriations subcommittee with oversight of independent scientific agencies, announced: "We hear that

[4] Abelson is quoted in Ralph E. Lapp, *The New Priesthood: The Scientific Elite and the Uses of Power* (New York, 1965), p. 30; Lilienthal, "Skeptical Look . . .," *New York Times Magazine*, Sept. 29, 1963, pp. 82, 84.

[5] Oppenheimer's remark is quoted in "The Scientists," *Fortune*, 38 (Oct. 1948), 112. It is interesting to note that Ernest Lawrence, who manifested grave reservations before Hiroshima about whether the bomb should be dropped, privately protested: "I am a physicist and I have no knowledge to lose in which physics has caused me to know sin." Quoted in Herbert Childs, *An American Genius: The Life of Ernest Orlando Lawrence* (New York, 1968), p. 405.

[6] Greenfield, "Science Goes to Washington," p. 26.

research money is running out of the ears of grant recipients. . . . It's time to pause and take a look."[7]

The look revealed that grant and contract money was used to increase scientific salaries, to purchase equipment unnecessary for the funded project, to underwrite a decidedly high standard of academic living. The Office of Naval Research found a disturbing "correlation between the specific European laboratories visited by American scientists and the tourist attractions of the area. . . ." At home, the system of granting research funds to professors for specific projects, rather than to institutions for general purposes, diminished the loyalty of the professors to their universities. Academic administrators found themselves at the mercy of star scientists—Gerard Piel, the editor of *Scientific American*, termed many of them "mercenaries"—who could easily threaten to leave for another university, taking their students, equipment, contracts, and overhead funds along.[8]

Academic officers, mindful of prestige and money, were hardly reluctant to see their leading scientists win federal grants. But the project-grant system undercut the ability of even the best-intentioned administrator to maintain balance in his university program, especially a balance between the sciences and the humanities. Echoing in reverse the complaint of scientists a century earlier, the critic David Boroff reported that students and professors of the humanities "find themselves inhabiting a dingy academic slum with modest research grants, or none, lower salaries, and few of the gaudy emoluments—travel and consulting—which have recently glamorized other branches of academic life." In the major universities, the project-grant system also intensified the long-developing ascendancy of research over teaching, the newer premium of entrepreneurial skill over university citizenship. "On some campuses," *The New York Times* editorialized, "a teacher's status is inversely proportional to the time he teaches, or to the amount of time he spends on university grounds at all."[9] Undergraduates in the large research-oriented universities increasingly complained that their professors, scientists and otherwise, were too busy with research, graduate students, or academic junketing to pay them any attention.

Complaints also mounted in the academic and industrial worlds about the distribution of federal R & D largess. In 1963 nine states on the east and west coasts received almost 71 percent of all such government contract dollars; California alone, with its large concentration of aerospace firms, absorbed almost 39 percent of the total. The geographical concentration was less marked in the distribution of federal funds for

[7] Quoted in *The New York Times*, Oct. 8, 1963, p. 26.

[8] The ONR finding is quoted in Greenberg, *Politics of Pure Science*, p. 289; Piel, in *The New York Times*, April 25, 1965, p. 52.

[9] Boroff, "A Plea to Save the Liberal Arts," *New York Times Magazine*, May 10, 1964, p. 18; *New York Times*, editorial, Jan. 3, 1964, p. 22.

science to universities. Still, almost 40 percent of such federal dollars went to only ten academic institutions, almost 60 percent, to twenty-five. The leading ten universities also drew a disproportionately large number of the graduate students holding federal fellowships.[1]

Best-science elitism accounted for a good deal of the geographical and institutional concentration. But in the mid-1960s, when the research and development cornucopia along Route 128 outside Boston was attributed to the scientific might of Cambridge, when in the words of an observer university science was "perceived as the great catalyst for the whole economy of a metropolitan district or geographical region," aspirants outside the favored regions were no longer inclined to accept a best-science rationale. "If all R & D grants are to be placed strictly where the greatest competence is . . .," Senator Gaylord P. Nelson of Wisconsin declared, "all the competence would become pretty well concentrated creating greater and greater imbalance in the Nation."[2]

Besides, whatever the best-science explanation, it was commonly noted that the National Aeronautics and Space Administration located a widely sought electronics research center in Cambridge after the election to the Senate of Edward M. Kennedy, whose campaign slogan had claimed that he could do more for Massachusetts. Few commentators doubted that the benefits heaped upon Texas by NASA had borne some relationship to the power of Senate Majority Leader, then Vice President, Lyndon B. Johnson or of Congressman Albert P. Thomas, whose sub-committee oversaw the space budget. Students of academic science also noted that the leading ten university recipients of federal funds supplied 40 percent of the advisers used by government to review research proposals. Scientists themselves, including some of the best, suspected that the peer review system sometimes operated with cliquish favoritism. Scientific research, the acrid wisecrack went, was the only pork barrel for which the pigs determined who got the pork.[3]

A disproportionately large fraction of the money for the physical sciences went to physics. Some observers charged that the reason was the dominance by physicists of the scientific community's governing elite. Whatever the reason, advocates of other disciplines wanted a larger portion of R & D expenditures. But now that the size of the federal R & D budget was itself becoming an issue, a growing number of analysts were coming to understand with Emmanuel G. Mesthene, the head of

[1] *New York Times*, Sept. 11, 1964, p. 11; Lapp, *The New Priesthood*, p. 22; Milton Lomask, *A Minor Miracle: An Informal History of the National Science Foundation* (Washington, D.C., 1976), p. 133.

[2] Donald R. Fleming, "The Big Money and High Politics of Science," in William R. Nelson, ed., *The Politics of Science: Readings in Science, Technology, and Government* (New York, 1968), pp. 301–2; Nelson is quoted in Anton G. Jachim, *Science Policy Making in the United States and the Batavia Accelerator* (Carbondale, Ill., 1975), pp. 92–93.

[3] Lapp, *The New Priesthood*, p. 20; Greenberg, *Politics of Pure Science*, p. 151.

the Harvard University Program on Technology and Society: "To spend a billion dollars on physics and a million on archaeology is not an indifferent choice . . ., even when the purely scientific claims and credentials of the archaeologist and the physicist are equally good." The more analysts recognized that the public treasury could not support all scientifically meritorious projects, the more they called for the "establishment of priorities," in the phrase of the day, not only within science but for science in relation to other national needs. "Insistence on the obligation of society to support the pursuit of scientific knowledge for its own sake," the University of Chicago economist Harry G. Johnson mused, "differs little from the historically earlier insistence on the obligation of society to support the pursuit of religious faith, an obligation recompensed by a similarly unspecified pay-off in the distant future."[4]

When Apollo 11 landed on the moon in July 1969, many Americans could understand the reaction of John Furst, a student at the University of Pennsylvania, who was "very proud" when he saw "that spaceship and the men with the flags on their sleeves. But I must confess that I also thought of all the people who live in the ghettos. . . . The flag may be flying on the moon, but it is also flying in their neighborhoods, where there are poverty, disease, and rats." Pollsters discovered that Americans ranked the space race far lower in importance as a national problem than water and air pollution or job training and poverty. Commentators also found a growing number of Americans worrying with Admiral Hyman G. Rickover that, uncontrolled, the exploitation of science might well turn into "a Frankenstein monster, destroying its creator."[5]

Although the United States signed the test ban treaty, the Atomic Energy Commission kept producing enough nuclear weapons to destroy the major cities of the world more than a hundred times over. Research and development might produce the technological marvels of the affluent society, but transistor radios made cacophony a commonplace in buses and parks. Together with miniaturized circuits, they also made possible the bugging revolution—"nobody is safe anywhere," said the Boston investigator Andrew J. Palermo—data banks, the threat of what Professor Arthur R. Miller of the University of Michigan Law School called a "dossier dictatorship." While computers eased the burden of record-keeping and paperwork, automation destroyed more than forty thousand blue- and white-collar jobs a week, and filled still more jobs with numbing boredom. William C. Jensen, who monitored a $160,000 machine that drilled, shaped, and capped solid metals into complex missile parts, sighed:

[4] Mesthene, "Can Only Scientists Make Government Science Policy?" in Nelson, ed., *The Politics of Science*, pp. 460–61; Johnson is quoted in *Science*, 148 (April 1965), 608–9.

[5] Furst is quoted in *Time*, 94 (July 25, 1969), 16; Rickover in *The New York Times*, Oct. 25, 1965, p. 39.

"You spend most of your time loading it, rewinding the tape, and sometimes you spend 45 minutes just watching it."[6]

In many major cities, it was often a trial to make a telephone call, that is, if one could get a dial tone. Detergents poisoned the waters; pesticides, the natural environment; auto exhausts, the air itself. The fragility of technological civilization was driven home when on November 9, 1965, the power failed in New York, blacking out eighty thousand square miles of the northeastern United States, and when summer after summer officials "browned" out sections of eastern cities because power supplies were insufficient to meet the air-conditioning demand. Experts raised apprehensive doubts whether technological society could continue to exploit natural resources, especially the resources that yielded energy, at a geometrically increasing rate. The answer to the energy crunch, various commentators and scientists declared, was nuclear power. But were reactors safe? Americans increasingly asked. Safe or not, would they not pollute the environment—air, water, and earth—with excess heat and radioactivity? In 1969 *Life* magazine concluded: "Perhaps the rush to go nuclear was premature."[7]

By the early 1970s many observers were wondering whether non-nuclear nations might go nuclear in a different way by diverting fuel from power plants into the production of atomic weapons. Despite pious promises of peaceful intentions, India acquired a nuclear power plant from Canada, then purified the fuel into the core of an explosive device. Terrorism also cast a pall over the once bright promise of nuclear power. By 1976 about 176 nuclear reactors were operating in the world; by 1980 the number was expected to reach 400. Richard Wilson, a Harvard physicist who specialized in nuclear safety, summarily warned: "One major problem of the nuclear age is how to prevent crazy people from making and exploding nuclear bombs."[8]

People, the critic John Leonard observed, tended to blame the perversions of modern technological society on single causes, but the cause most prevalently blamed, and with special virulence "among those of a literary or humanist sensibility," was science. "Science, which," in Leonard's itemization, "brought you technology . . .; science, which has steadily reduced the number of things for which God can be held accountable and thereby pinned the rap on man . . . that *science*. And those scientists." Pollsters found public confidence in scientists rapidly falling, down by 1971 to a "very favorable" rating of only 37 percent. President Lyndon B. Johnson reminded an audience gathered at the White

---

[6] Palermo is quoted in *Time*, 83 (March 6, 1964), 55; Miller in *The New York Times*, Feb. 24, 1971, pp. 1, 24; Jensen, in *Life*, 55 (July 19, 1963), 68 B.

[7] *Life*, 67 (Sept. 12, 1969), 32.

[8] Quoted in *The New York Times*, Jan. 3, 1976, p. 21. In 1976 John A. Phillips, a Princeton physics senior, designed a plutonium bomb using shrewd guesses and unclassified information. *Princeton Alumni Weekly*, Oct. 25, 1976, p. 6.

House for the National Medal of Science award ceremony: "An aggrieved public does not draw the fine line between 'good' science and 'bad' technology. . . . You and I know that Frankenstein was the doctor, not the monster. But it would be well to remember that the people of the village, angered by the monster, marched against the doctor."[9]

The humanists had their traditional responses to the hegemony of science. There was the distressed complaint of the university president that "the age of the cultivated man has passed; the age of the competent man is here"; the solemn calls for a renewed emphasis on values over techniques, ethics over engineering; the alarmed appeals for the liberal arts (without them, David Boroff warned, "we run the risk of producing a generation of mindless technicians, specialized boors or even sinister Dr. Strangeloves").[1] There were also the admirable learned treatises, including Lewis Mumford's eloquent *Pentagon of Power*. The trouble was, Mumford argued, that since 1940 a scientific-military-industrial machine had emerged whose purpose was to subordinate human purposes to its own mechanical aims.

Often borrowing humanist metaphors, other dissidents followed the lead of Mario Savio, the mathematics graduate student who, amid the Free Speech Movement at Berkeley in 1964, likened the university to a machine and cried to a throng occupying the administration building: "It becomes odious, so we must put our bodies against the gears, against the wheels . . . and make the machine stop until we're free." On the leading campuses by 1970, numerous students were opting to defy the machine by denying its culture. Astrology, mysticism, the occult, drugs, psychedelic art, acid rock, sexuality—all became staples of the counter-cultural wave. So did Charles Reich's *The Greening of America*, with its celebration of the new consciousness of love, flowers, and under-standing. The movement spread rapidly beyond the campuses, notably to trend-setting sectors of upper middle class suburbia. Some analysts declared the counterculture a form of neoromanticism, others found its roots in the communal offshoots of the transcendental movement. Whatever its precedents, in the apt observation of the sociologist Robert Nisbet, advocates of the counterculture implied both the existence and the value of a "kind of knowledge beyond the reach of science, indeed of reason, logic and research."[2]

The chief countercultural prosecutor of science was the young historian Theodore Roszak, who called high-energy accelerators an

[9] Leonard, "The Last Word: Should Science Be Shot?" *New York Times Book Review*, July 18, 1971, p. 31; *Public Papers of Lyndon B. Johnson, 1968–69*, p. 220.

[1] Boroff, "A Plea to Save the Liberal Arts," p. 78, which also quotes the university president, p. 75.

[2] Savio is quoted in William L. O'Neill, *Coming Apart: An Informal History of America in the 1960's* (Chicago, 1971), p. 280; Nisbet, "Knowledge Dethroned," *New York Times Magazine*, Sept. 28, 1975, p. 36.

example of "major scientific talent taking expensive advantage of the public gullibility." He indicted science for providing "the image of nature that invited the rape [of the environment] and . . . the sensibility that has licensed it." "Undeniably," Roszak conceded, "those who defend rationality speak for a valuable human quality. But they often seem not to realize that Reason as they honor it is the Godword of a specific and highly impassioned ideology," a heritage of the Enlightenment, a hand-maiden to the "aggressive . . . urban industrialization of the world . . . to the scientist's universe as the only sane reality . . . [to] an unavoidable technocratic elitism."[3]

Humanists and counterculturalists, protesters and political analysts—all found fault with the governing elite of post-World War II science. Advocates of science might long have claimed that the scientific pro-fessions would provide a corps of disinterested experts operating in the public interest. But "far from providing a counterforce to the business system," the cultural historian Leo Marx asserted, "the scientific and technological professions in fact have strengthened the ideology of American corporate capitalism, including its large armaments sector." Proponents of the National Academy might always have assumed that it institutionalized scientific disinterestedness. Former Secretary of the Interior Stewart Udall attacked the institution for functioning "all too often as a virtual puppet of the Government," and for failing to serve as an "independent voice." The Columbia sociologist Amitai Etzioni, normally a friend to the scientific community, took physicists to task for having "impeded" the treatment of social problems by restricting the access of social scientists to the ears and pocketbook of the government.[4]

In the government proper, the Atomic Energy Commission often seemed more an interested advocate of nuclear power than a disinterested guardian of the public safety. The Food and Drug Administration ap-peared at times to ignore evidence detrimental to drug manufacturers. NASA Administrator James E. Webb, following the flash fire that killed three astronauts, refused to disclose a pertinent report to Congress about his agency's agreements with North American Aviation on grounds that disclosure might destroy the "intimate and confidential" nature of NASA relationships with its contractors.[5] But nothing did more to raise questions about the power of the scientific elite in its relations with industry and government than the war in Vietnam.

The war threw a powerful spotlight on the enormous influence of the Defense Department in academic science. In 1964 defense agencies

[3] Roszak, "Science: A Technocratic Trap," *Atlantic Monthly*, 230 (July 1972), 58, 59; Roszak, "Some Thoughts on the Other Side of This Life," *New York Times*, April 12, 1973, p. 43.

[4] Marx, "American Institutions and Ecological Ideals," *Science*, 170 (Nov. 27, 1970), 948; Udall is quoted in *The New York Times*, Dec. 31, 1970, p. 6; Etzioni, in *ibid.*, Feb. 3, 1971.

[5] Quoted in *The New York Times*, editorial, April 19, 1967, p. 44.

supplied R & D funds to at least one hundred universities, fifty of which received more than $1,000,000, ten more than $5,000,000, three more than $10,000,000. MIT led the list with almost $47,000,000 worth of contracts; the California Institute of Technology was fifteenth with $4,232,000, which, while less than one-tenth the MIT figure, amounted to 20 percent of the school's budget. Not long after his antiwar presidential bid, Senator Eugene J. McCarthy of Minnesota decried the influence of defense-oriented research contracts on American science and technology. "They can determine what research shall be carried out. More subtle, but perhaps more important, is the danger that academic institutions may begin to tailor their whole direction and approach to court these research grants."[6]

Many critics of the war judged any scientist suspect who associated with the Defense Department, no matter in what capacity, or what policy he advocated. The indictment of scientists for guilt by such associations was particularly prominent among the group that identified itself as the New Left. Some of the New Left shared, while some did not, the counter-cultural antipathy to science. More important, distinct from liberals who opposed the war, the New Left was prone to ideological vision and Marxist rhetoric; they attributed most of the turmoil in the world to the military-industrial complex and its "exploitation" of science. New Left war critics typically focused special attention on a branch of the Institute of Defense Analysis, the Jason Division.

Named after the Jason of Greek mythology, the division had been established in 1958 to bring the imaginative talents of stellar young scientists, mainly physicists, to bear upon the problem of national defense. The Jason roster included Marvin L. Goldberger, a particle theorist at Princeton, and Murray Gell-Mann, as well as such experienced advisers from the Los Alamos generation as George Kistiakowsky and Jerome Wiesner. In addition to several short meetings each year, some forty Jasons, many with their families, would spend six or seven summer weeks pondering defense issues at Woods Hole, Massachusetts, or La Jolla, California. Initially they dealt with such strategic matters as antiballistic missile systems. As the war in Vietnam escalated, some, like the Harvard physicist Steven Weinberg, declined to contribute to that effort. Others turned to questions of counterinsurgency, insurrection, and infiltration. What most infuriated Jason's critics was its contribution to the development of an electronic barrier, a system of electronic sensing devices to be strung across a no-man's-land between North and South Vietnam that would blow up anyone, civilian or combatant, who ventured into the area.

[6] Ralph E. Lapp, *The Weapons Culture* (New York, 1968), pp. 196–97; McCarthy, "The Pursuit of Military Security," *Saturday Review*, 51 (Dec. 21, 1968), 10. A study of actual research projects by the Stanford Workshop on Political and Social Issues concluded: "The influence of the military has skewed the direction of research at Stanford." Quoted in *Science*, 175 (Feb. 25, 1972), 866.

Jason's defenders pointed to the group's considerable work in opposition to the strategic arms race. No matter, declared Fred Bramfman, the director of an antiwar research group in Washington, D.C. Those efforts only meant that the Jasons "are lesser, rather than greater, war criminals. They are dramatic examples of how it is possible to be a moderate, well-meaning, decent war criminal."[7]

The extravagant rhetoric was matched by impassioned action. For several days, students occupied the Pupin Physics Laboratory at Columbia to protest, among other things, the Jason membership of some of the professors. MIT demonstrators, including a number of the Institute's own students, demanded the end of weapons development at the instrumentation laboratory on the campus. At the University of Wisconsin, a bomb blew up part of a building shared by the physics department and the Army Mathematics Research Center, killing a thirty-three-year-old postdoctoral student. The next year two bombs ripped apart a section of the Stanford Linear Accelerator. In 1970, at the Chicago meeting of the American Association for the Advancement of Science, Herbert T. Fox, a forty-year-old physicist, rose in the name of "Science for the People" to indict Glenn T. Seaborg for playing "a self-serving and ruthless role" in shaping for "use by the ruling class" such institutions as the Atomic Energy Commission, the Department of Defense, and the University of California, where "the minds of the students are bent to the needs of Empire."[8] When the radicals crowded into the room where Seaborg was due to make his address as retiring president, the AEC chairman, following the advice of convention officials, fled through a side door.

Amid the melee, Mrs. Garret Hardin, the wife of the celebrated biologist, jabbed a heckling radical in the arm with her knitting needle, drawing a little blood. "It felt kind of good," she told reporters later. Numerous Americans may have sympathized with Mrs. Hardin's outrage at the frequent rudeness and violence of the radical protesters of the day. But many of the same people had also come together in a new revolt against science. It was a revolt made far more virulent and intense than that of the 1930s because so many outside the New Left believed that the nation's scientific elite was "an implement—and even a cause of war," in the early analysis of the journalist Evert Clark, "to be dismantled or at least blamed for all our current ills and punished in some way."[9]

The virulence of the revolt shook the leaders of the Los Alamos generation. Was there an advisory elite dominated by physicists and

---

[7] Steven Weinberg, "Reflections of a Working Scientist," *Daedalus*, 103 (Summer 1974), 35; Bramfman is quoted in "Jason Division . . .," *Science*, 179 (Feb. 2, 1973), 460.

[8] Quoted in the Los Angeles *Times*, Dec. 31, 1970, p. 4.

[9] Mrs. Hardin is quoted in *Science*, 171 (Jan. 8, 1971), 48; Clark, "The Scientific-Government-Industrial Machine," *New Republic*, 155 (Aug. 27, 1966), 32.

drawn from a narrow range of institutions? A matter of historical accident, retorted Harvey Brooks, the dean of engineering at Harvard and chairman of the National Academy Committee on Science and Public Policy. In any case, he added, it was justified if you wanted competent advice. When Peter B. Hutt, the onetime legal counsel of the Food and Drug Administration, called for public access to the deliberations of the Academy itself, President Philip Handler responded with blunt derision. "We choose the members of our committees with extreme care. We have no sense of participatory democracy. This *is* an elitist organization, sir."[1]

Were federal R & D funds unevenly distributed? "To achieve high quality results," Leland Haworth, the physicist who succeeded Alan Waterman as director of the National Science Foundation, understandably told a congressional committee, "requires going where the best capability exists. That capability is now quite concentrated geographically. . . ." Glenn Seaborg ridiculed the notion of a geographical distributional standard in which every congressional district was assumed to deserve a post office, a reclamation project, and a major scientific laboratory. To the rising insistence in an era of limited budgets upon the necessity for scientific choice, a panel of the President's Science Advisory Committee pronounced it impossible to assign relative priorities to different fields, adding that each should be funded according to its "needs." During his tenure as President Kennedy's Science Advisor, Jerome Wiesner believed that one of the most important functions of the Science Advisory Committee was to "protect the anarchy of science."[2]

The claim that science was not serving the social needs of the nation tended to elicit vociferous expressions of faith in a trickle-down approach reminiscent of Millikan—that the best way to deal with the problems of technological society was to invest in basic research. James R. Schlesinger, the RAND Corporation economist and future secretary of defense, only partly caricatured the prevailing rhetoric when he summarized the chief theses in a book of Wiesner's: "(1) Science is Good; (2) Progress is Good; (3) Money is Good for Science; (4) Science is Good for Peace; (5) Peace is Good for Science; (6) Money, Science, Peace, and Progress are Good for Impoverished Nations." When Charles C. Price, the president of the American Chemical Society, called for a crash research program to create life, *The New Yorker*'s normally impish "Talk of the Town" column declared the proposal "at least as much

[1] Brooks, "The Science Advisor," in Gilpin and Wright, eds., *Scientists and National Policy-Making*, pp. 82–83; Handler is quoted in *Science*, 191 (Feb. 13, 1976), 543.

[2] Haworth is quoted in Jachim, *Science Policy Making*, pp. 89–90; Seaborg, in *The New York Times*, March 15, 1964, p. 39; the panel and Wiesner, in Greenberg, *Politics of Pure Science*, pp. 231–32, 151.

concerned with salesmanship . . . as with high intellectual enterprise. To our layman's ear, his words hold a now familiar ring of scientific sell."[3]

To many lay Americans, the nation's scientists seemed to respond to the dissidence mainly with a panacea: more money. With more funds, one could create new "centers of excellence," to use the current best-science phrase, in regions where none existed, raise the humanists' academic standard of living, give university administrators institutional grants under their control, subvene every scientific field of merit and promise. Among the scientific statesmen of the mid-1960s, the argument was commonly advanced that federal funding of academic science should rise at least 15 percent a year, indefinitely. But if the leadership of the Los Alamos generation remained committed to political and best-science elitism, not to mention limitless growth, its response to the antiscientific revolt of the day differed significantly from that of Millikan's generation to the rebellion of the 1930s.

Among natural scientists, physicists constituted the most politically liberal group. Compared to the rank-and-file of the profession, too, the leading academic physicists tended to be more liberal still. Four out of five of the profession's elite disapproved of classified research on the campuses and opposed the administration's policies in Vietnam. Two out of three approved of the emergence of radical student activism.[4] Most, including some of the minority who belonged to the New Left, disapproved of violence and disruption, and many worried that the more extreme radicals would provoke a reaction destructive to the institutional structure of science. But if they advocated the ongoing advancement of basic knowledge, they rejected a trickle-down approach to social betterment. They endorsed governmental action to ensure that science was used to good social purpose and asserted that the social responsibility of the scientist properly included responsibilities in domestic as well as in foreign affairs. From the later 1960s on, many allied themselves with the movement against the wrongs in the prevailing society, determined all the while to maintain what they considered right in their traditional standards of professional responsibility, conduct, and autonomy.

The President's Science Advisory Committee, with its heavy complement of physicists, received briefings on the war every month. "We often gave the briefers a bad time," Herbert York recalled. There were "frequent and occasionally nasty arguments over the effectiveness of the bombing, over who really controlled what territory, over the idea of the stockade type villages, [over] where . . . the other side really got its weapons. . . . Perhaps half the PSAC was clearly and vocally though

---

[3] Schlesinger review, *The Reporter*, 33 (Aug. 12, 1965), 48; *New Yorker*, 41 (Oct. 9, 1965), 43.

[4] "Survey Finds Physicists on the Left," *Physics Today*, XXV (Oct. 1972), 61–62.

always in private opposed to the war. Some of us would even caucus occasionally about whether or not there would be any value in some sort of loud resignation."[5] George Kistiakowsky, the former Presidential Science Advisor and one of the most respected leaders of the Los Alamos generation, did resign, and, though refraining from denunciations, let it be known that his departure was impelled by the administration's policies in Vietnam. At the end of the 1960's, scientists in and out of the post-Hiroshima advisory elite formed an inconspicuous exodus from Washington committees.

At MIT early in 1969, some of the most distinguished professors of physics supported the proposal of a few of the graduate students in the department for a "strike" day on March 4. The professors were careful to explain publicly that this would not be a strike against the university. It would be a work stoppage to call attention, in the language of the faculty's ultimate announcement, to how the "misuse of scientific and technical knowledge presents a major threat to the existence of mankind." The March 4 strike quickly turned into a movement that spread to more than thirty universities. At MIT the day was highlighted by a rousing speech from George Wald, the Harvard Nobel laureate in biology, who won a standing ovation for denouncing as "criminally insane" Senator Richard B. Russell's recently expressed eagerness to ensure that after a nuclear war the new Adam and Eve would be Americans.[6] But for the most part, at MIT and elsewhere, the day was occupied with provocative debates about the responsibilities of intellectuals, the social uses of science, the academic community in its relationship to the federal government, and arms control. Especially notable everywhere was the degree to which senior professors of the Los Alamos generation made clear how misuse of science troubled them as much as it did their students.

The dissidence among the physicists had also become manifest in the affairs of the American Physical Society, initially through the efforts of Professor Charles Schwartz of Berkeley. In 1967 Schwartz, who would later propose that all his students sign a Hippocratic oath pledging not to use their science for the "harm of man," urged the Society to amend its constitution so that it could take a position on the war.[7] The amendment was resoundingly defeated by well over two to one. In 1971 Professor Robert March of the University of Wisconsin proposed another amendment—that the Society should "assist its members in the pursuit of . . . humane goals and . . . shall shun those activities . . . judged to contribute harmfully to the welfare of mankind." The members voted down this proposal by a solid majority, not least because they had no idea who would do the judging. Dr. Eugene J. Saletan of Northwestern University

---

[5] York to the author, May 8, 1974.

[6] The faculty statement of purpose and Wald's speech are in Jonathan Allen, ed., *March 4: Scientists, Students, and Society* (Cambridge, Mass., 1970), pp. xxxii, 113.

[7] Schwartz, letter to the editor, *Daily Californian*, April 1, 1970.

expressed the opposition common to both amendments. Though by his own description an "adamant 'extremist'" against the war in Vietnam, he believed that the Society "should remain pure."[8] But in the nationally tumultuous period between the Schwartz and March amendments, the Society did yield to demands for the discussion of controversial political issues. The governing council authorized the establishment of a Forum on Physics and Society, whose sessions attracted substantial crowds, and the council scheduled a debate for the April 1969 Washington meeting on the technical aspects of the antiballistic missile issue.

Over the preceding year, physicists had been taking an increasingly vociferous public stand against the deployment of any ABM, including the Nixon administration's proposed Safeguard system, which would be limited to the protection of American missile installations against a pre-emptive Soviet strike. At MIT on March 4, Hans Bethe had spoken for most of the Los Alamos generation when he explained concisely just why on technical grounds even Safeguard would not work. Now, in Washington, the American Physical Society session on ABM drew more than two thousand people and went far beyond strictly technical issues. Afterwards, 250 physicists marched from the hotel to the White House. A delegation was admitted to discuss the issue with President Nixon's Science Advisor, while on Capitol Hill other delegations visited more than sixty senators to petition against the deployment of Safeguard.[9]

With less vigor, prominent American physicists also responded to the gathering force of the women's rights movement. In part, they recognized the legitimacy of the complaint of Dr. Betsy Ancker-Johnson, a staff member of the Boeing Research Laboratory and professor at the University of Washington: "A woman in physics needs to be at least twice as determined as a man with the same competence to achieve as much."[1] In part, like academics in other fields, they felt a practical goad from the suit filed in 1970 by the Women's Equity Action League and the National Organization of Women. Charging that forty-three colleges and universities discriminated in employment against women, the two groups demanded that the government rectify the situation by using its civil rights enforcement powers, especially the power to cancel all federal contracts held by institutions found to practice illegal discrimination.

In 1971 the American Physical Society sponsored a session on women in physics that drew a standing-room crowd of some six hundred people. The traditional obstacles to enlarging the female contingent of the pro-

---

[8] The March amendment is quoted in *Physics Today*, XXV (Nov. 1972), 42; Saletan, in *New York Times*, July 14, 1968, p. 20.

[9] Joel Primack and Frank von Hippel, *Advice and Dissent: Scientists in the Political Arena* (New York, 1974), pp. 181, 187–88; Bethe's speech is in Allen, ed., *March 4*, pp. 142–50; Barry M. Casper, "Physicists and Public Policy: The 'Forum' and the APS," *Physics Today*, XXVII (May 1974), 31–33.

[1] Quoted in Gloria B. Lubkin, "Women in Physics," *Physics Today*, XXIV (April 1971), 24.

fession were clear to the panelists, not only discrimination but prevailing cultural standards and the inevitable conflicts between family and career. Ancker-Johnson conceded that she did not know how she would have managed without domestic help to care for her children in their preschool years. Still, the male as well as female participants agreed that, as a matter of simple justice, women deserved as much opportunity as men to be physicists. Less convincingly, Professor Chien-Shiung Wu of Columbia, one of the two female physicists in the National Academy, advanced a scientific version of the venerable yet still dubious argument of women's rights advocates. Men had "brought us to the gigantic brink of environmental ruin. . . . Women's vision and humane concern may be exactly what is needed . . . to warn us of chemical pollution . . . occupational health hazards . . . and drugs like thalidomide."[2]

Whatever women might—or might not—do, there was scarcely a technologically threatening issue on which scientists did not help to raise an alarm. In 1972 thirty-one of them, including Harold Urey, urged the Congress to deny the administration's request for funds to begin the construction of a model breeder reactor to generate electrical power. Scientists crowded professional gatherings on the problems of the environment. The biologist Paul Ehrlich warned: "We must realize that unless we are extremely lucky, everybody will disappear in a cloud of blue steam in 20 years." James D. Watson, the Nobel laureate in biology, predicted that when scientists managed to conceive a baby in a test tube —and the event, Watson asserted, was imminent—"all hell will break loose." Do we really want to do this? he asked. At Harvard, the young molecular biologist Jonathan Beckwith, bearded and dressed in flarebottomed trousers, told a press conference that the possibilities in genetic manipulation were "frightening—especially when we see work in biology used by our Government in Vietnam and in devising chemical and biological weapons."[3]

Of course, some prominent scientists, young and old, physicists and nonphysicists alike, opposed the March 4 movement, defended ABM, and encouraged the chief of a naval weapons laboratory to say: "We are having no trouble recruiting fine, serious young men who want to contribute to their nation's defense efforts. Maybe there is a hidden majority after all."[4] Still, the outspokenness of a Beckwith or a Bethe reflected a

[2] Quoted *ibid.*, p. 23.

[3] Ehrlich is quoted in Leo Marx, "American Institutions and Ecological Ideals," 946; Watson, in Los Angeles *Times*, Jan. 29, 1971, p. 6; Beckwith, in *The New York Times*, Nov. 23, 1969, p. 72.

[4] Quoted in *The New York Times*, Sept. 13, 1970, p. 74. During one of the protests at MIT, a conspicuously clean-shaven graduate student told a reporter that he felt no personal guilt about his work. "What I'm designing may one day be used to kill millions of people. I don't care. That's not my responsibility. I'm given an interesting technological problem and I get enjoyment out of solving it." Quoted *ibid.*, Nov. 9, 1969, p. 61.

trend away from an overly secretive approach to public policy, not only among younger, liberal and New Leftish scientists but among leaders of the Los Alamos generation itself. John W. Gofman, an AEC staff member at Berkeley who opposed nuclear reactors, declared: "This is not a matter to be decided in back rooms by a little group of 'experts.' . . . Every citizen in the United States has a right to know the risks first."[5] Yet the willingness of scientists ranging from Bethe to Beckwith to speak out revealed more than a commitment to open politics. It revealed how much so many devotees of the postwar scientific citadel sympathized with the rebellion against it or against developments with which it could be associated.

[5] Quoted in the Los Angeles *Times*, Jan. 27, 1970, p. 13.

XXV

A Degree of
Disestablishment

I n January 1966, Philip Abelson rightly lamented: "A twenty-year
honeymoon for science is drawing to a close."[1]

As the rebellion against science mounted, liberals grew less
willing to accept a degree of political elitism that insulated the
governance of science from accountability, channeled a lavish investment
of public money into pure instead of socially useful research, and permitted the Defense Department so large an influence in the academic
world. Conservatives quarreled with political elitism, too, not least
because so many members of the prevailing elite opposed the war in
Vietnam or the ABM and seemed too willing to tolerate disruptions on
the campuses. For Americans at most points of the political spectrum, the
time had come to halt the geometrical growth rate in the federal expenditure for R & D. If liberals wanted to shift money into such areas as
pollution control, conservatives simply wanted to spend less money. In
the scientific have-not regions of the country, no special political convictions were required to join against a system that operated in accord
with the tenets of best-science elitism. Withal, in the mid-1960s, in a way
reminiscent of the early 1890s or the 1930s, liberals, conservatives, some
of the New Left, and numerous other Americans formed a coalition committed to shifting the nation's scientific program from a politically
elitist to a politically responsive enterprise, from a best-science to a
geographically more even distribution, from luxury to leanness in level

[1] Quoted in John Fischer, "Why Our Scientists Are About to Be Dragged,
Moaning, into Politics," *Harper's*, 234 (Sept. 1966), 16.

of funding, from pure to applied research, from the goal of scientific advancement as such to that of socially purposeful utility.

The shift found an early presidential response when John F. Kennedy told the National Academy: "Scientists alone can establish the objectives of their research, but society, in extending support to science, must take account of its own needs."[2] But the new coalition gained considerably fuller expression in the White House of Lyndon B. Johnson, a reform President with a sharp eye for the dollar, a Texas teachers-college graduate who was, to say the least, ill at ease with the Northeastern establishment, scientific or otherwise, and a politico who never forgot the votes that could be corraled by being, as he liked to say, "truly national."

By 1965 Johnson had diversified the membership of the President's Science Advisory Committee, leaving only one member from Cambridge, Massachusetts. He also appointed the first black member of the Atomic Energy Commission—Samuel M. Nabrit, the president of Texas Southern University, a Ph.D. in biology from Brown University. In September 1965, the President, declaring that federal R & D funds were "concentrated in too few institutions in too few areas of the country," directed the federal establishment to share the wealth and enlarge the university administration's latitude in the use of grant money. The Johnson administration, elaborated Presidential Science Advisor Donald F. Hornig, hoped to strengthen academic capability throughout the country. It also aimed to eliminate the leverage of individual "prima donnas" whose independence detracted from the overall educational values of the university.[3]

Tell me what science can do for Grandma, Lyndon Johnson would query his science advisers. In June 1966, near the start of Medicare, he declared: "A great deal of basic research has been done. . . . I think the time has now come to zero in on the targets by trying to get this knowledge fully applied."[4] The last Johnson budgets kept the rate of growth for federal R & D expenditures down to a few percent annually, far lower than the post-1967 rate of inflation. According to the later report of William D. Carey, LBJ, his relations with the academic community steadily worsening as a result of the war, would peruse the budgetary documents, blue-penciling out items for university research. Scientists opposed to the war earned positions on a Department of Health, Education, and Welfare blacklist of people to be excluded from scientific advisory committees.[5]

When Richard Nixon took office in 1969, the scientific community

---

[2] *Public Papers of . . . John F. Kennedy, 1963*, p. 804.

[3] *Public Papers of . . . Lyndon B. Johnson, 1965*, p. 996; Hornig is quoted in *Science*, 149 (Sept. 24, 1965), 1484.

[4] Herbert York to the author, May 8, 1974; *Public Papers of . . . Lyndon B. Johnson, 1966*, p. 610.

[5] *Science*, 171 (March 5, 1971), 874; *New York Times*, Oct. 9, 1969, p. 1.

hailed his appointment as Presidential Science Advisor of Lee A. DuBridge, since 1946 the president of the California Institute of Technology. Nixon also won approval for his order increasing the budget of the National Science Foundation by $10,000,000. But with inflation, not to mention the Nixon impoundments of congressional appropriations, the Nixon budgets actually reduced the real value of the federal R & D budget. Within the budget, the Nixon program stressed applied over basic research, a war on cancer over the advancement of fundamental biology, technology over science. For business, there was the supersonic transport; for defense, the ABM; and for energy, the fast-breeder nuclear reactor. DuBridge, always a staunch advocate of basic research, himself warned his colleagues: "The day is past when scientists and other scholars can sit quietly in their ivory towers. . . . Scientists must carefully ponder the relevance of their work to the problems of human beings."[6]

And to politics, DuBridge said in the same breath. In the spring of 1969, the Nixon administration withdrew from Franklin A. Long its tender of appointment as director of the National Science Foundation. A respected chemist at Cornell, Long was a member of the Los Alamos generation and of the President's Science Advisory Committee. But he was on record against the ABM, and his appointment had been opposed by Senate Minority Leader Everett Dirksen and Congressman James G. Fulton of Pennsylvania, who had his own candidate. The leadership of the scientific community was outraged at this injection of "politics" into the choice to fill the traditionally apolitical directorship of the Foundation. In a remarkable reversal, the President apologized to the National Science Foundation board in a meeting at the White House, conceded that his staff had been wrong, and offered the job to Long, who declined.[7]

Despite Nixon's reversal on Long, DuBridge found himself increasingly caught between the mutual hostility of his scientific colleagues and the Nixon White House. In 1970 he resigned, amid stories that his influence with the Nixon White House had steadily waned. His successor was Edward E. David, a twenty-year-old graduate student at the time of Hiroshima, now a computer expert from Bell Telephone Laboratories, the first nonacademic appointed to the post, who had never been on the President's Science Advisory Committee and was hardly a household name in the world of science. The Committee itself harbored grave doubts about ABM, and it reported adversely on the supersonic transport. At the opening of 1973, not long after Nixon had won his smashing electoral victory, David resigned. "Ed feels less than useful," a White House source reportedly said.[8] A month later, Nixon announced the abolition of the

[6] Quoted in The New York Times, Nov. 2, 1969, p. 31.
[7] Science, 164 (April 25, 1969), 406–11; 164 (May 2, 1969), 532.
[8] Quoted in The New York Times, Jan. 3, 1973, p. 1. Philip Abelson remarked of the demise of the President's Science Advisory Committee that its members, "being

Science Advisory Committee altogether, along with the Office of Science and Technology. The duties of science-advising would pass to H. Guyford Stever, the current and prudent director of the National Science Foundation—in the middle of the Long imbroglio, Stever had said: "No administration can stand within itself an activist against itself"—who would have no responsibility for defense technology. After 1971 Nixon declined to award any National Medals of Science, and his staff also seriously discussed cutting off research funds to MIT, the president of which was Jerome Wiesner, an outspoken critic of the ABM whose name later appeared on the celebrated White House "enemies list."[9]

In the Congress, both houses had come to include a body of members who, after years of experience with legislation for research and development, were, in the remark of Congressman Charles A. Mosher, "no longer overawed in the presence of . . . famous scientists."[1] Few congressmen, experts or not, were overawed by the demands for more R & D appropriations. In 1963 the House had ordered a full-scale investigation into federally funded science, the first since the Allison Commission in the 1880s; it was not the last in the 1960s. The investigations revealed no scandals, but they did call a good deal of attention to the unevenness in the geographical distribution of funds and to the general lack of social purpose in the nation's scientific programs.

The impact of the investigations was evident when in 1965 National Science Foundation officials appeared before Congressman Albert P. Thomas' appropriations subcommittee to testify in defense of the agency's budget requests. Henry W. Riecken, Jr., the associate director for education, explained that it was policy to award fellowships to the top 5 percent of graduate students in science, mathematics, and engineering. Thomas exploded: "One of the defects in the program . . . [is] that you pick them from half a dozen universities to the exclusion of everybody else. . . . You give these [other] people an opportunity to spread their wings and fly and they will do it."[2] Thomas' subcommittee summarily recommended a 10 percent limit on the number of fellowship recipients in any one state. It also substantially reduced the Foundation's request for funds to subvene research projects while it awarded the agency's full

---

an elite group and occupying a lofty station in the scheme of things, fell victim to a common human disease: arrogance. . . . Behind the scenes, PSAC attempted to wield great influence on the decisions and policies of the various governmental agencies. In the process, the part-time committee made full-time enemies. The major political blunder, however, was that members of PSAC occasionally disagreed publicly with the President." *Science*, 182 (Oct. 5, 1973), 13.

[9] Stever is quoted in *Science*, 175 (March 31, 1972), 1443; Los Angeles *Times*, July 14, 1973, p. 23.

[1] Quoted in Ralph Sanders, "The Autumn of Power: The Scientist in the Political Establishment," *Bulletin of the Atomic Scientists*, 22 (Oct. 1966), 24.

[2] Quoted in *Science*, 148 (May 14, 1965), 929.

request for monies to institutions. Only twice before had the Senate voted to abide by the House cuts in the Foundation budget. Now, though it eliminated the 10 percent restriction, it went along with the other reductions.

Congressional liberals had been looking into the operations of the Foundation with the aim of nudging it toward supplying money for applied research of social value. In 1969, Congress amended the organic act of the agency, in part to put it into applied research, in part to direct it to fund the social sciences, those disciplines traditionally considered instruments for the social control of technology. The Foundation inaugurated further socially pertinent efforts, including the program Research Applied to National Needs. RANN was conceived by a young theoretical physicist, Joel Snow, who was now dealing with matters far removed from theoretical physics. The chief of the program, Alfred C. Eggers, Jr., an ex-NASA official, reflected: "A lot of us are involved in things today that are not our first choice in life."[3]

In the spring of 1965, the House subcommittee for military research and development had recommended lopping off almost 25 percent of the basic research section of the defense budget. In part, the reason was economy, the members said. In part, they added, "retrenchment might, so far as colleges and universities are concerned, have a corollary benefit of making the best faculty more available for the purpose of teaching students."[4] But for quite different reasons, not the least of them his concern about the dependence of academic science on the military, in 1968 Senate Majority Leader Mike Mansfield of Montana launched an effort to limit the largess that universities could receive from the Defense Department.

Unsuccessful that year, in 1969 the senator successfully pushed through a section in the military authorization bill, named the Mansfield amendment, that prohibited the Defense Department from financing any research not directly related to a specific military purpose. Mansfield, a liberal, was joined in the House by L. Mendel Rivers, chairman of the Armed Services Committee, an unquestioned conservative and all-out supporter of the Pentagon, who asked why universities should enjoy the benefits of defense dollars if they were so unwilling to perform defense research. When the Mansfield amendment passed, academic scientists and administrators cried foul. By now, Lee DuBridge complained, the Defense Department supplied only about one-fifth of the federal dollars for all science on the campus. But the fraction was a good deal larger, about 40 percent, for just the physical sciences, and like Harold Smith or even Vannevar Bush a quarter of a century earlier, Mansfield preferred

[3] Quoted *ibid.*, 172 (June 25, 1971), 1317.
[4] Quoted *ibid.*, 149 (July 16, 1965), 280.

the civilian National Science Foundation over the military as a major patron of academic science.[5] The Mansfield amendment was dropped from the military authorization bill the following year. Yet its statement of congressional intent had a lasting impact, especially in the defense bureaucracy's interpretation of its latitude in research funding.

To relieve the pressure on basic science created by the withdrawal of military support, the Congress increased appropriations to the National Science Foundation, by $85,000,000 in the year of the Mansfield amendment. But a number of meritorious basic projects were lost in the shuffle because part of the increase was for socially useful purposes. Expressing the wider shift to social concerns, the NASA electronics laboratory that Edward Kennedy had brought to Massachusetts went into transportation work; the Lawrence Radiation Laboratory initiated research in environmental problems; and Los Alamos set aside part of its resources for research in energy. Department of Defense expenditures for research on the campuses were reduced, especially on the basic research side. The campuses themselves dropped secret work for the Defense Department; at MIT the administration divested itself of the controversial instrumentation laboratory (while deciding to keep an electronics research laboratory which did all its business with the Defense Department but whose programs had not been attacked). By 1974 the Defense Department was spending only $108,000,000 on basic academic research, down in both real and absolute dollars from the $137,000,000 of 1965. Only 4 percent of the total was classified.[6]

As early as 1965 Don K. Price had recognized the historical import of the shift in White House and congressional policy for science. At the end of World War II, he recalled, "the Kilgore bill was defeated. Moreover, it was generally forgotten through almost a positive effort on the part of the scientists. . . . But its central notions are slipping up on us again rapidly. . . ."[7] And forcefully so, Price might have added had he written a few years later. By the 1970s Kilgore's, not Bush's, program held the greater sway on the federal scientific scene. By 1973 federal R & D budgetary obligations had risen to $17 billion, but inflation had driven their real value 18 percent below that of 1967.[8] Many members of the scientific elite, all too aware of the decline in their power and their

[5] "DuBridge and His Critics," *Science*, 169 (July 24, 1970), 356; National Science Foundation, *Federal Funds for Academic Science, Fiscal Year 1969* (NSF 71-7, 1971), pp. 9, 12.

[6] National Science Foundation, *Federal Funds for Research, Development . . . Fiscal Years 1964, 1965, and 1966* (NSF 65-19, 1966), p. 22; National Science Foundation, *Federal Funds for Research, Development . . . Fiscal Years 1972, 1973, and 1974* (NSF 74-300, 1974), p. 11; Dorothy Nelkin, *The University and Military Research: Moral Politics at M.I.T.* (Ithaca, N.Y., 1972), pp. 134-37.

[7] Price, "Federal Money and University Research," *Science*, 151 (Jan. 21, 1966), 289.

[8] *Federal Funds for Research, Development . . .* (NSF 74-300), p. 2.

purse, knew that the post-Hiroshima honeymoon had definitely ended— and some were anxious about the future of the marriage.

The end of the honeymoon was plain in physics, especially high energy physics. In 1960 high energy physics drew $53,000,000 in federal funds; in 1964, $135,000,000. Costs were expected to increase to $500,- 000,000 by 1970, particularly because of the American physics community's expectations of still more powerful accelerators. Their plans for the immediate future included a machine of 200 billion electron volts, then later one of 600 billion electron volts. Design studies on both machines were already under way at Berkeley and Brookhaven, where it was assumed the behemoths would be constructed. The estimated costs of the 200 billion electron volt machine were $280,000,000 to build and $50,000,000 a year to operate.

The high energy physicists justified such an enormous public investment by citing the necessity of maintaining American leadership in the field (in the mid-1960s the most powerful accelerator in the world was the 70 billion electron volt Soviet device at Serpukhov). Besides, basic research, however esoteric, had often produced practical results before, and now Luke L. C. Yuan of Brookhaven indulged in the extravagant suggestion that if certain elementary particles and antiparticles could be separated, then brought together, one might produce an unprecedentedly enormous explosion.[9] Even if no practical results ensued, did not advanced fields of research improve the cultural fiber of the society, enrich its educational tone? And high energy physics was both scientifically laudable and ripe for further progress. But whatever the merits of such arguments, both President and Congress were counting closely high energy physics costs.

Jerome Wiesner told high energy physicists that, on grounds of economy, they would be unwise even to ask for, let alone expect to get, everything they wanted. In the Congress, Representative Chet Holifield of California, the chairman of the Joint Committee on Atomic Energy, emphasized in 1964: "There is no end to scientific ambitions to explore, but there is an end to the public purse." High energy physics was squeezing out other parts of the Atomic Energy Commission's program, Holifield complained, especially in the applied, practical area. Presidential Science Advisor Donald Hornig contended that high energy physics was one of the most revolutionary fields in science. Holifield scoffed: "Every scientist that comes before us thinks his crow is the blackest."[1]

In and out of the Congress, informed Americans wondered with John Fischer, the editor of *Harper's*, what the nation might expect to earn from its investment in the huge machine. So far as Fischer could

[9] *Physics Today*, XVIII (May 1965), 100.
[1] Quoted in *The New York Times*, March 4, 1964, p. 30.

discover, "nobody really knows. . . . Why, then, should we be in such a hurry to build this vastly expensive piece of specialized equipment—especially at a time when the economy is overheated, the budget strained, and scientific talent in short supply? Why shouldn't it be downgraded on the priority list—to be considered again in a few years or decades?"[2]

Eugene Wigner, one of the physics community's respected conservative statesmen, testified that high energy accelerators unfairly and, from the point of view of the national interest, unwisely deprived other, potentially more practical fields of money and manpower. Half a billion dollars had already been spent on high energy physics, Philip Abelson, a geochemist by training, pointed out, but apart from a number of Nobel Prizes, no practical return had come from the investment. "Never, in the history of science have so many fine minds been supported on such a grand scale, and worked so diligently, and returned so little to society for its patronage." The trouble, Abelson asserted, was that in deciding how to spend money the Atomic Energy Commission had relied principally upon the advice of high energy physicists. "This is, of course, like asking a hungry cat to make recommendations about the disposition of some cream."[3]

The most arresting critique of the nation's high energy program came from the director at Oak Ridge, Alvin Weinberg, who in the early 1960s had given currency to the phrase "big science." Over the years Weinberg had thought deeply about the measures of merit for making choices among expensive scientific fields. His criteria were, in summary, "relevance to the science in which it is embedded . . . to human affairs, and . . . to technology." Weinberg rated high energy research "poorly" on the degree to which it might illuminate neighboring scientific disciplines, including other parts of physics. On the scales of technology or social merit, he counted its contributions "nil."[4] Weinberg had diversified the Oak Ridge program to stress basic research appropriate to such socially utilitarian purposes as cheap energy sources, desalination, and environmental problems. If he had likened the mighty accelerators of the modern era to the pyramids of the ancients or the cathedrals of the Middle Ages, he knew what pyramids and cathedrals cost ordinary human beings. In making choices between fields of science, he argued, we should "remember the experiences of other civilizations. Those cultures which have devoted too much of their talents to monuments which had nothing to do with the real issues of human well-being have usually fallen upon bad days. . . . We must not allow ourselves, by short-sighted seeking after

[2] Fischer, "Why Our Scientists Are About to Be Dragged, Moaning, into Politics," p. 27.

[3] *Science,* 147 (March 19, 1965), 1426; Abelson, "Are the Tame Cats in Charge?" *Saturday Review,* 49 (Jan. 1, 1966), 101–2.

[4] Alvin M. Weinberg, *Reflections on Big Science* (Cambridge, Mass., 1967), pp. 78, 80.

fragile monuments of Big Science, to be diverted from our real purpose, which is the enriching and broadening of human life."[5]

Or at least not to be diverted from achieving geographical equity, cried the growing chorus. People, in the language of Congressman Craig Hosmer of California, were likening the proposed machine to a "200 BEV bonanza," "the government's biggest free offer to all comers since opening the Cherokee Strip to the homesteaders." Apart from the money it would bring, the accelerator would employ some two thousand scientists and technicians. They were the "real gold mine," a senior editor of *Look* exclaimed. "Such human capital, many businessmen know, creates new wealth in today's educated economy. Around Boston's Route 128 and San Francisco's Bay Area, research brains spin off waves of new products and services practical enough to make money, lots of it." Out in the state of Washington, a consultant to the Tri-City Nuclear Industrial Council told the Pasco Kiwanis Club that the accelerator would bring "literally thousands of small industries to the Tri-Cities."[6]

Regional competition for the proposed machine was intense. The accelerator would have to be located on a large site of flat land, with access to sufficient electrical power and to the kind of academic or metropolitan center where physics research would flourish and Ph.D. scientists would want to live. "Our problem," Paul W. McDaniel, the director of the Atomic Energy Commission Division of Research, gibed, "is to find a site where scientists can continue their little girls in ballet school."[7] The St. Louis *Post-Dispatch* promptly announced that more than forty ballet dance studios were listed in the city telephone directory. In communities around the country, local leadership groups twisted the arms of electrical utility companies to promise the Commission favorable power rates. Educators issued brochures about the local school systems, cultural attractions, and artistic efforts. ("We're upgrading the whole country," a Commission official marveled.[8]) By August 1965 the Commission had received formal site proposals from 117 communities in forty-six states.

Brookhaven had the country's second-largest accelerator, Berkeley its first; the regional campaign for the 200 BEV machine was especially intense in the Midwest, which had unsuccessfully been demanding a high energy accelerator for a decade. In the early 1960s, enlisting the aid of their congressmen, midwestern physicists had almost won approval of a big machine. President Johnson killed the enterprise in 1964, but Senator William Proxmire of Wisconsin, then a new member of the Appropria-

[5] Weinberg, "Impact of Large-Scale Science on the United States," *Science*, 134 (July 21, 1961), 164.

[6] Hosmer is quoted in *The New York Times*, March 23, 1966, p. 38; George Harris, "The Great Atom Smasher Contest," *Look*, 29 (March 8, 1966), 37; the remark to the Kiwanis Club is quoted in *Science*, 149 (Aug. 13, 1965), 731.

[7] Quoted *ibid.*, p. 730.

[8] Quoted in Harris, "The Great Atom Smasher Contest," p. 38.

tions Committee, had warned Jerome Wiesner: "The failure to approve an accelerator for the Midwest would seriously compromise the prospects for approving a $250 million accelerator on the east or west coasts [in] a few years. . . ." Now Governor James A. Rhodes of Ohio declaimed: "We have been discriminated against continually. We're not asking for something here. We're entitled to it."[9] In the new accelerator competition, the major midwestern universities agreed to support whatever site in their region seemed to have the greatest chance after the initial screening by the Atomic Energy Commission.

Frederick Seitz, the physicist who headed the National Academy, proposed to defuse the geographical issue by forming University Research Associates, a nationally representative group, to manage the accelerator. All the same, Congressman Hosmer of the Joint Committee on Atomic Energy denounced Seitz's proffer of the group's services. "This is not . . . World War II; this is 1965. . . ." Neither the Congress nor the Atomic Energy Commission should "abandon responsibility to oversee the spending of the people's money . . . to some nongovernmental group of university presidents or physicists." "The high-energy fraternity," Hosmer added, "has created a model of the best of all possible worlds with everything in it that they would like. It is kind of like a bunch of business tycoons sitting down and figuring out what taxes they are going to pay and writing the laws to fit them."[1]

Meanwhile, at the request of the Atomic Energy Commission, a committee of the National Academy, distinguished and geographically representative, evaluated every potential location in light of its available land, electrical power, academic facilities, and transportation. In March 1966, the committee narrowed down the choice for the Atomic Energy Commission by recommending six sites: Brookhaven on the East Coast, Sacramento on the West, Denver in the mountain states, and Wisconsin, Michigan, and Illinois in the Midwest. President Seitz announced that University Research Associates would be happy with any of the choices. Crude political considerations had played no significant role in the deliberations of the Academy committee, one of the members recalled. "After all, we didn't recommend any site in Texas."[2]

The Atomic Energy Commission chose Illinois, where the preferred site had been South Barrington, an affluent suburb of Chicago. But South Barrington, fearing among other things that an influx of scientists would "disturb the moral fiber of the community," removed itself from the

---

[9] Proxmire is quoted in Daniel S. Greenberg, "When Pure Science Meets Pure Politics," *The Reporter*, 30 (March 12, 1964), 41; Rhodes, in the Washington *Post*, Sept. 21, 1965.

[1] Quoted in Anton J. Jachim, *Science Policy Making in the United States and the Batavia Accelerator* (Carbondale, Ill., 1975), pp. 103–4.

[2] Interview with the member, who would prefer to remain anonymous.

running.[3] It was replaced by Weston, a village not far from Chicago, with no full-time municipal employees and a budget of $2,500 but one that contained 6,800 acres of fairly flat farmland which the state of Illinois promised to purchase and donate to the Atomic Energy Commission. On the route of a high power electrical line, Weston was also close to Argonne Laboratory, the universities in Chicago, and O'Hare Airport, which put it within a few hours of scientists anywhere in the continental United States.

Despite the considerable merits of the choice, Senator Jacob Javits attacked the decision on behalf of his disappointed constituents at Brookhaven: "We cannot afford to horse around with a cornfield."[4] Or with racial discrimination, other eastern Senators warned. Housing in the Weston area was generally segregated, and the choice of the site was rendered more controversial by its embroilment in the struggle to pass a state open-housing law. To remove the accelerator from any tainted association with the open-housing issue, the managers renamed the site "Batavia" and titled the enterprise the Fermi National Laboratory. Still, even more objectionable than the question of segregated housing, *The New York Times* editorialized, was the expensive "irrelevance of a 200 billion electron volt accelerator to any real present national problem. The nation is engaged in a bloody war in Vietnam; the streets of its cities are swept by riots born of anger over racial and economic inequities; millions of Americans lack proper housing, adequate medical care, and essential educational opportunity."[5]

The normally friendly Joint Committee on Atomic Energy opted for a slower pace of construction by recommending smaller appropriations for design work than the high energy physics community had hoped for, then by reducing still further the budget for operating expenses and equipment. Moreover, in order to pay for the substantial cost of the machine, and for a low energy meson production installation at Los Alamos, the Atomic Energy Commission began phasing out some of its accelerators and cutting back funds for others. At Princeton, an accelerator run jointly with the University of Pennsylvania was closed down for lack of money. While the budget for high energy physics rose, most of the increment went to Batavia. By 1972, accelerators at the five other Atomic Energy Commission laboratories were working at only 60 to 70 percent capacity. At the same time, acting under the prod of budgetary restrictions and the Mansfield amendment, the Office of Naval

[3] Quoted in *Science*, 152 (April 15, 1966), 326. President Johnson, Glenn T. Seaborg recalled, "told me that he wanted not to be involved in the actual selection, and in fact he played no role in the selection of the final site and accepted it without question once the Atomic Energy Commission had made its choice." Seaborg to the author, April 21, 1976.

[4] Quoted in Jachim, *Science Policy Making*, p. 133.

[5] *New York Times*, July 16, 1967, p. 12.

Research withdrew from nuclear physics. There was no more talk about the 600 billion electron volt machine long assumed to follow Batavia. Between 1970 and 1974, inflation combined with the reduced budgetary growth rate to lower the high energy physics budget by some 50 percent.

Whatever the scientific field, the cutbacks, including a drastic decrease in graduate fellowships, created an employment squeeze reminiscent of the 1930s depression. In 1968, only seven out of ten physicists who received their Ph.D.'s found jobs; a year later, one out of ten remained unemployed, but the lower figure masked the large number—one out of three—of physicists who were earning their living as postdoctoral fellows. As early as 1966 the number of registrants for jobs at the annual meeting of the American Physical Society was almost double the number of openings; by 1970 the ratio of registrants to openings had climbed to more than ten to one. In 1971, among all the natural sciences, physics was the scientific discipline with the highest—4 percent—and best-publicized unemployment rate. By 1973 physics unemployment was falling back toward 1 percent, but a significant fraction of the employed remained in a "holding pattern" as postdoctoral fellows; one out of eight physicists wanted to change jobs. The unemployment rate for young physicists remained at almost 4 percent. In 1970 the number of physics Ph.D.'s awarded was 1,550; in 1973, 1,300; in 1976 about 1,200.[6]

At the Argonne National Laboratory, staff physicists expressed their demoralization by wearing buttons: "Argonne—love it and/or leave it." Younger physicists filled the pages of their journals with bitter complaints about a national sellout. "Here is a generation of people who have studied physics under the stimulus of Sputnik," Victor Weisskopf said. "As kids in school they were told this was a great national emergency, that we needed scientists. So they worked hard—it's not easy to become a physicist—and now they have maybe a wife and a child and they are out on the street and naturally they feel cheated."[7] Cheated or not, younger physicists, the promise of the profession, were less likely to obtain what research money remained available, since the grants system tended to favor established senior scientists. On their part, senior physicists worried whether the profession could sustain a sufficient influx of new talent.

Senior physicists also fretted over how to reconcile the budgetary squeeze with federal requirements for affirmative action, especially with

[6] *Science*, 166 (Oct. 31, 1969), 583; 168 (May 22, 1970), 934; 172 (May 21, 1971), 823; Arnold A. Strassenberg, "Supply and Demand for Physicists," *Physics Today*, XXIII (April 1970), 28; editorial, *ibid.*, XXVI (Dec. 1973), 92; Beverly F. Porter, Sylvia Barisch, and Raymond W. Sears, "A First Look at the 1973 Register," *Physics Today*, XXVII (April 1974), 23; Eugen Merzbacher, "Rethinking Graduate Education," *ibid.*, XXIX (June 1976), 88.

[7] The buttons are quoted in Ann Mozley, "Change in Argonne . . .," *Science*, 173 (Oct. 1, 1971), 36; Weisskopf is quoted in Spencer Klaw, "Letter from MIT," *Harper's*, 244 (May 1972), 24.

respect to women. The number of women who sought to become physicists remained at a low level, and the reduction in resources made it difficult to hire those who did take Ph.D.'s. Between 1971 and 1974, the fraction of women on the assistant professorial staffs of the ten leading physics departments tripled, but only to about one in fifty. By the mid-1970s no college or university had lost its federal funding for failure to comply with the civil rights laws. Still, an administrator of physics could never be sure whether *his* department's funds would be cut off tomorrow, and for reasons beyond his control. "If we are to receive support for physics . . .," President Kingman Brewster of Yale said in expressing the common anxiety, "we must conform to federal policies in the admission of women to the art school, in the provision of women's athletic facilities, and in the recruitment of women and minorities, not just in the federally supported field, but throughout the university."[8]

Physicists generally questioned, too, whether the small projects of little science could survive the power. of big science, especially high energy research, to monopolize the increasingly limited resources. Advocates of big or little science, the entire profession agreed with the forecast of Marvin L. Goldberger that the long-standing American lead in physics might "very likely" pass to the Soviet Union and Western Europe.[9] Yet *The New York Times*, aware of the "revolt against physics," chastised its practitioners for their "feelings of rejection and self-pity." In the judgment of the newspaper's editors, the gloom reflected more the dashing of extravagant expectations than anything else. Even at the height of the revolt, physics in America remained remarkably strong and its progress seemed little retarded.[1]

The explanation was in part that American physicists, despite their diminished funding, continued to enjoy a comparatively high degree of public and private subvention, at least as much as that in the rest of the Western world combined. The profession included more than one-third of the physicists on the face of the globe and published more than half the significant research in the discipline. The explanation was also that physics in America remained constructed along the lines of Rowland's pyramid; the practitioners at the top, who produced the bulk of the significant research, were still well supported. "In science as in art," George Kistiakowsky acknowledged, "there are Picassos and then there are people who make Pentagon briefing charts. They're both useful. In my career I published over 200 papers. Only a few of them could be described as really important, but the others made a contribution, too. In today's situation, it's the 'others' that are suffering." And the people

[8] *Physics Today*, XXVIII (July 1975), 62; Brewster is quoted in *Science*, 188 (April 11, 1975), 105.

[9] Quoted in the *New York Times*, April 25, 1968, p. 17.

[1] *The New York Times*, Feb. 12, 1968, p. 38.

who produced the other papers, Kistiakowsky might have added, not so much the really gifted young physicists.[2]

Similarly, if the financial squeeze shut down a number of less essential particle accelerators, the concentration of available funds provided sufficient means to develop and operate the few machines capable of experiments on the frontier of the discipline. The Batavia accelerator was completed on time, within the budget, and with a top energy of 500 instead of just 200 billion electron volts, making it the most powerful accelerator on earth. In November 1974, using the excellent accelerator equipment at Brookhaven and Stanford, two teams of high energy physicists independently detected a new particle—it was believed to be a "charmed" quark bound to an antiquark—whose discovery aroused major scientific interest and in 1976 earned the team leaders a joint Nobel Prize. Of the eleven Nobel awards in physics after 1965, this was the sixth won or shared by Americans.

All the same, measured on scales different from research achievements or international stature, physics in America, especially pure physics, had undergone an indisputable degree of disestablishment. Polls found that in the mid-1970's a rising fraction of Americans once again thought well of science, but what mainly compelled admiration were the contributions of the life sciences to medicine.[3] Applied physics helped set out a menu of technology to suit every taste, including digital watches and pocket electronic calculators; television photos transmitted from spacecraft on Mars; or the remarkable lasers exfoliating with myriad uses. Nevertheless, such marvels did nothing to establish a climate more favorable to basic physical research. Defense experts repeatedly warned that the Soviets were not only challenging but overcoming the nation's qualitative advantage in military technology. By 1977 the evidence was convincing enough to quiet most opposition to higher defense expenditures, particularly for research and development, but it stimulated no public outcry for higher-powered physics. Physics, as *The New York Times* had earlier conceded, was simply no longer the "glamor king" of the sciences.[4]

Neither were physicists any longer the leaders of the public forum. With the departure of Lee DuBridge from the White House, the policy

---

[2] Editorial, *Physics Today*, XXIV (Oct. 1971), 88; Jonathan R. Cole and Stephen Cole, "The Ortega Hypothesis," *Science*, 178 (Oct. 27, 1972), 374; Kistiakowsky is quoted in Daniel S. Greenberg, "Money Troubles in the Laboratory," Los Angeles *Times*, Nov. 28, 1971, p. 2; Klaw, "Letter from MIT," p. 24. Back in the halcyon post-Sputnik days, Jerome Wiesner said that the National Science Foundation and the National Institutes of Health were supporting "not only the unusual scientists . . . but . . . the average person as well." Quoted in Harold Orlans, "Developments in Federal Policy Toward University Research," *Science*, 155 (Feb. 10, 1967), 666.

[3] *The New York Times*, May 5, 1976, p. 15.

[4] *The New York Times*, Feb. 12, 1968, p. 38.

dominance of the Los Alamos generation had ended. The war in Vietnam, which produced no heroes, left in its wake no respected scientific elite comparable in stature to Bush and Conant or DuBridge and Oppenheimer. In 1976 President Gerald R. Ford signed into law a bill reestablishing a White House science advisory apparatus. But the passage of this measure evoked no special fanfare, and physicists were not particularly prominent among the candidates mentioned for either the presidential science advisory post or the staff. Physicists, to use Philip Abelson's assessment of scientists in general, were now regarded as "mortals—fairly intelligent, fairly well-meaning, but still merely mortals. . . . Their views are discounted just as those of any other group."[5]

Especially discounted were their budgetary views. Federal funds for basic physics in 1974 amounted to an estimated $412,000,000—somewhat less, allowing for inflation, than a decade earlier. While the Ford budget for fiscal 1977 proposed an increase in total R & D expenditures to more than $23 billion, including a sizable increase in the amount for academic physics, in constant dollars the figure was still lower than the peak of 1967. National defense accounted for about 50 percent of the total; the space program, especially the space shuttle, for another 15 percent; energy, including reactors, for about another 12 percent. Economy, military technology, and social purposefulness continued to dominate the budget. Practical politicians remained unpersuaded, to use Don K. Price's observation at the beginning of the revolt, that "what's good for science is good for the nation." Both scientists and politicians, Price had gone on to remark, "no longer rely on the assumption—which was acceptable enough to the general public when Dr. Bush presented his memorable report—that science and democracy are natural allies. Especially since some scientists have never believed it."[6] Whether they believed it or not, the assumption was, to say the least, riddled with troublesome questions.

The troublesome questions had always resided in the house of American science, and especially of physics—in the program of Henry Rowland, the ambitions of George Ellery Hale, the pieties of Robert Millikan, the zestful triumphs of an Ernest Lawrence or I. I. Rabi. If they had been pushed aside at the end of World War II, the revolt begun in the early 1960's had forced them back upon the attention of thoughtful Americans, physicists and nonphysicists alike. Now, more than a century of historical experience made the questions seem so impervious to easy answers that they amounted to full-scale dilemmas.

[5] *Science*, 180 (April 20, 1973), 259.

[6] *Federal Funds for Research, Development . . . Fiscal Years 1972, 1973, and 1974*, p. 19; *Science*, 191 (Feb. 6, 1976), 444–6; *Physics Today*, XXIX (March 1976), 85; Don K. Price, *The Scientific Estate* (Cambridge, Mass., 1965), pp. 4–5. In fiscal 1977, R & D took only about 6 percent of the federal budget compared to nearly 13 in 1965.

From the days of Joseph Henry, physicists had claimed as an article of faith that knowledge ought to be advanced and appreciated for its own sake. Certainly modern physics, one of the most magnificent intellectual achievements of human history, bolstered their argument. "This work," Stanford Professor Sidney D. Drell said of elementary-particle research, "is the difference between us and the people who came out of the forest a million years ago, dragging their knuckles on the ground." Yet in the era of particle physics, the science was even more incomprehensible to laymen than in Einstein's day, let alone Tyndall's. Physicists, of course, claimed that the advancement of science served the general welfare over the long run. In the short, socially pertinent run, Alvin Weinberg candidly conceded, "Isobaric analog states in nuclei won't resolve racial tension in Detroit or religious tension in Belfast."[7] Whatever its historic appeal to the social drives of philanthropists, physics had found its most effective claims to patronage in economic, then in defense arguments; its most faithful allies, in the sometimes popularly suspect institutions of big business and the military.

Physics no doubt flourished most vigorously under a system of self-governance that favored its own expert leadership. "There is no democracy in physics," the Nobel laureate Luis Alvarez once put it. "We can't say that some second-rate guy has as much right to [an] opinion as Fermi."[8] It could not be said that lay voters, congressmen, or presidents commanded comparable intellectual authority either. But they did command both the right and the power to control the public purse and purpose.

Physicists might believe that they could contribute to decisions of public policy with complete disinterestedness; some might even declare that, accustomed to the internationalism of science, they could help in some apolitical fashion to diminish the likelihood of military conflict. But historically, scientists had of course strongly tended to approach foreign policy from a posture of national self-interest. Now various scientific groups—including the long-dormant Federation of American Scientists, its membership of physicists down to about 25 percent—might actively help to maintain a vigorous public debate on such issues as the environment, nuclear power, and strategic weapons. Yet could their technical opinions be genuinely separated from their political and ideological convictions? More generally, how could physicists claim authority as disinterested experts if, like other human beings, they showed normal

[7] Drell is quoted in the Los Angeles *Times*, Nov. 15, 1976, p. 23; Weinberg, "In Defense of Science," *Science*, 167 (Jan. 9, 1970), 143. Benjamin W. Lee, the chief theorist at Batavia, remarked that the ultimate elementary-particle picture might prove as complex as Jackson Pollock's "Conversions." *The New York Times*, April 28, 1976, p. 18.

[8] Quoted in Daniel S. Greenberg, *The Politics of Pure Science* (New York, 1967), p. 43.

loyalty to the industrial corporations, particular universities, or federal agencies that gave them jobs, money, and status?

How was the ethic of pure science, its subject matter confined to the few, to be rendered persuasive in a nation traditionally most comfortable with achieving practical goals for the many?

How was physics to enjoy sustained support in identification with the needs of the economy and defense, yet avoid becoming their creatures or suffering the volatility of American attitudes toward big business and the military?

How was the scientific community's demand for political elitism to be reconciled with the principle of politically responsive public policy? How was best-science elitism to be accommodated to the geographical and institutional pluralism of the United States?

The more physicists had changed the world, the more had these dilemmas grown vexing. Now, the capacity of physicists to comprehend the universe far beyond the ordinary ken, their power to master nature yet unleash terrifying forces, their command of a body of esoteric knowledge essential to public decisionmaking, and their simultaneous dependence upon a political process of public choice—all promised to keep the dilemmas alive indefinitely. However far the nation's physicists had come in status, power, prestige, and professional accomplishment since Joseph Henry's day, as the republic moved ahead into its third century they remained a special establishment in American life, destined to function in uneasy tension with the democracy from which they derived vitality, sustenance, and purpose.

Glossary of
Manuscript Citations

The manuscript and records collections listed below are cited in the footnotes by the code letters on the left. Additional abbreviations are: NA, for National Archives, Washington, D.C.; LC, for Library of Congress, Washington, D.C.; RG, for Record Group.

AAF	Records of the Army Air Force, RG 18, NA
AC	Andrew Carnegie Papers, LC
AEF	Records of the American Expeditionary Forces, RG 120, NA
AGO	Records of the Adjutant General's Office, U.S. Army, RG 94, NA
AHC	Arthur Holly Compton Papers, Niels Bohr Library, American Institute of Physics, New York, New York
AIP	History of Physics Collection, Niels Bohr Library, American Institute of Physics, New York, New York
ALL	Abbot Lawrence Lowell Papers, Archives, Harvard University
BOB	Records of the Bureau of the Budget, RG 51, NA
BuM	Records of the Bureau of Mines, RG 70, NA
BuS	Records of the Bureau of Ships, RG 19, NA
CB	Carl Barus Papers, John Hay Library, Brown University, Providence, Rhode Island
CB–AIP	Carl Barus Papers, Niels Bohr Library, American Institute of Physics, New York, New York
CM	Charles Murphy Papers, Harry S Truman Library, Independence, Missouri
CNO–CIC	Records of the Chief of Naval Operations and Commander-in-Chief of the U.S. Fleet, Naval History Division, Washington Navy Yard, Washington, D.C.
COH	Oral History Collection, Oral History Research Office, Columbia University, New York, New York
CUP	Records of the Cornell Physics Department, Archives, Cornell University, Ithaca, New York
CWE	Charles W. Eliot Papers, Archives, Harvard University, Cambridge, Massachusetts
DCG	Daniel Coit Gilman Papers, Archives, Johns Hopkins University, Baltimore, Maryland
DF	Dana Family Papers, Stirling Library, Yale University, New Haven, Connecticut
DST	Records of the Department of State, RG 157, NA
EGC	Edwin Grant Conklin Papers, Rare Book Room, Princeton University, Princeton, New Jersey

EJK   Ernest J. King Papers, Naval History Division, Washington Navy Yard, Washington, D.C.

EM   Ernest Merritt Papers, Archives, Cornell University, Ithaca, New York

EOL   Ernest O. Lawrence Papers, Center for History of Science and Technology, Bancroft Library, University of California, Berkeley, Berkeley, California

EP   Edward C. Pickering Papers, Archives, Harvard University, Cambridge, Massachusetts

FB   Franz Boas Papers, American Philosophical Society, Philadelphia, Pennsylvania

FDR   Franklin D. Roosevelt Papers, Franklin D. Roosevelt Library, Hyde Park, New York

GBP   George B. Pegram Papers, Department of Physics, Columbia University, New York, New York

GEH   George Ellery Hale Papers, Archives, California Institute of Technology, Pasadena, California

HAR   Henry A. Rowland Papers, Archives, Johns Hopkins University, Baltimore, Maryland

HAS   H. Alexander Smith Papers, Rare Book Room, Princeton University, Princeton, New Jersey

HC   Henry Crew Papers, Niels Bohr Library, American Institute of Physics, New York, New York

HDS   Harold D. Smith Papers, Franklin D. Roosevelt Library, Hyde Park, New York

HGB   Harold G. Bowen Papers, Rare Book Room, Princeton University, Princeton, New Jersey

HH   National Research Endowment–National Academy of Sciences, Secretary of Commerce Files, Herbert C. Hoover Papers, Herbert Hoover Presidential Library, West Branch, Iowa

HLH   Harry L. Hopkins Papers, Robert Sherwood Collection, Franklin D. Roosevelt Library, Hyde Park, New York

HLS   Henry L. Stimson Papers, Stirling Library, Yale University, New Haven, Connecticut

HM   Alfred M. Mayer–Alfred G. Mayer–Alpheus Hyatt Papers, Rare Book Room, Princeton University, Princeton, New Jersey

HMK   Harley M. Kilgore Papers, University of West Virginia Library, Morgantown, West Virginia

HR   Henry Rowland Papers, Harriette Rowland Collection, Archives, Johns Hopkins University, Baltimore, Maryland

HST   Harry S Truman Papers, Harry S Truman Library, Independence, Missouri

IL   Irving Langmuir Papers, LC

JAF   Julius A. Furer Papers, Naval Historical Foundation Collection, LC

JAG   Records of the Judge Advocate General, U.S. Navy, RG 125, NA

JCM   John C. Merriam Papers, LC

JCS   Records of the Joint Chiefs of Staff, RG 218, NA

JD   Josephus Daniels Papers, LC

JF   James V. Forrestal Papers, Rare Book Room, Princeton University, Princeton, New Jersey

JFN   Records of James V. Forrestal, Secretary of the Navy, in General Records of the Navy Department, RG 80, NA

JM   James McCosh Papers, Rare Book Room, Princeton University, Princeton, New Jersey

JMC   James McKeen Cattell Papers, LC

JRO   J. Robert Oppenheimer Papers, LC

KTC   Karl T. Compton Papers, Rare Book Room, Princeton University, Princeton, New Jersey

LAD   Lee A. DuBridge Papers, Archives, California Institute of Technology, Pasadena, California

LL–AIP   Leonard Loeb Papers, Niels Bohr Library, American Institute of Physics, New York, New York

NACA   Records of the National Advisory Committee for Aeronautics, RG 255, NA

NBS   Records of the National Bureau of Standards, RG 167, NA

NCB   Records of the Naval Consulting Board, in General Records of the Navy Department, RG 80, NA

NCONR   Records of the Naval Coordinator of Research and Development, in Records of the Office of Naval Research, RG 298, Federal Records Center, Suitland, Maryland

NR   Naval Records Collection of the Office of Naval Records and Library, RG 45, NA

NRCM   Records of the National Academy of Sciences–National Research Council, Archives, National Academy of Sciences, Washington, D.C.

NRPB   Records of the National Resources Planning Board, RG 187, NA

OCE   Records of the Office of the Chief of Engineers, United States Army, RG 77, NA

OCM   Othniel C. Marsh Papers, Peabody Museum, Yale University, New Haven, Connecticut

ONR   Records of the Office of Naval Research, RG 298, Federal Records Center, Suitland, Maryland

OSC   Oscar S. Cox Papers, Franklin D. Roosevelt Library, Hyde Park, New York

OSR   Records of the Office of Scientific Research and Development, RG 227, NA

OSW   Records of the Office of the Secretary of War, RG 107, NA

OV   Oswald Veblen Papers, LC

OWMR   Records of the Office of War Mobilization and Reconversion, RG 250, NA

PSRB   Records of the President's Scientific Research Board, Harry S Truman Library, Independence, Missouri

RAM   Robert A. Millikan Papers, Archives, California Institute of Technology, Pasadena, California

RF   Records of the Rockefeller Foundation, Archives of the Rockefeller Foundation, Rockefeller Archive Center, Pocantico Hills, North Tarrytown, New York

RR   Rush Rhees Papers, Archives, University of Rochester, Rochester, New York

RTB–AIP   Richard T. Birge Papers, Niels Bohr Library, American Institute of Physics, New York, New York

SAG   Records of the Secretary of Agriculture, RG 16, NA

SHQP   Sources for History of Quantum Physics, Center for History of Science and Technology, Bancroft Library, University of California, Berkeley, Berkeley, California

SI   Records of the Smithsonian Institution, Archives, Smithsonian Institution, Washington, D.C.

SIT   Records of the Secretary of the Interior, RG 48, NA

SN   Simon Newcomb Papers, LC

SNV   Records of the Office of the Secretary of the Navy, in General Records of the Navy Department, RG 80, NA

SWS   Samuel Wesley Stratton Papers, Archives, Massachusetts Institute of Technology, Cambridge, Massachusetts

TCM   Thomas Corwin Mendenhall Papers, Niels Bohr Library, American Institute of Physics, New York, New York

UCBP   Records of the Department of Physics, University of California, Berkeley, Department of Physics, University of California, Berkeley, California

UCH   Office of the President Papers, University of Chicago, Special Collections, University of Chicago Library, Chicago, Illinois

UPP   Minutebooks of the Department of Physics, Princeton University, 1909–1941, Office of the Chairman, Department of Physics, Princeton University, Princeton, New Jersey

UWS   Records of the University of Wisconsin, Archives, University of Wisconsin, Madison, Wisconsin

VB   Vannevar Bush Papers, LC

VBM   Vannevar Bush Papers, Archives, Massachusetts Institute of Technology, Cambridge, Massachusetts

WDGS   Records of the War Department General Staff, RG 165, NA

WFGS   William F. G. Swann Papers, American Philosophical Society, Philadelphia, Pennsylvania

WFM   William F. Meggers Papers, Niels Bohr Library, American Institute of Physics, New York, New York

WW   Woodrow Wilson Papers, LC

Acknowledgments

The writing of this book would have been decidedly more difficult if not impossible without the help of numerous people and institutions. Many members and associates of the physics community took time to discuss with me their careers as well as important events; others kindly answered my queries by correspondence or in some cases made available to me important personal materials. For aid of this type my thanks go to the following individuals, including for the record those now deceased: Roger G. Alexander, Jr., Carl D. Anderson, Samuel K. Allison, Luis W. Alvarez, Robert F. Bacher, Raymond T. Birge, Walker Bleakney, Felix Bloch, Ira S. Bowen, Edward L. Bowles, Robert Brode, Gordon S. Brown, Vannevar Bush, Mrs. Karl T. Compton, James B. Conant, Edward U. Condon, H. Richard Crane, Karl K. Darrow, David M. Dennison, Lee A. DuBridge, Ora S. Duffendack, Jesse W. M. DuMond, Carl Eckart, Luther P. Eisenhart, Paul S. Epstein, Eugene Feenberg, Raymond B. Fosdick, James Franck, E. B. Fred, Samuel A. Goudsmit, David T. Griggs, George Harrison, Ralph Hayes, Lucy J. Hayner, Harold L. Hazen, Raymond G. Herb, Robert M. Hutchins, Donald W. Kerst, Paul H. Kirkpatrick, Dr. and Mrs. Benjamin O. Koopman, Otto Laporte, Charles C. Lauritsen, Harvey B. Lemon, Leonard B. Loeb, Joseph Mayer, Edwin M. McMillan, Burton J. Moyer, Robert S. Mulliken, Henry V. Neher, Lothar W. Nordeim, J. Robert Oppenheimer, Linus C. Pauling, Milton S. Plesset, I. I. Rabi, Harrison M. Randall, Samuel I. Rosenman, Ralph A. Sawyer, Herbert Schimmel, Allen G. Shenstone, William Shockley, Henry DeWolf Smyth, Julius A. Stratton, Hugh S. Taylor, Edward Teller, John C. Trump, Merle Tuve, George E. Uhlenbeck, Harold C. Urey, John F. Victory, Earnest C. Watson, Warren Weaver, Eugene P. Wigner, Charles Wiltse, Frederick W. Zachariasen, and Fritz Zwicky.

I greatly appreciate the assistance given me in my research by the staffs at the American Physical Society; Special Collections, Brown University Library; the Chelsea, Massachusetts, Public Library; Special Collections, the University of Chicago Library; the Connecticut Historical Society; the Archives, Harvard University Library; the Herbert Hoover Presidential Library; the Henry E. Huntington Library; the Indiana Historical Society; the International Council of Scientific Unions; the Iowa State Department of History and Archives; the Los Angeles Public Library; the Archives, Massachusetts Institute of Technology; the University of Michigan Historical Collection; the Naval History Division, Washington Navy Yard; Special Collections, Princeton University Library; the Research Corporation of New York; the Rockefeller Foundation Archives; the Franklin D. Roosevelt Library; the Tennessee State Library and Archives; the Harry S Truman Library; the Universalist Historical Society; the Archives, Washington University Library; the University of Wisconsin Archives; and the Stirling Library, Yale University. My thanks are also due the alumni records officers at American colleges and universities who kindly supplied biographical information about physicists among their graduates.

I am especially grateful to the people upon whom I made the sustained demands inevitable in a long project of research and writing. On a number of occasions I received extended hospitality and helpfulness from Joan Warnow and Charles Weiner at the Center for History and Philosophy of Physics, American Institute of Physics; Murphy D. Smith, American Philosophical Society Library; Arthur Norberg at the Bancroft Library, and Judy Fox at the Office for History of Science and Technology, University of California, Berkeley; the staffs of the Manuscripts Division, Library of Congress, and of the National Archives, especially Joseph Howerton, Sarah D. Jackson, and John Taylor; and Jean R. St. Clair, Archives of the National Academy of Sciences – National Research Council. The late Roger Stanton first introduced me to the manuscript materials at the California Institute of Technology, and Judith R. Goodstein, the Institute Archivist, unfailingly facilitated my use of the rich collections under her care. I could scarcely have pursued major parts of the research for this book without the assiduous support of the staff in the Institute's Millikan Library, including Janet Casebier, Roderick J. Casper, Helen Lyons, Donald McNamee, William Stanley, Erma Wheatley, Sophia Yen, and, in the Inter-Library Loan Division, Ruth Bowen along with her successor, Jeanne F. Tatro.

For generously sharing with me their material, ideas, or unpublished writings upon various topics covered in this history, I happily thank Jerold S. Auerbach, Robert Bruce, Albert Christman, Rexmond C. Cochrane, Robert D. Cuff, John Whitney Evans, James K. Flack, Richard G. Hewlett, Thomas P. Hughes, Robert Kargon, Milton Lomask, Gloria Lubkin, Robert Maddox, Albert E. Moyer, Nathan Reingold, Margaret Rossiter, Harvey M. Sapolsky, Brigitte Schroeder-Gudehus, Martin Sherwin, Michael Sherry, Henry Small, Roger Stuewer, William B. Tuttle, Spencer Weart, and Charles Weiner. I was also fortunate along the way to have the assistance in research of P. Thomas Carroll, Kenneth Garbade, Carolyn Harding, Jeffrey Ross, and Lawrence Shirley. For the care, patience, and enthusiasm they contributed to this project, I am deeply indebted to the secretarial staff of the Division of Humanities and Social Sciences at Caltech, including Joy Hansen and Edith Taylor, who typed the early drafts, Rita Pierson, who did and redid the final ones, and Karen Wales, who saw to the endless closing details. Major credit and thanks for the preparation of the index are due Carol Pearson.

A significant part of my work on this project was made possible by research and travel funds from the Caltech Division of Humanities and Social Sciences; the award of the Division's Old Dominion Fellowship at an early stage permitted me an intensive year of essential research in Washington, D.C. The completion of my archival studies was advanced significantly by grants from the National Science Foundation and the American Council of Learned Societies. For permission to republish material in modified form which appeared first in various scholarly collections, I am grateful to *Isis*, *Military Affairs*, *Minerva*, MIT Press, Northwestern University Press, *The Physics Teacher*, and *Technology and Culture*.

I have benefited immeasurably from the collective advice and suggestions of friends and colleagues, including Robert F. Bacher, Margaret Bates, Lance E. Davis, John A. Ferejohn, Charles C. Gillispie, R. Cargill Hall, Robert A. Huttenback, Byrd Jones, Alfred Kazin, Clayton R. Koppes, J. Morgan Kousser, Hugh and Marilyn Nissenson, Rodman W. Paul, and Robert A. Rosenstone. For taking time from their crowded lives to give me exceptionally detailed criticism of the work, both literary and substantive, I wish to thank with special warmth Paul Forman, John L. Heilbron, W. T. Jones, and particularly Bettyann Kevles, who read through more drafts than either of us cares to remember. I am equally grateful to my editors at Alfred A. Knopf, Inc.—the late Harold Strauss, who urged me to pare down the manuscript, and Ashbel Green, who, besides supplying much wise advice, tactfully made me do it.

My good friend Eric F. Goldman once warned me, with a look suggesting that I might consider doing something more enjoyable, like selling shoes, that writing books is a lonely and arduous task. He advanced the writing of this book in countless essential respects, especially by furnishing map and compass for a journey of high literary and historical endeavor. My wife and children, to whom this book is dedicated, know the magnitude of my debt to them, not least for their patience and faith.

Essay on Sources

This book draws on a wide range of sources, including a mass of periodical literature, both popular and scientific. The bibliographical notes that follow cover only magazines and articles of special importance and are also selective with respect to the government documents, biographies or autobiographies, and contemporary or historical works consulted. Since this book rests heavily on manuscript sources, the discussion of these materials is comprehensive. To save space, manuscript collections cited in the text are identified by the acronyms found in the Glossary of Manuscript Citations; the locations of all other manuscripts are supplied where they are mentioned. The abbreviation GPO is used for Government Printing Office.

CHAPTERS I–VII | 1865–1916

A contemporary introduction to the state of science in post-Civil War America is George Brown Goode, "The Beginnings of American Science: The Third Century," *Report of the United States National Museum*, Part II, 407–66, in *Annual Report of the Board of Regents of the Smithsonian Institution, 1897* (GPO, 1901). For many years the historiography of American science was highly influenced by Richard Shryock, "American Indifference to Basic Science during the 19th Century," *Archives Internationales d'Histoire des Sciences*, 5 (1948), 50–65. Revisions of Shryock's thesis are argued diversely in George H. Daniels, ed., *Nineteenth Century American Science: A Reappraisal* (Northwestern U. Press, 1972), which is usefully supplemented by Alexandra Oleson and Sanborn C. Brown, eds., *The Pursuit of Knowledge in the Early American Republic: American Learned and Scientific Societies from Colonial Times to the Civil War* (Johns Hopkins U. Press, 1976); Nathan Reingold, ed., *Science in Nineteenth-Century America: A Documentary History* (Hill and Wang, 1964); David Van Tassel and Michael G. Hall, eds., *Science and Society in the United States* (Dorsey Press, 1966). George H. Daniels, *Science in American Society: A Social History* (Knopf, 1971), is disappointing.

The life of science generally is further revealed through a small number of biographies, especially the thorough studies by A. Hunter Dupree, *Asa Gray* (Harvard U. Press, 1959), and Edward Lurie, *Louis Agassiz: A Life in*

Science (U. of Chicago Press, 1966). Also helpful are Frederick H. Getman, *The Life of Ira Remsen* (Easton, Pennsylvania: Journal of Chemical Education, 1949); Henry Fairfield Osborn, *Cope: Master Naturalist* (Princeton U. Press, 1931); David Starr Jordan, ed., *Leading American Men of Science* (Henry Holt, 1910); Bernard Jaffe, *Men of Science in America* (rev. ed.; Simon and Schuster, 1958). For the affairs of American science in their respective editorial periods, the Dana Family Papers (DF) and the James McKeen Cattell Papers (JMC) are surprisingly unrewarding. G. Brown Goode, ed., *The Smithsonian Institution 1846–1896* (GPO, 1897), is helpful, but particularly useful analyses for my purposes are Wilcomb E. Washburn, "The Influence of the Smithsonian Institution on Intellectual Life in Mid-Nineteenth Century Washington," *Records of the Columbia Historical Society, Washington, D.C., 1963–65*, 96–121, and "Joseph Henry's Conception of the Purpose of the Smithsonian Institution," in Whitfield J. Bell, ed., *A Cabinet of Curiosities* (U. Press of Virginia, 1967), 106–66. Paul H. Oehser, *Sons of Science, the Story of the Smithsonian Institution* (Henry Schuman, 1949), is anecdotal.

The starting point for any study of governmental research is A. Hunter Dupree's pioneering *Science in the Federal Government: A History of Policies and Activities to 1940* (Harvard U. Press, 1957). The special agencies are solidly treated in Donald R. Whitnah, *A History of the United States Weather Bureau* (U. of Illinois Press, 1965); T. Swann Harding, *Two Blades of Grass: A History of Scientific Development in the United States Department of Agriculture* (U. of Oklahoma Press, 1947); and A. C. True, *A History of Agricultural Experimentation and Research in the United States, 1607–1925* (U.S. Dept. of Agriculture, Misc. Pub. No. 251; GPO, 1937). The chief activity of late-nineteenth-century federal science is treated with masterful scope in William H. Goetzmann, *Exploration and Empire: The Explorer and the Scientist in the Winning of the American West* (Knopf, 1966). William Culp Darrah, *Powell of the Colorado* (Princeton U. Press, 1951), is an admiring treatment. The reformist qualities of Powell's program were early recognized in Walter Prescott Webb, *The Great Plains* (Ginn, 1931), and are the central theme in Wallace Stegner's adventurously written *Beyond the Hundredth Meridian: John Wesley Powell and the Second Opening of the West* (Houghton Mifflin, 1953). The political and administrative development of Powell's agency is perceptively treated in Thomas G. Manning, *Government in Science: The U.S. Geological Survey, 1867–1894* (U. of Kentucky Press, 1967), which is helpfully supplemented by Charles Schuchert and Clara Mae LeVene, *O. C. Marsh: Pioneer in Paleontology* (Yale U. Press, 1940).

A mine of information on science in late-nineteenth-century government is the record of the Allison investigation: Joint Commission to Consider the Present Organization of the Signal Service, Geological Survey, Coast and Geodetic Survey, and the Hydrographic Office of the Navy Department . . ., *Testimony*, March 16, 1886, 49th Cong., 1 Sess., Sen. Misc. Doc. 82 (Ser. 2345). Additional glimpses of federal science are contained in Simon Newcomb, *The Reminiscences of an Astronomer* (Houghton Mifflin, 1903); William Heally Dall, *Spencer Fullerton Baird* (Lippincott, 1915); G. R. Agassiz, ed., *Letters and Recollections of Alexander Agassiz* (Houghton Mifflin, 1913); and Charles G. Abbott, *Adventures in the World of Science* (Public Affairs Press, 1958). J. Kirkpatrick Flack, *Desideratum in Washington: The Intellectual Community in the Capital City, 1870–1900* (Schenkman, 1975), is a discerning portrait. A wide window onto the affairs of Washington science is provided by the Simon Newcomb Papers (SN), along with the Othniel C. Marsh Papers (OCM) and the Thomas Corwin Mendenhall Papers (TCM). The William B. Allison Papers, State Historical Building, Des Moines, Iowa, are devoid of information on the subject. George H. Daniels, "The Pure-Science Ideal and Democratic Culture," *Science*, 156 (June 30, 1967), 1699–1705, casts light on a key issue in the Allison investigation, as does A. Hunter Dupree, "Central Scientific Organization in the United States Government," *Minerva*, 1 (Summer 1963), 453–69. For the expansion of the government into the physical sciences in the early twentieth century, Rexmond G. Cochrane, *Measures for Progress: A History of the National Bureau of Standards* (National Bureau of Standards, U.S. Department of Commerce, 1966), is an informative official history. The Samuel Wesley Stratton Papers (SWS) are slightly helpful on the founding of the Bureau but not on much else. Background to the impulse given the physical sciences by federal regulation is supplied in James Harvey Young's admirable *The Toadstool Millionaires: A Social History of Patent Medicines in America Before Federal Regulation* (Princeton U. Press, 1961), and in Oscar E. Anderson, Jr.'s, wide-ranging *The Health of a Nation: Harvey W. Wiley and the Fight for Pure Food* (U. of Chicago Press, 1958).

Useful introductions to the development of technology in the post-Civil War decades are Roger Burlingame, *Engines of Democracy* (Scribner's, 1940), and J. W. Oliver's encyclopedic *History of American Technology* (Ronald Press, 1956). Nathan Rosenberg's incisive *Technology and American Economic Growth* (Harper & Row, 1972) stresses the emphasis that went to innovations in production and mechanical invention. The mid-nineteenth-century friendliness of scientists to technological enterprise is argued in Nathan Reingold, "Alexander Dallas Bache: Science and Technology in the American Idiom," *Technology and Culture*, 11 (April 1970), and Reingold, "Theorists and Ingenious Mechanics: Joseph Henry Defines Science," *Science Studies*, 3 (Oct. 1973). Monte A. Calvert, *The Mechanical Engineer in America, 1830–1910* (Johns Hopkins U. Press, 1967), spotlights the distinction in the factories between shop- and school-trained men. Catherine Mackenzie, *Alexander Graham Bell* (Houghton Mifflin, 1928), deals with the inventor of the telephone, while Robert V. Bruce's absorbing *Bell: Alexander Graham Bell and the Conquest of Solitude* (Little, Brown, 1973) captures the many sides of the man, including his serious interest in science. David O. Woodbury, *A Measure for Greatness: A Short Biography of Edward Weston* (McGraw-Hill, 1949), is in many ways a model study of a late-nineteenth-century inventor-entrepreneur. Matthew Josephson's rousing *Edison* (McGraw-Hill, 1959) detects an early shift from individual to organized invention, which is also clear in Thomas Parke Hughes's excellent *Elmer Sperry: Inventor and Engineer* (Johns Hopkins U. Press, 1971) and unpublished article, "Edison's Method." The corporate tension imposed on technical life is emphasized in Edwin T. Layton, *The Revolt of the Engineers: Social Responsibility and the American Engineering Profession* (Case Western Reserve U. Press, 1971). Irving G. Wyllie, *The Self-Made Man in America* (Rutgers U. Press, 1954), deals with the attitudes of businessmen toward higher education.

The evolution of the industry most closely related to physics is best approached through a number of studies, including Malcolm MacLaren, *The Rise of the Electrical Industry during the Nineteenth Century* (Princeton U. Press, 1943), and Harold C. Passer, *The Electrical Manufacturers, 1875–1900* (Harvard U. Press, 1953). The transition of the industry to its increasing reliance on scientifically trained personnel is evident from the pages of the trade journal *Electrical World*; Lee De Forest's boastful autobiography, *Father of Radio* (Chicago: Wilcox and Follett, 1950); John J. O'Neill, *Prodigal Genius: The Life of Nikola Tesla* (Ives Washburn, 1944); John Winthrop Hammond, *Charles Proteus Steinmetz* (Century, 1924); and W. Rupert Maclaurin, *Invention and Innovation in the Radio Industry* (Macmillan, 1949). A critical assessment of the birth of the Western Electric industrial research laboratory is N. R. Danielian, *A.T.&T.: The Story of Industrial Conquest* (Vanguard Press, 1939), while M. D. Fagen, ed., *A History of Engineering and Science in the Bell System: The Early Years, 1875–1925* (Bell Telephone Laboratories, 1975), is an encyclopedic technical account. A useful authorized study is Kendall Birr, *Pioneering in Industrial Research: The Story of the General Electric Research Laboratory* (Public Affairs Press, 1957), which can be supplemented by John T. Broderick's admiring *Willis Rodney Whitney: Pioneer of Industrial Research* (Albany, N. Y.: Fort Orange Press, 1945).

The popularization of science before the Civil War is assessed in Wyndham D. Miles, "Public Lectures on Chemistry in the United States," *Ambix*, 15 (1968), 130–53, and Margaret Rossiter, "Benjamin Silliman and the Lowell Institute: The Popularization of Science in Nineteenth-Century America," *New England Quarterly*, 44 (1971), 602–26. The standard work on one of the main subjects of the postwar popularization is Richard Hofstadter, *Social Darwinism in American Thought* (Beacon Press, 1955), which is helpfully supplemented by Arthur Schlesinger, Jr. and Morton White, eds., *Paths of American Thought* (Houghton Mifflin, 1963); Windsor Hall Roberts, "The Reaction of American Protestant Churches to the Darwinian Philosophy, 1860–1900" (unpublished Ph.D. dissertation, History, U. of Chicago, 1936), and Bert J. Loewenberg, "The Impact of the Doctrine of Evolution on American Thought, 1859–1900" (unpublished Ph.D. dissertation, History, Harvard U., 1934); and William G. McLoughlin, *The Meaning of Henry Ward Beecher: An Essay on the Shifting Values of Mid-Victorian America* (Knopf, 1970). *Popular Science Monthly* reveals the overall substance, purpose, and attitudes of the popularizers and their audience. A sympathetic biography of the founder is John Fiske, *Edward Livingston Youmans: Interpreter of Science for the People* (D. Appleton, 1894). Considerable material on the relationship between science and the reform of higher education is in *The Nation*, *The Educational Review* (1891–   ), and *Science* (1895–   ).

There is no better introduction to the history of the higher learning than Frederick Rudolph's sparkling and insightful *The American College and University: A History* (Knopf, 1962). Also useful are Richard Hofstadter and C. DeWitt Hardy, *The Development and Scope of Higher Education in the United States* (Columbia U. Press, 1952); Richard Hofstadter and Walter P. Metzger, *The Development of Academic Freedom in the United States* (Columbia U. Press, 1955); George P. Schmidt, *The Liberal Arts College: A*

Chapter in American Cultural History (Rutgers U. Press, 1957). For public higher education, the starting points are Edward D. Eddy, Jr., *Colleges for Our Land and Time* (Harper, 1956); Earle D. Ross, *Democracy's College: The Land Grant Movement in the Formative Stage* (Iowa State College Press, 1942); Allan Nevins, *The State Universities and Democracy* (U. of Illinois Press, 1962). Edwin E. Slosson, *Great American Universities* (Macmillan, 1910), rousingly reports on the entrance of the state universities into graduate work and research.

Of the numerous histories of educational institutions, particularly useful for my purposes were Samuel Eliot Morison, *Three Centuries of Harvard, 1636–1936* (Harvard U. Press, 1936); Samuel Eliot Morison, ed., *The Development of Harvard University since the Inauguration of President Eliot, 1869–1929* (Harvard U. Press, 1930); George Wilson Pierson, *Yale: College and University, 1871–1937* (2 vols.; Yale U. Press, 1952); Russell H. Chittenden, *History of the Sheffield Scientific School of Yale University, 1846–1922* (Yale U. Press, 1928); Ralph Henry Gabriel, *Religion and Learning at Yale: The Church of Christ in the College and the University, 1757–1957* (Yale U. Press, 1958); Thomas J. Wertenbaker, *Princeton, 1746–1896* (Princeton U. Press, 1946); Samuel C. Prescott, *When MIT Was 'Boston Tech': 1861–1916* (The Technology Press, 1954); Thomas LeDuc, *Piety and Intellect at Amherst College, 1865–1912* (Columbia U. Press, 1946); Richard J. Storr, *Harper's University: The Beginnings* (U. of Chicago Press, 1966); Merle Curti and Vernon Cartensen, *The University of Wisconsin: A History, 1848–1925* (2 vols.; U. of Wisconsin Press, 1949); Winton U. Solberg, *The University of Illinois, 1867–1894* (U. of Illinois Press, 1968); and Hugh Hawkins, *Pioneer: A History of the Johns Hopkins University, 1874–1889* (Cornell U. Press, 1960).

My own understanding of the perceptions and purposes of the university reformers was substantially advanced by Laurence R. Veysey, *The Emergence of the American University* (U. of Chicago Press, 1965). Burton J. Bledstein, *The Culture of Professionalism: The Middle Class and the Development of Higher Education in America* (Norton, 1976), analyzes the ideas of the reform university presidents in rich detail. The inaugural addresses and annual reports of the leading university presidents amply repay examination. A first-rate study of Eliot's reforms in the cultural context of his day is Hugh Hawkins, *Between Harvard and America: The Educational Leadership of Charles William Eliot* (Oxford U. Press, 1972). A discerning biography is Henry James, *Charles W. Eliot: President of Harvard University, 1869–1909* (2 vols.; London: Constable, 1930); and Henry H. Sanderson, *Charles W. Eliot: Puritan Liberal* (Harper, 1928), stresses the religious springs of Eliot's thought. Firsthand insight into Eliot's thought is to be gained from William Allan Neilson, ed., *Charles W. Eliot: The Man and His Beliefs* (2 vols.; Harper, 1926); Charles W. Eliot, *Educational Reform* (Century, 1898) and *American Contributions to Civilization* (Century, 1897); and the Charles William Eliot Papers (CWE). Eliot's ally at Cornell is perceptively assessed in Walter P. Rogers, *Andrew D. White and the Modern University* (Cornell U. Press, 1942), which can be usefully supplemented by White's *Autobiography of Andrew Dickson White* (2 vols.; Century, 1905) and Carl L. Becker,

Cornell University: Founders and the Founding (Cornell U. Press, 1943). An able biography is Fabian Franklin, *The Life of Daniel Coit Gilman* (Dodd, Mead, 1910).

William H. Sloane, *The Life of James McCosh* (Scribner's, 1896), is an uncritical biography. McCosh's ideas are best obtained from his own writings, including *The New Departure in College Education* (Scribner's, 1885), *Religion in a College: What Place It Should Have* (New York: A. C. Armstrong, 1886), *The Religious Aspect of Evolution* (Putnam's, 1888), *What an American University Should Be* (New York: J. K. Lees, 1885), and the James McCosh Papers (JM). The most coherent statement of the conservative position on science and educational reform is Noah Porter's eloquent *The American Colleges and the American Public* (Scribner's, 1878). Henry W. Bragdon perceives the educational conservatism of Princeton's latter-day president in *Woodrow Wilson: The Academic Years* (Harvard, 1967). In *Alma Mater* (Farrar and Rinehart, 1936), Henry Seidel Canby takes a rose-colored look at the old-time college, and in *University Control* (New York: The Science Press, 1913), James McKeen Cattell reports sympathetically upon the struggle for greater faculty power.

Stanley M. Guralnick, *Science and the Ante-Bellum American College* (American Philosophical Society, 1975), provokes a reconsideration of the strength of science in pre-Civil War higher education. Stimulating for its comparative approach is Joseph Ben-David, "The Universities and the Growth of Science in Germany and the United States," *Minerva*, 7 (Autumn 1968), 1–35, which should be read in the context of Ben-David's more sweeping assessment *The Scientist's Role in Society: A Comparative Study* (Prentice-Hall, 1971). Considerable information is to be gleaned from Frank Wigglesworth Clarke, "A Report on the Teaching of Chemistry and Physics in the United States" (U.S. Bureau of Education Information Circular #6, 1880; GPO, 1881); A. Riedler, "American Technological Schools," in *Report of the Commissioner of Education, 1892–93*, I, 657–86; Charles R. Mann, *A Study of Engineering Education* (Carnegie Foundation for the Advancement of Teaching, Bulletin #11, 1918); Donald Fleming, *Science and Technology in Providence, 1760–1914* (Brown U. Press, 1952); Maurice Caullery, *Universities and Scientific Life in the United States*, trans. James Haughton Woods and Emmet Russel (Harvard U. Press, 1922). Albert E. Moyer details the educational spillover of the reform movement in "The Emergence of the Laboratory Approach in the Teaching of Secondary-School Physics in Late Nineteenth-Century America" (unpublished master's thesis, History of Science, U. of Wisconsin, 1974). Helpful in an encyclopedic fashion is Merle Curti and Roderick Nash, *Philanthropy in the Shaping of American Higher Education* (Rutgers U. Press, 1965). Howard S. Miller, *Dollars for Research: Science and Its Patrons in Nineteenth-Century America* (U. of Washington Press, 1970), deals with the origin of the Carnegie Institution of Washington, which may be further studied in Joseph Frazier Wall, *Andrew Carnegie* (Oxford U. Press, 1970); David Madsen, "Daniel Coit Gilman at the Carnegie Institution of Washington," *History of Education Quarterly*, IX (Summer 1969), 158; Nathan Reingold, "National Science Policy in a Private Foundation: The Carnegie Institution of Washington, 1903–1920" (unpublished manuscript, 1975); and the Institution's Yearbooks.

Documentation, relatively sparse for this period, on the general life of university physicists includes a few biographies and autobiographies: Howard McClenahan, *Cyrus Fogg Bracket, 1833–1915: An Appreciation* (Princeton: The Guild of Brackett Lecturers and the Princeton Engineering Association, 1934); Michael Pupin, *From Immigrant to Inventor* (Scribner's, 1923); "Raymond T. Birge," Transcript of a tape-recorded interview conducted by Edna T. Daniel, Regional Cultural History Project, General Library, University of California, Berkeley, 1960; and Leonard B. Loeb, "The Autobiography of Leonard B. Loeb" (AIP). Also helpful are the Alfred M. Mayer–Alfred G. Mayer–Alpheus Hyatt Papers (HM), the George B. Pegram Papers (GBP), and the Ernest Merritt Papers (EM). Of only incidental value are the papers of Edward L. Nichols, Cornell University Archives; Bergen Davis, Physics Department, Columbia University; Edwin P. Adams and William F. Magie, respectively in the Physics Department and the Rare Book Room, Princeton University; Leonard Ingersoll, Archives of the University of Wisconsin; and Henry G. Gale, Special Collections, University of Chicago.

The Henry A. Rowland Papers (HR) are a valuable collection of letters written while he was a student, and the Rowland Papers (HAR) are a major source for the working physicist. Rowland's research articles and addresses are conveniently collected in *The Physical Papers of Henry Augustus Rowland* (Johns Hopkins U. Press, 1902). A summary treatment is Daniel J. Kevles, "Henry A. Rowland," *Dictionary of Scientific Biography*, XII, 577–79. John D. Miller, "Rowland and the Nature of Electric Currents," *Isis*, 63 (March 1972), 5–27, "Rowland's Magnetic Analogy to Ohm's Law," *ibid.*, 66 (June 1975), 230–41, and "Rowland's Physics," *Physics Today*, XXIX (July 1976), 39–45, are indispensable. Also informative are Samuel Rezneck, "The Education of an American Scientist: H. A. Rowland, 1848–1901," *American Journal of Physics*, 28 (1960), 155–62, and "An American Physicist's Year in Europe: Henry Rowland, 1875–1876," *ibid.*, 30 (1962), 877–86. Bernard Jaffe, *Michelson and the Speed of Light* (Anchor Books, 1960), must be used with care. Dorothy Michelson Livingston, *The Master of Light: A Biography of Albert A. Michelson* (Scribner's, 1973), is a personally informative biography by a daughter. Aspects of the Michelson-Morley experiment are revealed in the Edward W. Morley Papers, Library of Congress, which may be supplemented by Howard R. Williams, *Edward Williams Morley: His Influence on Science in America* (Easton, Pa.: Chemical Education Publishing Co., 1957). A solid summary of Michelson's scientific career is Loyd S. Swenson, Jr., "Albert A. Michelson," *Dictionary of Scientific Biography*, IX, 371–74. Helpful on a special aspect of Michelson's work is Robert A. Shankland, "Michelson and His Interferometer," *Physics Today*, XXVII (April 1974), 37–43.

Lynde P. Wheeler, *Josiah Willard Gibbs: The History of a Great Mind* (Yale U. Press, 1951), is an appreciative introduction, and Muriel Rukeyser, *Willard Gibbs* (Doubleday, Doran, 1942), deals sensitively with the difficulties of being a theorist in late-nineteenth-century America. Martin J. Klein, "Josiah Willard Gibbs," *Dictionary of Scientific Biography*, V, 386–93, is a masterful treatment of Gibbs's scientific work. Henry A. Bumstead and Ralph G. Van Name edited *The Collected Works of J. Willard Gibbs* (2 vols; Yale U. Press, 1948), which may be explored with F. G. Donnan and Arthur Haas, eds., *A*

Commentary on the Scientific Writings of J. Willard Gibbs (2 vols.; Yale U. Press, 1936). The Willard Gibbs Papers in the Beinecke Library, Yale University, contain little correspondence. In *The Autobiography of Robert A. Millikan* (Prentice-Hall, 1950), Millikan provided valuable accounts of his childhood, education, and work on the electronic charge. A comprehensive introduction to Millikan's scientific work is Daniel J. Kevles, "Robert A. Millikan," *Dictionary of Scientific Biography*, IX, 395–400, and a provocative assessment is Robert H. Kargon, "The Conservative Mode: Robert A. Millikan and the Twentieth Century Revolution in Physics," *Isis* (December, 1977). The Robert A. Millikan Papers (RAM) are a voluminous collection which contains little for the prewar period. William Seabrook, *Doctor Wood* (Harcourt, Brace, 1941), does no justice to Robert Wood's remarkable scientific research.

Further information on the institutional progress of physics in academia can be gleaned from the catalogues of the major universities. Also useful are Edwin H. Hall, "Physics Teaching at Harvard Fifty Years Ago," *The American Physics Teacher*, VI (1938); Allen G. Shenstone, "Princeton and Physics," *Princeton Alumni Weekly*, 61 (Feb. 24, 1961), 6–13, 20; L. R. Ingersoll, "The First Hundred Years of the Department of Physics of the University of Wisconsin" (AIP); Harley E. Howe and Guy Grantham, "Seventy Years of Physics at Cornell," copy in my possession; the minutebooks in the Princeton Physics Department; the reports to the president by the chairman of the Michigan Physics Department in the Michigan Historical Collection, University of Michigan; the minutes in the Physics Department at the University of Wisconsin; and the notes on the Columbia Physics Department of the late Professor Lucy J. Hayner. Little light on physics in the federal government is shed by Thomas Coulson, *Joseph Henry: His Life and Work* (Princeton U. Press, 1950) or by the autobiographies of federal scientists, including Harvey L. Curtis, *Recollections of a Scientist* (Bonn, Germany: Privately printed, 1958); William J. Humphreys, *Of Me—W. J. Humphreys* (Washington, D.C.: Privately printed, 1947); or William Coblentz, *From the Life of a Researcher* (Philosophical Library, 1951). More helpful is Donald L. Obendorf, *Samuel P. Langley: Solar Scientist, 1867–1891* (University Microfilms, 1969), and highly rewarding is Carl Barus, "The Life of Carl Barus," along with his correspondence (CB). David O. Woodbury, *Beloved Scientist: Elihu Thomson* (McGraw-Hill, 1944), is an able portrait of this physicist, inventor, and businessman. An outstanding treatment of the new industrial research scientist is Albert Rosenfeld, "The Quintessence of Irving Langmuir," in *The Collected Works of Irving Langmuir*, ed. G. Guy Suits and Harold E. Way (12 vols.; Pergamon Press, 1960–62), XII, 3–232.

The social composition of the scientific community at the opening of the period is treated in Donald de B. Beaver, "The American Scientific Community, 1800–1860" (unpublished Ph.D. dissertation, History of Science, Yale, 1966), and Clark A. Elliott, "The American Scientist, 1800–1863: His Origins, Career, and Interests" (unpublished Ph.D. dissertation, History of Science, Case Western Reserve, 1969). My analysis of the social composition of the physics community was made by drawing up a list of the leading physicists of the period, meaning those who were most productive in research, elected to the National Academy of Sciences or to offices in the American Physical Society, or identified as prominent by a star in the early editions of *American*

Men of Science. I then gleaned as much biographical information on these physicists as possible from the *Dictionary of American Biography*, the *National Cyclopedia of Biography*, the *National Academy of Sciences Biographical Memoirs*, college alumni records, and the biographical materials at the Center for the History of Physics, American Institute of Physics, in New York City. The resulting data are summarized in Daniel J. Kevles, "The Study of Physics in America" (unpublished Ph.D. dissertation, History, Princeton U., 1964), Appendices V to VII.

My conclusions about the distribution of research both by field and by publisher are based on a statistical analysis of the physics articles published in the *American Journal of Science* in the late nineteenth century and *The Physical Review* in the twentieth. The results are summarized in Daniel J. Kevles and Carolyn Harding, "The Physics, Mathematics, and Chemical Communities in the United States, 1870–1915: A Statistical Survey," California Institute of Technology, *Social Science Working Paper No. 136*, March 1977. Indispensable for setting the size and resources of American physics in international perspective is Paul Forman, John L. Heilbron, and Spencer Weart, *Physics circa 1900: Personnel, Funding, and Productivity of the Academic Establishments* (Vol. 5, *Historical Studies in the Physical Sciences*; Princeton U. Press, 1975). An essential source for the development of Ph.D. work is M. Lois Marckworth, comp., *Dissertations in Physics: An Indexed Bibliography of All Doctoral Theses Accepted by American Universities, 1861–1959* (Stanford U. Press, 1961). The Baconian tradition in American science is emphasized in George H. Daniels, *American Science in the Age of Jackson* (Columbia U. Press, 1968). The elitist, hierarchical nature of the scientific enterprise is suggested in Derek J. deSolla Price, *Little Science, Big Science* (Columbia U. Press, 1963), and forms one of the major themes in the remarkable work of the pioneer sociologist of science Robert Merton, whose papers, some of the more important of which for my purpose were done in collaboration with Harriet Zuckerman, have been collected in *The Sociology of Science: Theoretical and Empirical Investigations*, ed. Norman W. Storer (U. of Chicago Press, 1973). Other essential works in the Merton school are Norman W. Storer, *The Social System of Science* (Holt, Rinehart, Winston, 1966); Diana Crane, *Invisible Colleges: Diffusion of Knowledge in Scientific Communities* (U. of Chicago Press, 1972); Jonathan R. Cole and Stephen Cole, *Social Stratification in Science* (U. of Chicago Press, 1973); and Warren O. Hagstrom, *The Scientific Community* (Basic Books, 1965).

The early difficulties of American scientists in establishing professional standards institutionally are dealt with in Sally Gregory Kohlstedt, "The Geologists' Model for National Science, 1840–47," *Proceedings of the American Philosophical Society*, 118 (April 1974), 179–95, "A Step Toward Scientific Self-Identity in the United States: The Failure of the National Institute, 1844," *Isis*, 62 (Fall 1971), 339–62, and more fully in her able *The Formation of the American Scientific Community: The American Association for the Advancement of Science, 1848–1860* (U. of Illinois Press, 1976). Also useful are John D. Holmfield, "From Amateurs to Professionals in American Science: The Controversy over the Proceedings of an 1853 Scientific Meeting," *Proceedings of the American Philosophical Society*, 114 (1970), 22–36, and Stephen Goldfarb, "Science and Democracy: A History of the Cincinnati Observatory,

1842–1872," *Ohio History*, 78 (1969), 172–78. Robert V. Bruce, "Democracy and American Scientific Organizations in the Mid-Nineteenth Century," unpublished manuscript, perceptively discusses the conflicts generated by scientific elitism. The eagerness for a genuinely elitist organization is evident from A. Hunter Dupree, "The Founding of the National Academy of Sciences: A Reinterpretation," *American Philosophical Society Proceedings*, 101 (1957), 434–40. The inadequacies of the Academy and of the American Association for the Advancement of Science in setting standards of scientific merit are clear respectively from the annual *Reports* and *Proceedings* of the two organizations. The origins and early history of the physicists' own organization are recalled in Frederick Bedell, "What Led to the Founding of the American Physical Society," *Physical Review*, 75 (May 15, 1949), 1601–4; the "Minutes of the Council," May 20, 1899–Dec. 30, 1920, Pupin Laboratory, Columbia University; *Bulletin of the American Physical Society*, Vols. I–III (1899–1903), after which the publication was discontinued; and Ernest Merritt, "Early Days of the Physical Society," *Review of Scientific Instruments*, 5 (April 1934), 146–47.

The attachment to mechanical explanations of nature among late-nineteenth-century physicists is explored in David R. Topper, "Commitment to Mechanism: J. J. Thomson, the Early Years," *Archives for History of Exact Sciences*, 7 (1971), 393–410; Robert Kargon, "Model and Analogy in Victorian Science: Maxwell's Critique of the French Physicists," *Journal of the History of Ideas*, 30 (July–Sept. 1969), 423–36; P. M. Heimann, "The Unseen Universe: Physics and Philosophy of Nature in Victorian Britain," *British Journal of the History of Science*, 6 (1972), 73–79. The strength of the mechanical view in the Anglo-American school is manifest in J. J. Thomson, *Recollections and Reflections* (London: G. Bell, 1936); Arthur Schuster, *The Progress of Physics During Thirty-Three Years* (Cambridge, England: University Press, 1911); A. S. Eve and C. H. Creasy, *Life and Work of John Tyndall* (London: Macmillan, 1945); Richard T. Glazebrook, *James Clerk Maxwell and Modern Physics* (Macmillan, 1896). Kenneth F. Schaffner, *Nineteenth-Century Aether Theories* (Pergamon Press, 1972), deals ably with its subject. Stillman Drake, "John B. Stallo and the Critique of Classical Physics," in Herbert M. Evans, ed., *Men and Moments in the History of Science* (U. of Washington Press, 1959), is a thoughtful essay. An alternative to the mechanical view is arrestingly studied in Russell McCormmach, "H. A. Lorentz and the Electromagnetic View of Nature," *Isis*, 61 (1970), 459–97. In "The Completeness of Nineteenth-Century Science," *Isis*, 63 (1972), 48–58, Lawrence Badash finds little evidence that physicists thought the conceptual development of their discipline was finished.

Charles C. Gillispie deals with physics through relativity in his discerningly interpretive *The Edge of Objectivity: An Essay in the History of Scientific Ideas* (Princeton U. Press, 1960). The origins of the revolution in modern physics are treated in Otto Glasser, *Wilhelm Conrad Roentgen and the Early History of Roentgen Rays* (Springfield, Ill.: Charles C Thomas, 1934); Robert W. Nitske, *The Life of Wilhelm Conrad Roentgen: Discoverer of the X Ray* (U. of Arizona Press, 1971); Lawrence Badash, "The Early Developments in Radioactivity, with Emphasis on Contributions from the United States" (unpublished Ph.D. dissertation, History of Science and

Medicine, Yale, 1964); and Arthur S. Eve, *Rutherford* (Macmillan, 1939). To be used with care are Max Planck, *Scientific Autobiography*, trans. Frank Gaynor (Philosophical Library, 1949); Planck, *The Origin and Development of the Quantum Theory*, trans. H. T. Clarke and L. Silberstein (Oxford: The Clarendon Press, 1922); and Banesh Hoffmann, *The Strange Story of the Quantum* (2d ed.; Dover, 1959). Much to be preferred are Martin J. Klein, "Max Planck and the Beginnings of the Quantum Theory," *Archive for History of Exact Sciences*, I (1960–62), 459–79, "Einstein's First Paper on Quanta," *The Natural Philosopher*, II (1963), 59–86, and "Thermodynamics and Quanta in Planck's Work," *Physics Today*, XIX (Nov. 1966), 23–32. Roger H. Stuewer, "Non-Einsteinian Interpretations of the Photoelectric Effect," in Roger H. Stuewer, ed., *Historical and Philosophical Perspectives of Science* (U. of Minnesota Press, 1970), pp. 246–63, is helpful. Authoritative introductions to the increasingly central arena of quantum studies are Clifford L. Maier, "The Role of Spectroscopy in the Acceptance of an Internally Structured Atom, 1860–1920" (unpublished Ph.D. dissertation, History of Science, University of Wisconsin, 1964); John L. Heilbron, "A History of the Problem of Atomic Structure from the Discovery of the Electron to the Beginning of Quantum Mechanics" (unpublished Ph.D. dissertation, History, University of California, Berkeley, 1964); and William McGucken, *Nineteenth-Century Spectroscopy: Development of the Understanding of Spectra, 1802–1897* (Johns Hopkins U. Press, 1969). The historiography of atomic physics has gained increasing knowledge and insight from Sigeko Nisio, "From Balmer to the Combination Principle," *Japanese Studies in the History of Science*, 5 (1966), 50–74; Russell McCormmach, "Henri Poincaré and the Quantum Theory," *Isis*, 58 (Spring 1967), 37–55; McCormmach, "The Atomic Theory of John William Nicholson," *Archive for History of Exact Sciences*, 3 (1966), 160–84; John L. Heilbron, "The Scattering of Alpha and Beta Particles and Rutherford's Atom," *ibid.*, 4 (1968), 247–307; Heilbron and Thomas S. Kuhn, "The Genesis of the Bohr Atom," *Historical Studies in the Physical Sciences*, I (1969), 211–90; Tetu Hirosige and Sigeko Nisio, "The Genesis of the Bohr Atom Model and Planck's Theory of Radiation," *Japanese Studies in the History of Science*, 9 (1970), 35–47; Paul Forman, "The Discovery of the Diffraction of X-Rays by Crystals: A Critique of the Myths," *Archive for History of Exact Sciences*, 6 (1969), 38–71; Sigeko Nisio, "X-Rays and Atomic Structure at the Early Stages of the Old Quantum Theory," *Japanese Studies in the History of Science*, 8 (1969), 55–75. An exquisite study is John L. Heilbron, *H. G. J. Moseley: The Life and Letters of an English Physicist, 1887–1915* (U. of California Press, 1974). Max Jammer, *The Conceptual Development of Quantum Mechanics* (McGraw-Hill, 1966), provides a major technical account. Ruth Moore, *Niels Bohr: The Man, His Science, and the World They Changed* (Knopf, 1966), is a model popular biography.

The origins of Einstein's special revolution are perceptively explored in Russell McCormmach, "Einstein, Lorentz, and the Electron Theory," *Historical Studies in the Physical Sciences*, II (1970), 41–87; Stanley Goldberg, "The Lorentz Theory of Electrons and Einstein's Theory of Relativity," *American Journal of Physics*, 37 (Oct. 1969), 982–94; Tetu Hirosige, "Origin of Lorentz' Theory of Electrons and the Concept of the Electromagnetic Field," *Historical Studies in the Physical Sciences*, I (1969), 151–209; Gerald Holton,

"Einstein, Michelson, and the 'Crucial' Experiment," *Isis*, 60 (Summer 1969), 133–97. Lewis S. Feuer, "The Social Roots of Einstein's Theory of Relativity," *Annals of Science*, 27 (1971), 277–98, 313–44, is extravagantly imaginative. Banesh Hoffmann, *Albert Einstein: Creator and Rebel* (Viking, 1972), is disappointing, and Ronald W. Clark, *Einstein: The Life and Times* (World Publishing Co., 1971), is encyclopedic. Philipp Frank, *Einstein: His Life and Times*, trans. George Rosen (Knopf, 1953), remains rewarding. The resistance to Einstein's theory is dealt with thoroughly in Stanley Goldberg, "The Early Response to Einstein's Special Theory of Relativity, 1905–1911: A Case Study in National Differences" (unpublished Ph.D. dissertation, Education, Harvard, 1968). The seminal work on its subject, which draws heavily on the upheavals of quanta and relativity, is Thomas S. Kuhn, *The Structure of Scientific Revolutions* (U. of Chicago Press, 1962). Many of the important papers in the early history of atomic physics are conveniently gathered in Henry A. Boorse and Lloyd Motz, eds., *The World of the Atom* (2 vols.; Basic Books, 1966).

CHAPTERS VIII–X | WORLD WAR I

There is no general history of science in the war. The memoirs, autobiographies, and biographies of key figures, military and civilian, scarcely mention the role of scientists and neither for the most part do the major histories of the conflict, including the following helpful introductions: Cyril Falls, *The Great War* (Putnam's, 1959); Walter Millis, *Arms and Men: A Study in American Military History* (Putnam's, 1956); Frederick L. Paxson, *American Democracy and the World War: Pre-War Years, 1913–1917* (Houghton Mifflin, 1936); Paxson, *Post-War Years: Normalcy, 1918–1923* (U. of California Press, 1948); Preston William Slosson, *The Great Crusade and After, 1914–1928* (Macmillan, 1930); William S. Sims, with Burton J. Hendrick, *The Victory at Sea* (London: John Murray, 1920); Elting E. Morison, *Admiral Sims and the Modern American Navy* (Houghton Mifflin, 1942); Edward C. Coffman, *The Hilt of the Sword: The Career of Peyton C. March* (U. of Wisconsin Press, 1966); John J. Pershing, *My Experiences in the World War* (2 vols.; New York: Frederick A. Stokes, 1931); Frederick Palmer, *Newton D. Baker: America at War* (2 vols.; Dodd, Mead, 1931); Josephus Daniels, *The Wilson Era—Years of War: 1917–1923* (U. of North Carolina Press, 1946); and Joseph L. Morrison, *Josephus Daniels: The Small-d Democrat* (U. of North Carolina, 1966).

Helen Wright, *Explorer of the Universe: A Biography of George Ellery Hale* (Dutton, 1966), stresses Hale's career as a practitioner and entrepreneur of science. The major sources for the history of the National Research Council are the George Ellery Hale Papers (GEH), together with the Edwin Grant Conklin Papers (EGC), the Robert A. Millikan Papers (RAM), the Edward C. Pickering Papers (EP), and the Records of the National Academy of Sciences–National Research Council (NRCM). Further information on the Council is in Simon Flexner and James Thomas Flexner, *William Henry Welch and the Heroic Age of American Medicine* (Viking, 1941); the reports of the National Academy of Sciences for the war years; Daniel J. Kevles, "George

Ellery Hale, the First World War, and the Advancement of Science in America," *Isis*, 59 (Winter 1968), 427–37; the successive annual reports of the Council of National Defense; and Robert D. Cuff, "Business, Government, and the War Industries Board" (unpublished Ph.D. dissertation, History, Princeton, 1966). The voluminous records of the Council of National Defense in the National Archives contain virtually no materials on the mobilization of science, but the relations of the Research Council with the White House are documented in the Woodrow Wilson Papers (WW), along with the diaries in the Papers of Edward M. House, Stirling Library, Yale University.

The background to the venture of the International Research Council is explored in Brigitte Schroeder, "Caractéristiques des relations scientifiques internationales, 1870–1914," *Cahiers d'Histoire Mondiale*, 13 (1966), 161–77, and in Gavin R. De Beer, *The Sciences Were Never at War* (Nelson, 1960). A detailed account of Hale's international efforts is Daniel J. Kevles, " 'Into Hostile Political Camps': The Reorganization of International Science in World War I," *Isis*, 62 (1970), 47–60. The relation of the government to the scientific attachés and the Council is suggested in the Records of the Department of State (DST). The Millikan Papers are particularly important for the involvement of the Research Council in the movement to establish engineering experiment stations, and so are the successive government documents: U.S. Congress, Senate Committee on Agriculture and Forestry, *Hearings, Experiment Stations in Connection with Land-Grant Colleges*, 64th Cong., 1st Sess., June 24, 1916; Senate, *Industrial Research Stations*, 64th Cong., 1st Sess., Sen. Doc. No. 446, May 18, 1916; House Committee on Education, *Hearings, Engineering Experimental Stations for War Services*, 65th Cong., 2d Sess., June 11, 1918. The cooperation between Phineas V. Stephens and Samuel Wesley Stratton is evident in the Records of the National Bureau of Standards (NBS). A detailed account of the matter is Daniel J. Kevles, "Federal Legislation for Engineering Experiment Stations: The Episode of World War I," *Technology and Culture*, 12 (April 1971), 182–89. Helpful on the establishment of the National Research Council fellowships are Raymond B. Fosdick, *The Story of the Rockefeller Foundation* (Harper, 1952); the Records of the Rockefeller Foundation (RF); the John C. Merriam Papers (JCM); and Nathan Reingold, "World War I: The Case of the Disappearing Laboratory," unpublished manuscript in the possession of the author.

Peripherally useful for the mobilization of civilian science are Parke R. Kolbe, *The Colleges in War Time and After* (D. Appleton, 1919), and Charles F. Thwing, *The American Colleges and Universities in the Great War, 1914–1918* (Macmillan, 1920). The technological challenges presented by the war are detailed in National Bureau of Standards, *War Work of the Bureau of Standards* (GPO, 1921); Benedict Crowell, *America's Munitions, 1917–1918* (GPO, 1919); Frederick E. Wright, *The Manufacture of Optical Glass and Optical Systems: A War-Time Problem* (GPO, 1921); Frank Parker Stockbridge, *Yankee Ingenuity in the War* (Harper, 1920); and Robert M. Yerkes, *The New World of Science: Its Development During the War* (Scribner's, 1920). The special field of aeronautics may be approached through Archibald Turnbull and Clifford L. Lord, *A History of United States Naval Aviation* (Yale U. Press, 1949); Henry H. Arnold, *Global Mission* (Harper's,

1949); Alfred Goldberg, ed., *A History of the United States Air Force, 1907–1957* (Van Nostrand, 1958); and I. B. Holley, Jr., *Ideas and Weapons: Exploitation of the Aerial Weapon by the United States during World War 1* (Yale U. Press, 1953).

The general inattention given the technological side of mobilization before 1917 is evident in U.S. Congress, Senate Committee on Military Affairs, *Hearings, Preparedness for National Defense*, 64th Cong., 1st Sess., 1916, and House Committee on Military Affairs, *Hearings, To Increase the Efficiency of the Military Establishment of the United States*, 64th Cong., 1st Sess., 1916. An introduction to the new militarily purposeful research agency is Arthur L. Levine, "United States Aeronautical Research Policy, 1915–1958: A Study of the Major Policy Decisions of the National Advisory Committee for Aeronautics" (unpublished Ph.D. dissertation, Political Science, Columbia, 1963). The Records of the National Advisory Committee for Aeronautics (NACA) are a mine of information enriched by study of the Charles D. Walcott Papers in the Records of the Smithsonian Institution (SI). The Records of the Army Air Force (AAF) for the war are enormous, but the materials for the relations among the military service, the Advisory Committee, and Millikan's Science and Research Division are centered in entries 81, 82, and 166.

Otto L. Nelson, *National Security and the General Staff* (Washington, D.C.: Infantry Journal Press, 1946), is a helpful introduction to its subject. The Records of the War Department General Staff (WDGS), especially the entries for the Chief of Staff, the War College Division, and the War Plans Division, repay imaginative use of the indexes with considerable material on scientific and technological matters. Also useful are the Records of the Office of the Chief of Engineers (OCE), but the Records of the Office of the Chief Signal Officer and of the Office of the Chief of Ordnance in the National Archives were, for my purposes, disappointing, as were the Records of the Secretary of War (OSW). Paul W. Clark, "Major General George Owen Squier: Military Scientist" (unpublished Ph.D. dissertation, Division of Special Interdisciplinary Studies, Case Western Reserve University, 1974), is an able account of Squier's technical achievements. For flash and sound ranging, the Records of the American Expeditionary Forces (AEF) are rich with material. Also essential are "The War Letters of Augustus Trowbridge, August 28, 1917 to January 19, 1919," in *Bulletin of the New York Public Library*, 43 (1939), 591–617, 645–66, 725–38, 830–44, 901–14; 44 (1940), 8–35, 117–29, 331–50. Jesse R. Hinman, *Ranging in France with Flash and Sound* (Portland, Ore.: Press of Dunham Printing Co., 1919), is a spirited memoir by a participant. John R. Innes, comp., *Flash Spotters and Sound Rangers: How They Lived, Worked and Fought in the Great War* (London: George, Allen & Unwin, 1935), tells the British story. Effective official histories that cover World War I are Leo P. Brophy and George J. B. Fisher, *The Chemical Warfare Service: Organizing for War* (United States Army in World War II, The Technical Services; Washington, D.C.: Office of the Chief of Military History, 1959), and Leo P. Brophy et al., *The Chemical Warfare Service: From Laboratory to Field* (United States Army in World War II, The Technical Services; Washington, D.C.: Office of the Chief of Military History, 1959). Both are to be supplemented by Daniel P. Jones's informative monograph "The Role of Chemists in Research on War Gases in the United States during World War I" (unpublished Ph.D. dissertation, History of Science, University of Wisconsin, 1969), and by Frederick J. Brown's thoughtful *Chemical Warfare* (Princeton U. Press, 1968).

A contemporary introduction to its subject is Lloyd N. Scott, *Naval Consulting Board of the United States* (GPO, 1920). The Board's and the Navy's attitudes on technological preparedness find expression in U.S. Congress, House Committee on Naval Affairs, *Hearings, Estimates Submitted by the Secretary of the Navy, 1916*, 64th Cong., 1st Sess., 1916. The Josephus Daniels Papers (JD) reveal the secretary's faith in Edison and contain materials on the Board's activities, as does E. David Cronon, ed., *The Cabinet Diaries of Josephus Daniels, 1913–1921* (U. of Nebraska Press, 1963). The Records of the Board (NCB) include some materials on the submarine detection effort, and the Office of Inventions section of the Records of the Judge Advocate General (JAG) suggests the Board's difficulties in mobilizing the inventive genius of the people.

Used with the subject index cards, the Records of the Office of the Secretary of the Navy (SNV) yield considerable information on the scientific mobilization. The Records of the Bureau of Ships (BuS), especially entry 988, are the basic source for activities in submarine detection research, some of which are summarized in *History of the Bureau of Engineering, Navy Department, During the World War* (Office of Naval Records and Library, Historical Section, Publication No. 5; GPO, 1922). Also helpful on the technical side are Richard D. Fay, "Underwater-Sound Reminiscences: Mostly Binaural," *Sound: Its Uses and Control*, 2 (Nov.–Dec. 1963), 37–42; Harvey C. Hayes, "World War I: Submarine Detection," *ibid.*, 1 (Sept.–Oct. 1962), 47–48; Walter G. Cady, "Piezoelectricity and Its Uses," *ibid.*, 2 (Jan.–Feb. 1963), 46–52; A. B. Wood, "Reminiscences of Underwater Sound Research, 1915–1917," *ibid.*, 1 (May–June 1962), 8–17; Max Mason, "Submarine Detection by Multiple Unit Hydrophones," *The Wisconsin Engineer*, 25 (Feb.–April 1921), 75–77, 99–102, 116–20; and Frederick V. Hunt, *Electroacoustics: The Analysis of Transduction and Its Historical Background* (Harvard Monographs in Applied Science, No. 5; Harvard U. Press, 1954). The operational uses of the listening devices are documented in Naval Records Collection of the Office of Naval Records and Library (NR) and vividly described in Ray Millholland, *The Splinter Fleet of the Otranto Barrage* (Bobbs-Merrill, 1936). The British effort in detection devices is told from an administrative point of view in Roy MacLeod and E. Kay Andrews, "Scientific Advice in the War at Sea, 1915–1917: The Board of Invention and Research," *Journal of Contemporary History*, 6 (1971), 3–40.

Postwar attitudes of the army and navy toward research and development may be drawn from Clyde S. McDowell, "Naval Research," *U.S. Naval Institute Proceedings*, 45 (June 1919), 895–908; U.S. Congress Subcommittee of the Committee on Naval Affairs, *Hearings, Naval Investigation*, 66th Cong., 2d Sess., 1920; House Committee on Military Affairs, *Hearings, Army Reorganization*, 66th Cong., 1st Sess., 1919; and Senate Subcommittee of the Committee on Military Affairs, *Hearings, Reorganization of the Army*, 66th Cong., 1st and 2d Sess., 1919.

CHAPTERS XI–XVIII | 1920–1939

The conservative thrust of the popularization of science in the 1920s is a central thesis of Ronald C. Tobey, *The American Ideology of National Science, 1919–1930* (U. of Pittsburgh Press, 1971), which also deals authoritatively with the founding of Science Service. The strong opinions of the Service's original sponsor are set forth in Charles R. McCabe, ed., *Damned Old Crank: A Self-Portrait of E. W. Scripps* (Harper, 1951). Maurice Holland, *Industrial Explorers* (Harper, 1928), is a typical celebration of the industrial scientist. Frederick Lewis Allen, *Only Yesterday* (Bantam Books, 1957), captures the flavor of the popularization, but its content is best obtained from an examination of the popular magazines of the day. Charles A. Beard, ed., *Whither Mankind* (Longmans, Green, 1928) and *Toward Civilization* (Longmans, Green, 1930) can be used together as an introduction to the debate between humanists and scientists. A special aspect of the discussion is dealt with in Carroll Pursell, Jr., " 'A Savage Struck by Lightning': The Idea of a Research Moratorium, 1927–37," *Lex et Scientia*, 10 (Oct.–Dec. 1974), 146–61. Henry Elsner, Jr., *The Technocrats: Prophets of Automation* (Syracuse U. Press, 1967), includes useful information on the origins of this symptom of the discontent with scientific civilization in the 1930s. Typical of various writings on science in its relationship to society during the depression are Thomas A. Boyd, *Research: The Pathfinder of Science and Industry* (Appleton-Century, 1935); Clifford C. Furnas, *The Next Hundred Years: The Unfinished Business of Science* (Reynal and Hitchcock, 1936); Jesse E. Thornton, comp., *Science and Social Change* (Brookings Institution, 1939). Rogers D. Rusk, *Atoms, Men and Stars* (Knopf, 1937), unhappily matches the wonders of modern physics against the worries of modern society. Arthur Holly Compton celebrated free will as an implication of quantum mechanics in *The Freedom of Man* (Yale U. Press, 1935).

The studies on AT&T and General Electric cited in the bibliographical notes for Chapters I–VII are helpful for the ongoing development of industrial research. A mine of information on the subject is *Research—A National Resource* (3 vols.; National Resources Committee, 1938–41). The patronage of academic science by industrial corporations awaits its historian. The story of the National Research Fund is extensively documented in the papers of Hale (GEH), Millikan (RAM), and Herbert Hoover (HH). In "The National Research Fund: A Case Study in the Industrial Support of Academic Science," *Minerva*, XII (April 1974), 207–20, Lance E. Davis and Daniel J. Kevles analyze the episode with the aid of a theory of economic institutions. The tax and legal background of corporate giving is discussed in F. Emerson Andrews, *Corporation Giving* (Russell Sage Foundation, 1952). Thomas F. Devine, *Corporate Support for Education: Its Bases and Principles* (Catholic U. of America Press, 1956), contains information on the passage of the law allowing corporate donations up to 5 percent of profits.

Abraham Flexner, *Funds and Foundations* (Harper, 1952), and Ernest V. Hollis, *Philanthropic Foundations and Higher Education* (Columbia U. Press, 1938), are useful introductions to their subjects. Helpful for my purposes were Robert M. Lester, *Forty Years of Carnegie Giving* (Scribner's, 1941), and *Summary of Grants Primarily for Research in Biological and Physical Science, 1911–1931* (Carnegie Corporation of New York, 1932). The scientific program of the Rockefeller philanthropies in the 1920s is recounted in Raymond B. Fosdick, *Adventure in Giving: The Story of the General Education Board* (Harper & Row, 1962). George W. Gray, *Education on an International Scale: A History of the International Education Board, 1923–38* (Harcourt, Brace, 1941), is essential. Wickliffe Rose's attitudes are revealed in the annual reports of the International Education Board and General Education Board, and are treated along with those of Guggenheim Foundation officials in Stanley Coben, "Foundation Officials and Fellowships: Innovation in the Patronage of Science," *Minerva*, XIV (Summer 1976), 225–40. The eventual shift to science for social purposes may be discerned in the annual presidential reports of the Carnegie Corporation and the Rockefeller Foundation, whose records (RF) are rich with material on the development of Warren Weaver's program. Weaver's *Scene of Change: A Lifetime in American Science* (Scribner's, 1970), generally neglects the author's social concerns of the depression era; his *U.S. Philanthropic Foundations: Their History, Structure, Management and Record* (Harper & Row, 1967) is weak on historical detail. In *Chronicle of a Generation* (Harper, 1958), Raymond B. Fosdick expressed his troubled concern about the course of technological civilization.

Many of the general works and university histories cited in the notes for Chapters I–VII are useful for the history of higher education between the wars. The annual reports of the presidents at the major universities yield considerable material on attitudes and institutional developments. Also helpful in a statistical way are Floyd W. Reese et al., *Trends in University Growth* (U. of Chicago Press, 1933); Malcolm M. Willey, ed., *Depression, Recovery and Higher Education: A Report of Committee Y of the American Association of University Professors* (McGraw-Hill, 1937); and Ernest V. Hollis, *Toward Improving Ph.D. Programs* (American Council on Education, 1945). In *No Friendly Voice* (U. of Chicago Press, 1936), Robert M. Hutchins advances his humanist idea of education, which evoked the important response of Harry Gideonse, *The Higher Learning in a Democracy* (Farrar and Rinehart, 1937).

For the activities of the National Academy of Sciences–National Research Council, the papers of Hale (GEH), Millikan (RAM), and John C. Merriam (JCM) are essential. So are the annual reports of the National Academy itself. The unhappy outcome of Hale's international venture is analyzed in Paul Forman, "Scientific Internationalism and the Weimar Physicists: The Ideology and Its Manipulation in Germany after World War I," *Isis*, 64 (June 1973), 151–80; Brigitte Schroeder-Gudehus, "Challenge to Transnational Loyalties: International Scientific Organizations after the First World War," *Science Studies*, 3 (1973), 93–118, and "Les professeurs allemands et la politique du rapprochement," *Annales d'études internationales*, 1970, 1–22. The doings of the Academy and Research Council at home were frequently discussed in *Science* and the *Scientific Monthly*, which are a rich source for the social attitudes of the nation's scientific leadership. In the 1920s, the proindustrial posture of federal science was manifest in the annual reports of the secretary of commerce and is detailed in Cochrane's *Measures for Progress*.

In the 1930s, the history of the Science Advisory Board and the fate of its programs require close study of the Millikan Papers (RAM); the Franklin D. Roosevelt Papers (FDR); the Records of the National Resources Planning Board (NRPB), which include those of its predecessors, the National Resources Board and the National Resources Committee; the Records of the Department of the Interior, Office of the Secretary (SIT); and the Records of the Office of the Secretary of Agriculture (SAG). Little about the Board is revealed in Harold Ickes, *The Secret Diary of Harold L. Ickes* (2 vols.; Simon and Schuster, 1953), and the Ickes Papers at the Library of Congress were closed as of the time of this writing. The career of the Board's chairman is ably summarized in Richard G. Hewlett, "Karl Taylor Compton," *Dictionary of Scientific Biography*, III, 372–73; Compton's political views are best obtained from his popular speeches and writings. Other materials of considerable use are the two reports of the Science Advisory Board, 1934 and 1935; Carroll W. Pursell, Jr., "The Anatomy of a Failure: The Science Advisory Board, 1933–1935," *Proceedings of the American Philosophical Society*, 109 (Dec. 1965), 342–51; and Lewis E. Auerbach, "Scientists in the New Deal . . .," *Minerva*, III (Summer 1965), 457–82. The general attitudes of the National Resources Committee are exemplified in its *Technological Trends and National Policy* (GPO, 1937). The story of the Briggs bill is well documented in the Records of the National Bureau of Standards (NBS), together with the Records of the Department of Commerce in the National Archives. The failure of the bill is insightfully analyzed in Carroll W. Pursell, Jr., "A Preface to Government Support of Research and Development: Research Legislation and the National Bureau of Standards, 1935–1941," *Technology and Culture*, 9 (April 1968), 145–64. Also helpful on other aspects of science and government in the 1930s are Pursell, "The Farm Chemurgic Council and the United States Department of Agriculture, 1935–1939," *Isis*, 60 (Fall 1969), 307–17, and "The Administration of Science in the Department of Agriculture, 1933–1940," *Agricultural History*, 42 (July 1968), 231–40; and Donald C. Swain, "The Rise of a Research Empire: NIH, 1930–1950," *Science*, 138 (Dec. 14, 1962), 1233–37.

Material on the financial and institutional progress of physics in academia can be gleaned from the Millikan Papers (RAM); the Karl T. Compton Papers (KTC); the Abbott Lawrence Lowell Papers (ALL); the Papers of the Office of the President, University of Chicago (UCH), especially the file "University and Developmental Campaigns . . ."; the Rush Rhees Papers (RR); the Records of the Berkeley Physics Department (UCBP); and the Records of the University of Wisconsin (UWS). Also helpful are Sheridan W. Baker, Jr., *The Rackham Funds of the University of Michigan, 1933–53* (U. of Michigan Press, 1955); Tenney L. Davis and H. M. Goodwin, *A History of the Department of Chemistry and Physics at MIT, 1865–1933* (The Technology Press, 1933); Theodore W. Lyman, "The Future of the Department of Physics," *Harvard Alumni Bulletin*, 31 (June 13, 1929), 1056–58; and Raymond T. Birge, "History of the [Berkeley] Physics Department, 1868–1932" (AIP), which must be used with care, and Charles Weiner, "Physics in the Great Depression," *Physics Today*, XXIII (Oct. 1970), 31–38. The university catalogues for the period indicate the evolution of the faculty and the subjects of teaching and research.

My discussion of the development of the American physics community between the wars rests in significant part upon interviews with physicists of the day; their names are listed in the acknowledgments. I also drew considerable information from the autobiographical recollections and the interviews conducted by Charles Weiner in the History of Physics Collection at the American Institute of Physics (AIP), and from the interviews that comprise part of the extensive materials gathered by the Project on the Sources for the History of Quantum Physics (SHQP). Especially helpful for my purposes were the Project correspondence files of John H. Van Vleck, Edwin C. Kemble, and Samuel Goudsmit. For the complete holdings of the Project, see Thomas S. Kuhn, John L. Heilbron, Paul Forman, and Lini Allen, *Sources for History of Quantum Physics: An Inventory and Report* (Memoirs of the American Philosophical Society, vol. 68; American Philosophical Society, 1967). Incidentally informative are the Percy W. Bridgman Papers, Lyman Physics Laboratory, Harvard, and the William F. G. Swann Papers (WFGS). A pioneering introduction to the ascendancy of American physics in the postwar years is Stanley Coben, "The Scientific Establishment and the Transmission of Quantum Mechanics to the United States, 1919–32," *American Historical Review*, 76 (April 1971), 442–66. John H. Van Vleck, "American Physics Comes of Age," *Physics Today*, XVII (June 1964), 21–26, raises but does not satisfactorily answer the question of why physics in the United States became so good. Useful in many respects is Spencer R. Weart, "The Physics Business in America, 1919–1940: A Statistical Reconnaissance," to be published in Nathan Reingold, ed., *The Sciences in the United States: A Bicentennial Perspective* (Princeton, forthcoming). Robert Jungk, *Brighter than a Thousand Suns*, trans. James Cleugh (Penguin, 1960), catches the glowing excitement of physicists abroad in the 1920s and the collapse of their world in the 1930s.

The interviews and recollections form the basis of my conclusions about the evolving social composition of the American physics community. That the trends among physicists were common is evident from R. H. Knapp and H. B. Goodrich, *Origins of American Scientists* (U. of Chicago Press, 1952). Logan Wilson, *The Academic Man: A Study in the Sociology of a Profession* (Oxford U. Press, 1942), brings explicitly to light the status drives of the professoriat. Indispensable for its subject are Margaret Rossiter, "Women Scientists in America before 1920," *American Scientist*, 62 (May-June 1974), 312–23, and her unpublished paper, "Quantitative History of Women Scientists in the United States, 1920–1950." A useful case-study supplement is Helen Wright, *Sweeper in the Sky: The Life of Maria Mitchell, First Woman Astronomer in America* (Macmillan, 1950). A mine of statistical information on women and work is *Recent Social Trends in the United States: Report of the President's Research Committee on Social Trends* (2 vols.; McGraw-Hill, 1933). For my purposes, especially useful are Louella Cole Pressey, "The Women Whose Names Appear in 'American Men of Science' for 1927," *School and Society*, 29 (Jan. 19, 1929), 96–100; "Preliminary Report of Committee W, on Status of Women in College and University Faculties," *Bulletin of the American Association of University Professors*, VII (1921), 21–32; "Second Report of Committee W . . .," *ibid.*, X (1924), 563–71; Jessie Bernard, *Academic Women* (Pennsylvania State U. Press, 1964); and Mabel Newcomer, *A Century of Higher Education for Women* (Harper, 1959). Madame Curie's visit to the United States is described briefly in Eve Curie, *Madame Curie*, trans. Vincent Sheean (Doubleday, Doran, 1938), and Robert Reid, *Marie Curie* (Saturday Review Press, 1974). General attitudes about women, careers, and science are advantageously explored in the popular magazines of the interwar period.

Among the numerous works in recent years to pay attention to the history of women beyond the fight for suffrage, I found especially helpful William L. O'Neill, *Everyone Was Brave: The Rise and Fall of Feminism in America* (Quadrangle, 1969).

The small number of Catholics in American science in this period is reported in Harvey C. Lehman and Paul A. Witty, "Scientific Eminence and Church Membership," *Scientific Monthly*, 33 (Dec. 1931), 544–58, and in Knapp and Goodrich, *Origins of American Scientists*. The disappointing quality of scientific work in Catholic colleges is evident from the *Bulletin of the American Association of Jesuit Scientists*. Contemporary concern about the problem among Catholics is revealed in Roy J. Deferrari, "Catholics and Graduate Study," *Commonweal*, 14 (June 24, 1931), 203–5, and Deferrari, ed., *Vital Problems of Catholic Education in the United States* (Catholic U. of America Press, 1939); John A. O'Brien, ed., *Catholics and Scholarship: A Symposium on the Development of Scholars* (Huntington, Ind.: Our Sunday Visitor, 1938); and Karl Herzfeld, "Scientific Research and Religion," *Commonweal*, 9 (March 20, 1929), 560–62. A leading cause of the small number of Catholic scientists is best treated in Edward J. Power, *A History of Catholic Higher Education in the United States* (Bruce, 1958); Philip Gleason, "American Catholic Higher Education: A Historical Perspective," in Robert Hassenger, ed., *The Shape of Catholic Higher Education* (U. of Chicago Press, 1967); it is also touched upon with historical perception in Christopher Jencks and David Riesman, *The Academic Revolution* (Doubleday, 1968). The economic and cultural conditions discouraging the development of Catholic scientists are penetratingly explored in Philip Gleason, "Immigration and American Catholic Intellectual Life," *Review of Politics*, 26 (April 1964), 147–73; Gerhard Lenski, *The Religious Factor: A Sociological Study of Religion's Impact on Politics, Economics, and Family Life* (Doubleday, 1961); James W. Trent, with Jeanette Golds, *Catholics in College: Religious Commitment and the Intellectual Life* (U. of Chicago Press, 1967). An exceptionally good introduction to the subject is John Tracy Ellis, "American Catholics and the Intellectual Life," *Thought: Fordham University Quarterly*, 30 (Fall 1955), 351–88.

The cultural propensity of Jews for careers in scholarship is suggested in Mark Zborowski and Elizabeth Herzog, *Life Is with People: The Culture of the Shtetl* (Schocken Books, 1962), and sensitively revealed in the physicist Leopold Infeld's autobiographical novel *Quest: The Evolution of a Scientist* (Doubleday, Doran, 1941). There is no thorough study of academic anti-Semitism. Helpful in a general way are John Higham, "Social Discrimination Against Jews in America, 1830–1930," in Abraham J. Karp, ed., *The Jewish Experience in America* (5 vols.; American Jewish Historical Society; New York: Ktav, 1969), V, 349–81; Carey McWilliams, *A Mask for Privilege: Anti-Semitism in America* (Little, Brown, 1948); Abram L. Sachar, *Sufferance Is the Badge: The Jew in the Contemporary World* (Knopf, 1939); and Charles B. Sherman, *The Jew Within American Society: A Study in Ethnic Individuality* (Wayne State U. Press, 1961). A rich mine of information on its subject is Nathan Goldberg, "Occupational Patterns of American Jews," *Jewish Review*, 3 (Oct.–Dec. 1945; Jan.–March 1946), 3–23, 161–86, 262–89. My estimates of the number of Jews entering physics between the wars are based on an analysis of Harry Cohen and Itzhak J. Carmin, eds., *Jews in the World of Science: A Biographical Dictionary of Jews Eminent in the Natural and Social Sciences* (Monde, 1956). Einstein's visit to America is described in Peter Michelmore, *Einstein: A Profile of the Man* (Dodd, Mead, 1962), and in Clark, *Einstein*.

The starting point for understanding the flight of the refugees to the United States is Charles Weiner, "A New Site for the Seminar: The Refugees and American Physics in the Thirties," in Donald Fleming and Bernard Bailyn, eds., *The Intellectual Migration* (Harvard U. Press, 1969), pp. 190–234, which may be profitably supplemented by Charles John Wetzel, "The American Rescue of Refugee Scholars and Scientists from Europe, 1933–1945" (unpublished Ph.D. dissertation, History, University of Wisconsin, 1964); Stephen Duggan and Betty Drury, *The Rescue of Science and Learning* (Macmillan, 1948); Donald Peterson Kent, *The Refugee Intellectual: The Americanization of the Immigrants of 1933–1941* (Columbia U. Press, 1953); and Laura Fermi's encyclopedic *Illustrious Immigrants: The Intellectual Migration from Europe* (U. of Chicago, 1968). The Oswald Veblen Papers (OV) are rich with information on the problems of the refugees, in addition to material on the general affairs of physics and mathematics in America. Light on the job situation for natives and refugees alike is cast by "Academic Unemployment," *Bulletin of the American Association of University Professors*, 19 (1933), 354–55; F. K. Richtmeyer and H. M. Willey, "The Young College Instructor and the Depression," *ibid.*, 22 (Dec. 1936), 507–9; Edward J. v. K. Menge, *Jobs for the College Graduate in Science* (Bruce, 1932). Karl T. Compton et al., *Physics in Industry* (New York: American Institute of Physics, 1937), suggests the positions opening up in the 1930s. The Clinton J. Davisson Papers in the Library of Congress reveal little about the life of an industrial physicist, but the William F. Meggers Papers (WMF) suggest the difficulties of working at the Bureau of Standards between the wars.

The remarkable childhood of the Compton brothers is portrayed in James R. Blackwood, *The House on College Avenue: The Comptons at Wooster, 1891–1913* (MIT Press, 1968). Arthur Holly Compton's autobiographical *Atomic Quest: A Personal Narrative* (Oxford, 1956) relates little about the working scientist. Roger H. Stuewer, *The Compton Effect: Turning Point in Physics* (New York: Science History Publications, 1975), explores the development of its topic in exhaustive technical detail. The public Compton is advanced in Marjorie Johnston, ed., *The Cosmos of Arthur Holly Compton* (Knopf, 1967), and some of his private thoughts are set down in twenty-three volumes of notebooks, 1919 to 1940, in the Compton files (AIP). The dispute over the nature of cosmic rays with Millikan is advantageously approached through the Millikan Papers (RAM) and the Arthur Holly Compton Papers, Washington University, St. Louis. Nuel Pharr Davis' vividly written *Lawrence and Oppenheimer* (Simon and Schuster, 1968) is insightful but often inaccurate. Herbert Childs, *An American Genius: The Life of Ernest Orlando Lawrence* (Dutton, 1968), is the authorized biography; it can be usefully supplemented with the Ernest O. Lawrence Papers (EOL). A good deal of information about the young Oppenheimer is in Denise Royal, *The Story of J. Robert Oppenheimer* (St. Martin's Press, 1969), and Oppenheimer the prewar physicist is affectionately recalled in I. I. Rabi et al., *Oppenheimer* (Scribner's,

1969). Also helpful are Peter Michelmore, *The Swift Years: The Robert Oppenheimer Story* (Dodd, Mead, 1969), and the interviews with Oppenheimer and Harold Urey in the Sources for History of Quantum Physics (SHQP).

An engaging popular introduction to the development of quantum mechanics is Barbara Lovett Cline, *The Questioners: Physicists and the Quantum Theory* (Thomas Y. Crowell, 1965). Though marred by its tone of ineluctability, Jammer, *The Conceptual Development of Quantum Mechanics*, is a masterful technical overview. Leon Rosenfeld, "Men and Ideas in the History of Atomic Theory," *Archive for History of Exact Sciences*, 7 (1971), 69–90, discerningly captures the groping quality of the approach to the difficulties of the old quantum theory at Bohr's institute. Paul Forman analyzes the struggle in a brilliant series of articles: "Alfred Landé and the Anomalous Zeeman Effect, 1919–1921," *Historical Studies in the Physical Sciences*, 2 (1970), 153–261; "The Doublet Riddle and Atomic Physics *circa* 1924," *Isis*, 59 (Summer 1968), 156–74; "Weimar Culture, Causality, and Quantum Theory, 1918–1927: Adaptation by German Physicists and Mathematicians to a Hostile Intellectual Environment," *Historical Studies in the Physical Sciences*, 3 (1971), 1–115; and, with V. V. Raman, "Why Was It Schroedinger Who Developed de Broglie's Ideas?" *ibid.*, 1 (1969), 291–314.

The state of quantum physics at the opening of the 1920s is revealed in Niels Bohr, *The Theory of Spectra and Atomic Constitution* (Cambridge University Press, 1922). The scientific reasons for a sense of crisis by 1925 are reported in John H. Van Vleck, "Quantum Principles and Line Spectra," *Bulletin of the National Research Council*, 10 (No. 54, March 1926). Many of the original papers leading up to matrix mechanics are conveniently collected in B. L. Van der Waerden, ed., *Sources of Quantum Mechanics* (Dover, 1968). A complementary collection is Gunter Ludwig, ed., *Wave Mechanics* (Pergamon Press, 1968). Helpful recollections by participants include Werner Heisenberg, *Physics and Beyond: Encounters and Conversations*, trans. Arnold J. Pomerans (Harper & Row, 1971); M. Fierz and V. F. Weisskopf, eds., *Theoretical Physics in the Twentieth Century* (New York: Interscience, 1960); Edward U. Condon, "60 Years of Quantum Physics," *Physics Today*, XV (Oct. 1962), 37–49; John H. Van Vleck, "Reminiscences of the First Decade of Quantum Mechanics," *International Journal of Quantum Chemistry*, 5 (1971), 3–20; and the interviews in the Sources for History of Quantum Physics (SHQP).

Ruth Moore, *Niels Bohr*, supplies a reliable popular introduction to the debate over the interpretation of quantum mechanics. Schrödinger's views are advanced in K. Przibram, ed., *Letters on Wave Mechanics*, trans. Martin J. Klein (Philosophical Library, 1967), and Erwin Schrödinger, "An Undulatory Theory of the Mechanics of Atoms and Molecules," *Physical Review*, 28 (Dec. 1926), 1049–70. Bohr's position is well stated in his *Atomic Theory and the Description of Nature* (Cambridge U. Press, 1934). Also helpful are Werner Heisenberg, "The Development of the Interpretation of the Quantum Theory," in W. Pauli, ed., *Niels Bohr and the Development of Physics* (McGraw-Hill, 1955), pp. 12–29; Gerald Holton, "The Roots of Complementarity," *Daedalus*, 99 (Fall 1970), 1015–55; Martin J. Klein, "The First Phase of the Bohr-Einstein Dialogue," *Historical Studies in the Physical*

Sciences, 2 (1970); and S. Rozental, ed., *Niels Bohr: His Life and Work as Seen by His Friends and Colleagues* (John Wiley, 1967).

Daniel J. Kevles, "Towards the Annus Mirabilis: Nuclear Physics before 1932," *The Physics Teacher*, 10 (April 1972), 175–81, provides background to the increasingly central subject of physics in the 1930s. The puzzlement about nuclear structure before 1932 is evident in "Discussion on the Structure of Atomic Nuclei," *Proceedings of the Royal Society, A*, 123 (April 6, 1929), 373–90; *ibid.*, 136 (June 1, 1932), 735–62; and *Convegno di Fisica Nucleare Ottobre 1931–IX* (Rome: Reale Accademia d'Italia, 1932–X). The surge of confidence after 1932 is manifest in *International Conference on Physics, London, 1934 . . . Papers and Discussions* (2 vols.; London: The Physical Society, 1935), and the shift into nuclear studies is sketched in Charles Weiner, "Moving into the New Physics," *Physics Today*, *XXV* (May 1972), 40–49. A number of the important original papers in nuclear physics are collected in Robert T. Beyer, ed., *Foundations of Nuclear Physics* (Dover, 1949). Joan Bromberg, "The Impact of the Neutron: Bohr and Heisenberg," *Historical Studies in the Physical Sciences*, 3 (1971), 307–41, sets Chadwick's achievement in theoretical context.

Emilio Segrè, *Enrico Fermi: Physicist* (U. of Chicago Press, 1970), details its subject's seminal contributions to the discipline. Charles Weiner, ed., *Exploring the History of Nuclear Physics* (AIP Conference Proceedings, No. 7; American Institute of Physics, 1972), provides a guide to many of the important historical issues and original publications in the field. M. Stanley Livingston, ed., *The Development of High-Energy Accelerators* (Dover, 1966), collects the fundamental papers of Lawrence, Cockroft and Walton, and their successors. Bruno Rossi, *Cosmic Rays* (McGraw-Hill, 1964), surveys the technical issues with considerable historical sensitivity. Norwood R. Hanson's *The Concept of the Positron* (Cambridge U. Press, 1963) explores the preparation of the intellectual ground for the ultimate recognition of Anderson's particle. Hideki Yukawa recalls "The Birth of the Meson Theory" in *American Journal of Physics*, 18 (March 1950), 154–56.

CHAPTERS XIX–XXV | WORLD WAR II TO THE PRESENT

The interviews and recollections cited for Chapters XI–XVIII make clear the early shift to preparedness of American physicists, as do the pages of *Science* for the scientific community in general. The attempt of such bodies as the Science Advisory Board to invigorate the relationship between the military and civilian science may be followed in the Millikan Papers (RAM) and the annual reports of the National Academy of Sciences. Some attention to research is paid in Mark Skinner Watson, *Chief of Staff: Prewar Plans and Preparations* (United States Army in World War II; GPO, 1950). The story of the founding of the National Defense Research Committee rests on Vannevar Bush, *Pieces of the Action* (Morrow, 1970); James B. Conant, *My Several Lives: Memoirs of a Social Inventor* (Harper & Row, 1970); the Vannevar Bush Papers (VB); the Franklin D. Roosevelt Papers (FDR); and the Harry L. Hopkins Papers (HLH). The story is cast in long-range perspective in A. Hunter Dupree's thoughtful "The Great Instauration of

1940," in Gerald Holton, ed., *The Twentieth-Century Sciences* (Norton, 1972), pp. 443–67. On Bush himself, the autobiographical *Pieces of the Action* must be supplemented by the transcript of Eric Hodgins' interview with Bush, on which the book is based, copy of which is in the MIT Archives. The Vannevar Bush Papers at MIT limn his early development as an electrical engineer and administrator.

The essential source for the wartime scientific mobilization is the Records of the Office of Scientific Research and Development (OSR). Especially useful are the documents in Entry 1, which contains the office files of Bush and Conant; in Entry 2, which includes material on such reports to the President as *Science—The Endless Frontier*; Entry 13, the general records of the agency, which are rich with material for postwar planning and legislation; and Entry 39, which contains the office files of Frank Jewett, Karl Compton, and the other civilian members of NDRC. A useful wartime account is George W. Gray, *Science at War* (Harper, 1943). The overall official history is James Phinney Baxter, *Scientists Against Time* (MIT Press, 1968). Informative official histories on special subjects include: Irvin Stewart, *Organizing Scientific Research for War: The Administrative History of the Office of Scientific Research and Development* (Little, Brown, 1948); C. Guy Suits and George R. Harrison, eds., *Applied Physics* (Little, Brown, 1948), which deals with radar countermeasures; John Herrick, *Subsurface Warfare: The History of Division 6, NDRC* (Department of Defense, 1951); John E. Burchard, ed., *Rockets, Guns, and Targets* (Little, Brown, 1948); John E. Burchard and Lincoln R. Thiesmeyer, *Combat Scientists* (Little, Brown, 1947); and Joseph C. Boyce, ed., *New Weapons for Air Warfare* (Little, Brown, 1947). A stimulating and lively study of one of the major wartime physics projects is Albert Christman, *Sailors, Scientists, and Rockets: Origins of the Navy Rocket Program and of the Naval Ordnance Test Station, Inyokern* (Vol. I, History of the Naval Weapons Center, China Lake, California; GPO, 1971). Also helpful is John Burchard, *Q.E.D.: MIT in World War II* (John Wiley, 1948).

The best official history is Henry Guerlac's massive, unpublished "Radar," which details the development and work of the Radiation Laboratory, along with the operational challenges and uses for the Laboratory's products. Robert Watson-Watt, *The Pulse of Radar: The Autobiography of Sir Robert Watson-Watt* (Dial, 1959), stresses the British contributions. Harold G. Bowen, *Ships, Machinery, and Mossbacks: The Autobiography of a Naval Engineer* (Princeton U. Press, 1954), emphasizes the importance of the prewar development work of the Naval Research Laboratory. Helpful for Lee DuBridge's prewar career are the Lee A. DuBridge Papers (LAD) and the Rush Rhees Papers (RR). Interviews flesh out life at the Rad Lab, as do the engaging picturebook *Five Years at the Radiation Laboratory* (MIT, 1946) and "The Reminiscences of Norman F. Ramsey" (COH).

The relative ease with which scientists achieved a role at the highest army levels as advisers on tactical and strategic issues is suggested in Henry L. Stimson and McGeorge Bundy, *On Active Service in Peace and War* (Harper, 1948), and Elting E. Morison, *Turmoil and Tradition: A Study of the Life and Times of Henry L. Stimson* (Houghton Mifflin, 1960). Stimson's friendliness to Bush as an adviser is evident from his Diary in the Henry L. Stimson Papers (HLS). Materials essential for the important advisory role achieved by physicists with the air forces in the European theater include Lee DuBridge's diaries of the Compton Mission, in DuBridge's possession; John G. Trump, "A War Diary, 1944–45," in the possession of Trump; the files of Edward L. Bowles in the Records of the Office of the Secretary of War (OSW), entries 77, 78, 80, 81, and the file of Harvey H. Bundy, entry 82. The files of the Advisory Specialist Group are in Records of the Army Air Force (AAF), entry 60A. A critique of blind bombing via radar is in Wesley Frank Craven and James Lea Cate, eds., *The Army Air Forces in World War II* (7 vols.; U. of Chicago Press, 1948–58). Bush's difficulties with Admiral King over the change in antisubmarine warfare doctrine are recorded in the diary of Admiral Furer, Julius A. Furer Papers (JAF). Also useful on this issue are Morison's *Turmoil and Tradition;* the Stimson Diary; the Ernest J. King Papers (EJK); the Records of the Chief of Naval Operations and Commander-in-Chief of the U.S. Fleet (CNO–CIC); Ladislas Farago's lively *The Tenth Fleet* (New York: Ivan Obolensky, 1962); Ernest J. King and Walter Muir Whitehill, *Fleet Admiral King* (Norton, 1952); and Samuel Eliot Morison, *History of United States Naval Operations in World War II* (15 vols.; Little, Brown, 1947–62).

Useful for special aspects of wartime research and development are the series on the Technical Services in the official history of the United States Army in World War II, especially: Dulany Terrett, *The Signal Corps: The Emergency* (GPO, 1956); George Raynor Thompson et al., *The Signal Corps: The Test* (GPO, 1957); Thompson and Dixie R. Harris, *The Signal Corps: The Outcome* (GPO, 1966); and Constance McLaughlin Green, Harry C. Thomson, and Peter Roots, *The Ordnance Department: Planning Munitions for War* (GPO, 1955). Scientific manpower problems may be approached initially through Albert A. Blum, *Drafted or Deferred: Practices Past and Present* (U. of Michigan, 1967), but sorting out the draft and deferment policies for scientists in particular requires a close study of U.S. Selective Service, *Occupational Bulletins Nos. 1–44 and Activity and Occupational Bulletins, Nos. 1–35* (GPO, 1944). Also helpful is *Industrial Deferment* (U.S. Selective Service System, Special Monograph No. 6; 3 vols.; GPO, 1948). Attempts to enlarge the pool of scientifically trained manpower during the war are reported in Henry H. Armsby, *Engineering, Science and Management War Training: Final Report* (U.S. Office of Education, Federal Security Agency, Bulletin No. 9, 1946; GPO, 1946), and in William B. Tuttle, Jr., "Higher Education and the Federal Government: The Lean Years, 1940–42," *The Record,* 71 (Dec. 1969), 297–312 and "Higher Education and the Federal Government: The Triumph, 1942–1945," *ibid.,* 71 (Feb. 1970), 485–99.

In the planning of postwar military research policy, the conviction that German scientists were too hamstrung by their own military is a theme in Samuel Goudsmit, *Alsos* (Henry Schuman, 1947), the record of his mission to ascertain the state of German war research, and of the army officer Leslie E. Simon's *German Research in World War II* (John Wiley, 1947). *The War Reports of General of the Army George C. Marshall, General of the Army H. H. Arnold, Fleet Admiral Ernest J. King* (J. B. Lippincott, 1947), acknowledges the heavy debt of the armed services to science. The degree of the army and navy's willingness to give civilian scientists something of an independent hand in postwar military research and development is evident in U.S. Congress, House, *Select Committee on Post-war Military Policy, Hearings, Surplus Material—Research and Development*, 78th Cong., 2d Sess., Nov. 1944–Jan. 1945, and in the Records of the War Department General Staff (WDGS), New Developments Division entries 486 and 487 and the files of the Special Planning Division. The attitudes of the navy on the issue are further revealed in the Furer Diary (JAF); the Records of James V. Forrestal, Secretary of the Navy (JFN), especially files 39–1–8 and 70–2–18; the files of the Naval Coordinator of Research and Development (NCONR); U.S. Congress, Senate Committee on Naval Affairs, *Hearing, Establishing a Research Board for National Security*, 79th Cong., 1st Sess., June 20, 1945. The failure to establish an independent civilian agency for peacetime military research is discussed in detail in Daniel J. Kevles, "Scientists, the Military, and the Control of Postwar Defense Research: The Case of the Research Board for National Security, 1944–46," *Technology and Culture*, 16 (Jan. 1975), 20–47.

The origins of the Office of Naval Research are outlined in University of Pittsburgh Historical Staff at ONR, "The History of U.S. Naval Research and Development in World War II," unpublished ms., copy in Office of Naval History, Washington, D.C. The Records of the Office of Naval Research (NCONR) illuminate little of the crucial history of the Office of Research and Inventions; neither do the Harold G. Bowen Papers (HGB), Secretary Forrestal's Papers (JFN), or the James V. Forrestal Papers (JF). More helpful are The Bird Dogs, "The Evolution of the Office of Naval Research," *Physics Today*, XIV (Aug. 1961), 30–35; John E. Pfeiffer, "The Office of Naval Research," *Scientific American*, 180 (Feb. 1949), 11–15; Barton Nanus, "The Evolution of the Office of Naval Research" (unpublished master's thesis, Sloan School of Industrial Management, MIT, 1959); and Harvey M. Sapolsky, *ONR: Science and the Navy* (in preparation).

On the civilian side of postwar research policy, the evolution of Harley Kilgore's program must be traced through U.S. Senate, Subcommittee of the Committee on Military Affairs, *Hearings, Technological Mobilization*, 77th Cong., 2d Sess., Oct.–Dec. 1942; *Hearings, Scientific and Technical Mobilization*, 78th Cong., 1st Sess., March 1943 to May 1944; and *Report, The Government's Wartime Research and Development, 1940–44*, 79th Cong., 1st Sess., Part I (Jan. 23, 1945); Part II (July 23, 1945); and the Harley M. Kilgore Papers (HMK). The best statement of Bush's position is of course *Science—The Endless Frontier* (GPO, 1945), the origins of which are explored in detail in Daniel J. Kevles, "The National Science Foundation and the Debate over Postwar Research Policy, 1942–1945: A Political Interpretation of *Science—the Endless Frontier*," *Isis*, 68 (March 1977), 5–26. The joining of the legislative battle between the programs of Bush and Kilgore is recorded in the U.S. Congress, Senate, Subcommittee of the Committee on Military Affairs, *Hearings on Science Legislation*, 79th Cong., 1st Sess., Oct. 1945, and in Senate Committee on Military Affairs, *Report, National Science Foundation . . .*, 79th Cong., 2d Sess., April 9, 1946.

Indispensable for the struggle over both military and civilian research policy from the war years are the Franklin D. Roosevelt Papers (FDR); the Records of the Office of War Mobilization and Reconversion (OWMR); the Harold D. Smith Papers (HDS); the Records of the Bureau of the Budget

(BOB); the Records of the President's Scientific Research Board (PSRB); the Harry S Truman Papers (HST); the Charles Murphy Papers (CM); the Records of the Office of Scientific Research and Development (OSR); the Bush Papers (VB); the H. Alexander Smith Papers (HAS); and the Millikan Papers (RAM). John R. Steelman, *Science and Public Policy* (5 vols.; President's Scientific Research Board; GPO, 1947), advances the Truman administration's position. Don K. Price, *Government and Science* (New York U. Press, 1954), takes advantage of Price's special knowledge of the issues. Robert P. McCune, *Origins and Development of the National Science Foundation and Its Division of Social Sciences* (University Microfilms, 1971), relies heavily on the *Congressional Record*, and Kenneth M. Jones, "Science, Scientists, and Americans: Images of Science and the Formation of Federal Science Policy, 1945–1950" (unpublished Ph.D. dissertation, History, Cornell, 1975), contains a wealth of information drawn from the periodical literature of the era.

The development of the atomic bomb and of atomic energy policy has received extensive treatment. Essential first-hand accounts of the Manhattan Project are Arthur Holly Compton, *Atomic Quest: A Personal Narrative* (Oxford U. Press, 1956), and Leslie R. Groves, *Now It Can Be Told: The Story of the Manhattan Project* (Harper, 1962). The official history is Richard G. Hewlett and Oscar E. Anderson, Jr.'s first-rate *The New World, 1939/1946* (*A History of the United States Atomic Energy Commission*, vol. I; Pennsylvania State U. Press, 1962), which covers the Project, the decision to drop the bomb, the drive for international control of atomic energy, and the legislative battle over the McMahon bill. Margaret Gowing, *Britain and Atomic Energy, 1939–1945* (Macmillan, 1964), rightly stresses the importance of the British contributions early in the war and is provocatively reflective on key issues. The industrial role is emphasized in Stephane Groueff, *Manhattan Project: The Untold Story of the Making of the Atomic Bomb* (Little, Brown, 1967).

Lansing Lamont, *Day of Trinity* (Atheneum, 1965), portrays its subject dramatically but is not entirely reliable. Life at the Metallurgical Laboratory and at Los Alamos is variously recalled in Jane Wilson, ed., *All in Our Time: The Reminiscences of Twelve Nuclear Pioneers* (Bulletin of the Atomic Scientists, 1975); Richard S. Lewis and Jane Wilson, eds., *Alamogordo Plus Twenty-five Years* (Viking, 1971); Laura Fermi, *Atoms in the Family: My Life with Enrico Fermi* (U. of Chicago Press, 1954); Victor F. Weisskopf, "The Los Alamos Years," in I. I. Rabi et al., *Oppenheimer* (Scribner's, 1969); and "The Reminiscences of Kenneth T. Bainbridge" (COH). Helpful for human details is William L. Laurence, *Dawn over Zero: The Story of the Atomic Bomb* (Knopf, 1946), a treatment by the *New York Times* science reporter who witnessed the Trinity test. The J. Robert Oppenheimer Papers (JRO) reveal something of the anxiety and tension at Los Alamos.

An outstandingly thorough and arresting study of the policy that led to the dropping of the bomb is Martin J. Sherwin, *A World Destroyed: The Atomic Bomb and the Grand Alliance* (Knopf, 1975). Equally penetrating are Barton J. Bernstein's "Roosevelt, Truman, and the Atomic Bomb: A Reinterpretation," *Political Science Quarterly*, 90 (Spring 1975), 23–69, and "The Quest for Security: American Foreign Policy and International Control of Atomic Energy, 1942–1946," *Journal of American History*, LX (March

1974), 1003–44. Both Sherwin and Bernstein are a corrective to Gar Alperovitz, *Atomic Diplomacy: Hiroshima and Potsdam* (Vintage, 1965). Also helpful is Arthur Steiner, "Scientists, Statesmen, and Politicians: The Competing Influences on American Atomic Energy Policy, 1945–46," *Minerva*, XII (Oct. 1974), 469–509. Szilard's petitions against the dropping of the bomb on Japan are in Leo Szilard, "Reminiscences," in Fleming and Bailyn, eds., *The Intellectual Migration*. The Irving Langmuir Papers (IL) document Langmuir's impressions of the strength of Russian science as a result of his visit to the anniversary celebration of the Soviet Academy of Sciences in 1945.

The struggle over the domestic control of atomic energy at the end of the war is told from the scientists' point of view in Alice Kimball Smith, *A Peril and a Hope: The Scientists' Movement in America, 1945–1947* (U. of Chicago Press, 1965). A useful supplement is Donald A. Strickland, *Scientists in Politics: The Atomic Scientists Movement 1945–1946* (Purdue U. Studies, 1968), and William R. Nelson, *Case Study of a Pressure Group: The Atomic Scientists* (University Microfilms, 1966). The wide range of issues at stake is discussed in James R. Newman and Byron S. Miller, *The Control of Atomic Energy: A Study of Its Social, Economic, and Political Implications* (McGraw-Hill, 1948). Morton Grodzins and Eugene Rabinowitch, eds., *The Atomic Age: Scientists in National and World Affairs* (Basic Books, 1963), is a litmus paper indicator of the policy opinions from the war on of scientists associated with the *Bulletin of the Atomic Scientists*. Atomic energy policy in the postwar decades can be followed in Robert Gilpin, *American Scientists and Nuclear Weapons Policy* (Princeton U. Press, 1962), and Richard G. Hewlett and Francis Duncan, *Atomic Shield, 1947/1952* (*A History of the United States Atomic Energy Commission*, Vol. II; Pennsylvania State U. Press, 1969). A retrospective look is *The New York Times*, prep., *Hiroshima Plus 20* (Delacorte Press, 1965). Herbert F. York, *The Advisors: Oppenheimer, Teller, and the Superbomb* (W. H. Freeman, 1976), argues Oppenheimer's side on the issue of the H-bomb. Philip M. Stern, *The Oppenheimer Case: Security on Trial* (Harper & Row, 1969), is by far the best study of its subject. Lewis L. Strauss defended the proceedings in *Men and Decisions* (Doubleday, 1962). Stanley A. Blumberg and Gwinn Owens, *Energy and Conflict: The Life and Times of Edward Teller* (Putnam's, 1976), while informative on Teller's early life, strains credulity on most key issues after 1940. The transcript of the Gray board hearings, U.S. Atomic Energy Commission, *In the Matter of J. Robert Oppenheimer* (GPO, 1954), is a mine of information. Important for the test ban treaty are Theodore C. Sorensen, *Kennedy* (Harper & Row, 1965), and Arthur M. Schlesinger, Jr., *A Thousand Days: John F. Kennedy in the White House* (Houghton Mifflin, 1965). The considerable groundwork done for the treaty under Eisenhower is evident from George B. Kistiakowsky, *A Scientist at the White House: The Private Diary of President Eisenhower's Special Assistant for Science and Technology* (Harvard U. Press, 1976).

On the general relations of science and government, a useful source of documents is James L. Penick, Jr., et al., eds., *The Politics of American Science, 1939 to the Present* (rev. ed.; MIT Press, 1972), which can be profitably supplemented by William R. Nelson, ed., *The Politics of Science: Readings in Science, Technology, and Government* (Oxford, 1968). Also helpful are J. Stefan Dupré and Sanford Lakoff, *Science and the Nation: Policy and*

Politics (Prentice-Hall, 1962); Robert Gilpin and Christopher Wright, eds., *Scientists and National Policy-Making* (Columbia U. Press, 1964); Sanford A. Lakoff, ed., *Knowledge and Power: Essays on Science and Government* (Free Press, 1966); Donald W. Cox, *America's New Policy Makers: The Scientists' Rise to Power* (Chilton Books, 1964). The Committee on Science and Public Policy, National Academy of Sciences, *Federal Support of Basic Research in Institutions of Higher Learning* (National Academy of Sciences–National Research Council, 1964), endorses the postwar structure of federal science. A somewhat critical view is advanced in Milton Lomask, *A Minor Miracle: An Informal History of the National Science Foundation* (National Science Foundation, 1976). A knowledgeably acerbic critique by the former news editor of *Science* magazine is Daniel S. Greenberg, *The Politics of Pure Science* (New American Library, 1967).

A major source of information on the life and welfare of the postwar physics community is *Physics Today* (1948–   ). There is also a wealth of data in Physics Survey Committee, National Research Council, *Physics in Perspective* (National Academy of Sciences, 1972). M. Stanley Livingston, *Particle Accelerators: A Brief History* (Harvard U. Press, 1969), covers the technical advances. Anton G. Jachim, *Science Policy Making in the United States and the Batavia Accelerator* (Southern Illinois U. Press, 1975), is uncritical but useful. Robert Jungk, *The Big Machine* (Scribner's, 1968), deals romantically with CERN. Kenneth W. Ford, *The World of Elementary Particles* (Blaisdell, 1963), is a helpful introduction. Information on attitudes and occupational patterns is to be found in William H. Chafe, *The American Woman: Her Changing Social, Economic, and Political Role, 1920–1970* (Oxford U. Press, 1972). Jacquelyn A. Mattfeld and Carol G. Van Aken, eds., *Women and the Scientific Professions: The MIT Symposium on American Women in Science and Engineering* (MIT Press, 1965), is highly useful, and so is Harriet Zuckerman and Jonathan R. Cole, "Women in American Science," *Minerva*, XIII (Spring 1975), 82–102. Julius H. Taylor, ed., *The Negro in Science* (Morgan State College Press, 1955), and Aaron E. Klein, *The Hidden Contributors: Black Scientists and Inventors in America* (Doubleday, 1971), only scratch the surface of the subject. Essential statistical information is in James M. Jay, *Negroes in Science: Natural Science Doctorates, 1876–1969* (Detroit: Balamp, 1971). Thomas F. O'Dea, *American Catholic Dilemma: An Inquiry into the Intellectual Life* (Sheed and Ward, 1958), is typical of the rising self-critical concern among American Catholics in the 1950s about their low production of scholars and scientists. The increasing number of Catholics choosing to enter science is documented in Andrew M. Greeley, *Religion and Career: A Study of College Graduates* (Sheed and Ward, 1963), and in Stephen Steinberg, *The Academic Melting Pot: Catholics and Jews in American Higher Education* (Carnegie Commission on Higher Education; McGraw-Hill, 1974).

For the mid-1960s revolt against science and its aftermath, I have relied heavily on newspapers, general periodicals, *Physics Today*, and *Science* magazine. Don K. Price, *The Scientific Estate* (Harvard U. Press, 1965), explores the political issues raised by the requirement of scientific autonomy. Typical of the muckraking books of the period are William Gilman, *Science: U.S.A.* (Viking Press, 1965), and Harold L. Nieburg, *In the Name of Science* (Quadrangle, 1966). Insightfully critical are Ralph E. Lapp, *The New Priest-*

hood: The Scientific Elite and the Uses of Power (Harper & Row, 1965), and Spencer Klaw, *The New Brahmins: Scientific Life in America* (Morrow, 1968). Alvin M. Weinberg's thoughtful series of articles is summarized in his *Reflections on Big Science* (MIT Press, 1967). The reentrance of scientists into public debates on technological policy is treated in Joel Primack and Frank von Hippel, *Advice and Dissent: Scientists in the Political Arena* (Basic Books, 1974), and Jonathan Allen, ed., *March 4: Scientists, Students, and Society* (MIT Press, 1970). Dorothy Nelkin, *The University and Military Research: Moral Politics at M.I.T.* (Cornell U. Press, 1972), reports in broad context the turmoil on that campus. The politics of physicists and professors in other disciplines are explored in the strongly statistical Everett Carll Ladd, Jr., and Seymour Martin Lipset, *The Divided Academy: Professors and Politics* (Carnegie Commission on Higher Education; McGraw-Hill, 1975).

Index